Lectins
Biology, Biochemistry, Clinical Biochemistry
Volume 5

Lectins

Biology, Biochemistry Clinical Biochemistry Volume 5

Proceedings of IUB Symposium No. 144
The Seventh International Lectin Meeting
Bruxelles, Belgium, August 18–23, 1985

Editors
T. C. Bøg-Hansen · E. van Driessche

Walter de Gruyter · Berlin · New York 1986

Editors
Thorkild Christian Bøg-Hansen, cand. scient., lic. techn.
The Protein Laboratory
University of Copenhagen
Sigurdsgade 34
DK-2200 Copenhagen N

Edilbert van Driessche
Protein Chemistry Laboratory
Free University of Bruxelles
Paardenstraat, 65
B-1640 Sint-Genesius-Rode

CIP-Kurztitelaufnahme der Deutschen Bibliothek

Lectins, biology, biochemistry, clinical biochemistry : proceedings of the ... Internat. Lectin Meeting. – Berlin ; New York : de Gruyter
Bis Vol. 4, Kongressname: Lectin Meeting

NE: International Lectin Meeting; Lectin Meeting

Vol. 5 = 7. Bruxelles, Belgium, August 18 – 23, 1985. – 1986.
(IUB symposium ; No. 144)
ISBN 3-11-010699-X (Berlin)
ISBN 0-89925-102-1 (New York)

NE: International Union of Biochemistry: IUB symposium

3 11 010699 X Walter de Gruyter · Berlin · New York
0-89925-102-1 Walter de Gruyter, Inc., New York

 Printing: Gerike GmbH, Berlin. – Binding: D. Mikolai, Berlin. Printed in Germany.

PREFACE

Since their discovery by Stillmark, nearly 100 years ago, lectins have fascinated scientists working in many different fields such as immunology, cell biology, biochemistry, histochemistry and physiology. Lectins have been devised and selected by nature and are isolated, characterized and manipulated by man to be used as tools.

The idea of bringing together people sharing a common interest i. e. lectins, was born at the Protein Laboratory of the University of Copenhagen, Sigurdsgade 34, 2nd floor, in May 1978; about 60 researchers mainly working in Copenhagen met during Lec 1 to discuss their findings and ideas on reactions of lectins with cells and proteins. Guests from Poland and Canada gave the second Copenhagen Lectin Meeting an international flavour. In 1980, during Lec 3 which may be considered to be a real international meeting with about 100 lectinologists, it was decided to publish the original results in a proceedings volume. The initiative has been repeated 4 times then. The fourth Copenhagen Meeting was organized in June 1981. By then, and due to their great success, it had become clear that future meetings should be organized on a collaborative basis. Interlec 5 was held in Bern, Switzerland with Dr. Georg Spengler as host and for Interlec 6 we were the guests of Dr. Jan Breborowicz in Poznan, Poland.

Interlec 7 was held in Brussels, Belgium from August 18-23, 1985 at the Faculty of Medicine and Pharmacy of the Vrije Universiteit Brussel. In this Meeting about 150 lectinologists from 27 countries participated. Our dream was to achieve a lectin meeting where participants would communicate their results, hopes and ideas and make plans for the future. The main characteristics of Interlec 7 were the high scientific standard of the communications and a spirit of collegiality and friendship, during both the scientific sessions and the social events. At the end of the congress we all returned home, satisfied but tired by the hand work during the constructive days we had spent in Brussels.

Following the tradition of previous meetings, this proceedings volume has been complied in order to make the most important and valuable information from Interlec 7 available to the scientific community. The book contains 85 selected papers covering important current topics in lectinology. We are grateful to the contributing authors for their efforts in helping us to achieve fast publication.

The first two chapters deal with the physiological function, biosynthesis, isolation and characterization of plant lectins. Although more than 100 plant lectins have been purified and characterized, the search for new lectin sources continues. There is no doubt that within a few years, for most lectins, a detailed description at the molecular level of their interaction with carbohydrates will be available. Investigations on endogeneous lectinreceptors will ultimatively lead to an answer to the question: What are plant lectins made for?

The next chapter consists of papers devoted to the function of animal lectins in normal and malignant tissues and to their implication in development. Chapter IV contains papers dealing with microbial lectins and their importance in pathogenesis.

Chapter V consists of papers describing the many and diverse biological effects of lectins both in experimental systems and when present in the diet. Finally the papers of chapter VI and VII clearly demonstrate how precious lectins are for the characterisation and histochemical localisation of glycoconjugates.

We would like to thank all our colleagues and friends who participated in Interlec 7, the members of the International Organizing Committee: Dr. S. Beechmans, Prof. R. Dejaegere, Prof. L. Kanarek, Dr. J. Breborowicz, Prof. A. Burny, Prof. M. Etzler, Dr. A. Faure, Dr. D. L. J. Freed, Dr. A. Lihme and Dr. G. A. Spengler, as well as the staff of the laboratories of Protein Chemistry and Plant Physiology (Vrije Universiteit Brussel) and of the Protein Laboratory (University of Copenhagen) for their joint efforts to achieve a successful Meeting. Last but not least we sincerely acknowledge the support given by: Vrije Universiteit Brussel (Belgium), the Belgian National Fund for Scientific Research (NFWO), the Belgian Ministry of Education, the Belgian Ministry of Health and Family, the International Union of Biochemistry, the Generale Bank of Belgium, the Olivetti Computer Company and the Boehringer-Mannheim Company.

During Interlec 7 plans were made for future Interlec meetings: in May 1986 Dr. Jose Ochoa will host Interlec 8 in La Paz in Mexico, and in 1987 Dr. David Freed will be responsible for the organization of Interlec 9 to be held in Cambridge, England. To celebrate the centenary year of Stillmark's discovery of lectins, we hope Interlec 10 to be in collaboration with our colleagues of the University of Tartu, USSR.

Copenhagen, January 1986

Edilbert van Driessche
Thorkild C. Bøg-Hansen

TABLE OF CONTENTS

I. PHYSIOLOGY AND BIOSYNTHESIS OF PLANT LECTINS

II. ISOLATION AND CHARACTERIZATION OF PLANT LECTINS

V. BIOLOGICAL EFFECTS OF LECTINS

VI. Histochemistry with Lectins

VII. Lectins as Tools

PART 1

BIOSYNTHESIS AND PHYSIOLOGY OF PLANT LECTINS

THE PHYSIOLOGICAL FUNCTION OF PLANT LECTINS.

Jan W. Kijne, Clara L. Díaz and Robert Bakhuizen,

Dept. of Plant Molecular Biology, Botanical Laboratory,
Nonnensteeg 3, 2311 VJ LEIDEN, The Netherlands.

The specific carbohydrate-binding properties of plant lectins are generally believed to determine their role inside and outside plant cells. Unfortunately, data on the biological activity of endogenous or exogenously applied lectins in plant cells or tissues are absent or, if available, inconclusive. Moreover, as long as the identity and biology of the endogenous complementary carbohydrate moieties (lectin receptors) have not been elucidated, the physiological function of plant lectins will remain unknown.

On the other hand, the molecular properties and sugar-binding characteristics of an increasing number of plant lectins are well-known, and the study of lectin genetics and biosynthesis is making good progress. These data, in combination with suggestive observations on lectins in other areas of cell biology (e.g. cell-cell adhesion in slime molds (1), induction of mitosis in T-lymphocytes (2), various endocytotic processes in human and animal cells (Kolb Bachofen et al., These Proceedings Vol 5)) have resulted in the existence of several hypotheses on plant lectin function. With regard to their presence in different plant tissues and their ability to bind dissimilar sugar receptors *in vitro*, albeit with different affinity, it has been proposed that plant lectins actually might have several functions. The discovery of specific high-affinity binding of adenine-derivatives by legume lectins (3) has even put the speculations across the frontier of sugar binding, and has introduced plant lectins into the complicated area of plant hormonal regulation.

We will not present an exhaustive review of the various existing hypotheses, facts and argumentations about plant lectin funct-

Lectins, Vol. V

ion(s); the reader is referred to instructive and stimulating review papers (e.g. 4,5) and books (e.g. 6). Rather, we intend to extract some essential elements from one exemplary case, and to see if an extension to other observations might lead to an acceptable generalization about the framework in which the study of plant lectin function might fit. Enzymes with lectin-like properties will not be considered (7).

LEGUME ROOT HAIRS

Lectin isolation

Root hairs are tubular outgrowths of plant root epidermal cells, functioning in the absorption of water and nutrients. From the capped tip of a young root, four successive zones can be distinguished: a "no root hair zone" with dividing and elongating hairless epidermal cells, a "developing root hair zone", a "mature root hair zone" and finally a zone with senescent and broken root hairs. Developing root hairs extend rapidly by a growth mechanism called "tip growth" (8), a general morphogenetic phenomenon in various plants and plant-like organisms (e.g. fungal hyphae, moss caulonemata, pollen tubes, algal rhizoids). In tip-growing plant cells new cell wall material is secreted by numerous vesicles (9) and membrane-associated enzyme-systems (10). The plasmamembrane is concomitantly enlarged. The root hair cell wall components are assembled extracellularly (external to the plasmamembrane) and a gradual cross-linking process, leading to cell wall rigidification, ultimately fixes the tubular cell shape. Within a few days the hairs reach their maximal length (determined by plant genotype and growth conditions), tip growth stops, the tip cell wall becomes rigid and the hairs are called "mature".

Growing root hairs of several leguminous plants have been shown to secrete lectins at the root hair tip (clover, 11; *Lotononis*, 12; pea, Díaz et al. in preparation). Also root slime (13) and root exudate (14,15) contain lectin. Apparently, root hair lectins are secretory proteins. The final location of secreted lectin molecules, apart from dispersed molecules in the rhizosphere, is unclear. Are they incorporated into the developing cell wall? Do some lectin molecules remain associated with externally exposed sugar-moieties of plasmamembrane-bound glycoconjugates? Are some lectin molecules inserted into the plasmamembrane after fusion of

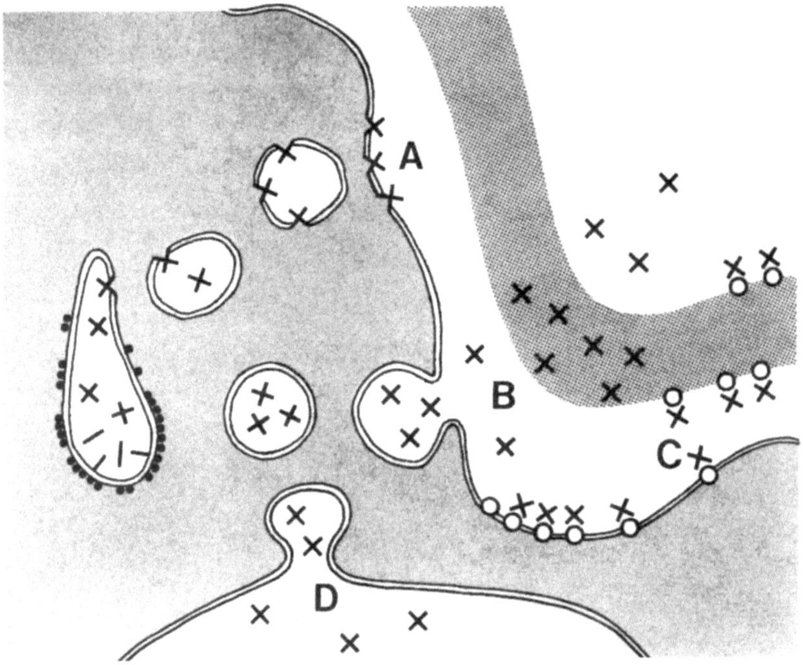

Figure 1: Possible location of plant lectin molecules (✖) after secretion:
A: incorporation in the plasmamembrane (experimental evidence: absent),
B: incorporation in the extracellular matrix and/or release in the cellular environment (experimental evidence: available, but incomplete),
C: association with membrane- or cell wall-receptors (experimental evidence: only indirect),
D: secretion into protein bodies/vacuoles (experimental evidence: conclusive, but lacking for root hairs).

secretory vesicles? Are they simultaneously stored in root hair vacuoles like seed lectins? Figure 1 shows a schematic representation of these possibilities.

Some fragmentary data give an indication of the probability of these suggestions. Pea lectin has been found in saline extracts of pea root cell walls (13), which points to non-covalent lectin-incorporation into the cell wall matrix. Cell wall localization has also been demonstrated for *Dolichos*-lectin (16). Hapten-elutable lectin is present on clover roots (11), and hapten-inhibitable adhesion of clover root hair tips to each other suggests the presence of lectin receptors on the root hair surface (17).

However, it is unknown if the root hair plasmamembrane contains externally exposed lectin receptors, as it has been demonstrated for soybean protoplasts (18). Any indication that plant lectins are integral plasmamembrane-constituents, not only in root hairs, is lacking. Root hairs are attractive model cells for the study of these problems. The possibility of collecting pure root hair preparations by breaking the hairs from the epidermal cells opens the way to a combined biochemical and (immuno)microscopical analysis.

Developmental control

Pea root hair cells constitute a heterogeneous population of cells as shown by lectin localization (Díaz et al. in preparation); only a part of the growing hairs produces lectin in a detectable amount. If lectins are involved in tip growth as future important cell wall constituents, one might expect lectin to be present at the tip of all growing hairs.

Two possibilities are open, and intuitively we favour the second:

- some root hairs produce lectins, others do not.
- some root hairs overproduce lectins, while others produce lectin at a low level, in non-detectable amounts or in a non-detectable form.

This brings us to the following central questions:

- do all plant cells produce lectin, or only special cells?
- is detectable lectin production actually an overproduction?

Interestingly, lectin-producing pea root hairs are positioned following a similar xylem-determined pattern (Díaz et al. in preparation) as found by Libbenga et al. (19) for the dedifferentiation of pea root cortical cells by an exogenously applied plant hormone regime. This observation suggests that gradients of plant growth substances and cell division factors might determine lectin excretion by pea root epidermal cells.

Developmental control of lectin production is also illustrated by the observation of Dazzo and co-workers (20,21) that nitrate-supply to clover seedlings significantly reduces lectin accumulation on the root surface. High-level nitrate stimulates root-growth, accentuates apical dominance of the root tip (own observations) and leads to a reduction of root hair formation by epidermal cells (22), in comparison with N-limited legume plants. In our opinion, nitrate influences the hormone balance in legume roots, and thereby lectin secretion.

Several authors have noticed the presence of serologically related but inactive lectin in legume root and root exudate preparations (21,23,24). Lectin inactivation might represent another regulatory mechanism, albeit at present completely unclear at the molecular level.

Another important element in this respect is the special status of the root hair itself. Several characteristics of root hair cells (e.g. a continuously mobile nucleus together with a high synthetic activity) suggest that root hair forming epidermal cells show inhibition of cytokinesis (25,26). The possibility should be considered that detectable lectin synthesis is correlated with a prolonged G2-stage of the cell cycle (Bakhuizen et al. in preparation).

In summary: which conditions favour lectin production by legume root epidermal cells? The picture emerges of developmentally arrested cells (a condition probably stimulated by nitrogen limitation) which secrete significant amounts of lectin under hormonal control (Fig.2). No need to say that incomplete sketches easily develop into surprisingly different finished products.

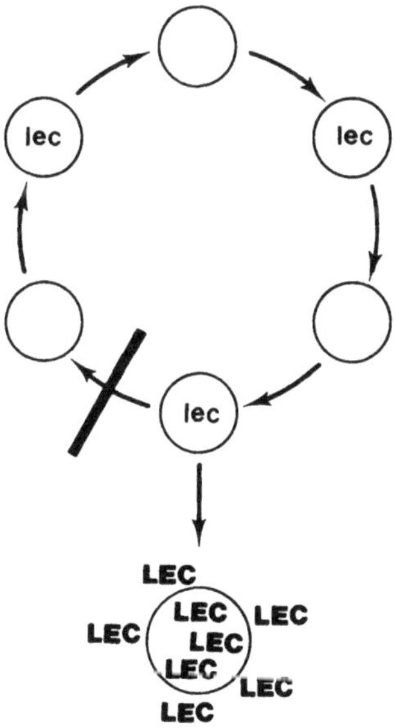

Figure 2: Developmental arrest might induce overproduction and excretion of lectin by plant cells. Possible growth limitations are nitrogen (20), hormones (16) and, in the case of seed lectins, water.

Root hair lectin function

The simultaneous secretion of plant lectin and cell wall components suggests that lectins are part of the extracellular matrix. Kauss and Glaser (27) proposed a specific role of plant lectins in cell wall extension, due to the acid-lability of lectin-sugar binding. However, the first strutural lectin receptor in plant cell walls still awaits discovery.

Much attention has been paid to host-specific recognition of the symbiotically nitrogen-fixing bacterium *Rhizobium* by legume root hair lectins (28,29). Supporting observations are the specific distribution of different lectins over the different host plant cross-inoculation groups, the location of lectins at the infection sites (growing root hair tips), production of lectin receptors by *Rhizobium*, and, more specifically, the inhibition of rhizobial attachment to root hair tips by haptenic monosaccharides (30,31,32,33). Recent collaboration between the research groups of Dazzo and Rolfe has shown that nodulation mutants of *R. trifolii* are impaired in clover-lectin binding (34).

The "lectin-recognition hypothesis" has been challenged by other observations: e.g. promiscuous rhizobial infections (35), lectin-less host plants (36), lectin-binding to non-host rhizobia (37), non-specific attachment of *Rhizobium* to root hairs (38). Several factors, however, explain these contradictory results: insensitive lectin-detection methods (39); measurement of low-affinity binding; interference of a non-specific rhizobial attachment mechanism (33); presence of lectins with different sugar specificity in the same host plant (40); another rhizobial infection mechanism bypassing the root hair (41). Nevertheless, as long as the growth conditions for host plant and rhizobia in soil and rhizosphere are not exactly known, the lectin-recognition theory for host-specific root hair infection will remain attractive but hypothetical.

Recent results of Halverson and Stacey (15) suggest that lectins also promote the virulence of *Rhizobium* ; the molecular mechanism of lectin- *Rhizobium* interaction is not known. Also the link between the infection stages following attachment (root hair curling, infection thread formation) and lectin-mediated attachment is not understood. It is tempting to suggest that lectin-binding triggers the next infection steps (42), but evidence is lacking. Minute quantities of clover-lectin binding polysaccharides of *R. trifolii* enhance rhizobial clover root hair infection; this biological activity is inhibited by combined nitrogen in the clover culture medium (34). Again, the molecular mechanism of this presumably lectin-mediated phenomenon is not known.

Experiments on the influence of well-characterized lectin receptor molecules on root hair growth and cell wall assembly are worth trying; rhizobial lectin receptors might function as model probes, as long as endogenous plant lectin receptors are not available.

Secondary functions

We consider recognition and binding of microorganisms as being secondary lectin functions, because the complementary receptors are non-self. Secondary functions can be stabilized in evolution when the ultimate effect on the plant is benificial; positive selection in agriculture may subsequently amplify these functions in crop plants.

Three preliminary conclusions can be drawn from the legume root hair case, each with successively less experimental evidence:

- lectins are secretory proteins, to be found in the extracellular matrix of plant cells and/or in the cellular environment.
- lectins are produced in significant amounts under conditions of developmental arrest, in combination with a special hormonal regime and a nutritional limitation.
- secondary functions of lectins are a consequence of overproduction and excretion.

OTHER EXAMPLES

Data from other plant lectin studies generally are consistent with the above-mentioned conclusions.

Legume (seed) lectins

Legume seed lectins are produced in large amounts during seed maturation, evidently an example of developmental arrest with an increasing nutritional limitation: water. Interestingly, for soybean this overproduction seems to be restricted to agricultural lines; close relatives of *Glycine max* do not accumulate the classical soybean seed lectin (SBA)(43). This might point to a beneficial function of SBA in soybean seeds, positively selected for in agriculture. It has been suggested that SBA, leaking from germinating seeds, has a protective function against certain microbial pathogens (44). Under laboratory conditions SBA is apparently redundant as seed-lectin-less lines (36) germinate and grow normally. It might however be interesting to compare germination physiology and soil microbiology of SBA^+ and SBA^- genotypes under field conditions. On the other hand, selection for SBA^+ lines might very well be coincidental within a general selection procedure for larger seeds containing more protein (Peumans, personal communication). Furthermore, a significant negative correlation was found between field emergence of legume seeds and seed exudation (45).

Study of the biosynthetic pathway of legume lectins (Chrispeels et al., These Procedings Vol 5) as well as *Ricinus* -lectin (a non-legume)(46,47) has underlined their secretory character. Seed lectins are secreted into protein bodies and might serve as an additional reserve protein (or protein binder: Einhoff et al., These Proseedings Vol 5), which is illustrated by their disappearance from the cotyledons during germination (e.g. Rougé et al., These Proceedings Vol 5).

Non-seed lectin of *Dolichos* (CRM) can be extracted from cell walls of cultured cells, and is also found in minute quantities in the medium (Etzler and Linsefors, These Proceedings Vol 5). Cell cultures of *Dolichos* only produce significant amounts of CRM after hormone depletion, which again suggests a relationship between lectin secretion and developmental arrest.

Wheat germ agglutinin

Wheat germ agglutinin (WGA) is different from legume seed lectins in that it is not a storage protein. WGA is produced in significant amounts by certain parts of the embryo, and is considered to be a dormancy-specific protein (48). Dormancy is a state of developmental arrest, of which the molecular biology is not well understood. The synthesis of WGA is under hormonal control (Raikhel and Palevitz, These Proceedings Vol 5).

A lectin indistinguishable from WGA is synthesized in wheat roots (49), and can be found in surface tissue of the root and, hapten-elutable, on the root surface (50). A role in resistance against chitin-containing soilborne pathogenic microorganisms has been suggested (51,52), after leakage from germinating seeds (53) or excretion by the roots. Which regulatory mechanism in wheat cells underlies the choice between storage of WGA in vacuoles (50) or excretion in the cellular environment is unknown. It also remains to be determined if the biosynthetic pathway of WGA, including the participating organelles, shows the characteristics of secretory-protein biosynthesis.

The specific binding of epiphytic bacteria from the rice rhizosphere by the WGA-related rice lectin (54) is an interesting observation in comparison with the legume-*Rhizobium* symbiosis.

Cucurbitaceae- and Robinia-lectins

Phloem exudate from several agriculturally important *Cucurbitaceae* (pumpkin, cucumber, watermelon) contains a N-acetyl-glucosamine specific lectin (55), which together with P-proteins might function in immobilization of invading microorganisms. Chemical gellation of phloem sap might be required as an emergency-barrier, in view of the large sieve pore diameter in some of these genera. Presence of lectin in phloem exudate again points to active secretion, but in this case data on biosynthesis and cell biology are lacking.

Also *Robinia* -phloem lectin is believed to function as an inhibitor of microorganisms (56). Earlier assumptions about a role of phloem lectins in sugar-transport have been challenged by negative results in the search for complementary sugars in sieve tube sap.

Solanaceae-lectins

Several members of the *Solanaceae* family (potato, tomato, thorn apple) have been shown to produce related lectins in several plant organs. These lectins are distinct from other plant lectins in their resemblance to hydroxyproline-rich cell wall glycoproteins (HRGP)(57) and a role as soluble HRGP-precursor is under discussion (58). Conclusive evidence for the characterization of these lectins as cell wall constituents is lacking. The presumed role of potato lectin in immobilization of avirulent pathogenic Pseudomonads (59) has been cancelled after the finding that the agglutination was caused by another protein, a HRGP (60,61).

Even for speculation on their function, more data are needed on biosynthesis and cell biology of these interesting lectin molecules.

PLANT LECTIN FUNCTION

The only function of plant lectins which gains experimental support is the interaction with microorganisms, for the good of the plant. We believe that this function is secondary and has evolved as a result of overproduction and excretion of lectins under special developmental conditions. A primary function of plant lectins is unknown. Its definition awaits the characterization of endogenous plant lectin receptors, the production of relevant plant mutants and a further development of the methodology in the study of plant membranes.

Depending on their concentration plant lectins might function as a local cytokinin-buffering system (3). A biologically relevant binding-release mechanism in lectin-cytokinin interactions is however still to be discovered.

A role of plant lectins in cell membrane mediated processes is doubtlessly an attractive suggestion on the analogy of lectin function in animal cell systems (Kolb-Bachofen et al., These Proceedings Vol 5). The cell wall matrix of plant cells might be a reservoir of non-covalently bound lectin, or can immobilize lectin molecules at the cell wall-plasmamembrane interface. Matrix changes may lead to release of lectin molecules and resulting changes at the membrane surface, provided that certain membrane constituents are lectin receptors. In this concept lectins act as signals. Wang et al.(18) presented some promising results on lectin-membrane interactions with soybean protoplasts, showing lectin influence om membrane receptor mobility in relation to cytoskeletal function. It should be remembered that the plant extracellular matrix is an essential regulatory element in the plant cell cycle: wall-less plant cells can not divide.

ACKNOWLEDGEMENTS

The authors wish to thank Albert N. van der Werf for his contribution to the review of lectin literature, Peter Hock for drawing the figures and June den Hartog for proof-reading of the manuscript.

REFERENCES

1. Ray,J., T.Shinnick, R.A.Lerner (1979) Nature 279, 215-221.
2. Nowell, P.C. (1960) Cancer Res. 20, 462-466.
3. Roberts,D.D., I.J.Goldstein (1983) see ref.6, pp.131-141.
4. Barondes,S.H. (1981) Annu.Rev.Bioch. 50, 207-231.
5. Etzler,M.E. (1985) Annu.Rev.Plant Physiol. 36, 209-234.
6. Goldstein,I.J., M.E.Etzler (eds.)(1983) Chemical taxonomy, molecular biology and function of plant lectins. Alan R.Liss Inc., New York.
7. Goldstein,I.J., R.C.Hughes, M.Monsigny, T.Osawa, N.Sharon (1980) Nature 285, 66.
8. Sievers,A., E.Schnepf (1981) in: O.Kiermayer (ed.)Cytomorphogenesis in plants, Springer Verlag, pp.265-299.
9. Dashek,W.V., W.G.Rosen (1966) Protoplasma 61, 192-204.
10. Mueller,S.C., R.M.Brown jr.(1980) J.Cell Biol. 84, 315-326.
11. Dazzo,F.B., W.E.Yanke, W.J.Brill (1978) Biochim.Biophys.Acta 539, 276-286.
12. Law,I.J., B.W.Strijdom (1984) Plant Physiol. 74, 779-785.
13. Díaz,C.L., P.Lems-van Kan, I.A.M.van der Schaal, J.W.Kijne (1984) Planta 161, 302-307.
14. Dazzo,F.B., E.M.Hrabak (1981) J.Supramol.Struct.Cell.Biochem. 16, 133-181.
15. Halverson,L.J., G.Stacey (1985) Plant Physiol. 77, 621-625.
16. James,D.W. jr., M.Ghosh, M.E.Etzler (1985) Plant Physiol. 77, 630-634.
17. Dazzo,F.B., G.L.Truchet, J.W.Kijne (1982) Physiol.Plant. 56, 143-147.
18. Wang,J.L., T.N.Metcalf III, M.Schindler (1983) see ref.6, pp. 273-276.
19. Libbenga,K.R., F.van Iren, R.J.Bogers, M.F.Schraag-Lamers (1973) Planta 114, 29-39.
20. Dazzo,F.B., W.J.Brill (1978)Plant Physiol. 62, 18-21.
21. Sherwood,J.E., G.L.Truchet, F.B.Dazzo (1984) Planta 162, 540-547.
22. Thornton,H.G. (1936) Proc.Roy.Soc.London B 119, 474-492.
23. Gatehouse,J., D.Boulter (1980) Physiol.Plant. 49, 437-442.
24. Kato,G., Y.Maruyama, M.Nakamura (1981) Plant Cell Physiol. 22, 759-771.
25. Cutter,E.C., L.J.Feldman (1970) Amer.J.Bot. 57, 190-201.
26. Cutter,E.C., L.J.Feldman (1970) Amer.J.Bot. 57, 202-211.
27. Kauss,H., C.Glaser (1974) FEBS Lett. 45, 304-307.
28. Bohlool,B.B., E.L.Schmidt (1974) Science 185, 269-271.
29. Dazzo,F.B., D.H.Hubbell (1975) Appl.Microbiol. 30, 1017-1033.
30. Dazzo,F.B., C.A.Napoli, D.H.Hubbell (1976) Appl.Environ. Microbiol. 32, 166-171.
31. Zurkowski,W. (1980) Microbios 27, 27-32.
32. Stacey,G., A.S.Paau, W.J.Brill (1980) Plant Physiol. 66, 609-614.
33. Kijne,J.W., G.Smit, C.L.Díaz, B.J.J.Lugtenberg (1986) Proceedings 6th Int.Symp.Nitrogen Fixation, Corvallis, in press.
34. Dazzo, F.B., R.I.Hollingsworth, J.E.Sherwood, M.Abe, E.M. Hrabak, A.E.Gardiol, H.S.Pankratz, K.B.Smith, Hong Yang (1986) Proceedings 6th Int.Symp.Nitrogen Fixation, Corvallis, in press.
35. Robertson,J.G., P.Lyttleton, C.E.Pankhurst (1981) in: A.H.Gibson, W.E.Newton (eds.) Current perspectives in nitrogen fixation, Aust.Acad.Sci., pp. 280-291.

36. Pull,S., S.Pueppke, T.Hymowitz, J.Orf (1979) Science 200, 1277-1279.
37. Van der Schaal,I.A.M., J.W.Kijne, C.L.Díaz, F.van Iren (1983) in: T.C.Bøg-Hansen, G.A.Spengler (eds.) Lectins Vol 3, W.de Gruyter, Berlin, pp. 531-538.
38. Badenoch-Jones,J., D.J.Flanders, B.G.Rolfe (1985) Appl.Env. Microbiol. 49, 1511-1520.
39. Tsien,H., M.Jack, E.Schmidt, F.Wold (1983) Planta 158, 128-133.
40. Dombrink-Kurtzman,M., W.Dick, K.Burton, M.Cadmus, M.Slodki (1983) Biochem.Biophys.Res.Comm. 111, 798-803.
41. Hrabak, E., G.Truchet, F.Dazzo (1984) In: C.Veeger, W.Newton (eds.) Advances in nitrogen fixation research, Nijhoff/Junk, Pudoc, p.419.
42. Kamberger,W. (1979) FEMS Microbiol.Lett. 6, 361-365.
43. Pueppke,S. (1983) In: T.C.Bøg-Hansen, G.A.Spengler (eds.) Lectins Vol 3, W.de Gruyter, Berlin, pp. 513-521.
44. Causse,H., P.Rougé (1983) In: T.C.Bøg-Hansen, G.A.Spengler (eds.) Lectins Vol 3, W.de Gruyter, Berlin, pp. 559-572.
45. Matthews,S., W.T.Bradnock (1968) Hort.Res. 8, 89-93.
46. Lord,J.M. (1985) Eur.J.Biochem. 146, 403-409.
47. Lord,J.M. (1985) Eur.J.Biochem. 146, 411-416.
48. Peumans,W., H.M.Stinissen (1983) see ref. 6, pp. 99-116.
49. Stinissen,H.M., M.J.Chrispeels, W.J.Peumans (1985) Planta 164, 278-286.
50. Mishkind,M.L., N.V.Raikhel, B.A.Palevitz, K.Keegstra (1983) see ref.6, pp. 163-176.
51. Mirelman,D.E., E.Galun, N.Sharon, R.Lotan (1975) Nature 256, 414-416.
52. Mishkind,M.L., N.V.Raikhel, B.A.Palevitz, K.Keegstra (1982) J.Cell Biol. 92, 753.
53. Mishkind,M., K.Keegstra, B.A.Palevitz (1980) Plant Physiol. 66, 950.
54. Tabari,F., J.Balandreau, R.Bourrilon (1984) Biochem.Biophys. Res.Comm. 119, 549-555.
55. Read,S.M., D.H.Northcote (1983) Planta 158, 119-127.
56. Gietl,C., H.Kauss, H.Ziegler (1979) Planta 144, 367-371.
57. Allen,A.K. (1983) see ref.6, pp. 71-85.
58. Van Driessche,E., S.Beeckmans, B.Dejaegere, L.Kanarek (1985) In: T.C.Bøg-Hansen, J.Breborowicz (eds.) Lectins Vol 4, W.de Gruyter, Berlin, pp. 567-582.
59. Sequeira,L., T.L.Graham (1977) Physiol.Plant Pathol. 11, 43-54.
60. Leach,J. (1981) PhD Thesis, Univ. Wisconsin, Madison.
61. Mellon,J.E., J.P.Helgeson (1982) Plant Physiol. 70, 401-405.

PEA LECTIN BINDING ACTIVITY OF PEA ROOT EXTRACTS

Tove M.I.E Christensen, Clara L. Díaz and Jan W. Kijne,

Dept. of Plant Molecular Biology, Botanical Laboratory, University of Leiden, Nonnensteeg 3, 2311VJ Leiden, The Netherlands.

Lectins of leguminous plants, localized on growing root hair tips have been suggested to recognize the bacterial root-nodule symbiont *Rhizobium*, and to mediate specific attachment of rhizobia to the host plant roots (1,2). Pea root lectin has been isolated and partially characterized; its sugar-binding specificity is similar to that of the pea seed isolectins (3,4,5). Presence of lectin receptors on the pea root surface is suggested by the specific adsorption of pea seed lectin (5). Similarly, surface lectin receptors seem to be involved in hapten-reversible root hair tip adhesion of white clover seedlings (6). Endogenous lectin receptors can play a regulatory role in the first rhizobial infection stages as bacterial and plant receptors may compete for lectin binding sites.

Most information about legume lectins and lectin binders concerns seed components. Endogenous seed lectin receptors have been found in *Glycine max*, *Canavalia ensiformis*, *Pisum sativum*, *Vicia sativa*, *Vicia faba* and *Ricinus comunis* seeds (7,8). Besides, a pea seed lectin binder was also found in other parts of the plant, and is said to be a cytoplasmatic protein (9).

This report shows preliminary results of the screening of pea root slime and root cell wall extracts for lectin binding activity with the use of an enzyme-linked assay (10). This technique allows the semi-quantitative determination of lectin binding by monovalent and polyvalent sugar-containing compounds, provided that the potential receptors are water soluble. The effect of seedling age (4 and 7 days old plants were used) and nitrogen supply (growth with 20 mM NO_3^- inhibits root nodulation, 11) were investigated.

Lectins, Vol. V

MATERIALS AND METHODS

Pea seeds (cultivar "Finale", CEBECO, Rotterdam) were surface sterilized, imbibed overnight with sterile water and planted in medium-soaked gravel (Raggio-medium, 12, supplied with 20 mM NO_3^- if required). Roots were collected after 4 or 7 days growth in darkness at 20°C. Cell walls were prepared according to (13), including extensive washings with phosphate buffer, 1:1 chloroform: methanol and acetone. Cell wall preparations were weighed immediately after lyophilization. Initially, cell walls were extracted with distilled water at 80°C during 2h. However, additional lectin binding activity could be obtained after prolonged extraction. Ultimately, for receptor isolation, cell walls of 4 days old seedlings grown in presence of 20 mM NO_3^- were extracted three successive times, the extracts being combined. All extracts were separated into 60% ethanol-soluble (S) and insoluble (P) fractions. P fractions were recovered by centrifugation (30 min 23,000 g at 2°C) and lyophilized after evaporation of residual ethanol. The ethanol-soluble fractions were concentrated under vacuum and either lyophilized (crude S) or separated into fractions with components greater or smaller than 10,000 D (S>10,000 and S<10,000, obtained by filtration through a CX-10 Millipore Immersible Unit with a cut-off of 10,000 D). Filtrates were either kept frozen at -20°C or in a lyophilized state until use.

Water-washed root slime (RS) and cell wall extracts performed with buffered 1.5 M NaCl (CWE) were prepared from a different set of plants, as previously described (11).

All extracts and fractions were tested with an enzyme-linked pea lectin binding assay, ELBA (10). In this test, the activity of unknown receptors is compared with known quantities (equivalents) of the hapten mannose needed to achieve a similar inhibition of the binding of labelled pea lectin to a solid phase-bound receptor (ovalbumin). Pea root lectin was determined with an immuno-sandwich ELISA (11).

Ion-exchange chromatography of S>10,000 filtrates was performed using a 14X1 cm Sephadex DEAE A-25 column equilibrated with water. After elution of unbound material with 100 ml water, a salt gradient (0-0.6 M NaCl in water, 140 ml) was applied. The elution pattern was followed by absorption at 280 nm and by estimation of total sugar content. Peaks were pooled, lyophilized, dissolved in water, desalted with Trisacryl GF 05 (10 ml columns) and again freeze-dried. Dry weight was determined immediately.

Total sugars wre analyzed with a micro phenol-sulfuric acid test (14) using glucose as a standard. Protein was estimated with a modified Coomassie Blue G-dye reagent (0.01% Serva Blue G in 1.6 M phosphoric acid/0.8 M ethanol) as described in (15) with BSA as a standard. Gas liquid chromatography was performed according to (16), after hydrolysis of the samples during 1h at 120°C in 2N trifluoracetic acid, with the use of a Becker 420 GC equipped with a 3% ECNSS/M Gaschrom Q column.

RESULTS

Different amounts of pea lectin binding activity were found in every type of root cell wall extract and in root slime (see table 1). The relatively low binding activity of root slime and cell wall extracts obtained with buffered salt might be correlated with the interfering presence of root lectin in these fractions. Higher receptor activity was observed in the hot water extracts, in which cell wall-associated lectin could not be detected. The latter result might be due either to the hot water treatment or to the firm association of the lectin to the cell wall.

Table 1. *Lectin binding activity (mannose equivalents) and lectin content of root extracts from 4 and 7 days old pea seedlings grown with or without 20 mM NO_3^-.*

	ELBA µg mannose equivalents/ mg sample	ELISA µg pea lectin/ mg sample
4 P[a]	125	0[e]
4 P NO_3^-	240	0
7 P	157	0
7 P NO_3^-	300	0
4 S[b]	185	0
4 S NO_3^-	175	0
7 S	150	0
7 S NO_3^-	95	0
4 CWe[c]	89	0.61[f]
4 CWe NO_3^-	54	0.28
7 CWe	140	0.50
7 CWe NO_3^-	94	1.01
4 RS[d]	55	0.47
4 RS NO_3^-	78	0.75
7 RS	60	0.06
7 RS NO_3^-	80	0.06

a: 60% ethanol precipitates, cell walls extracted with H_2O, 2h, 80°C.
b: 60% ethanol-soluble fractions, extracted as described in a.
c: cell walls extracted with 1.5 M NaCl, as in (11).
d: root slime, prepared as in (11).
e: below detection level (20 ng).
f: data from (11).

Table 2. *Total protein, sugar content and lectin binding acitvity of the ethanol soluble fractions (S) of hot water extracts obtained from 3 sets of cell wall preparations after different times of extraction. Results presented in mg/gram dry weight root cell walls from 4 days old plants grown with 20 mM NO_3^-.*

	mg protein	mg sugar	ELBA-activity
S>10,000 D			
2 h extraction	0.71	7.13	2.24
4 h extraction	1.01	9.88	3.54
6 h extraction	3.09	8.36	9.82
S<10,000 D			
2 h extraction	traces	1.28	0.42
4 h extraction	traces	1.51	0.74
6 h extraction	traces	2.94	4.31

Seedling age and growth in a NO_3^--rich medium significantly increased the lectin binding activity of the ethanol-insoluble P fractions. The S fractions also showed a relatively high lectin receptor activity when tested with the ELBA; lectin binding decreased when the plants were grown with NO_3^- for 7 days.

Gas liquid chromatography showed that glucose and mannose (pea lectin haptens) as well as rhamnose, arabinose, xylose and galactose were constant components of the root extracts tested. The distribution of these sugars presents characteristic patterns in relation to extract type, seedling age and growth in a NO_3^--rich medium (data not given). No direct positive correlation between hapten content and lectin binding activity could be found.

Since the S-fractions were water soluble and their yield was at least 2 times higher than that of the P-fractions, the S-fractions were chosen for further analysis. Table 2 resumes a time-scaled extraction of lectin binding activity of S-extracts from cell walls prepared from 4 days old plants grown with 20 mM NO_3^-. Most of the lectin binding activity could be obtaines from the set of cell walls subjected to 6 h extraction. Binding to lectin seems to follow carbohydrate contnet in the small molecular weight fraction. However, in the high molecular weight fraction binding activity appears to be correlated with the increased extraction of protein, as most of the sugar-containing compounds are extracted after 2 h.

Subjection of the S>10,000 fractions (4 h extraction) to ion exchange chromatography resulted in a neutral and several "acid" peaks. The elution pattern was similar for extracts prepared from 4 days old plants grown with or without NO_3^- (figure 1). When the pooled, desalted peaks were tested with the ELBA, all of them were found to have lectin binding activity. No correlation between sugar content and binding to lectin could be found. However, pea lectin receptor activity and protein content might be positively correlated, as the ELBA-activity roughly follows protein content.

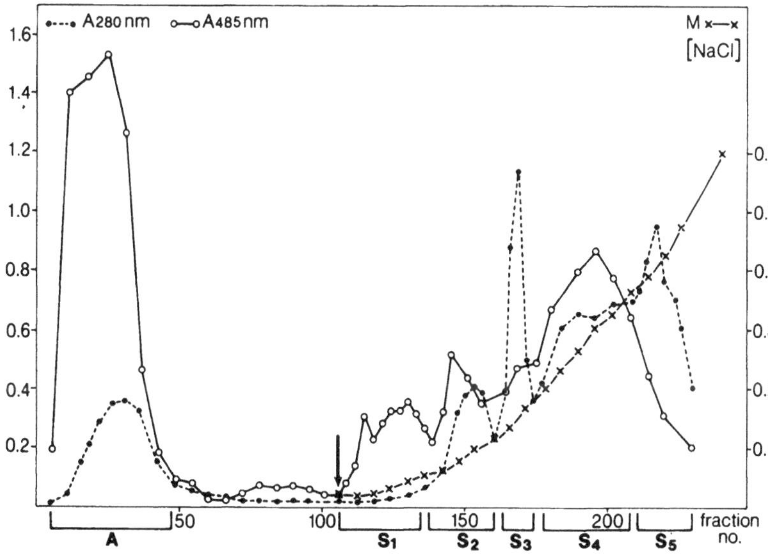

Figure 1. *Elution pattern of S>10,000 D after Sephadex DEAE-A25 chromatography. Carbohydrate content (o—o) was estimated with the Phenol-H_2SO_4 test (100 l per tube) and is expressed as absorption at 485 nm. Protein (•--•) is a function of the absorption at 280 nm. The application of the salt gradient (x—x) is indicated by the arrow. Unbound components are pooled as peak A; peaks S_1, S_2, S_3, S_4 and S_5 represent pooled components eluted at different salt concentrations.*

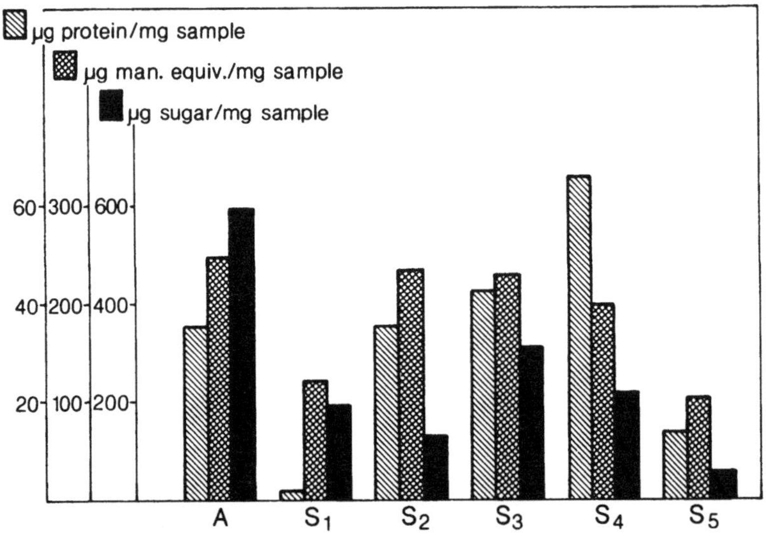

Figure 2. *Lectin binding activity, protein and sugar content per mg pooled fractions resulting from the ion exchange chromatography of S 10,000 D pea root cell wall extracts. (See Fig. 1). Pea lectin binding activity* ▧ *is expressed as μg mannose equivalents, carbohydrates* ■ *as μg glucose (Phenol-H_2SO_4 test) and μg protein* ▩ *as determined with a modified Coomassie Blue-dye reagent.*

DISCUSSION

These preliminary results show that it is possible to extract pea lectin-binding compounds from cell wall preparations obtained from 4 and 7 days old pea seedlings. The lectin binders are either secreted into the root slime or bound in the cell wall matrix. Extraction of the cell wall-associated lectin binders can be carried out using chaotropic agents (1.5 M NaCl) or with extended extraction at high temperature (6 h, water at 80°C). Although the possibility that one type of plant cell wall component might be extracted by different means (17) cannot be set aside, it is probable in this case that different components come into solution with the different extraction methods (18).

One of the characteristics of the ELBA is that the binding of complex receptors to pea lectin is compared to that of mannose. It is known that the lectin binding constant of a specific sugar might be several orders of magnitude lower than the binding constant of a glycocojugate containing the same sugar (19). That is also why we are using the GLC data only to confirm the presence of haptenic sugars in pea lectin binding extracts. Unfortunately the ELBA does not give information about the binding affinity of the lectin receptors tested.

These results suggest that pea root lectin may interact with various root cell wall-associated receptors rather than with one type, as it seems to be the case in cotyledons (7,8,9). It remains to be studied if pea lectin binding activity is restricted to pea root cell walls or has a wider distribution in leguminous plants. Speculations on the role of cell wall associated lectin receptors are not meaningful at this stage of study. Characterization of a lectin receptor molecule from these fractions should prove that the ELBA is a valuable tool for the screening of crude extracts for pea lectin binding activity in a purification procedure.

Acknowledgements: We thank Han Bernard Jansen for his help. T.M.I.E.C. was recipient of grants from EMBO and the Dutch Ministry of Education.

REFERENCES

1. Bohlool, B.B., Schmidt, E.L. (1974) Lectins: a possible basis for specificity in the *Rhizobium*-legume root nodule symbiosis. Science 185: 269-271.
2. Dazzo, F.B., Hubbell, D.H. (1975) Cross-reactive antigens and lectin as determinants of symbiotic specificity in the *Rhizobium*-clover association. Appl. Microbiol 30:1018-1033.
3. Hosselet, M., Van Driessche, E., Van Poucke, M., Kanarek,L. (1983) Purification and characterization of root lectins from *Pisum sativum* L. In: Lectins, Biology, Biochemistry, Clinical Biochemistry, vol 3, (Bøg-Hansen, T.C., Spengler, G.A., eds.) De Gruyter, Berlin-New York, pp 549-558.
4. Kijne, J.W., Van der Schaal, I.A.M.,De Vries, G.E. (1980) Pea lectins and the recognition of *Rhizobium leguminosarum*. Plant Sci. Letters 18: 65-74.
5. Kijne, J.W., Van der Schaal, I.A.M., Díaz, C.L., Van Iren, F. (1983) Mannose specific lectins and the recognition of pea roots by *Rhizobium leguminosarum*. In: Lectins, Biology, Biochemistry, Clinical Biochemistry, vol 3, (Bøg-Hansen, T.C., Spengler, G.A., eds.) De Gruyter, Berlin, New York, pp 521-529.
6. Dazzo, F.B., Truchet, G.L., Kijne, J.W. (1982) Lectin involvement in root hair tip adhesion as related to the *Rhizobium*-clover symbiosis. Physiol. Plant. 56: 143-147.
7. Bowles, D.J., Marcus, S. (1981) Characterization of receptors for the endogenous lectins of soybean and jackbean seeds. FEBS Letters 129: 135-138.

8. Gansera, R., Schurz, H., Rüdiger, H. (1979) Lectin-associated proteins from the seeds of Leguminosae. Hoppe-Seyler's Z. Physiol. Chem. 360: 1579-1585.
9. Rüdiger, H. Gebauer, G., Schurz, J. (1983) Lectins and lectin binders (receptors) from plants. In: Chemical Taxonomy, Molecular Biology and Function of Plant Lectins (Goldstein, I.J., Etzler, M.E. eds.) Alan R. Liss, Inc., New York, pp 237-248.
10. Van der Schaal, I.A.M., Logman, T.J., Diáz, C.L., Kijne, J.W. (1984) An enzyme-linked lectin binding assay for quantitative determination of lectin receptors. Anal. Biochem. 140:48-55.
11. Díaz, C.L., Lems-van Kan, P., Van der Schaal, I.A.M., Kijne, J.W. (1984) Determination of pea (*Pisum sativum* L.) root lectin using an enzyme-linked immunoassay. Planta 161: 302-307.
12. Raggio, N., Raggio M. (1956) Relación entre cotiledones y nodulación y factores que la afectan. Phyton 7: 103-119.
13. Buchala, A., Franz, G. (1974) A hemicellulosic β-glycan from the hypocotyls of Phaseolus aureus. Phytochem. 13: 1887-1889.
14. Klis, F., Rootjes, M., Groen, S., Stegwee, D. (1983) Accelerated accumulation of wall-bound hydroxyproline in artificially induced regions of bean hypocotyl sections. Z. Pflanzenphysiol. 110: 301-307
15. Read, S.M., Northcote, D.H. (1981) Minimization of variation in the response to different proteins to the Coomassie Blue G dye-binding assay for protein. Anal. Biochem. 116: 53-64.
16. Albersheim, P., Nevins, D.J., English, P.D., Karr, D. (1967) A method for the analysis of sugars in plant cell wall polysaccharides by gas liquid chromatography. Carbohydr. Res. 5: 340-345.
17. Hass, D., Frey, R., Thiesen, M., Kauss, H. (1981) Partial purification of a hemagglutinin associated with cell walls from hypocotyls of *Vigna radiata*. Planta 151: 490-496.
18. Kauss, H., Glaser, C. (1974) Carbohydrate-binding proteins from plant cell walls and their possible involvement in extension growth. FEBS Letters 45: 304-307.
19. Kornfeld, R., Feris, C. (1975) Interaction of immunoglobulin glycopeptides with Concanavalin A. J. Biol. Chem. 250: 2614-2619.

SOYBEAN LECTIN AND THE INTERACTION OF RHIZOBIUM WITH CULTURED SOYBEAN CELLS.

Siu-Cheong Ho, Shahnaz Malek-Hedayat and Sally Meiners.

Department of Biochemistry, Michigan State University, East Lansing, MI 48824, USA

Rhizobium infect root hairs of leguminous plants as an early step in the nitrogen-fixing root nodule symbiosis. This process is specific, since most Rhizobia will nodulate only one host (1). Bohlool and Schmidt (2) suggested that the attachment of Rhizobium to legume roots could be mediated by specific lectin-polysaccharide interactions. However, the presence of lectin at the site of Rhizobium-root cell interaction is still a subject of controversy (3,4). Attempts to detect the lectin (Soybean agglutinin, SBA) on the surfaces of soybean roots and root hairs by immunological techniques have yielded conflicting results (5,6).

We now report the identification and isolation of a lectin from cultures of the SB-1 cell line, which is derived originally from soybean roots. This lectin, designated soybean lectin (SB-1), can be found on the cell wall of the cultured cells. Moreover, this lectin may be responsible for our observed binding of Rhizobium japonicum to the cultured SB-1 cells.

Materials and Methods

Antibodies Directed Against Seed SBA

Seed SBA was isolated from soybean meal (Glycine max) by affinity chromatography on Sepharose-N-caproyl-galactosamine columns (7). Antibodies directed against seed SBA were raised in New Zealand rabbits. Monospecific anti-SBA antibodies were isolated by affinity chromatography on an SBA-Sepharose 4B column. The isolated lectin and the antibody preparation were characterized by polyacrylamide gel electrophoresis in sodium dodecyl sulfate (SDS) and immunoblotting techniques.

Identification and Localization of Soybean Lectin (SB-1)

The SB-1 cell cultures, derived from soybean (Glycine max) roots, were grown in liquid cell suspension in 1B5C medium as reported

Lectins, Vol. V

previously (8). Protoplasts were prepared by digestion of cell wall with cellulase and pectinase (8). The presence of soybean lectin SB-1 on the surface of SB-1 cells or protoplasts was detected by indirect immunofluorescent technique using rabbit anti-SBA antibody and fluorescein-conjugated goat anti-rabbit IgG as primary and secondary antibodies, respectively. Normal rabbit IgG was used as primary antibody in control experiments.

Soybean lectin (SB-1) was purified from the medium derived from 4-day old culture of SB-1 cells. The procedures used were similar to those carried out in the purification of seed SBA. To insure that the material isolated from SB-1 cells was not contaminated by seed SBA, affinity columns previously not exposed to seed SBA were used.

Rhizobium Culture and Binding to SB-1 Cells
Rhizobium japonicum (R110d) was cultured in liquid yeast extract-mannitol-gluconate medium at 30°C (9). SB-1 cells (2-day old cultures) in 1 ml suspension were co-cultured with 0.1 ml of 4-day-old Rhizobium culture. The co-cultures were incubated at 26°C for 24 h in the dark. At the end of the incubation, the cells were washed three times before observing under the Leitz microscope with phase contrast optics.

Results

Identification of Soybean Lectin (SB-1) in Cultured Cell Line

Soybean lectin (SB-1) was purified from the culture medium of SB-1 cells by affinity chromatography. Figure 1 shows the polyacrylamide gels of the purified seed SBA and soybean lectin (SB-1) analyzed by silver staining and immunoblotting techniques. The soybean lectin (SB-1) was characterized by cross-reactivity with anti-seed SBA antibody and its similar mobility to seed SBA under

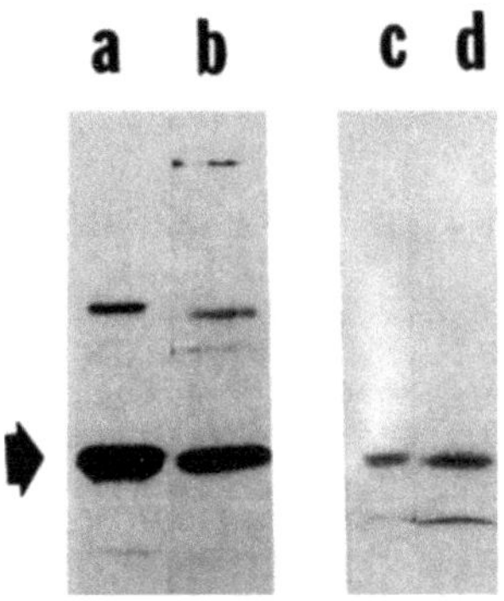

Figure 1. SDS-polyacrylamide gel electrophoresis of purified SBA (a and c) and soybean lectin (SB-1), (b and d). Immunoblotting analysis with rabbit anti-seed SBA antibody (a and b); silver staining (c and d).

SDS-gel electrophoresis. These results showed that the molecular weight is identical for the soybean lectin (SB-1) and seed SBA and that they share antigenic determinants.

Localization of Soybean Lectin (SB-1) on the Cell Surface

Indirect immunofluorescent staining of SB-1 cells using rabbit anti-seed SBA antibody showed fluorescence around the outer periphery of individual cells, while normal rabbit IgG failed to show similar staining (Figure 2a,b). Similarly, specific immunological staining was observed on the protoplast surface (Figure 2c,d). These results suggest the localization of a soybean lectin (SB-1) on the surfaces of the plasma membrane and the cell wall.

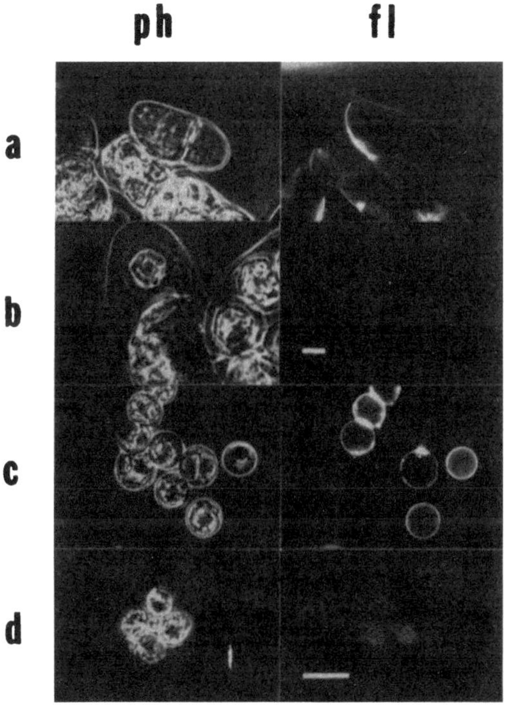

Figure 2. Immunofluorescent staining patterns of SB-1 cells and protoplasts derived from SB-1 cells. Cells were treated with rabbit anti-seed SBA (a and c) or normal rabbit immunoglobulins (b and d), followed by fluorescein conjugated goat anti-rabbit immunoglobulin. ph, phase contrast microscopy; fl, fluorescent microscopy, bar = 5 μm.

Binding of Rhizobium to SB-1 Cells
When SB-1 cells were co-cultured with Rhizobium japonicum (R110d), the bacteria attached to only certain cells (Figure 3a). These bacteria adhered to the plant cells in a polar fashion.

The binding of the Rhizobium to the soybean cells was specific. First, in a parallel experiment under the same conditions, Escherichia coli failed to bind to the plant cells. Second, the Rhizobium did not bind to protoplasts derived from SB-1 cells after removal of cell wall, even though the plasma membrane of SB-1 protoplasts contained sobyean lectin (SB-1) as detected by immunofluorescence studies. Third, the binding of Rhizobium was specifically inhibited by galactose (Figure 3c), but not by glucose (Figure 3b). Fourth, when the Rhizobium was incubated with the soybean cells in the presence of anti-seed SBA, inhibition of bacterial adhesion to the plant cells was observed. The lowest concentration that exhibited inhibition of Rhizobium binding was 20 µg/ml (Figure 3d). Normal rabbit immunoglobulin showed no inhibition of binding at a concentration of 1 mg/ml. These combined results strongly suggest that Rhizobium binds

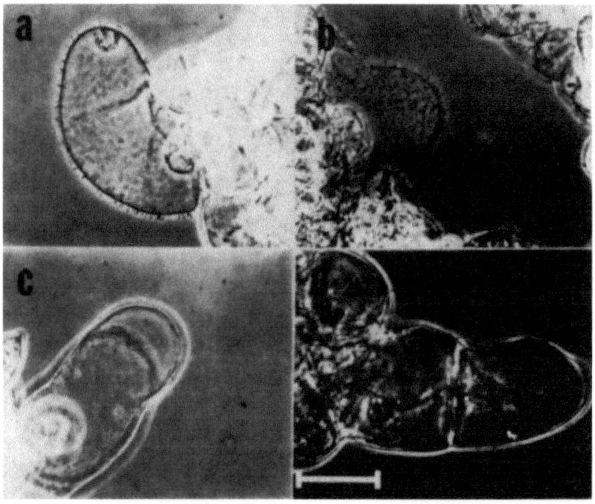

Figure 3. Representative photographs showing adhesion of Rhizobium japonicum (R110d) to SB-1 cells after co-culture for 24 h. (a) co-culture; (b) co-culture in the presence of 0.1 M glucose; (c) co-culture in the presence of 0.1 M galactose; (d) co-culture in the presence of 20 µg/ml of rabbit anti-seed SBA. Bar = 10 µm.

specifically to SB-1 cells, possibly via a lectin-mediated carbohydrate recognition.

Discussion

In this paper, we clearly demonstrate that SB-1 cells, a cultured cell line derived from roots of Glycine max, contain the soybean lectin (SB-1) with biochemical and immunological characteristics similar to seed SBA. This lectin immunologically cross-reacted with anti-seed SBA antibody and showed a distinct polypeptide band at 30,000 molecular weight in SDS-polyacrylamide gel. This lectin also showed a specific carbohydrate recognition to the affinity column by which its purification was achieved. In contrast to seed SBA, which is known to be a storage protein found inside the protein bodies in the cytoplasm of the cell (10), the soybean lectin (SB-1) was found located on the plasma membrane and on the cell wall surfaces. Moreover, it seems to be released into the culture medium, which provided a convenient source for its purification. Whether the biochemical nature of the released lectin and the cell-surface lectin is the same remains to be investigated. However, they both react with anti-seed SBA antibody. The present identification of the soybean lectin (SB-1) from the cultured cell line originally derived from soybean roots is consistent with the notion that the roots endogenously produce such a lectin. This finding fulfills the prerequisite for the lectin recognition hypothesis, that a lectin is involved at the site of Rhizobium recognition.

Indeed, our data strongly support the lectin recognition hypothesis (Figure 4). The arguments are listed as follows: (a) Rhizobium specifically binds to SB-1 cells but E. coli does not; (b) this binding appears to be mediated via carbohydrate recognition, inasmuch as galactose can inhibit the heterotype adhesion, whereas glucose failed to inhibit; (c) the likely candidate that mediates such an interaction is a lectin identified on the cell wall. This notion is supported by the observation that rabbit anti-seed SBA blocked the Rhizobium-soybean cell adhesion; (d) the cell surface lectin, produced endogenously by SB-1 cells, contained a polypeptide immunologically cross-reactive

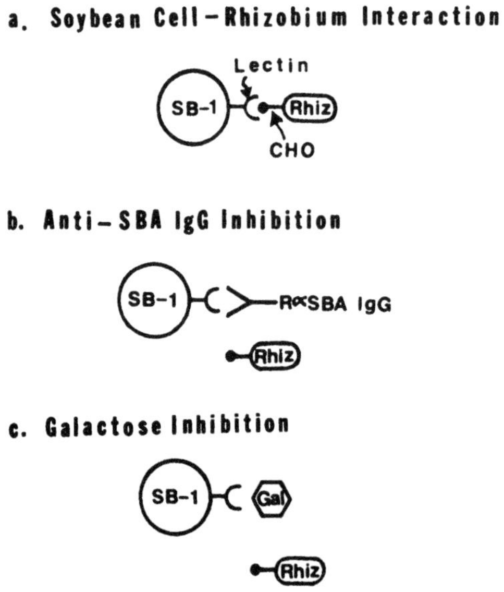

Figure 4. Evidences for soybean lectin (SB-1) in mediating Rhizobium recognition and binding. CHO, represents carbohydrate component in Rhizobium; SB-1, soybean cell line; Rhiz, Rhizobium japonicum; and Gal, galactose.

with seed SBA. However, our monospecific polyclonal anti-seed SBA IgG could not distinguish the two lectins.

It has been previously reported that a soybean lectin can be found in roots or root exudates (11,12). Gade et al. have extracted a soybean lectin from roots and showed its homology to seed lectin (11). The involvement of the soybean lectin in the initiation of the nodulation process has been clearly demonstrated by Halverson and Stacey (12,13). They further suggest that the lectin found in soybean root exudate is genetically distinct from the seed lectin. Two soybean lectin genes have been identified through molecular cloning techniques (14). Possibly, one of these genes is expressed during seed development and the other is expressed during root development.

There have been several previous reports on the binding of Rhizobium to cultured cells derived from callus of soybean roots (15-17). In these systems, the interaction of Rhizobium with soybean cells ultimately led to infection of the plant cells and the generation of a nitrogen-fixing symbiosis. It remains to be demonstrated that our present Rhizobium adhesion to SB-1 cells will lead to a symbiosis and activation of nitrogenase. Nevertheless, this root cell culture would provide a valuable system to further characterize the soybean lectin (SB-1), in terms of its molecular nature, its biosynthetic regulation, and its involvement in specific Rhizobium recognition.

Acknowledgement

This work was supported by Grant 83-CRCR-1-1288 from the U.S. Department of Agriculture, Grant PCM-8011736 from the National Science Foundation, and a McKnight Award in Plant Biology.

References

1. Bauer, W.D.: Annu. Rev. Plant Physiol. 32, 407-449 (1981)
2. Bohlool, B.B. and Schmidt, E.L.: Science 185, 269-271 (1974)
3. Pueppke, S.G. and Bauer, W.D.: Plant Physiol. 61, 779-784 (1978)
4. Su, L.C., Pueppke, S.G. and Friedman, H.P.: Biochim. Biophys. Acta 629, 292-304 (1980)
5. Stacey, G., Paan, A.S. and Brill, W.F.: Plant Physiol. 66, 609-614 (1980)
6. Pueppke, S.G., Friedman, H.P. and Su, L.C.: Plant Physiol 68, 905-909 (1981)
7. Allen, A.K. and Neuberger, A.: FEBS Lett. 50, 362-364 (1975)
8. Metcalf, T.N., Wang, J.L., Schubert, K.R. and Schindler, M.: Biochemistry 22, 3969-3975 (1983)
9. Bhuvaneswari, T.V., Turgeon, B.G. and Bauer, W.D.: Plant Physiol. 66, 1027-1031 (1980)
10. Horrisberger, M. and Vonlanthen, M.: Histochemistry 65, 181-186 (1980)
11. Gade, W., Jack, M.A., Dahl, J.B., Schmidt, E.L. and Wold, F.: J. Biol. Chem. 256, 12905-12910 (1981)
12. Halverson, L.J. and Stacey, G.: Plant Physiol. 77, 621-624 (1985)
13. Halverson, L.J. and Stacey, G.: Plant Physiol. 74, 84-89 (1984)
14. Vodkin, L.O., Rhodes, P.R., Goldberg, R.B.: Cell 34, 1023-1031 (1983)
15. Child, J.J. and LaRue, T.A.: Plant Physiol. 53, 88-90 (1974)
16. Phillips, D.A.: Plant Physiol. 53, 67-72 (1974)
17. Reporter, M., Raveed, D. and Norris, G.: Plant Science Lett. 5, 73-76 (1975)

GRAMINEAE LECTINS : A SPECIAL CLASS OF PLANT LECTINS

Willy J. Peumans, Bruno P.A. Cammue
Laboratorium voor Plantenbiochemie, Katholieke Universiteit Leuven, Kardinaal Mercierlaan 92, B-3030 Leuven, Belgium

Lectins are present in a wide variety of plant species spanning almost every taxonomical classification in the plant kingdom (1, 2, 3). At present over 100 phytohemagglutinins have been purified and partially characterized with respect to their biochemical properties, molecular structure and carbohydrate binding specificity. Although most information is available on legume lectins which have been proven an excellent source of material because of their abundance in the seeds of these plants, it is becoming increasingly clear that different classes of plants contain lectins with different biochemical properties and molecular structures. Moreover, as more information becomes available on lectins from different taxonomic groups, striking similarities between lectins from related plant species become obvious. Indeed, many legume lectins have homologous amino acid sequences (4) and are related serologically (5) whereas all Solanaceae lectins known at present have a common molecular structure and related serological determinants (6). Similarly, lectins found in Gramineae species appear to be closely related to each other both structurally and immunologically (7). These observations support the idea that plant lectins, or at least the greater part of them, represent several natural groups of related proteins which most likely are derived from common evolutionary ancestors (5, 6, 7). In this contribution an overview is given of the lectins which occur in different representatives of the plant family Gramineae. Thereby, the emphasis is put on the most recent advances made in this field especially with respect to Gramineae lectins from vegetative tissues.

Lectins, Vol. V

A. Gramineae seed lectins

A1. Wheat germ agglutinin (WGA) : The first lectin which was isolated from a member of the plant family Gramineae was purified from commercial wheat germ and hence designated as wheat germ agglutinin or WGA (8). At present it is generally accepted that WGA is a dimeric protein composed of 2 identical, not covalently bound subunits. The lectin monomers are polypeptides of 171 amino acids which do not contain covalently bound carbohydrate and are particularly rich in glycine and cysteine (42 and 32 residues respectively per monomer)(9). As could be concluded from the complete amino acid sequence of WGA, its subunits are build up of 4 strikingly homologous domains of 43 amino acids each (9). Although WGA is usually considered as a N-acetylglucosamine-specific lectin, oligomers of this sugar are much more inhibitory in hapten inhibition assays. Based on this different inhibitory effect of N-acetylglucosamine-oligomers of increasing chain length, it was proposed that the binding site of WGA consists of 3 or 4 subsites with differing specificity (8).

A2. WGA-like lectins from other Triticeae species : Although originally considered as a unique lectin, it soon became evident that WGA represented a mixture of different molecular forms (8). A logic explanation for the occurrence of the so-called WGA-isolectins followed after it turned out that in the cells of allohexaploid wheat (genome AABBDD) each genome (i.e. A, B and D) directs the synthesis of its own lectin (10, 11). Taking into consideration that the 3 related genomes of wheat (which are in fact derived from three different diploid wheat species) code for lectins with strikingly similar biochemical,physicochemical and biological properties, the question arose whether other species with genomes related to those of wheat also contain such lectins. The isolation of 2 lectins from rye and barley respectively, which were structurally and serologically indistinguishable from WGA confirmed the occurrence of WGA-like lectins (further referred to as 'cereal lectins') in at least these 2 cereal species (12). Later, the number of species containing cereal lectins was extended to not less than 77 when it became evident that embryos of all species of the Triticeae tribe (which comprises the genera

Aegilops, Agropyrum, Elymus, Eremopyrum, Haynaldia, Heteranthelium, Hordeum, Psathyrostachys, Secale, Triticosecale and *Triticum*) do contain lectins with the same molecular structure, carbohydrate-binding specificity and serological properties as WGA (13).

A3. WGA-related lectins in Brachypodium and Oryza (rice) species
The observation that WGA-like lectins occur in such a large number of Triticeae species justified a search for similar lectins in grass species from different taxonomic origin. A survey of such a search within the Gramineae family resulted in the detection of two types of WGA-related lectins. A first type, which occurs in the embryos of *Brachypodium* species has the same molecular structure and sugar specificity as WGA but is serologically different (14). The second type, wich is found in embryos of species belonging to the genus *Oryza* (rice) is rather complex as several subtypes can be discerned. Basically, all rice lectins have the same molecular structure and carbohydrate-binding specificity as WGA and are serologically related to but not identical with the latter lectin. However, 3 subtypes can be distinguished on the basis of the presence or absence of 'cleaved' lectin monomers. In the first subtype (further referred to as the *O. sativa* type) most of the lectin monomers are cleaved into 2 polypeptides with MW of 10 000 and 8 000 daltons respectively, whereas lectins of the second subtype (referred to as *O. alta* type) contain exclusively uncleaved monomers with a MW of 18 000 daltons. The third subtype (further referred to as the *O. minuta* type) takes an intermediary position as it contains about equal amounts of cleaved and uncleaved monomers. It should be mentioned that the cleavage of the monomers is essentailly a post-translational modification due to the proteolytic processing of the 18 000 dalton polypeptides (15). At present, it is not well understood why some rice lectins are processed and others are not. A likely explanation is that some genomes encode lectin polypeptides that have sequences which are recognized by the processing enzyme whereas other genomes direct the synthesis of lectin monomers that have no such sequence.

A4. Conclusion : From the observations described above, it may be concluded that within the Gramineae family one major type of seed- or embryo lectin occurs. It can be divided, however, into 3 subtypes namely cereal, *Brachypodium* and rice lectins. Since these 3 subtypes are both structurally and serologically related to each other, they probably have evolved from a common ancestor.

B. Gramineae lectins in vegetative tissues.

B1. WGA and barley lectin in vegetative tissues : Although cereal lectins were considered as typical seed lectins because of their location in the embryos there is no doubt anymore that they occur also in different vegetaive tissues. WGA has indeed been found in root tips, stem bases and leaves of wheat plants (16, 17, 18). Similarly, field-grown barley plants contain and synthesize lectin in root tips, leaves and developing ears (19). In both cases, the lectins which were present in the vegetative tissues were by no means distinguishable from the corresponding embryo lectins so that the hypothesis was put forward that at least in cereals like wheat and barley the same lectin genes are expressed in the embryo as well as in the different vegetative tissues (19).

B2. Agropyrum repens (couch grass) leaves contain a leaf-specific variant type of cereal lectin : As has been described in the previous section, typical cereal embryo lectins such as WGA and barley lectin are also present in and are synthesized by various vegetative tissues including leaves. Recently, however, a variant type of cereal lectin has been detected in leaves of couch grass (which is a wild perennial grass species closely related to wheat). Like all Triticeae species, couch grass contains in its embryo a typical cereal embryo lectin which by no means can be distinguished from other cereal embryo lectins (3). In addition to this embryo lectin, leaves of this plant contain relatively high levels of a lectin which is comparable to but not identical with cereal embryo lectins (20). The *Agropyrum repens* leaf lectin (further referred to as ARLL) has been isolated and characterized (20). It is a dimer composed of two monomers with a MW of 19 500 daltons (i.e. 15 amino acid residues longer than these of the embryo lectins). Structural differences between couch grass leaf lectin and cereal embryo lectins were inferred from *in vitro* subunit exchange experiments and serological analyses. Whereas the couch grass embryo lectin readily formed heterodimers with embryo lectins from other cereal species ARLL did not do so with the same embryo lectins (20). Moreover, ARLL and embryo lectins appeared to be different serologically as their precipitin lines did not fuse in immunodiffusion against WGA-antiserum (20). In addition to these structural differences, ARLL exhibited specificity towards N-acetylgalactosamine over N-acetylglucosamine and preferentially agglutinated blood-group A erythrocytes whereas the embryo lectin was not inhibited by N-acetylgalactosamine and exhibited no blood-group specificity (20). Obviously, ARLL and the embryo lectin from the same species are more different than the embryo lectins from different cereal species.

Although there is no doubt that both couch grass lectins are two distinct, tissue-specific proteins, it is not possible yet to decide whether they are products of different lectin genes or different lectin products of the same gene. One can imagine for instance, that the primary translation products of the lectin mRNAs are processed differently in embryo and leaf cells.

B3. Phragmites australis (common reed) contains 2 leaf-lectins : Despite many efforts to detect leaf-specific lectins in other cereal, rice and *Brachypodium* species no such lectin could be detected hitherto. Very recently, however, we observed that leaves of common reed (which is a wild grass species taxonomically far distant from the other lectin-containing Gramineae) contain relatively high concentrations of lectin. The lectin was isolated and appeared to be a mixture of two isolectins, which could readily be separated by ion-exchange chromatography and will further be referred to as *Pragmites australis* leaf lectin I ($PALL_I$) and $PALL_{II}$. From the preliminary experiments which have been carried out until now, several interesting conclusions can be drawn. First, $PALL_I$ strongly resembles ARLL. Indeed, like ARLL, $PALL_I$ is build up of monomers with MW of 19 500 daltons, exhibits specificity towards N-acetylgalactosamine over N-acetylglucosamine and preferentially agglutinates blood-group A erythrocytes. On the contrary, $PALL_{II}$ behaves much more like WGA as it is build up of monomers of MW of 18 000 daltons and exhibits specificity for N-acetylglucosamine. In addition, $PALL_I$ and $PALL_{II}$ are serologically different. They both cross-react with WGA-antiserum

but the precipitin lines obtained with both reed lectins do not fuse. It appears, therefore, that reed leaves contain 2 distinct but related lectins. One of them ($PALL_{II}$) strongly resembles the Gramineae embryo lectins whereas the other ($PALL_{I}$) behaves much more like the leaf-specific ARLL. Again the question arise whether both reed lectins are products of different genes or different protein products of the same gene.
A final remark to make concerns the location of the reed lectins. Indeed, although they are described here as leaf-lectins they occur in the shoot meristems as well and are present there in much higher concentrations than in the leaves.

C. Conclusions

From the data described above it can be concluded that the Gramineae lectins occur in at least 4 different taxonomic groups, namely the tribe of the Triticeae, the genera *Brachypodium* and *Oryza* and the species *Phragmites australis*. Although all these lectins are undoubtly related to each other(and hence are most likely derived from a common ancestor), it is evident that several subtypes can be distinguished on the basis of structural and serological criteria. In addition, it appears that subtypes (for instance ARLL) have to be considered as tissue-specific lectins since they are found exclusively in leaves and shoot meristems.

Acknowledgements

This work is supported in part by grants of the National Fund for Scientific Research (Belgium), of which W. Peumans is Senior Research Associate. B. Cammue receives a grant of the Belgian Instituut tot Aanmoediging van het Wetenschappelijk Onderzoek in Nijverheid en Landbouw.

References

1. Liener, I.E. 1976. Ann. Rev. Plant Physiol. 27, 291.

2. Goldstein, I.J. Hayes, C.E. 1975. Adv. Carboh. Chem. Biochem.

35, 127.

3. Etzler, M.E. 1985. Ann. Rev. Plant Physiol. 36, 209.

4. Strosberg, A.D., Lauwereys, M., Foriers, A. 1983. In : Chemical taxonomy, molecular biology, and function of plant lectins. (Goldstein, I.J., Etzler, M.E., eds.). Alan R. Liss, New York, pp. 7.

5. Hankins, C.N., Kindinger, I.J., Shannon, L.M. 1979. Plant Physiol. 64, 104.

6. Kilpatrick,D.C., Jeffree, C.E., Lockhart, C.M., Yeoman, M.M. 1980. FEBS Lett. 113, 129.

7. Peumans, W.J., Stinissen, H.M. 1983. In : Chemical taxonomy, molecular biology, and function of plant lectins. (Goldstein, I.J., Etzler, M.E., eds.). Alan R. Liss, New York, pp. 99.

8. Allen, A.K., Neuberger, A., Sharon, N. 1973. Biochem. J. 131, 155.

9. Wright, C.S., Gavilanes, F., Peterson, D.L. 1984. Biochemistry 23, 280.

10. Rice, R.H. 1976. Biochym. Biophys. Acta 444, 175.

11. Peumans, W.J., Stinissen, H.M., Carlier, A.R. 1982. Planta 154, 562.

12. Peumans, W.J., Stinissen, H.M., Carlier, A.R. 1982. Biochem. J. 203, 239.

13. Stinissen, H.M., Peumans, W.J., Carlier, A.R. 1983. Planta 159, 105.

14. Peumans, W.J., Spaepen, C., Stinissen, H.M., Carlier, A.R. 1982. Biochem. J. 205, 635.

15. Stinissen, H.M., Peumans, W.J., Carlier, A.R. 1983. Plant Molec. Biol. 2, 33.

16. Mishkind, M., Raikhel, N.V., Palevitz, B.A. 1982. J. Cell Biol. 92, 753.

17. Raikhel, N.V., Mishkind, M.L., Palevitz, B.A. 1984. Planta 162, 55.

18. Stinissen, H.M., Chrispeels, M.J., Peumans, W.J. 1985. Planta 164, 278.

19. Cammue, B., Stinissen, H.M., Peumans, W.J. 1985. Plant Physiol. 78, 384.

20. Cammue, B., Stinissen, H.M., Peumans, W.J. 1985. Eur. J. Biochem. 148, 315.

RICE GERM LECTIN: LOCALIZATION, DEVELOPMENT AND CELL AGGLUTINATION

Tang Xi-Hua, Shen Rui-Juan, Sun Ce, Zu Ji-Hua

Shanghai Institute of Plant Physiology and
Shanghai Institute of Biochemistry
Academia Sinica, 320 Yue Yang Road,
Shanghai, China

At present, the physiological functions of plant lectins remain unclear and investigations dealing with this topic have been confined largely to legume lectins. However, lectins have been shown to be present throughout the plant kingdom and most probably lectins perform a number of different functions (3,4). Within the graminae family, only WGA has been studied in some detail with respect to the physiological function. It was suggested by Mirelman and coworkers that WGA protect wheat against chitin-containing phytopatogens during seed imbibition, germination, and early seedling growth (5). The finding that the lectin is mainly located in the tissues which eventually make contact with the soil (6-8) in both the embryo and the adult plant is in aggreement with a defensive function.

Based on the fact that the accumulation of WGA (9) and rice embryo lectin (10,11) is governed by abscissic acid, Peumans and Stinissen suggested that both lectins might have a regulatory function in embryogenesis, i.e. they might be dormancy factors inhibiti cell division and precausious germination of the embryo (12).

In the present paper we describe the location of rice germ lectin at different stages of seed development as well as the agglutination of embryo cells by the lectin.

Lectins, Vol. V

Materials and Methods

Rice (*Oryza sativa* L. subsp. *japonica*, cv. Shuang Feng No. 1) plants were grown in a phytotron at 25°C with 12 h light/dark cycles. For the cell agglutination experiments, rice embryos were collected between 7 (immature) and 30 (mature) days post anthesis (DPA).

The method for isolation and purification of rice germ lectin (RGL) was as previously (2).

In both developmental and localizational experiments, detection of RGL was achevied using the indirect double immuno-binding method. The first antibody was rabbit anti-RGL, and the second antibody was sheep anti-rabbit immunoglobulin G (IgG). The sheep anti-rabbit IgG was conjugated with HRPO and detected with diaminobenzidine (DAB). Antibody binding sites in embryo sections were visualized by the deep color, control sections labelled with preimmune rabbit serum showed no reaction.

Mesophyll protoplasts of barley (*Hordeum vulgare* var. *nudum*) and of tobacco (*Nicotiana tabacum*) were compared with those from rice.

Results

RGL-induced agglutination was found in resuspended cells from both mature and immature rice embryo cells. Isolated cells from either leaves or roots of rice seedlings and also mesophyll protoplasts from tobacco and barley were agglutinated by RGL. All these agglutinations were lectin induced, ie they were specifically inhibited by N-acetyl glucosamine, except for that with cultured cells from the radicle of the immature embryos (negligable inhibition). These results (Table 1) suggest that common sites for RGL-specific binding exist on the cell surface of these plants. The glycoprotein molecules of the radicle callus show some differences from the plumule callus cells (data not shown).

These results indicate that some kind of similarity exists between either the lectins or cell-surface glycoconjugates of these plant species (Gramineae and Solanaceae).

Table 1. RGL-induced Cell Agglutination

Cell sources	*agglutination*	*inhibition*
rice (Oryza sativa)		
mature embryos	++++	+
immature embryos	++++	ND
seedling, roots	++++	ND
plumule	+++	+
leaves	++++	+
mature embryo callus	++++	+
immature embryo callus	++++	ND
plumule callus	++++	+
radicle callus	++++	NA
rud. panicle callus	++++	ND
barley (Hordeum vulgare)		
protoplasts	++++	+
chloroplasts	++++	ND
tobacco (Nicotiana tabacum)		
protoplasts	+++	ND

++++ = *75-100 % agglutination*, +++ = *50-75 % agglutination*, NA = *not appreciable*, ND = *not determined*.

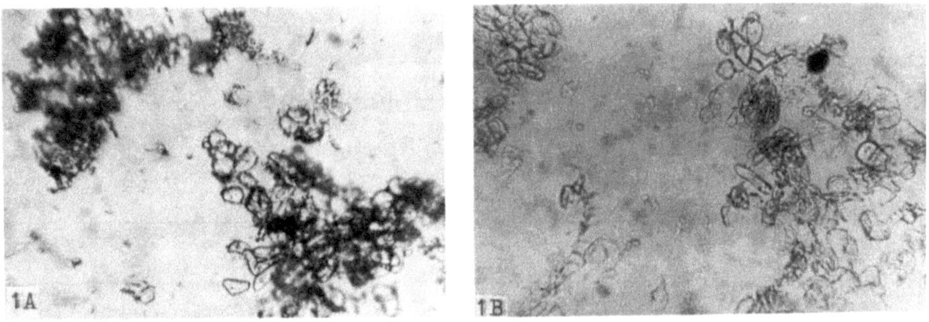

Figure 1. RGL-induced agglutination of suspended cells from rice immature embryo radicle callus.
A: in RGL, 250 µg/ml.
B: in RGL plus N-acetyl-glucosamine, 0.5 M.
x 200.

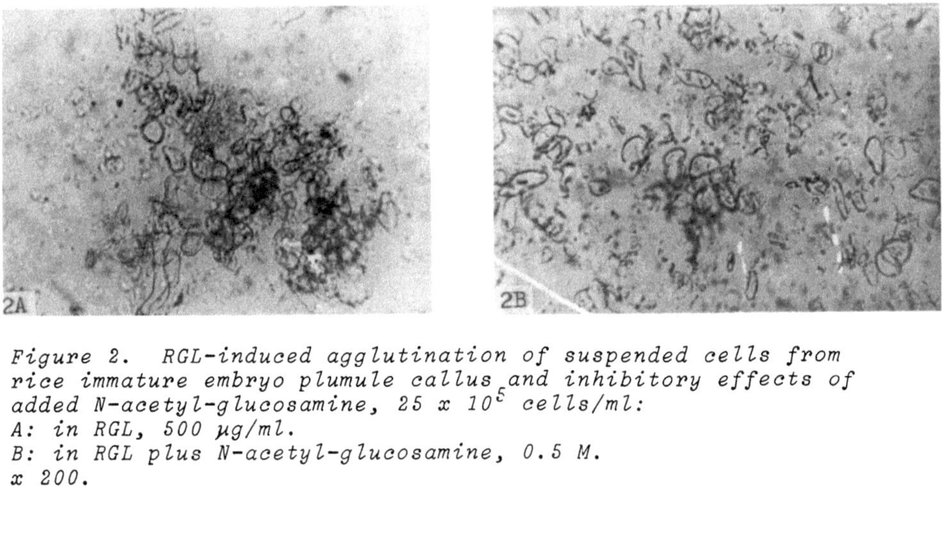

Figure 2. RGL-induced agglutination of suspended cells from rice immature embryo plumule callus and inhibitory effects of added N-acetyl-glucosamine, 25 x 10^5 cells/ml:
A: in RGL, 500 µg/ml.
B: in RGL plus N-acetyl-glucosamine, 0.5 M.
x 200.

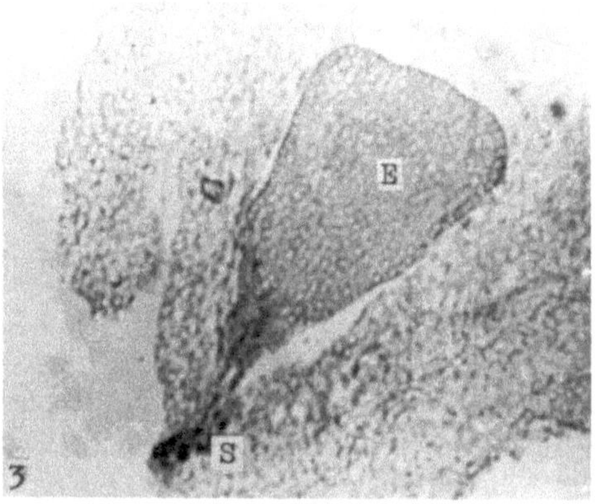

Figure 3. Longitudinal section through a 5 DPA rice embryo showing the embryo proper (E), suspensor (S) and its surrounding tissues stained for RGL-reaction product. x 200.

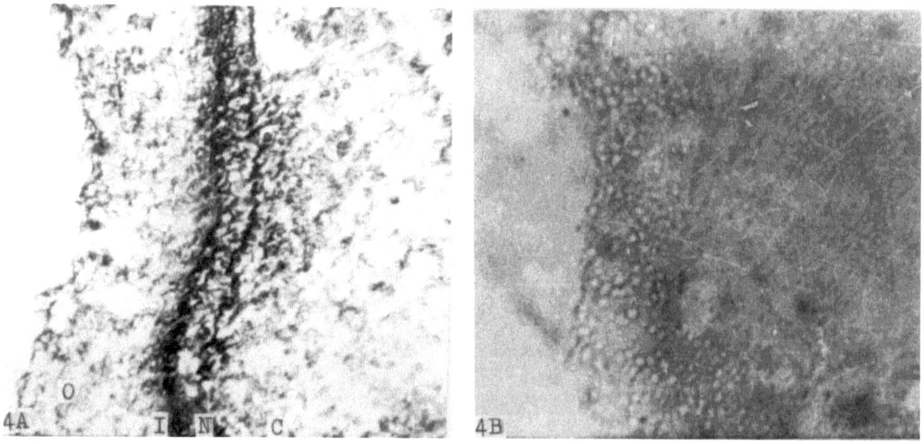

Figure 4. Longitudinal section of 4 DPA rice kernel near the embryo, showing A: the reaction product located in the inner layer of the ovary wall (O), integuments (I), nucellus (N) and cellular endosperm (C). B: the control section. x 200.

Immunocytological evidence indicates that in the early developmental stages (4-6 days after anthesis) RGL is uniformly distributed throughout the embryo proper, whilst the suspensor was heavily stained (Fig. 11). Single isolated cells, as well as cells induced from the plumule and the radicle primordia (both from matured and immatured embryo, 9-11 DPA) were also lectin positive. As the rice embryo differentiates and develops (about 21 DPA), RGL levels increase dramatically to the level present in the mature embryo. In the near mature embryo, RGL is located throughout all layers of the coleoptile to a greater extent than in surrounding tissues (scutellum epiblast, coleorhiza and radicle). Only a few cells in the embryonic leaves display the reaction product. It is also interesting to note that the reaction product is located in the cells of the inner layer of the ovary wall, integuments, nucellus, and cellular endosperm for early embryos (4-6 days) (Fig. 2). It is thus tempting to postulate that the existence of interrelationships between the RGL in the early nonembryonic tissues, and embryo differintiation.

References

1. Sun, C., X.H. Tang. 1981. Acta Phytophysiol.Sin. 7, 385.
2. Shen, Z.W., C. Sun, Z. Shu, X.H. Tang, R.J. Shen. 1984. Can.J.Biochem. 62, 1027.
3. H.Lis and N.Sharon in Biochemistry of Plants, Vol.6, 371-447, Academic Press, 1981.
4. H.Rüdiger, Bioscience 34,22, 95-99 (1984).
5. D.Mirelman, E.Galun, N.Sharon, R.Lotan, Nature 256, 414-416 (1975).
6. M.Mishkind, K.Keegstra, B.Palevitz, Plant Physiol. 66, 950-955 (1980).
7. M.Mishkind, N.Raikel, B.Palevitz, K.Keegstra, J.Cell Biol. 92, 753-564 (1982).
8. N.Raikel, M.Mishkind, B.Palevitz, Planta 162, 55-61 (1984).
9. B.Triplett, R.S.Quatrano, Developm. Biol. 90, 491-496 (1982).
10. H.M.Stinissen, W.Peumans, A.Carlier, Planta 159, 105-111 (1983).
11. H.M.Stinissen, W.Peumans, E. de Langhe, Plant Cell Reports 3, 55-59 (1984).
12. W.Peumans, H.M.Stinissen in (I.Goldstein and M.Etzler, eds.) Proceedings Chemical Taxonomy Molecular Biology and Function of Plant Lectins, Allan R.Lis, New York 1983, pp. 99-116.

INTERACTION OF LEGUMINOUS SEED LECTINS WITH SEED PROTEINS - LECTINS AS PACKING AIDS OF STORAGE PROTEINS

W. Einhoff, G. Fleischmann, T. Freier, H. Kummer and H. Rüdiger
Institut für Pharmazie und Lebensmittelchemie der Universität, Am Hubland, D-8700 Würzburg, W. Germany

The biological role of plant seed lectins still remains an unsolved question. In the present paper, some recent experiments aiming at this problem are summarized and conclusions drawn are presented. We dealt primarily with leguminous seed lectins because they represent the most important lectin group studied until now. They are localized in the cotyledons of the seeds, with a particularly high concentration during the resting state. Inside the cells, lectins are found in the protein bodies (1). The lectins share this location with the storage proteins (2, 3) and with some glycosidases (4). Only very little lectin is found outside the cotyledons (5).
In asking for the pysiological function of Leguminosae lectins one should primarily regard the site where these proteins are localized. Functions at other places would require the lectins to migrate from the protein bodies or to be synthesized at the place where they act. In the latter case, they are no longer seed lectins and thus not within the scope of our work.
As a precondition of exploring the function of lectins, we looked for lectin binding components at the natural site of the lectins, namely the resting cotyledons and their protein bodies.

Isolation of lectin binding proteins from whole seed extracts

Our technique consists in affinity chromatography. After immobilizing a lectin to an insoluble matrix, an extract of that plant material which had been the source of the particular lectin

Lectins, Vol. V

is passed over the lectin column. In one of the predecessors of this Conference series in Copenhagen, earlier results had been presented (6). In the meantime, we have extended our studies to many different plant species in order to get a more general view. From our experience, commercially available lectins are neither very pure nor can be obtained at a reasonable price. We therefore had to isolate as many different lectins as possible. By immobilizing glycoproteins that are easily available (hog gastric mucin, ovomucoid) we prepared general affinity adsorbents that are both inexpensive and simple to work with. Thus far, we succeeded in isolating about 30 lectins from various sources, mostly but not exclusively from Leguminosae (7, 8). Capacities of these gels are in the order of 1 - 4 mg/ml which allows to isolate nearly gram-amounts on columns of a reasonable size.
The next step was to immobilize the lectins to Sepharose. We used a new method introduced by Kohn and Wilchek (9) which allows to load gels in a reproduceable manner up to high protein densities, i. e. 8 - 11 mg/ml, measured by recording the derivative spectra (10).
With the lectin gels prepared by this means, the following species were studied:

Arachis hypogaea	*Pisum sativum*
Canavalia ensiformis	*Robinia pseudoacacia*
Glycine max	*Sophora japonica*
Griffonia simplicifolia	*Vicia faba*
Lens culinaris	*Wisteria floribunda*
Phaseolus vulgaris	

In any case, if seed extracts or their globulin fractions were applied to lectin columns, most of the material passed through. A small but defined amount was released on changing to higher ionic strength - as a little as 12 mM NaCl will often suffice. In some cases *(Lens, Pisum)* a further amount is desorbed by eluting with the binding sugars, mannose, glucose or their α-glycosides. This is not the case with plant species which contain lectins of other sugar specificities.

Identification of lectin binding proteins

Electrophoreses of the lectin binding fractions were performed in 15% polyacrylamide gels under dissociating conditions according to (11). By close inspection of the band patterns we suspected that lectin binding proteins might be related to the storage protein groups legumin and vicilin. Storage proteins from some species have been studied very intensively during the last few years (2, 3). Unfortunately, electrophoretical patterns are only available for those few species.
Before being able to assign lectin binding proteins to storage proteins we therefore had to screen the plants which we studied for their storage protein compositions. We used a zonal isoelectric precipitation method similar to that described by Casey (12). This method is very effective in separating the two storage protein groups, legumin from vicilin, taking advantage of their different solubilities: On a column of Sephadex G-50, equilibrated with a citrate buffer pH 4.7, the sample which had been dissolved in a Tris buffer at pH 8 is applied, and the column first developed with citrate buffer, followed by Tris buffer. This makes the vicilin run through nearly unretarded whereas legumin leaves the column only at the pH jump.
By this means, we could assign electrophoretical patterns to either storage protein group. Common to all vicilins is a group of about 50 kDa (50, 48, and 46 for *Lens*), common to the legumins are multiple bands at about 36 and 22 kDa. Some minor side bands presumably are proteolytic degradation products, others may be larger precursors.
In all species studied, the protein fractions isolated from cotyledon extracts by affinity chromatography on lectin Sepharose columns display typical storage proteins patterns in electrophoresis. As examples, from *Lens* extracts, proteinaceous material is desorbed at high ionic strength which contains legumin and vicilin bands (ratio 1 : 0.5). Additional material appears if a sugar (mannose of glucose) is added to the eluting buffer. This material belongs exclusively to the vicilin type. From *Glycine* extracts, legumin and vicilin type proteins (ratio 1 : 0.6) are isolated at high ionic strength by this method. No material is desorbed if the binding sugar galactose or any

other sugar is added to the eluant.
It should, however, be stressed at this point that the binding components are not the storage proteins in general but only a certain, nevertheless well defined subfraction, in the case of *Lens* only a few percent. Thus it is evident that binding and non-binding storage proteins though identical in their electrophoretical patterns are different. In contrast, some other proteins from the extract do not bind at all to the lectin. As an example, the trypsin inhibitor from *Glycine* passes the lectin column under all conditions used.

Isolation of lectin binding proteins from protein bodies

In further experiments, we tried to isolate the lectin binding material not only from the seeds but from the protein bodies, i. e. from that cell compartment where the lectins are localized. Protein bodies from the species

Canavalia ensiformis	*Pisum sativum*
Glycine max	*Sophora japonica*
Lens culinaris	*Wisteria floribunda*
Phaseolus vulgaris	

were prepared by a density gradient method similar to that described by van der Wilden et al. (13), extracted with buffer in the same way as whole cotyledons and subjected to affinity chromatography using immobilized lectins. In electrophoresis, protein body extracts displayed exclusively the storage protein and lectin band patterns, in contrast to cotyledon extracts which in addition contain other proteins. The material obtained by affinity chromatography, however, yielded the same band patterns, irrespective whether it had been isolated from protein body extracts, from whole seed extracts or from the globulin fraction. If protein bodies were the starting material, however, the relative amount of material bound to lectins was higher by a factor of 3 - 7. This corresponds well with the general enrichment of storage proteins in protein bodies.
Lectin binding material desorbed at high ionic strength is common to all species. This material belongs either to the legumin type or to the vicilin type or to both. As a specialty

of the mannose/glucose binding lectins from *Lens* and *Pisum*, a second fraction desorbed with the binding sugar is obtained. This material exclusively belongs to the vicilin type. This finding corresponds with the glycoprotein nature of vicilins, in contrast to legumins which are devoid of carbohydrates (14). We found that in *Lens* the vicilin-like lectin binder desorbed with sugar is a glycoprotein with 3 - 4% carbohydrate whereas the vicilin-like material desorbed at high ionic strength and associated with the legumin-like lectin binder does not contain carbohydrate. This shows that *Lens* vicilins are glycosylated only in part.

We also found that some glycosidases are specifically bound by lectins. So far it seems that this binding is restricted only to enzymes localized in protein bodies. Though this aspect is essential for the following discussion it would go beyond the space available to present these results here.

Discussion

Binding of proteins to lectins in an ion strength dependent manner may seem to be a totally unspecific process. However, simple overall ion exchange cannot be the underlying principle of binding to the immobilized lectin since at pH 8 which was used in most experiments, all lectins and lectin binding proteins are above their isoelectric points and therefore should bear an overall negative charge. If a salt gradient is applied in order to elute the loaded column instead of a jump in ionic strength, legumin- and vicilin-like lectin binders leave the column in this order, contrary to the order in which the storage proteins elute if subjected to ion exchange chromatography on DEAE cellulose. Moreover, the lectin binding proteins make up only a certain fraction of the total of storage proteins.

Common occurrence of lectins and lectin-binding proteins in cotyledon protein bodies and the specificity of their interaction compared with other cell components lead us to propose that lectins serve as packing aids for storage proteins in protein bodies of Leguminosae. This may be true for other plant species as well since in plants other than Leguminosae, lectins often are found in storage tissues, mostly in seeds as is the case with the Leguminosae, but also in hibernating rhizomes or bulbs (15 - 17). During protein body assembly, lectins may form association centres around which the storage proteins, primarily the lectin binding subfraction, may accumulate in an orderly manner. Since storage proteins strongly tend to undergo self-association (12, 18), further packing may take place spontaneously. Thus, low amounts of lectin in seeds will suffice to pack high amounts of storage proteins.

At the time of germination, water imbibes the seeds and penetrates the cell wall. It is known that this leads to an overall proton extrusion (19) which in consequence might raise the ionic strength within the cell. As seen from our results, lectin-storage protein aggregates dissociate at elevated ionic strengths. *In vivo* the storage proteins after being released from the complex with lectins may be degraded by proteases, and the cleavage products may leave the cell whereas the lectins remain intact for a while. Indeed, lectins are much more resistant to degradation than storage protein both *in vivo* and *in vitro* (20). At the same time, also glycosidases may be released from their connection with the lectins. They now may degrade oligosaccharide side chains of glycoproteins, e. g. of vicilins, which in turn may become more sensitive to proteolytic degradation (21). Glycosidases may also degrade storage oligosaccharides of the seeds (22). The resulting monosaccharides now may on the one hand further help to dissociate lectins from their binding partners and on the other hand may contribute to energy metabolism.

At the present stage, our proposal does not take into account the sugar binding site of most lectins which is only operative in lectin binders of the vicilin type in plants which contain mannose/glucose specific lectins. Lectins with other sugar specificities will demand other binding partners. As a working hypothesis, we propose that lectins may interact also with protein body membranes.

Protein body membranes may be glycosylated as are other membranes. Since they originate from the endoplasmatic reticulum, oligosaccharide side chains should be located inside, thus facing the protein body interior. The lectins after having entered the developing protein body might bind to the membrane by their sugar

binding sites. This interaction in consequence may be reinforced by hydrophobic binding to other membrane constituents - a well known phenomenon with lectins (18, 23) - and even more strengthened by additional lectin molecules that combine with the membrane bound ones by self-association - also well known for lectins (e. g. 18, 24, 25). At the same time, enzymes and lectin binding storage proteins may associate with the membrane-fixed lectins. By deposition of the bulk of storage proteins to these complexes, the assembly of protein bodies is finished. Though preferential deposition of lectins near the protein body membrane is not seen in all microscopic pictures published, some recent papers (26, 27) may be interpreted in that way. Another paper (28) describes a close association of kidney bean lectin with the protein body membrane.
In summary, the interaction between lectins on the one hand and lectin binding storage proteins and glycosidases on the other hand suggests that seed lectins may participate in combining and in degrading the protein body constituents in an orderly manner.

Acknowledgements

The support from the Deutsche Forschungsgemeinschaft and from the Fonds der Chemischen Industrie is gratefully acknowledged.

References

1. Bollini, R., M.J. Chrispeels. 1978. Planta 142, 291.

2. Boulter, D.. 1981. Adv. Bot. Res. 9, 1.

3. Boulter, D.. 1983. Qual. Plant. Foods Hum. Nutr. 32, 247.

4. Dey, P.M.. 1984. Phytochemistry 23, 257.

5. Borrebaeck, C.A.K., B. Matthiasson. 1983. Physiol. Plant. 58, 29.

6. Rüdiger, H., R. Gansera, G. Gebauer, H. Schurz. 1981. In: Lectins - Biology, Biochemistry, Clinical Biochemistry (T.C. Bøg-Hansen, ed.) Vol. 1, De Gruyter, Berlin, p. 135.

7. Freier, T., G. Fleischmann, H. Rüdiger. 1985. Biol. Chem. Hoppe-Seyler 366, in press.

8. Fleischmann, G., I. Mauder, W. Illert, H. Rüdiger. 1985. Biol. Chem. Hoppe-Seyler 366, in press.

9. Kohn, J. & M. Wilchek. 1982. Biochem. Biophys. Res. Commun. 107, 878.

10. Schurz, H., H. Rüdiger. 1982. Anal. Biochem. 123, 174.

11. Laemmli, U.K.. 1970. Nature (London) 227, 680.

12. Casey, R., 1979. Biochem. J. 177, 509.

13. Van der Wilden, W., E. Herman, M.J. Chrispeels. 1980. Proc. Natl. Acad. Sci. (US) 77, 428.

14. Gatehouse, J.A., R.R.D. Croy, D. Boulter. 1980. Biochem. J. 185, 497.

15. Allen, A.K., A. Neuberger. 1971. Methods Enzymol. 50, 343.

16. Stinnissen, H.M., W.J. Peumans. 1985. Biochem. Physiol. Pflanzen 180, 85.

17. Peumans, W.J., M. Nsimba-Lubaki, B. Peeters, W.F. Broekart. 1985. Planta 164, 75.

18. unpublished results from this laboratory.

19. Marré, E., P. Lado, F. Rasi-Caldogno, R. Colombo, M.I. De Michelis. 1974. Plant Sci. Letters 3, 365.

20. Goldstein, I.J., C.E. Hayes. 1978. Adv. Carbohydr. Chem. Biochem. 35, 127.

21. Dey, P.M. & E. Del Campillo. 1984.Adv. Enzymol. Related Areas Mol. Biol. 56, 141.

22. Bewley, J.D., M. Black. 1983. Biology and Biochemistry of Seeds in Relation to Germination, Vol. 1, Springer Berlin.

23. Ochoa, J., T. Kristiansen, S. Påhlman. 1979. Biochim. Biophys. Acta, 577, 102.

24. Gansera, R., H. Schurz, H. Rüdiger. 1979. Hoppe-Seyler's Z. Physiol. Chem. 360, 1579.

25. Schurz, J., H. Rüdiger. 1985. Biol. Chem. Hoppe-Seyler 366, 367.

26. Etzler, M.E., S. McMillan, S. Scates, D.M. Gibson, W.J. Douglas, D. Cole, S. Thayer. 1984. Plant Physiol. 76, 871.

27. Van Driessche, E., G. Smets, R. Dejaegare, L. Kanarek. 1981. Planta 153, 287.

28. Pusztai, A., R.R.D. Croy, J.S. Stewart, W.B. Watt. 1979. New Phytol. 83, 371.

MICROCALORIMETRIC EFFECTS OF LECTIN INTERACTIONS

W. Einhoff, T. Freier, H. Kummer and H. Rüdiger
Institut für Pharmazie und Lebensmittelchemie der Universität Würzburg, Am Hubland, D-8700 Würzburg, W. Germany

Introduction

By means of immobilized lectins as affinity adsorbents lectin binding proteins have been found in the seeds of several leguminous plants. The isolation and some properties of one of them have already been described (1, 2). By electrophoretical techniques (3, 4), lectin binding proteins were shown to belong to the storage proteins (5).

In mature seeds, lectins and storage proteins occur together in the protein bodies (6 - 8). Since both are found in the same compartment and interact with each other, we suggest that this interaction might also be important *in vivo* (5).

The present paper deals with the interaction between the lectin from the garden pea *(Pisum sativum)* with the lectin binding proteins from the same plant. The interaction is measured by microcalorimetry (9, 10), a method that leads to reliable and quantitative data for any type of chemical reaction.

Lectins, Vol. V

Material and Methods

Pea seeds were purchased from a local supplier, Sephadex G-100 and Sepharose 4B from Deutsche Pharmacia (Freiburg). All other reagents were from Sigma Chemie (Taufkirchen).
The pea seed lectin was isolated by affinity chromatography using Sephadex G-100 (1). The lectin binding proteins were isolated from pea cotyledon extracts by affinity chromatography using pea lectin immobilized to Sepharose 4B (2). Two types of lectin binding proteins were found:

- Type I, eluted at high ionic strength, representing non-glycosylated storage proteins (legumin and vicilin) in the molar ratio of 2 : 1. The legumin portion was separated from the vicilin by zone isoelectric precipitation (11).
- Type II, eluted with a glucose containing buffer, representing the glycosylated portion of the storage protein vicilin.

Measurements were carried out with a Bioactivity Monitor from LKB Instrument (Bromma, Sweden), which is a twin cup heat flux calorimeter (Tian-Calvet method) working under isoperibol conditions to measure a local temperature difference (12, 13).
All experiments were carried out in 50 mM Tris/acetate buffer pH 8.0 at 36.88 °C. A stainless steel ampoule was filled with 3.6 ml of a lectin solution and slowly lowered into the measuring cup of the calorimeter. After temperature equilibrium had been established, 100 µl of the lectin binder solution was added through a long steel capillary at 1 µl/s under constant stirring. In order to avoid neutralisation or dilution heat effects, both protein solutions were dialyzed twice against the same buffer prior to the measurements. In preliminary control runs no measurable effects had been observed if the buffer (0.05 M Tris/acetate pH 8) was mixed with buffer containing glucose, NaCl or bovine serum albumin as an inert protein (2) under the conditions used.

Results

The lectin was reacted with α-methylmannoside (Fig. A) which is the monosaccharide with the highest affinity for this lectin (14), and with the lectin binders of type I (Fig. B) and type II (Fig. C).

Since the reaction shown in Fig. A is very fast, the response of the system does not allow to calculate a velocity constant for the lectin - sugar complex formation. On the other hand, from the slow decline of the curve in Fig. A, the thermal equilibration constant of the system was calculated to be 0.183 min^{-1}.
The curve shown in Fig. B could be simulated very precisely by the kinetic equation of consecutive reactions (15) by employing a complex formation constant of 0.083 min^{-1}.

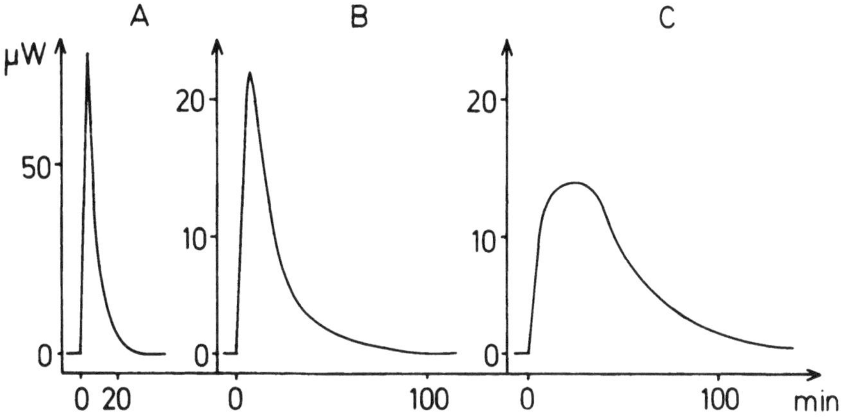

Fig. : Reaction between 10.8 mg pea lectin and
A: 5 x 10^{-7} mol α-methylmannoside,
B: 1.7 mg lectin binder (legumin from type I), and
C: 1.4 mg lectin binder type II.

With the binder type II (Fig. C) the reaction apparently is composed of more than one step which does not allow to apply the equation of ref. (15). Nevertheless, it could be estimated that the initial reaction between lectin and binder type II proceeds slower by a factor of 40 compared with that of binder type I. Thus it is seen that the reaction of the lectin with the sugar occurs at a fast rate whereas protein - protein interactions are slow.

From the areas under the curves, enthalpies were estimated. From Fig. A, an enthalpy of about 70 kJ/mole was obtained. Since the exact mol masses of the native lectin binding proteins are not known, values based on their masses were calculated. The values from Figs. B and C were 19 and 15 J/g, respectively.

Conclusion

Our results show that microcalorimetry is a valuable tool for measuring the interaction between lectins and their ligands. Quantitative data of interactions with potentially natural binding partners will help to elucidate the physiological function of seed lectins *in vivo*.

Acknowledgements: The support from the Deutsche Forschungsgemeinschaft and from the Fonds der Chemischen Industrie is gratefully acknowledged. We want to express our gratitude to Prof. Dr. P. Fürst and Dr. J. Nittinger, University of Hohenheim, for kindly permitting us to use their measuring equipments.

References

1. Gansera, R., H. Schurz, H. Rüdiger. 1979. Hoppe-Seyler's Z. Physiol. Chem. 360, 1579.

2. Schurz, J., H. Rüdiger. 1985. Biol. Chem. Hoppe-Seyler 366, 367.

3. Laemmli, U. K. 1970. Nature (London) 227, 680.

4. Campbell, W. P., C. W. Wrigley, J. Margolis. 1983. Anal. Biochem. 129, 31.

5. Einhoff, W., G. Fleischmann, T. Freier, H. Kummer, H. Rüdiger. 1985. These Proceedings.

6. Varner, J. E., G. Schidlovsky, 1963. Plant Physiol. 38, 139.

7. Pernollet, J.-C.. 1978. Phytochemistry 17, 1473.

8. Van Driessche, E., G. Smets, R. Dejaegere, L. Kanarek. 1981. Planta 153, 287.

9. Hinz, H.-J.. 1983. Ann. Rev. Biophys. Bioeng. 12, 285.

10. Hinz, H.-J.. 1983. In: Topics in Molecular Pharmacology (A.S. Burgen and G.C.K. Roberts, eds.). Elsevier Amsterdam, p. 71.

11. Casey, R.. 1979. Biochem. J. 177, 509.

12. Bioactivity Monitor, Instruction Manual, LKB Instrument.

13. Hemminger, W., G. Höhne. 1984.Calorimetry, Verlag Chemie Weinheim, p. 224.

14. Goldstein, I. J., C.E. Hayes. 1978. Adv. Carbohydr. Chem. Biochem. 35, 127.

15. Moore, W. J.. 1972. Physical Chemistry 5th edition, Longman London, p. 345.

LECTIN RELEASE FROM SEEDS OF DATURA STRAMONIUM AND INTERFERENCE OF THE DATURA STRAMONIUM LECTIN WITH BACTERIAL MOTILITY.

W.F. Broekaert, W.J. Peumans
Laboratorium voor Plantenbiochemie, Katholieke Universiteit Leuven, Kardinaal Mercierlaan 92, B-3030 Leuven, Belgium

Among the divergent possible functions of plant lectins that have been proposed and discussed in the vast literature on this subject (for reviews, see 1,2), involvement in host defence mechanisms represents one of the most challenging hypotheses. For such an hypothesis to be supported by experimental evidence, three primary conditions have to be evaluated. First, the lectin must be present in the area where plant tissue is naturally in contact with microorganisms. Second, the lectin must be shown to cause a negative effect on growth, competitivity or pathogenicity of the microorganisms. Third, the concentration of the lectin in the contact area must be high enough for exhibition of the effect. The rhizosphere is undoubtedly the best studied example of an interface area between plant and microorganisms. Evidence is now accumulating on the release of lectins into the rhizosphere, either by roots (3,4,5) or by germinating seeds (6,7) of some species. In this communication, we report on the highly specific release of lectin from seeds of *Datura stramonium*. In addition, a study of the effect of this lectin on the motility of bacteria is presented. The relevance of these experiments in view of a possible role of the lectin in host defence will be discussed.

Materials and Methods

Analysis methods: Agglutination assays were performed with trypsin-treated human blood-group A erythrocytes as previously described (8). Protein samples were analysed by SDS-PAGE by using a discontinuous system (9) on 12.5-25% (w/v)-acrylamide gradient gels. Samples were carboxyamidated (10) before electrophoresis. The gels were stained with Coomassie Brilliant Blue.

Lectin purification: Chitin-binding proteins were isolated from seed extract of *Datura stramonium* by affinity chromatography on a chitin column. *Datura stramonium* lectin (DSL) was further purified from this fraction by ion-exchange chromatography (11).

Capillary assay: Pseudomonas fluorescens was grown on a glycerol-salts medium (12) at 30 °C with rotary shaking (100 rpm), until an optical density of 0.5 at 590 nm was reached. The bacteria were labeled for 20 min by addition of 10 µCi of ^{3}H-leucine (128 Ci/mmol) per ml of suspension. Cells were harvested by centrifugation (1500 x*g* for 10 min) and washed three times with chemotaxis medium (60 mM potassium phosphate buffer, pH 6.8/ 0.1 mM EDTA). Finally, the cells were gently resuspended in chemotaxis medium and cell density was adjusted to 10^{8} cells per ml, as determined by optical density. Aliquots (150 µl) of labeled bacteria suspension containing DSL at different concentrations, were placed in a chemotaxis device (13), and 5 µl microcapillaries (filled with chemotaxis medium) were dipped into the suspension. The assemblies were incubated at 30 °C for 50 min. The capillaries were then removed, wiped free of adhering bacteria, and the contents blown out and counted. All experiments were performed at least 6 times.

Channeled chamber assay: A *P. fluorescens* suspension with a cell density of $4x10^{8}$ per ml, was prepared as described above, with the exception of the labeling step. The channeled chambers (14) were filled with chemotaxis medium containing DSL at different concentrations. Aliquots (20 µl) of the bacteria suspension were injected in the source wells. After 4 h of incubation at 30 °C, 1 µl aliquots were sampled from the target wells and cell numbers determined by microscope counting in a Thoma chamber. All experiments were performed at least 4 times.

Results

1.Lectin release from seeds of Datura stramonium: Seeds of *Datura stramonium* are surrounded by a hard and impermeable seedcoat. When intact seeds were imbibited in phosphate-buffered saline for 24 h, no or very little agglutinating activity could be detected in the imbibition solution. However, after perforation or partial removal of the seedcoat, relatively high agglutination titres were observed in the imbibate after a few minutes upon imbibition. Fig. 1 represents the release of lectin from pealed seeds (complete removal of the seedcoat) into different imbibition solutions. Agglutinating

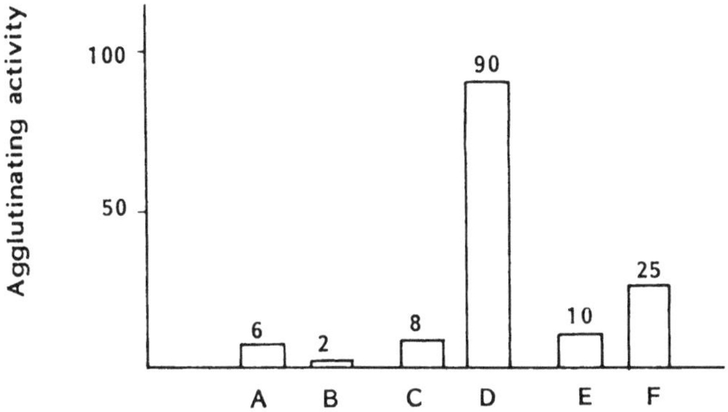

Fig. 1: Lectin release from seeds of Datura stramonium in different imbibition solutions. Lots of 8 aseptically decoated seeds were incubated at 15 °C in 200 µl of sodium- 4-morpholineethanesulphonic acid (Na-Mes) buffer, pH 6.2 (A,B), 2 mM Na-Mes buffer/10 mM EDTA (C,D) and 2mM Na-Mes/100 mM NaCl (E,F). After 15 minutes of incubation, the imbibates were taken off, and stored at -20 °C until agglutination assay. Another 200 µl of imbibition solution was added to the seeds, and they were further incubated for a period of 15 days. Agglutinating activity was determined of the 0-15 min imbibates (A,B,C) and of the 15 min-15 days imbibates(D,E,F). Data are expressed relative to the total agglutinating activity of the seeds.

activity was measured either after 15 minutes of imbibition (short term release)or after prolonged imbibition for a period of 15 days (long term release). From these data it can be seen that a reasonable fraction of the total lectin content of the seeds (about 50 µg per seed), is released after 15 minutes of imbibition. This fraction varies from 6 to 10% of the total lectin content, dependent on the type of imbibition medium used. However, the composition of the imbibition solution obviously has a greater impact on the long term elution. Indeed, whereas lectin release during the long term period is almost negligible in 2 mM Na-Mes (pH 6.2), 90% of the total lectin content is released when 10 mM EDTA is included in the same buffer. The addition of 100 mM NaCl to the Na-Mes buffer also stimulated lectin release (25% release of the total lectin content during the long term period), but the stimulating effect was obviously lower than that observed for the buffer containing the chelating agent EDTA.

In fig. 2, the SDS-PAGE analysis is shown of a PBS solution in which pealed *Datura stramonium* seeds had been imbibited for 1 h. By comparing lane 1 to lane 2, it appears that the major part of the proteins released by the seeds are chitin-binding proteins.

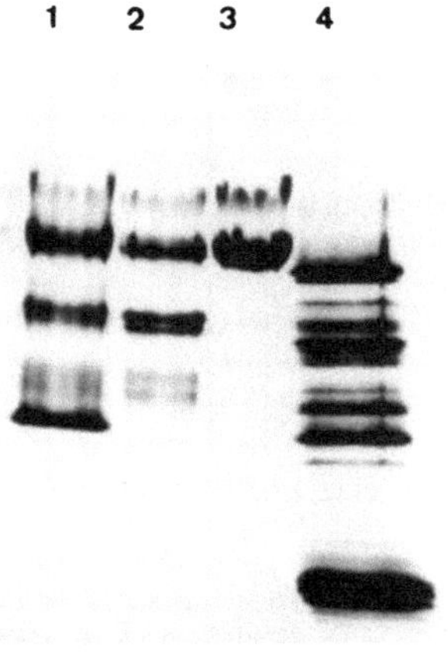

Fig.2: SDS-PAGE analysis of the imbibate from Datura stramonium seeds (lane 1), affinity-purified chitin-binding proteins (lane 2), purified DSL (lane 3) and total soluble seed protein (lane 4)

In a previous report (11) we demonstrated that the chitin-binding proteins from *Datura stramonium* seeds consist of DSL (shown in lane 3) and two lower molecular weight proteins which do not exhibit haemagglutinating activity. When the imbibition solution (lane 1) is compared to total soluble seed protein (lane 4), it becomes evident that the process of lectin release is highly specific. Indeed, whereas DSL is hardly detectable in the pattern of total soluble protein, it clearly is a major component of the imbibition solution.

2. Interference of DSL with the motility of bacteria: Suspensions of motile cells of *Pseudomonas fluorescens* were examined under the light microscope in the presence or absence of purified DSL. It was observed that the lectin caused a dramatic block of all bacterial movements at the air- water interface, where motion was noted to be most intense in the control droplets. Swimming of the bacteria within the drop was also affected but not completely stopped by the lectin. Boiling or digestion with pronase of the lectin preparation destroyed its particular activity on bacterial motion, ruling out the possibility that other components but proteins were responsible for the observed effects. Furthermore, the immobilisation at the air-water interface of lectin-treated cells could be reversed by addition of fetuin, a glycoprotein that binds to DSL (15). The agglutination-inhibiting oligosaccharide chitotriose;

however, was unable to reverse immobilisation in the concentration range tested (1-100 µg). It is noteworthy to mention here that chitotriose also fails to desorb DSL from the affinity-matrix fetuin-agarose.

To assess to what extent the interference of DSL with the motility of *P. fluorescens* is a general phenomenon, we investigated the action of this lectin on 10 different species of motile bacteria. As shown in Table 1, eight of the 10 species tested were immobilised by the lectin, and in all cases immobilisation could be reversed by addition of fetuin.

To obtain a more quantitative image of the immobilisation effect, two different tests were performed, that were originally developed for the assay of chemotaxis. The first method used, was the capillary assay, which measures the number of bacteria that enter a

Table 1. Activity of DSL on different bacterial species. DSL was added to 20 µl drops of a bacteria suspension (10^8 cells/ml) to a final concentration of 50 µg/ml, and motility was examined at the air-water interface by light microscopy. Reversion of motility inhibition was examined after a second addition of fetuin to a final concentration of 5 mg/ml.

Species	Immobilisation at the air-water interface	Reversion by fetuin
Escherichia coli	yes	yes
Pseudomonas fluorescens	yes	yes
P. aeruginosa	yes	yes
P. alcaligenes	yes	yes
Proteus vulgaris	yes	yes
Serratia marcescens	yes	yes
Bacillus subtilus	yes	yes
B. megaterium	yes	yes
Agrobacterium tumefaciens	no	/
Azospirillum brasiliense	no	/

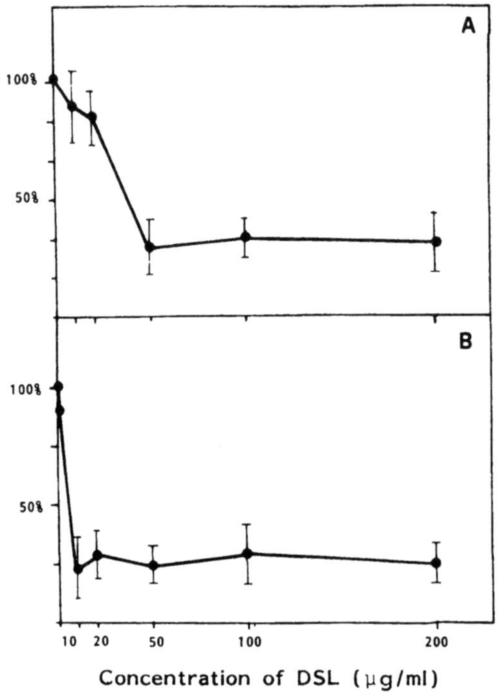

Fig.3: (A) Migration of P. fluorescens measured by the capillary assay, with DSL included in the bacteria suspension at the different concentrations indicated. Results are expressed as percentages of the control (suspension without DSL) value. (B) Migration of P. fluorescens measured by the channeled chamber assay, with DSL included in the chemotaxis medium at the different concentrations indicated. Results are expressed as percentages of the control (chemotaxis medium without DSL) value.

capillary dipped into a bacteria suspension. The influence of DSL on the migration of *P. fluorescens* is shown in Fig.3A. This figure reveals that concentrations of 50 µg/ml and higher caused a 3-fold reduction in the number of bacteria that reached the inside of the capillaries. Alternatively, the interference of DSL with the motility of *P. fluorescens* was examined by the channeled chamber assay. This method measures the migration of bacteria through an open channel system from one well to another. As shown in Fig.3B, the inhibitory effect of DSL on bacterial migration is even more pronounced when assayed in channeled chambers. Indeed, the number of migrating bacteria dropped by 4-fold relative to the control, at a lectin concentration of 10 µg/ml. Higher lectin concentrations did not bring about a further increase of the inhibitory effect.

Discussion

In this paper we present evidence for a rapid and highly specific release of lectin from seeds of *Datura stramonium*. The seedcoat acts as a physical barreer limiting the export of lectin, but once this structure is disrupted, lectin release occurs readily. During the initial phase of imbibition (first 15 minutes) about 10 % of the total lectin content is released in the surrounding imbibate. Most probably, this fraction is released by the partly resorbed integument layer, which is a particularly rich source of lectin (Broekaert, unpublished observations). Further release of lectin occurs more slowly and requires the presence of EDTA in the imbibition solution. At present, the action of EDTA on lectin release is not fully understood. It may be that EDTA affects the structural organisation of cell membranes (16) in such a way that it causes an increased leakiness of the cells. Alternatively, it is possible that DSL is associated with the cell walls by divalent cation bridges, which can be disrupted in presence of EDTA. Further knowledge of the intracellular distribution of DSL is required to distinguish between both possibilities. Specificity of lectin release is clearly demonstrated by comparison of the SDS-PAGE pattern of the imbibate to that of the total soluble seed protein. In previous reports on lectin release (3, 4, 5, 6, 7), specificity was never found to be so high as in the case of *Datura stramonium* seeds. Our results also suggest that lectin release occurs in concert to seed germination, since it is during the latter process that the seedcoat is perforated. In view of our results on the interference of DSL on the motility of bacteria, the release of lectin during seed germination might be physiologically relevant. Indeed, the released lectin might counteract the chemotactic approach of soil bacteria to germinating seeds. The importance of motility in the infection process has been demonstrated by several investigators (17, 18, 19). Leben (20) demonstrated that species that first succeed in reaching the rhizosphere of a seedling, will have the greatest chance to establish themselves and either to infect or colonize the roots of the adult plant. Hence, chemotaxis is considered to be an important determinant in the competitivity of pathogenic and non-pathogenic soil bacteria.

Our quantitative results obtained by the capillary and the channeled chamber assays, support the idea that DSL might have an influence on chemotaxis. Indeed, a 4-fold reduction of bacterial migration was noted by the channeled chamber technique at a lectin concentration as low as 10 µg/ml. The fact that DSL mainly affects bacterial motion at the air-water interface, as could be observed by light microscopy, might explain why the capillary assay needs higher lectin concentrations (50 µg/ml and higher) to evidence an effect on bacterial migration. Indeed, whereas the channeled chamber assay measures the migration of bacteria through the air-water interface as well as through the bulk of aqueous solution, only the latter migration is accounted for by the capillary assay. It looks worthwhile to mention here that movement of soil bacteria is essentially restricted to thin water film layers, which suggests that the influence of DSL on bacterial migration *in vivo* might be even more pronounced as can be measured by the assays used in our study.

Acknowledgements

This work was supported in parts by grants from the National Fund for Scientific Research (Belgium), of which W. Broekaert is Research Assistant and W. Peumans Senior Research Associate.

References

1. Pusztai, A., Croy, R.R.D., Grant, G., Stewart, J.C. 1982. In: Seed Proteins (Doussant, J., Mosse, J., Vaughen, J., eds.) New York, Academic pp. 53.

2. Rudiger, H. 1984. BioSci. 34, 95.

3. Anderson, A.J. 1983. Can. J. Bot. 61, 3438.

4. Gietl, C., Ziegler, H. 1979. Naturwissenschaften 66, 161.

5. Dazzo, F.B., Yanke, W.E., Brill, W.J. 1978. Biochym. Biophys.

Acta 539, 276.

6. Fountain, D.W., Foard, D.E., Replogle, W.D., Yang, W.K. 1977. Science 197, 1185.

7. Mishkind, M., Raikhel, N.V., Palevitz, B.A., Keegstra, K. 1982. J. Cell Biol. 92, 753.

8. Broekaert, W.F., Nsimba-Lubaki, M., Peeters, B., Peumans, W.J. 1984. Biochem. J. 231, 524.

9. Laemmli, U.K. 1970. Nature (London) 227, 680.

10. Raikhel, N.V., Mishkind, M.L., Palevitz, B.A. 1984. Planta 162, 55.

11. Broekaert, W.F., Peumans, W.J., Allen, A.K. 1985. In: Lectins, Vol 4 (Bøg-Hansen, T.C., Breborowicz, J., eds.), Walter de Gruyter Co., Berlin, New York, pp. 481.

12. Adler, J. 1973. J. Gen. Microbiol. 74, 77.

13. Galluci, K.K., Paerl, H.W. 1983. Appl. Envir. Microbiol. 45, 557.

14. Barak, R., Nur, I., Okon, Y. 1983. J. Appl. Bacteriol. 53, 399.

15. Singer, S.J., Nicolson, G.L. 1972. Science 175, 720.

16. Crowley, J.F., Goldstein, I.J., Arnarp, J., Lönngren J. 1984. Arch. Biochem. Biophys. 231, 524.

17. Scroth, M.N., Hildebrand, D.C. 1964. Ann. Rev. Phytopathol. 2, 101.

18. Panopoulos, N.J., Scroth, M.N. 1974. Phytopathology 64, 1389.

19. Chet, I., Zilberstein, Y., Henis, Y. 1973. Physiol. Plant Pathol. 3, 473.

20. Leben, C. 1983. Phytopathology 73, 577.

ARE THE MISTLETOE LECTINS STORAGE PROTEINS ?

D. Neumann, U. zur Nieden
Institut für Biochemie der Pflanzen der Akademie der Wissenschaften der DDR, 4010 Halle (Saale), German Democratic Republic

P. Ziska, H. Franz
Staatliches Institut für Immunpräparate und Nährmedien
DDR - 1120 Berlin, German Democratic Republic

Relatively large amounts of lectins occur in seeds, especially those of legumes. Nevertheless no physiological role is known with certainty for any plant lectin (15). In the last decade many different functions for lectins have been suggested (e.g. defense against pathogens (7), a role in the symbiosis between legumes and *Rhizobium* (3, 14), cell-cell recognition (8), a physiological role in the leaf growth and development (15)) little, if any, conclusive existence has been found for any of them (1, 10, 16). Certainly for some plant lectins physiological functions will be found in the future. However plants produce several substances (e.g. alkaloids), physiologically highly active but without any known function in plant cells. These situation could be true for some lectins, too. Besides a physiological role of plant lectins in the sense of the influence on development processes, other functional aspect should not be overlooked. Many plants accumulate proteins as C- and N-source for particular developmental stages. Such storage proteins are localized, in all cases investigated so far, in vacuolar compartments (11, 12, 17). The localisation of some legume lectins in (storage protein containing) protein bodies (2, 6, 9, 15) supports the hypothesis that some lectins may play a role as storage protein in plant cells. The aim of this work was to collect data about a possible storage function of lectins in mistletoe (*Viscum album* L.) especially of mistletoe lectin I (ML I).

Material and methods

Plant material for electron microscopy

Leaves of *Viscum album* (grown on *Acer carpinifolium*) were harvested in February and June from the same plant.

Electron microscopy

Small pieces of the leaves were prefixed with 6 % phosphate buffered glutaraldehyde (2 h), fixed in a series of acetone and embedded in Araldite. Ultrathin sections were stained with lead and viewed in B 5500 transmission electron microscope (TESLA, Brno).

Immuno electron microscopy

For immuno electron microscopy the leaf tissue was fixed only with freshly prepared 3 % phosphate buffered formaldehyde (1 h) and embedded in Araldite. Ultrathin sections were collected on Ni-grids and treated as following:

1. Anti-ML I antibody (diluted 1:1 with PBS/6 % human serum albumin (HSA)/1 % Tween 20) 12 h at 4°C
2. Washing with PBS (3 % HSA/0,5 % Tween 20 for 3 x 10 min)
3. Protein A-Au_{16} diluted 1:10 with PBS (0.02 % polyethyleneglycol) 1 h at 22°C
4. Washing with PBS (3 % HSA/0.5 % Tween 20 for 3 x 10 min)
5. Washing with H_2O for 10 min
6. Staining with 5 % aqueous uranylacetate for 20 min

Colloidal gold (particle 16 nm) was prepared by reduction of $HAuCl_4$ with sodium citrate (4). The protein A-Au_{16}-complex was prepared according to Roth et al. (13). Control incubations

Table 1

Time study of the lectin content of mistletoe grown on an apple-tree

month of harvest		µg lectin / g material
November	1982	600
January	1983	1000
March	1983	400
April	1983	450
September	1983	340
October	1983	340
December	1983	480

were carried out with normal rabbit serum. Estimation of the lectin content: 1 g of dried mistletoe was extracted with 10 ml PBS for 16 h at 4°C. Dilutions of the filtrates were used for lectin determination by the ELISA technique (18).

Results and discussion

Leaf cells of *Viscum album* L. exhibit a normal ultrastructure with well structured chloroplasts, mitochondria and numerous vacuoles (Fig. 1). Most of cell volume is occupied by the nucleus. Endoplasmatic reticulum (ER) occurs predominantly in tubular form. Vacuoles contain well contrasted protein clumps, mostly attached to the plasmalemma (Fig. 2) with specific deposits in the vacuoles of leaf parenchyma cells (Fig. 3, 4). The protein clumps were moderately labelled, indicative of the presence of other proteins. Whether the lectin is dissolved in the vacuole sap or exists in the form of small protein clumps cannot be decided by these experiments.

It may be possible that the dissolved lectin becomes bound to the protein clumps during the fixation process. The lack of contrast in the immuno-labelled objects arises from technical grounds, namely that immuno-labelling of ML I is only possible by omitting OsO_4 during fixation because OsO_4 seems to destroy the antigenic determinants of ML I. The result is diminished contrast. The localization of ML I inside the vacuoles corresponds in general to the accumulation of storage proteins in plant cells. Storage proteins are localized inside the vacuoles or corresponding aqueous compartments (e.g. protein bodies). Efflux of intact protein molecules out of these compartments could not be observed. During particular developmental stages of plants (e.g. germination) these proteins are degraded and the degradation products are metabolised. Indications for a storage function of ML I came also from the seasonal change of the lectin content in mistletoe (Table 1). The highest lectin content in the leaves could be observed during wintertime. In this period a large number of protein clumps are found in the vacuoles (Fig. 2). The lectin content decreases during spring in old leaves (Fig. 5). In very young leaves (Fig. 6) only a few protein deposits can be observed.

In summary, mistletoe lectin I (ML I) was localized in vacuolar

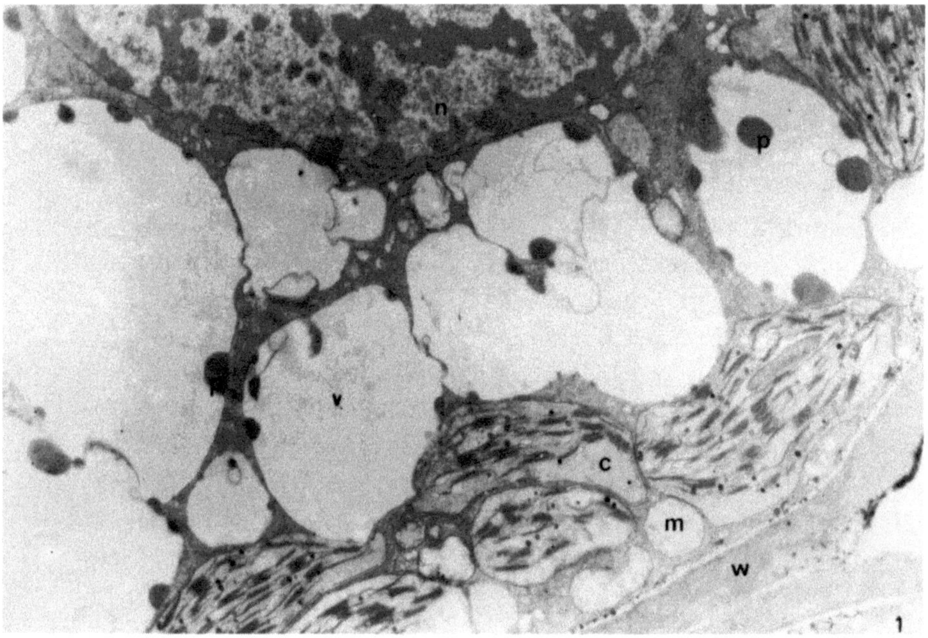

Fig. 1. Part of a leaf parenchyma cell of Viscum album (collected in February). Note the protein lumps (p) in the vacuoles (v); chloroplast (c), mitrochondrium (m), nucleus (n), cell wall (w); 7500x.

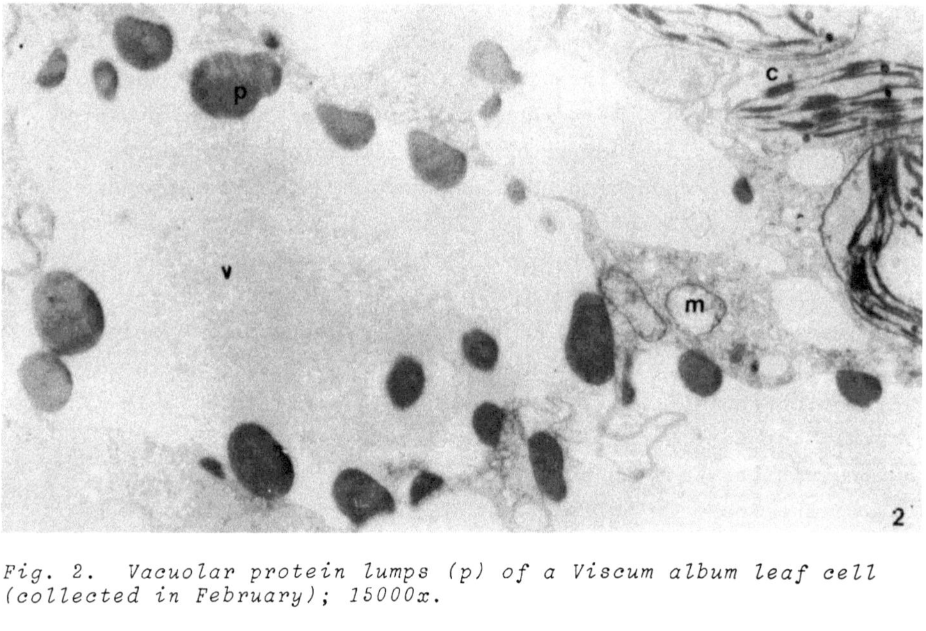

Fig. 2. Vacuolar protein lumps (p) of a Viscum album leaf cell (collected in February); 15000x.

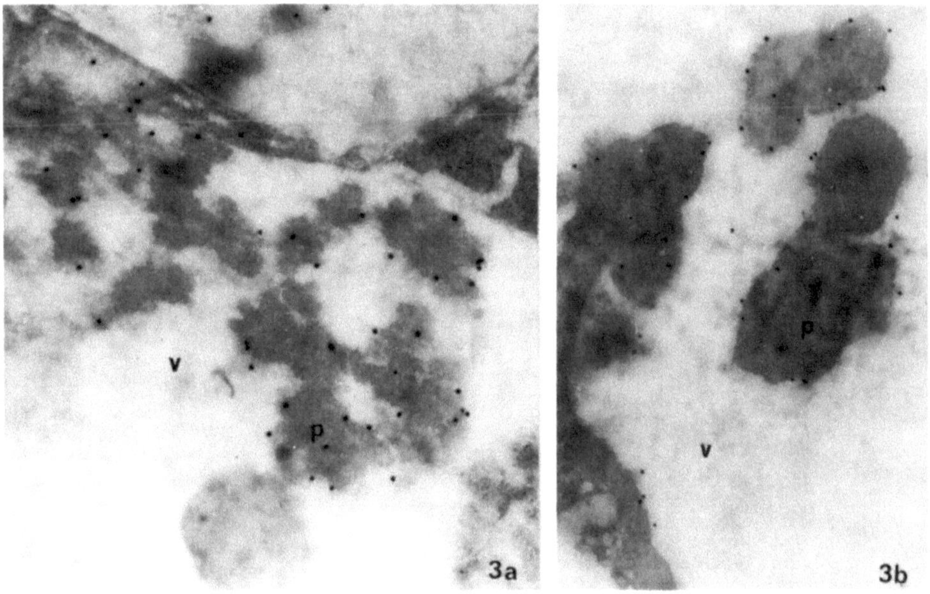

Fig. 3a, b. Protein lumps (p) in vacuoles (v) of Viscum album leaf cells after labeling with antimistletoe lectin I antibody/protein A-/Au_{16}; 31500x (a=, 27000x (b).

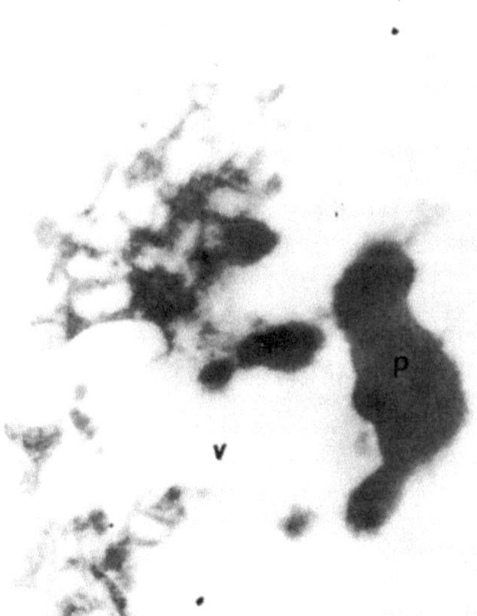

Fig. 4. Control to Fig. 3a, b. Normal rabbit serum was used in place of the specific antibody; 21000x.

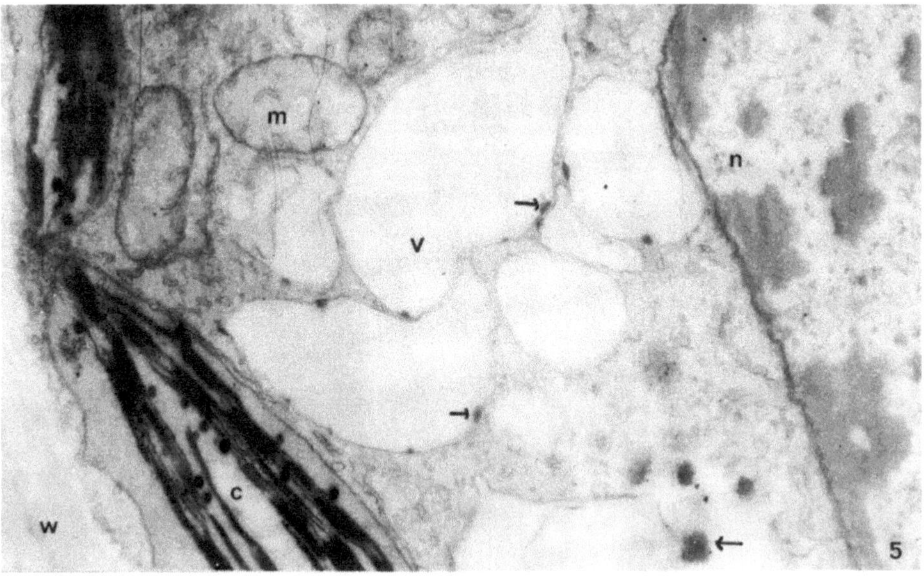

Fig. 5. Part of a leaf parenchyma cell of Viscum album (collected in June, old leaf of the foregoing year). Only small protein lumps (arrow) are visible inside the vacuoles (v); 18000x.

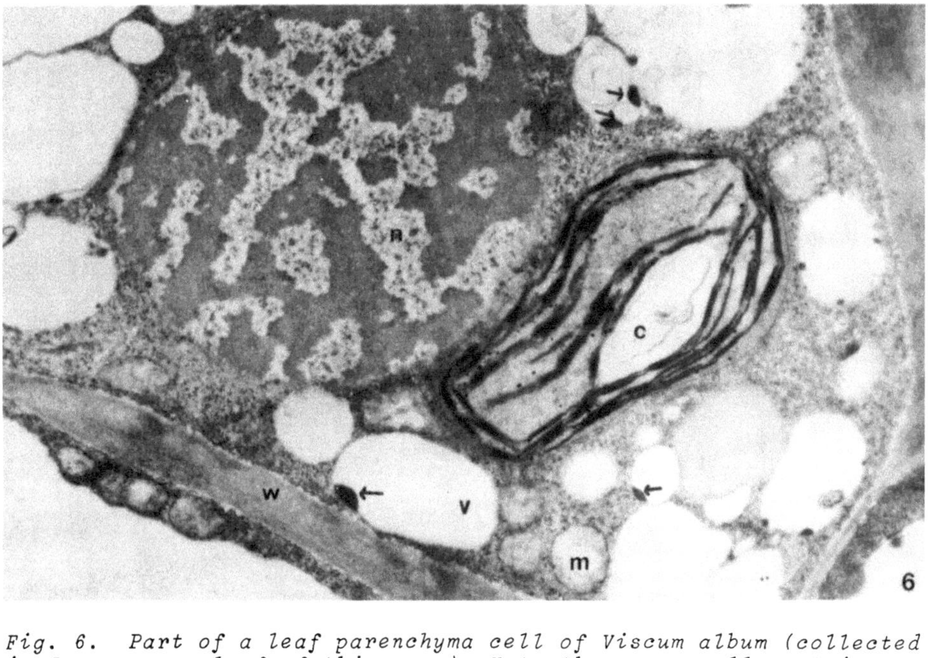

Fig. 6. Part of a leaf parenchyma cell of Viscum album (collected in June, young leaf of this year). Note the very small protein deposits (arrow) inside the vacuole (v); 8000x.

protein deposits in leaves of *Viscum album* L. by a specific antibody/protein A/Au-technique. The seasonal changes in the lectin content and the localization of the lectin in the plant indicate that ML I may play a role as a storage protein.

References

1. Barondes, S.H. 1981. Ann.Rev.Biochem. **50**, 207.

2. Chrispeels, M.J. 1983. Planta **158**, 140.

3. Dazzo, F.B. 1980. In: Advances in Legume Science (eds. A.H. Buntiny and R.J. Summerfield). p. 49.

4. Frens, G. 1973. Nature **241**, 20.

5. Herman, E.M., L.M. Shannon. 1984. Planta **161**, 97.

6. Horrisberger, M., M. Volanthen. 1980. Histochemistry **65**, 181.

7. Janzen, J.M., H.B. Juster. 1976. Science **192**, 795.

8. Marx, J.L. 1977. Science **196**, 1429.

9. Manen, J.F., A. Pusztai. 1982. Planta **155**, 328.

10. Mirelman, J., E. Galien, N. Sharon, R. Lotan. Nature

11. Pernollet, J.-C. 1978. Phytochemistry **17**, 1473.

12. Pernollet, J.-C. 1985. Physiol.Veg. **23**, 45.

13. Roth, J., M. Bendayan, L. Orci. 1978. J.Histochem. Cytochem. **26**, 1074.

14. Saint-Paul, M. 1961. Bull.Soc.Nat.Trans.Sang. **4**, 3.

15. Stinissen, H.M., W.J. Peumans. 1985. Biochem.Physiol. Pflanzen **175**, 85.

16. Toms, G.C. 1981. In: Advances in Legume Systematics (eds. R.M. Polhill and P.H. Raven). p. 561

17. Weber, E., J. Neumann. 1980. Biochem.Physiol.Pflanzen **175**, 279.

18. Ziska, P., H. Franz. 1985. In: Lectins. Biology, Biochemistry, Clinical Biochemistry IV (eds. T.C. Bog-Hansen and J. Breborowicz). W. de Gruyter, p. 475.

PATTERN OF WHEAT GERM AGGLUTININ ACCUMULATION IN DIFFERENT GENOTYPES OF WHEAT.

N. V. Raikhel[1], B. A. Palevitz[1] and R. S. Quatrano[2]

[1]Dept. of Botany, University of Georgia, Athens, GA 30602, U.S.A.
[2]Dept. of Botany and Plant Pathology, Oregon State University, Corvallis, OR 97331, U.S.A.

Wheat germ agglutinin (WGA) isolated from commercial wheat germ consists of 3 closely related polypeptides derived from each of the 3 genomes of *Triticum aestivum* (AA, BB, DD; 1, 2). The subunits are organized into homo- and heterodimeric isolectins (AA, BB, DD, AB, AD, BD).

Examination of the diploids *Triticum monococcum* (AA) and *Triticum tauschii* (*Aegilops squarrosa*, DD) shows that each contains only a single homodimer composed of polypeptides that appear to be homologous with the respective A and D subunits in the hexaploids (2). Furthermore, the AA, AB and BB isolectins of *Triticum aestivum* seem to be identical to the 3 dimer types found in tetraploids of the *Triticum turgidum* group (genome AABB). These data are compatible with the proposed origin of the A and B genomes of *Triticum aestivum* from AABB tetraploids and the D genome from *Triticum tauschii* (3). They further support the ultimate origin of the A genome from *Triticum monococcum*. However, the diploid source of the B genome remains unclear (3). Most speculation centers around the S genome group consisting of *Triticum speltoides* (SS), *Triticum sharonensis* and *Triticum longissimum* (S^lS^l), *Triticum bicorne* (S^bS^b) and *Triticum searsii* (S^sS^s), but considerable modification would have had to occur during the evolution of the B genome from these progenitors. Alternatively, an as yet unidentified diploid may have been the source of the B chromosomes.

Lectins, Vol. V

Previous work has shown that WGA is confined to the embryo portion of the wheat grain (4). However, a similar lectin is found in the roots and shoot base of seedlings or adult plants (5, 6, 7). Within the embryo, WGA is restricted to several organs and tissues, most notably the seminal roots, coleorhiza, epiblast and coleoptile (5). Recent evidence indicates that the accumulation of lectin is sequential during embryogenesis, appearing first in the coleorhiza and roots around day 10 (stage II) and lastly in the coleoptile around day 20-25 postanthesis (stage III) (Raikhel, N. V., R. S. Quatrano and B. A. Palevitz, in preparation). In barley (*Hordeum vulgare*), on the other hand, WGA never appears in the coleoptile (8). It is also of interest that abscisic acid, which seems to regulate WGA synthesis in embryos and adult plants (7, 9, 10), can induce WGA accumulation in lectin-minus coleoptiles.

Because of the differences in the occurrence of coleoptile lectin, we wished to further probe the relationship between genome, tissue differentiation and lectin expression using various wheat genotypes. The results of this study reported here may also bear on the origin of hexaploid wheat.

Materials and Methods

Various wheat genotypes were obtained from the National Small Grains Collection, United States Department of Agriculture, Beltsville, MD. The amphiploid hybrids AASS, obtained from a cross between *Triticum monococcum* and *Triticum bicorne*, and AADD, resulting from a cross between *Triticum monococcum* and *Triticum tauschii*, were provided by Dr. G. Kimber, University of Missouri, Columbia, MO.

Twenty embryos of each genotype were dissected from imbibed grain as described previously (4). Crude extracts were assayed directly be enzyme-linked immunosorbent assay essentially as described earlier (6). Experiments were repeated 3 times for each genotype.

Embryos obtained from imbibed grain were also processed for immunocytochemistry (5, 11) using formaldehyde fixation and treatment of frozen sections with rabbit anti-WGA followed by rhodamine-conjugated second antibody. Control sections were treated with non-immune rabbit IgG. Slides were viewed with epiflourescence optics using video enhancement (12).

Mature embryos excised from grain of *Triticum monococcum* were treated with 10^{-4}M abscisic acid for 2 days in the dark at 26°C and then fixed for WGA localization.

Results

WGA Levels in Wheat Extracts. Lectin levels in crude extracts of different wheat genotypes are shown in Table 1. Among the diploid wheats, *Triticum monococcum* has the lowest level of WGA. Diploids with S genomes have up to 2 times more lectin than *Triticum monococcum*, while lectin levels are even higher (up to 4 times) in *Triticum tauschii*.

All of the polyploids examined have greater levels of lectin than any of the diploids. It is interesting, however, that the highest level is present in a tetraploid (*Triticum turgidum*) and not a hexaploid.

Immunolocalization of WGA. In all of the wheats, lectin is found in the seminal roots, coleorhiza and epiblast. The embryos differ, however, in the presence of lectin in the coleoptile. While all of the AABBDD hexaploids, including different cultivars of *Triticum aestivum* (Marshall, Era and Chinese Spring) have lectin in the coleoptile (5), *Triticum zhukovskyi* lacks it (Table 1). Coleoptile WGA is also absent in the diploid *Triticum monococcum*, including 2 wild representatives, *Triticum monococcum boeticum* and *Triticum monococcum urartu*, although it's present elsewhere in the embryo (Fig. 1, A-C). On the other hand, lectin is present in the coleoptile of the S and D diploids (Table 1; Fig. 2, A,B). Coleoptile WGA is absent in the 3 AABB tetraploids, as well as *Triticum timopheevii*. However, both amphiploid hybrids AASS and AADD contain WGA in this organ (Fig. 2C).

Abscisic acid Effects on Lectin Accumulation in the Coleoptile. When mature embryos of barley, which normally lack coleoptile lectin, are treated with abscisic acid, lectin appears in this organ in some varieties. Likewise, *Triticum monococcum* embryos also accumulate lectin in the coleoptile when treated with this hormone (Fig. 1D). However, embryos of *Triticum turgidum* and *Triticum timopheevii* do not respond in this manner.

Discussion

Our evidence shows that the presence of coleoptile lectin varies

Table 1.
Amount and location of WGA in different wheat genomes (standard error indicated).

	Genome	Amount of WGA (ng/seed)	Localization in radicle, coleorhiza, epiblast	Localization in coleoptile
		DIPLOID WHEATS		
T. monococcum	AA	33.5 ± 3.1	+	-
T. longissimum	S^lS^l	63.7 ± 6.0	+	+
T. bicorne	S^bS^b	44.6 ± 5.3	+	+
T. speltoides	SS	64.5 ± 8.7	+	+
T. tauschii	DD	135.0 ± 13.6	+	+
		TETRAPLOID WHEATS		
T. timopheevii	AAGG	364.5 ± 11.5	+	-
T. turgidum	AABB	1126.2 ± 60.6	+	-
T. dicoccoides	AABB	173.4 ± 25.8	+	-
T. dicoccum	AABB	336.6 ± 19.5	+	-
		HEXAPLOID WHEATS		
T. aestivum (CV Marshall)	AABBDD	1010.7 ± 108.2	+	+
T. vavilovi	AABBDD	165.9 ± 26.5	+	+
T. compactum	AABBDD	538.2 ± 101.7	+	+
T. spelta	AABBDD	366.8 ± 2.7	+	+
T. zhukovskyi	AAAAGG	241.3 ± 1.1	+	-

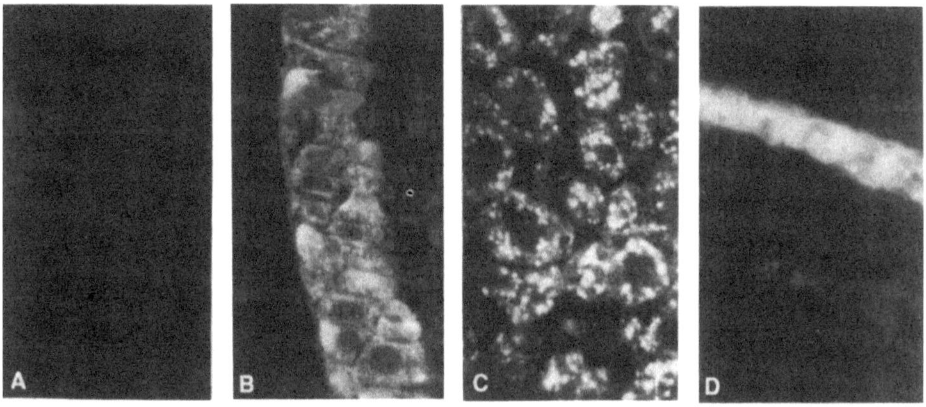

Fig. 1. A-D. Immunocytochemistry of a Triticum monococcum embryo incubated with anti-WGA. Reaction product is absent in the coleoptile (A) but present in the radicle and coleorhiza (B, C). Panel D shows lectin induction in the coleoptile of Triticum monococcum. X 525

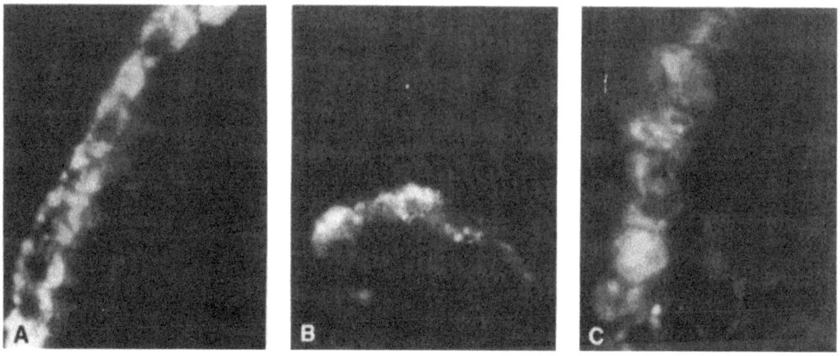

Fig. 2. A-C. Immunolocalization in the coleoptile of diploids Triticum bicorne (A) and Triticum tauschii (B) and hybrid AADD (C). X 525

among wheat genotypes. Although this difference may be useful in future studies on the mechanism and significance of organ-specific protein synthesis, its basis remains unclear. For example, we do not know whether lectin synthesis or degradation (or both) are altered in the WGA-minus coleoptiles.

Previous studies have indicated that abscisic acid regulates lectin accumulation in wheat and rice embryos as well as in the roots of older plants (7, 9, 10, 13). Because exogenous abscisic acid can induce WGA accumulation in some coleoptiles where it normally does not occur (*Triticum monococcum* and barley), it is possible that an alteration in endogenous abscisic acid distribution is responsible for the differences reported here. However, a change in abscisic acid receptors must also be considered. It is unclear why the embryos of other wheats without coleoptile lectin do not respond to abscisic acid. We are now quantitating and localizing abscisic acid in wheat in order to clarify these issues.

Because *Triticum monococcum* and the AABB tetraploids lack WGA in the coleoptile, it is tempting to speculate that the presence of WGA in this organ in *Triticum aestivum* is determined by the D genome. The reported absence of B lectin polypeptides in many *Triticum aestivum* varieties (2) and the presence of WGA in the coleoptile of the AADD hybrid reinforce this hypothesis. We realize, however, that such a proposal may be simplistic and that interactions between genomes are undoubtedly complex. More needs to be learned about such interactions as well as the possibility that WGA comprises a multigene family.

It has been speculated that *Triticum turgidum* and perhaps *Triticum timopheevii* arose from the hybridization of *Triticum monococcum* with one or more of the S genome species (reviewed in 3). However, considerable modification of the S chromosomes must have occurred during the subsequent development of the B and G genomes. Our evidence supports this hypothesis. While coleoptile lectin is absent in the tetraploids and *Triticum monococcum*, all of the S diploids have it. Thus, if an S diploid similar to those examined here was a progenitor of the tetraploids, lectin expression in the coleoptile was subsequently lost. It is noteworthy that the amphiploid hybrid AASS does have lectin in the coleoptile. It is also of interest that coleoptile lectin has been restored in the AABBDD hexaploids. The simplest explanation is that the hybridization of AABB tetraploids with DD (3) was responsible for this change. This assumption is indirectly supported by the lack of coleoptile WGA in *Triticum zhukovskyi*, which contains 4 A and 2 G chromosomes, but none of D. Direct evidence is provided by the presence of coleoptile lectin in the AADD hybrid. We realize that inferences about plant origins based on single characteristics are risky at best. Our comments are meant to encourage further work in the hope that new data on WGA and its genes can more definitively aid in our understanding of wheat systematics.

Acknowledgements We thank Dr. M. Smith and Mss Asha Wise, Laurie Hanna and Roswitha Hopkins for technical assistance. Supported by National Science Foundation grant DMB83-14374 to N.V.R. and B.A.P. and U.S.D.A. competitive grant (84-CRCR-1380) to R.S.Q. We are grateful to Dr. G. Kimber and the U.S.D.A. Small Grains Collection for providing various wheat grain.

References

1. Rice, R. H., Etzler, M. E.: Biochemistry *14*, 4093-4099 (1975).
2. Peumans, W. J., Stinissen, H. M., Carlier, A. R.: Planta *154*, 562-567 (1982).
3. Feldman, M. In: Evolution of Crop Plants, ed. N. W. Simmonds, pp. 120-128, Longman, London (1976).
4. Mishkind, M., Keegstra, K., Palevitz, B. A.: Plant Physiol. *66*, 950-955 (1980).
5. Mishkind, M., Raikhel, N. V., Palevitz, B. A., Keegstra, K.: J. Cell Biol. *92*, 753-764 (1982).
6. Raikhel, N. V., Mishkind, M., Palevitz, B. A.: Planta *162*, 55-61 (1984).
7. Stinissen, H. M., Chrispeels, M. J., Peumans, W. J.: Planta *164*, 278-286 (1985).
8. Mishkind, M., Palevitz, B. A., Raikhel, N. V., Keegstra, K.: Science *220*, 1290-1292 (1983).
9. Triplett, B. A., Quatrano, R. S.: Dev. Biol. *91*, 491-496 (1982).
10. Raikhel, N. V., Palevitz, B. A., Haigler, C. H.: Plant Physiol. in Press (1985).
11. Raikhel, N. V., Mishkind, M., Palevitz, B. A.: Protoplasma *121*, 25-33 (1984).
12. Palevitz, B. A., O'Kane, D. J., Kobres, R. E., Raikhel, N. V.: Protoplasma *109*, 23-55 (1981).
13. Stinissen, H. M., Peumans, W. J., DeLanghe, F.: Plant Cell Rep. *3*, 55-59 (1984).

PART 2

ISOLATION AND CHARACTERIZATION OF PLANT LECTINS

STUDIES ON LECTINS FROM INDIAN PLANTS.

Rajindar D. Sandhu, Jatinder S. Arora, Sushil K. Chopra and Sukhdev S. Kamboj.

Department of Biology, Guru Nanak Dev University, Amritsar-143005, India.

The occurrence of lectins as seed proteins is fairly well established for a number of plant species (1, 2, 15). Lectins are readily revealed by the hemagglutinating property associated with crude seed extracts using a battery of erythrocyts. In this way we have earlier detected and characterised 50 new phytolectins from India (3).

In a continuing survey of the Indian flora 450 plant species were thoroughly screened revealing 54 new lectin sources. These lectins have been characterized with respect to their biological action spectra and sugar specificity.

MATERIALS AND METHODS

Seed collection: Seeds of 450 different wild plant species were collected during the two main ripening seasons, that is, April to June and October to December in the years 1982-1984, from the following places: Kumaon Hills, Nepal, Siliguri, Darjeeling, Shillong, Rani forest in Assam, Pachmarhi, Western Ghats including Valpoi, Sirigere, Hanigere, Medikeri and deciduous Balle forest in Mysore. Seeds of species of *Crotalaria*, *Erythrina* and *Amaranthus* were procured from CSIR Regional Research Laboratory, Jammu, and National Botanic Gardens, Lucknow.

The procedures adopted in the preparation of seed extracts and the bioassay of lectin activity employing various normal and neuraminidase treated erythrocytes, have already been described (3, 4). The lowest inhibitory concentration of a sugar in the hemagglutionation assay was worked out by addition of serial dilutions of 50 mM stock solutions to the bioassay mixture.

RESULTS AND DISCUSSION

As many as 47 plant species were found to possess lectin activity against one or more types of erythrocytes. Seven additional seed lectins agglutinated exclusively the neuraminidase-treated cells. The former represented 21 different genera belonging to 11 families. The largest number of lectins were contributed by the family Fabaceae (20 species) mostly belonging to the genera *Crotalaria* and *Erythrina*. Among the non-fabaceous dicots the lectin-rich genera included *Amaranthus*, *Artocarpus* and *Michaelia*. The present study also brings out the occurrence of potent lectins in as many as 9 species belonging to 6 different genera of family Araceae, thus revealing new sources of lectins among the monocots. In at least two of these species, that is, *Arisaema intermedium* and *Sauromatum guttatum* the lectin activity was present in turbers but not in seed extracts. Data on the overall prevalence of phytolectins are summarized in Table 1.

The biological and sugar specificities of hemagglutinins encountered in this study are given in Table 2. In their biological action spectra a majority of the hemagglutinins proved to be non-specific, agglutinating 2-5 different types of erythrocytes. Eight species belonging to the genus *Amaranthus*, were able to agglutinate all 9 types of erythrocytes employed in the assay. Also several earlier studies have shown the preponderance of non-specific phytolectins (1, 3, 5), probably indicative of the availability of similar lectin binding sites on a variety of erythrocyte types.

Nevertheless, 2 lectins turned out to be specific for human A, 4 for rabbit and one for rat erythrocytes. The total absence of blood group B and O specific lectins in the present study is again reflective of extreme paucity of such lectins in nature (6, 7).

The rabbit erythrocytes were most prone to agglutination with manifest sensitivity to 37 plant species as compared to human erythrocytes which were agglutinated by 28 different lectins. The present study corroborates our earlier findings on the refractory nature of erythrocytes of ruminants which were agglutinated by not more than 8-11 lectins (3). The biological significance of this finding remains to be identified.

Neuraminidase treatment of target cells facilitated the detection of 7 additional phytolectins in *Crotalaria barbata*, *C. candicans*, *C. sericea*, *Clerodendron colebrookianum*, *C. indicum*, *C. roseus* and *Tephrosia purpurea*. The similarity of these phytolectins with peanut agglutinin insofar as they exclusively bind to desialized erythrocytes is quite obvious (8), and warrants their further investigation as potentially useful lectins for the selective isolation of immature hemopoietic cells (9). Generally, the biological action spectra of various lectins was broadened considerably when desialized erythrocytes were employed. Neuraminidase removes terminally located sialic acid which masks galactose moieties on the acceptor molecules besides reducing surface negative charge on the target cells which then become much more susceptible to agglutination (10).

A perusal of the data on minimal erythrocyte agglutinating protein concentration (MEAPC) for various lectins would suggest that seeds of *Artocarpus* spp. possess exceptionally potent lectins which may turn out to be preferred affinity ligands. Phytolectins being proteinaceous in nature are susceptible to denaturation although they may withstand incubation at elevated temperatures to varying degrees as shown in this study. The majority of the lectins exhibited a gradual loss of hemagglutinating activity when incubated at different temperatures ranging between 60°-$75^{o}C$ (Table 2). Among some of the structural features that might be contributing to improved thermo-

TABLE 2.

Biological action spectra and other characteristics of lectins from Indian plants.

	Agglutination of erythrocytes Human	Rb	Gp	Rt	Sh	Gt	*MEAPC µg/ml	Thermo-stability °C	Sugar inhibitor (mM)
Fabaceae									
1. *Abrus pulchellus*	(ABO)	+	(+)	(+)	(+)	(+)	236	55	Afet (25)
2. *Crotalaria anagyroides*	ABO*	(+)	(+)	(+)	(+)	(+)	2656	75	GalNAc (1.25)
3. *C. barbata*	(A*)	–	–	–	–	–	480	70	"(5.0)
4. *C. brevifolia*	(ABO)	+	–	+	(+)	–	35	70	"(5.0)
5. *C. brownei*	AB*	–	(+)	(+)	–	–	886	60	"(2.5)
6. *C. candicans*	(ABO)*								
7. *C. incana*	(AB)	+	(+)	+	–	–	156	65	"(2.5)
8. *C. laburnifolia*	(ABO)	(+)	(+)	+*	(+)	–	935	75	"(1.25)
9. *C. medicaginea*	ABO*	(+)	(+)	(+)	(+)	(+)	806	75	"(1.25)
10. *C. saltiana*	A*(B)	–	–	(+)	–	–	145	75	"(0.62)
11. *C. sericea*	(ABO)	(+)	(+)	(+)	(+)	–	2810	70	"(1.25)
12. *C. striata*	AB(0)	+	(+)	+	–	–	73	65	"(0.31)
13. *C. usaramoensis*	(ABO)	+	(+)	+	(+)	–	35	75	"(2.5)
14. *C. walkeri*	ABO	+	(+)	+	(+)	(+)	209	70	"(0.62)
15. *C. wightiana*	ABO*	(+)	(+)	(+)	(+)	(+)	781	65	"(0.62)
16. *Dalbergia lanceolaria*	ABO	+	(+)	+	(+)	–	3155	50	"(5.0)

TABLE 2 contd. ....

17. *Derris robusta*	ABO	+	+	+	(+)	-	647	60	Gal Nac(25)
18. *Dumasia villosa*	A*(BO)	-	-	(+)	(+)	-	136	65	Gal Nac(2.5)
19. *Erythrina fusea*	AB*O	(+)	(+)	(+)	(+)	(+)	113	65	Lac (10)
20. *E. resuparcellii*	AB*O	+	(+)	(+)	(+)	(+)	58	60	" (2.5)
21. *E. resupinata*	AB*O	+	(+)	(+)	(+)	(+)	384	65	" (1.25)
22. *E. stricta*	AB*O	(+)	(+)	(+)	(+)	(+)	512	65	" (1.25)
23. *E. verspertilio*	AB*O	(+)	(+)	(+)	(+)	(+)	768	60	" (2.5)
24. *Tephrosia purpurea*	(ABO)	(+)	(+)	(+)	(+)	-	961	50	Gal Nac(10)
Mimosoideae									
25. *Parkia roxburghii*	-	+	(+)	(+)	(+)	-	430	55	Undetermined
Amaranthaceae									
26. *Amaranthus caudatus*	ABO	+	+	+	+	+	490	65	Afet (125)
27. *A. cruentus*	ABO	+	+	+	+	+	223	65	" (75)
28. *A. dubius*	ABO	+	+	+	+	+	812	60	" (175)
29. *A. graecizans*	ABO	+	+	+	+	+	1070	65	" (50)
30. *A. hybridus*	ABO	+	+	+	+	+	178	60	" (125)
31. *A. hypochondriacus*	ABO	+	+	+	+	+	462	65	" (50)
32. *A. lividus*	ABO	+	+	+	+	+	555	65	" (125)
33 *A. tricolor*	ABO	+	+	+	+	+	1962	60	" (150)
Moraceae									
34. *Artocarpus hirsuta*	ABO	+	+	+	+	(+)	.006	60	Meb (20)
35. *A. lakoocha*	ABO	+	+	+	+	(+)	.021	55	" (2.5)
Convolvulaceae									
36. *Argyrea hypocratoriformis*	(ABO)	+	+	+	-	-	61	55	Undetermined

TABLE 2 contd. ....

	Verbenaceae									
37.	*Clerodendron colebrook-ianum*	(ABO)	(+)	(+)	(+)	(+)	(+)	69	50	Gal NAc(0.31)
38.	*C. indicum*	(ABO)	(+)	(+)	(+)	(+)	(+)	142	55	" " (1.25)
39.	*C. roseus*	(ABO)	(+)	(+)	(+)	(+)	(+)	101	50	" " (2.5)
	Tiliaceae									
40.	*Corchorus capsularis*	A(B)O	+	(+)	(+)	(+)	(+)	80	65	Gal (2.5)
	Fumariaceae									
41.	*Dicentra thalicitifolia*	-	+	-	+	-	-	322	55	GalNAc(10)
	Magnoliaceae									
42.	*Michaelia cartcatic*	-	+	(+)	(+)	(+)	-	2855	55	" (5)
43.	*M. linogiosa*	-	+	(+)	(+)	-	-	2221	55	" (2.5)
	Cucurbitaceae									
44.	*Trichosanthes sp.*	ABO	+	+	+	(+)	(+)	25	70	Lac (2.5)
	Vitaceae									
45.	*Vitis repanda*	ABO	+	(+)	+	(+)	-	390	50	GalNAc(2.5)
	Araceae									
46.	*Amorphophallus bulbifer*	-	+	+	+	+	-	301	65	Afet (100)
47.	*Arisaema consanguineum*	-	+	+	+	+	-	32	70	" (100)
48.	*A. flavum*	-	+	(+)	+	(+)	(+)	163	65	" (75)
49.	*A. intermedium (Tuber)*	-	+	+	+	+	-	26	65	" (125)
50.	*A. tortuosum*	-	+	+	+	(+)	-	122	75	" (125)
51.	*Pothos scandans*	-	+	(+)	+	(+)	-	1250	60	" (50)
52.	*Remusatia vivipara*	-	+	+	+	(+)	-	9	70	" (75)
53.	*Sauromatum quttatum (Tuber)*	-	+	+	+	(+)	-	12	65	" (50)

TABLE 2 contd. .....

54.	*Scindopsus officinalis*	ABO	+	+	+	+	-	155	75	Afet	50

Positive agglutination test is depicted by blood group symbol or + sign; parenthesis indicate use of neuraminidase treated erythrocytes: Rb = rabbit, Gp = guinea pig, Rt = rat, Sh = sheep, and Gt = goat.

MEAPC = minimal protein concentration for erythrocyte agglutination with respect to rabbit RBC or otherwise indicated by asterisk mark.

Thermostability = the highest temperature (°C) at which lectin is stable after 15 min incubation.

Afet = asialofetuin, the concentration is given in µg/ml.

TABLE 1.

Overall prevalence of phytolectins in Indian plants.

Taxon	Number positive/Number investigated			
	Family	Genera	Species	
			Number	Per cent
Dicots				
Fabaceae	1/1	7/25	24/75	32
Other legumes	1/2	1/23	1/77	1.29
Non-legumes	9/75	9/190	20/259	7.72
Monocots	1/7	6/20	9/27	33.33
Gymnosperms	0/4	0/9	0/12	0
Total	12/89	23/267	54/450	12

stability of a lectin are: Presence of extensive β-pleated structure (11), glycocomponents and sulphur containing amino acids (12, 13).

Based on the data pertaining to the most effective sugar or glycoconjugate inhibitor, the new lectins fall into two large categories. Twenty-five of the lectins responded best to N-acetylgalactosamine. Of the remainig, 18 lectins were exclusively inhibited by asialofetuin or fetuin indicating extended binding sites. Only two lectins failed to respond to any of the test sugars. The overwhelming preponderance of lectins specific for galactose and its derivatives is in conformity with the results of earlier studies on phytolectins (3, 14).

ACKNOWLEDGEMENTS

This study has been supported by the financial assistance from the Department of Science and Technology, New Delhi. JSA and SSK have been the Senior Research Fellows of Council of Scientific and Industrial Research and Department of Atomic Energy, New Delhi, respectively. We are greatful to Drs. C.K. Atal (RRL, Jammu), Mohinder Pal and G.S. Srivastava (NBRI, Lucknow), for donating samples from their seed collection. The authors thank Mr. Hjalmar Johansen for typing the manuscript.

REFERENCES

1. Gold, E.R., Balding, P.: Receptor-specific proteins - Plant and Animal Lectins, American Elsevier Publishing Company Inc., New York, pp. 151-236 (1975).
2. Sandhu, R.S., Reen, R.S.: In: Lectins, Biology, Biochemistry, Clinical Biochemistry (T.C. Bøg-Hansen, ed.) W. de Gruyter, Berlin, Vol. 2, pp. 113-150 (1982).
3. Sandhu, R.S., Reen, R.S., Singh, S., Singh, J., Chopra, S.K. In: Lectins, Biology, Biochemistry, Clinical Biochemistry (T.C. Bøg-Hansen, ed.) W. de Gruyter, Berlin, Vol. 2, pp. 679-692 (1982).
4. Sandhu, R.S., Chopra, S.K., Arora, J.S., Reen, R.S. In: Lectins, Biology, Biochemistry, Clinical Biochemistry (T.C. Bøg-Hansen and J. Breborowicz, eds.) W. de Gruyter, Berlin, Vol. 4, pp. 547-557 (1985).
5. Toms, G.C., Western, A.: Chemotaxonomy of Leguminosae (J.B. Harbourne, D. Boulter and B.L. Turner, eds.) Academic Press, London and New York, pp. 367-462 (1971).
6. Bird, G.W.G. Brit. Med. Bull. *15*, 165-168 (1959).
7. Allen, N.K., Brilliantine, L. J. Immunol. *102*, 1295-1299 (1969).
8. Bird, G.W.G. Vox. Sang. *9*, 748 (1964).
9. Reisner, Y., Sharon, N.: Trends Biochem. Sci. *5*, 29-31 (1980).
10. Nicolson, G.L., Locobiere, M., Eckhart, W.: Biochemistry *14*, 172-179 (1975).
11. Reeke, G.N., Becker, J.W., Edelman, G.M.: J. BIol. Chem. *250*, 1525-1547 (1975).
12. Burger, M.M., Goldberg, A.R.: Proc. Natl. Acad. Sci. *57*, 359-365 (1967).
13. Nagata, Y., Burger, M.M.: J. Biol. Chem. *249*, 3116-3122 (1974).
14. Mankins, C.N., Kindinger, J.I., Shannon, L.M.: Plant Physiol. *64*, 104-107 (1979).
15. Rüdiger, H. In: Lectins, Biology, Biochemistry, Clinical Biochemistry (T.C. Bøg-Hansen, ed) W. de Gruyter, Berlin, Vol. 1, pp. 1-10 (1981).

SCREENING FOR LECTINS IN COMMON FOODS BY LINE-DIVE IMMUNOELECTROPHORESIS AND BY HAEMADSORPTION LECTIN TEST

Mette M. Andersen and Kirsten Ebbesen

Nutrition Laboratory, Kommunehospitalet, DK-1399 Copenhagen K, Denmark

Lectins in food can affect the uptake of nutritional substances and impair the integrity of the intestinal mucosal surfaces in various animal species (1, 2, 3, 4). It is possible that food lectins can produce similar effects in humans (5, 6).

We have screened the edible parts of common vegetables and spices consumed in Denmark for the presence of lectins. The Line-Dive immunoelectrophoresis (7) and the haemadsorption lectin test (HALT) (8) were chosen since crude extracts of foods can be analysed directly in both techniques and because of the possibility of testing at least 20 items per plate. In the Line-Dive technique lectins are detected by their capacity to precipitate or bind human serum glycoproteins, thereby retarding their electrophoretic mobility. In the HALT technique lectins are adsorped to microtitre plates and detected by their capacity to bind human erythrocytes.

Lectins, Vol. V

Materials and Methods

All purified lectins (con A, ECG, GS1, GS2, LCA, PHA, PNA, PSA, SBA, STA, UEA, VAA, VFA and WGA) were obtained from Kem-en-Tec, Hellerup, Denmark. Purified orosomucoid and its con A-binding fraction (the latter isolated from a con A-Sepharose column) were gifts from T.C.Bøg-Hansen. Vegetables and spices were bought locally. All antibodies were obtained from Dakopatts A/S, Denmark.

Fresh vegetables were crushed in a juicer. Dry vegetables were crushed in a blender, spices in a mortar, and extracted in phosphate buffered saline (PBS) pH 7.2 (1:4 W/V) on a rotator for 1 hour. After centrifugation (4.000 rpm, 20 min) of the PBS extracts and the vegetables juices, the supernatants were used directly or frozen (-20 ^{o}C) until use. Coffee and tea were brewed as for drinking. Three members of the laboratory staff (blood types A, B and 0) provided the erythrocytes for the HALT.

The Line-Dive immunoelectrophoresis was carried out according to Krøll and Andersen (7). The antigen used was the solubilized precipitate of human serum treated with polyethylenglycol 6000 (PEG) (125 g + 100 ml serum). By this procedure the major part of albumin and of the lipoproteins are removed. Routinely this antigen was run against polyspecific antihuman antiserum. The purified lectins or the food extracts to be tested were placed in 5 µl and 10 µl wells anodically to the antigen trough. For quantitative purposes, the same antigen, mixed with serum (1:1) was run against each of 17 monospecific antibodies against the major serum glycoproteins. In these cases dilutions of the respective monospecific antibody were placed in wells in the same row as the lectins.

The haemadsorption test, HALT, was carried out according to Kint (8) with the following modifications: The erythrocytes were treated with neuraminidase (½ hour, 37 ^{o}C acetate buffer, pH 5.0) instead of with trypsin. The duration of coating was 18 - 24 hours at 4 ^{o}C. The concentration of erythrocytes used was 0.225%.

Results

In all, 75 food items including coffee and tea were tested in both techniques. Among 30 fresh vegetables tested, 11 were positive in the Line-Dive and 8 were positive in the HALT (Table I). Among 21 dry vegetables 13 were positive in Line-Dive and 6 in HALT (Table II). The greatest number of positive results came from spices where 16 of 24 were positive in Line-Dive and 15 in HALT (Table III). All teas tested, including green tea and herb-tea were positive in both techniques. A total of 40 items were positive in Line-Dive and 29 items in HALT.

The Line-Dive was positive with 8 of the 14 purified lectins (con A, GS1, LCA, PHA, PSA, VAA, VFA and WGA). GS1 could not be precisely evaluated since it moved faster anodically than the serum proteins, but when placed cathodically to the antigen it reacted with many glycoproteins. UEA reacted with α_2-macroglobulin by producing a rocket on this line and this could not be inhibited by the specific sugar. Nine of the 14 lectins were tested as Sepharose conjugated lectins, and all had very weak or negative reactions.

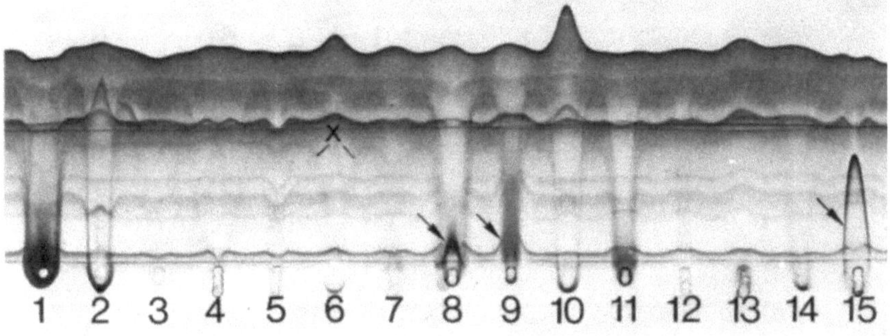

Fig. 1. Line-Dive immunoelectrophoresis for screening of food items. The reference antigen was human serum (solubilized PEG-precipitated serum corresponding to 22.5 µl native serum) and the reference antibody was polyspecific anti-human antiserum (250 µl polyspecific in the upper gel and 100 µl polyspecific plus 75 µl anti-IgM in the lower gel). The wells contained 2 - 10 ul of the following items or monospecific antibodies: 1) cinnamon, 2) anti-IgA, 3) paprika, 4) cayenne pepper, 5) cardamom, 6) anti-α_1-lipoprotein, 7) rhubarb, 8) clove, 9) thyme, 10) anti-β-lipoprotein, 11) sweet gale, 12) ginger, 13) curry, 14) anti-IgG and 15) garlic.
Note the precipitate rockets by clove, thyme and garlic (arrows). The thin line 'x' represents the remainings of α_1-lipoprotein.

In Fig. 1 the screening of food items in Line-Dive is illustrated. Garlic gave rise to a rocket precipitate and this was not a lectin reaction as it also appeared when neither antigen nor antibody were present in the gel. Clove and thyme also gave rise to rocket precipitates, but these were based on reactions with the serum proteins (e.i. line-rockets).

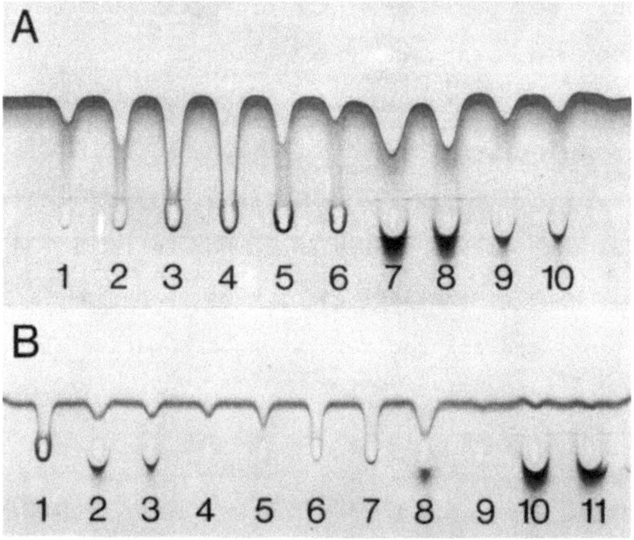

Fig. 2. Line-Dive immunoelectrophoresis for quantitative purposes. The reference antigen was as in Fig. 1. A: antibody 50 µl anti-IgA. The wells contained as follows: 1-4) anti-IgA 0.1 µl, 0.2 µl, 0.4 µl and 0.6 µl respectively, 5-6) VAA 10 µl and 5 µl, 7-8) con A 5 µl and 2 µl, 9-10) PHA 10 µl and 5 µl. B: antibody 150 µl anti-orosomucoid. The wells contained as follows: 1) VAA 5 µl, 2-3) PHA 10 µl and 5 µl, 4-7) anti-orosomucoid 0.05 µl, 0.1 µl, 0.3 µl and 0.5 µl, 8) WGA 10 µl (11 mg/ml), 9) WGA 10 µl, a different batch (0.2 mg/ml), 10-11) con A 5 µl and 2 µl.

In A note that for con A and PHA the line-dives produced by decreasing amounts of lectin are nearly equal, indicating that the lectins are in surplus, while this is not the case for VAA.
In B note that the amount of WGA in well 9) is not sufficient for a positive reaction. Also note the "waves" on the line produced by con A.

When the positive lectins were tested for their quantitative reaction with individual glycoproteins it could be seen that the binding capacity of some, but not all lectins were in surplus (Fig. 2 A). Most lectins reacted strongly with α_2-macroglobulin, β-lipoprotein and IgM (Table IV). All 7 lectins reacted with antithrombin III, ceruloplasmin and IgA, but none were positive with Gc-globulin. The reaction of con A with unpurified (Fig. 2 B) or purified orosomucoid was to produce "waves" on the line. Only when the con A-binding fraction of orosomucoid was used alone, did con A produce a line-dive reaction (not shown).

Many food items reacted with a thin, weak precipitate line, representing the remainings of α_1-lipoprotein after PEG-precipitation 'x' in Fig. 1. With 0.5% detergent in the gel this line was not precipitated, demonstrating the presence of hydrophobic sites on this protein. Three purified lectins (LCA, VAA and PSA) reacted with this protein, despite their very modest reactions with the native α_1-lipoprotein (Table IV).

In the HALT all 14 purified lectins were positive. None of the lectins were strictly blood group specific but there were differences in the lectin concentrations required. Similarly, the food items, if positive, reacted with all blood types. The HALT did not detect all purified lectins when trypsinized erythrocytes were used. Thus GS1, GS2, LCA, PNA, PSA, VFA and UEA reacted weakly or not at all

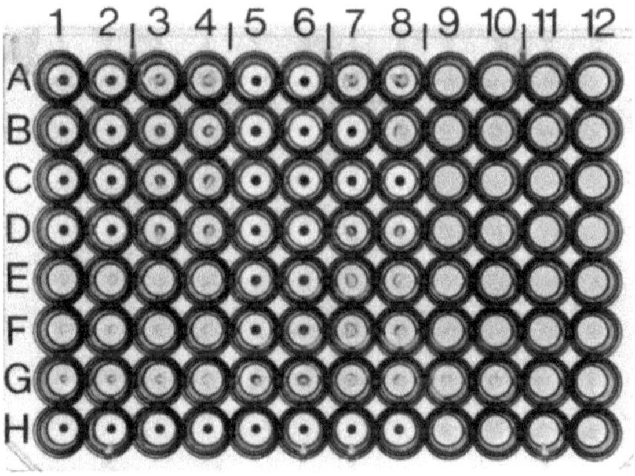

Fig. 3. Haemadsorption lectin test for screening of food items. Rows 1-2, 5-6, and 9-10 blood type 0. Rows 3-4, 7-8, and 11-12 blood type A. Serial dilutions (1:2) of each item were tested in duplicate starting with the crude extract in well A or E.

$A\text{-}D_{1-4}$ = pea pod	*$A\text{-}D_{5-8}$ = spinach*	*$A\text{-}F_{9-12}$ = brown bean*
$E\text{-}G_{1-4}$ = tomato	*$E\text{-}G_{5-8}$ = radish*	*$G\text{-}H_{9-12}$ = water from overnight soaking of beans*
H_{1-8} = buffer only		

Note the disturbed negative reactions by pea pod, spinach and radish.

with these cells. The general background staining of the wells, caused by adherence of erythrocytes to the plastic material, made it difficult to distinguish the weakly positive from the negative reactions. Moreover, some food items left a sediment (after wash) at the bottom of the wells, preventing the erythrocytes from settling properly (Fig. 3). Although their respective lectins were positive in HALT, broad bean, lentil, pea and soy bean extracts were negative in this technique.

Both Line-Dive and HALT could detect about 0.2 μg purified lectin, except for WGA where 4 μg was necessary (Fig. 2 B), and in both techniques positive reactions could be inhibited by specific sugars. In the Line-Dive the glycoprotein-lectin reactions were inhibited one-by-one with increasing sugar concentration, demonstrating differences in the strength of affinity to the lectin. Except for β-lipoprotein, the reactions of all the glycoproteins with 1 mole con A were inhibited by 380 moles of α-methyl-D-glycopyranoside (Fig. 4).

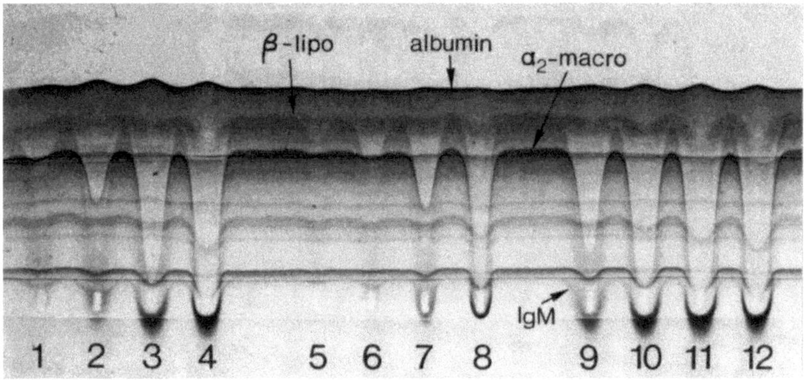

Fig. 4. Line-Dive immunoelectrophoresis with sugar inhibition. The reference antigen and antibody as in Fig. 1. The wells contained as follows: 1-4) and 9-12) con A 1 μg, 5-8) con A 0.1 μg. Wells 1 + 5), 2 + 6) and 3 + 7) contained α-methyl-D-glycopyranoside 5 μl (50%), (12.5%) and (3.1%) respectively. Wells 9), 10) and 11) contained glucose 5 μl (50%), (12.5%) and (3.1%) respectively. Note the sugar inhibition at well 5) is complete except for β-lipoprotein.

Discussion

The Line-Dive technique was originally described for identification of individual line-precipitates in a complicated pattern, among these the con A-binding proteins (7). The technique has since been applied by others for screening various seeds for lectins (9). The fact that only glycoproteins produce line-deflections is an assurance of the specificity of the lectin reaction, with the non-glycoproteins (e.g. albumin) providing a built-in negative reference. The main draw-back of this method is that relatively few purified lectins exhibit positive reactions. This could be due to the small quantity of lectin (10 µl) in the wells (as in the case for one of the batches of WGA) or the pH of the system not being optimal for some lectins. One lectin (GS1) was shown to have an anodic mobility. Most likely however, the carbohydrate structures of the serum proteins (10) are not complex enough for reaction with a number of lectins. An advantage of the technique is that the lectin reactions do not seem to be disturbed by competitive proteins in the crude extracts.

Many food items and three purified lectins reacted with the re-solubilized part of α_1-lipoprotein after PEG precipitation. This could be a lectin-reaction with a glycoprotein associated with α_1-lipoprotein (e.g. arylesterase, known to bind con A) or a non-glycoconjugate reaction with the hydrophobic part of this protein - or a combination of both. A non-lectin reaction was seen with UEA which produced a rocket on the α_2-macroglobulin line. Since the rocket precipitate was less dense than the line precipitate and since it could not be inhibited by specific sugar, this reaction was most likely due to degrading activity in the lectin preparation. Also clove and thyme gave rise to line-rockets, but these were very

Table I

Fresh vegetables with detectable lectin activity (14 of 30)

Name	Latin name	Line-Dive*	HALT*
Avocado	Persea americana	(x)	0
Beetroot	Beta vulgaris	(x)	(x)
Carrot	Daucus carota	(x)	0
Green bean (haricot)	Phaseolus vulgaris	X	X
Jerusalem artichoke	Helianthus tuberosus	(x)	(x)
Leek	Allium porrum	(x)	X
Mushroom	Agaricus bisporus	(x)	0
Pea	Pisum sativum	X	0
Potato	Solanum tuberosum	0	X
Red cabbage	Brassica oleracea	0	X
Red pepper	Capsicum annuum	(x)	0
Rhubarb	Rheum rhaponticum	X	0
Spring cabbage	Brassica oleracea	0	X
Tomato	Lycopersicum exculentum	X	X

Negative items: Asparagus, cauliflower, celery, chinese cabbage, cucumber, eggplant, fennel, garlic, green pepper, maize, onion, pea pod, radish, salad, spinach, squash.

* X : *strong positive reaction*
(x): *weak positive reaction*
0 : *no reaction*

dense, suggesting a lectin-protein complex. This could lead to masking of antigenic sites, see ref. 11. In all cases of questionable reactions, the final proof of a lectin reaction is its inhibition by specific sugars.

In the present study we attempted to quantify the area of the line-deflections caused by lectins by two means: a) by comparing the re-

Table II

Dry vegetables with detectable lectin activity (13 of 21)

Name	Latin name	Line-Dive	HALT
Bean, black	Phaseolus vulgaris	X	X
Bean, brown	Phaseolus vulgaris	X	X
Bean, chili	Phaseolus vulgaris	X	X
Bean, Pinto	Phaseolus vulgaris	X	X
Bean, white	Phaseolus vulgaris	X	X
Broad bean	Vicia faba	X	0
Buckwheat	Fagopyrum esculentum	(x)	0
Chick pea	Cider arietinum	X	0
Lentil (red)	Lens culinaris	X	0
Lentil (white)	Lens culinaris	X	0
Lima bean	Phaseolus limensis	X	X
Soy bean	Glycine max	(x)	0
Split pea	Pisum sativum	X	0

Negative items: Green gram, linseed, millet, potatoflour, rice, sesame, sunflower-seed, wheat flour.

spective line-deflections produced by specific antibodies, and b) by measuring the amount of specific sugar necessary to inhibit the reactions.

Ad a) it is required that the lectin in question has the same electrophoretic mobility and strength of affinity to the glycoproteins as the monospecific antibody it is compared with. The strength of affinity of the glycoprotein-lectin complex has to be strong enough for precipitation or retardation. Moreover, the lectin should be in surplus to prevent competition for the lectin from other glycoproteins in the antigen sample. Probably these requirements are

Table III
*Spices with detectable lectin activity (18 of 24)**

Name	Latin name	Line-Dive	HALT
Basil	Ocimum basilicum	(x)	0
Bay-leaf	Laurus nobilis	X	X
Cardamom	Elettaria cardamomum	(x)	(x)
Cayenne pepper	Capsicum annuum	X	(x)
Chive	Allium shoenoprasum	(x)	(x)
Cinnamon	Cinnamomum ceylanicum	X	X
Clove	Eugenia aromatica	X	X
Dutch kreuter		X	(x)
Ginger (chinese)	Zingiber officinale	X	0
Oregano	Origanum vulgare	(x)	X
Parsley	Petroselinum crispum	(x)	0
Rosemary	Rosmarimus officinalis	X	X
Savory	Satureja hortensis	X	(x)
Sweet gale	Myrica gale	X	X
Thyme	Thymus vulgaris	X	X
Turmeric	Curcuma longa	0	(x)
Coffee		0	(x)
Tea		X	X

Negative items: Curry, paprika, pepper, mace, almond, hazelnut.
**including coffee, tea and nuts.*

fulfilled only in some cases. Thus the area of the line-dive can only be a relative measure of lectin activity. When orosomucoid reacted with con A no line-dive occurred, probably due to the lectinprotein complexes being able to "keep up" with the immunocomplexes forming the precipitate line during electrophoresis. Only the

Table IV

*Reactivity of lectins with serum proteins in Line-Dive using monospecific antibodies as references**

Lectin (μg) / Protein	con A 46	LCA 34	PHA 12	PSA 60	VAA 14	VFA 14	WGA 110
orosomucoid	w	w	0.9	w	≥ 5	w	2.4
α_1-lipo	w	w	-	w	w	w	-
α_1-AT	2	w	-	-	3	w	-
α_1-ACT	1.5	0.6	1.5	w	≥ 8	1	3
Gc-glob	w	w	-	w	-	w	-
$Inter_{\alpha}TI$	2	0.9	-	w	1.5	1	1
α-Tromb. III	≥ 1	0.5	0.2	0.2	≥ 1	0.5	0.3
Cerulo	1.5	0.6	0.2	1	≥ 5	1	0.7
α_2HS	1	w	-	w	1	w	w
Hapto	4	w	1	w	≥ 7	0.5	2
α_2m	≥ 8	3	6	2	6	4	3
β-lipo	≥ 10	1	7	6	≥ 10	4	4
hemopexin	≥ 8	w	w	w	≥ 8	w	3
C_3	6	w	-	w	-	w	-
transferrin	2	w	w	w	w	w	-
IgA	3	1	1	0.5	2	3	0.5
IgM	≥ 5	≥ 5	≥ 5	≥ 5	2	≥ 5	2

≥ : *larger or the same as the highest amount of antibody, where the line dives around the antibody-well.*
w : *"waves" on the line.*
\- : *no reaction (undisturbed, straight line precipitate).*

**The reactivity of each lectin is expressed as the amount in μl (x 0.1) of serum protein antibody giving the same area of line-dives.*

purified con A-binding fraction of orosomucoid exhibited a line-dive. "Waves" on the lines may thus be due to non-precipitating lectin reactions, but theoretically, migration disturbances of non-lectin origin may also play a role.

Ad b) the sugar used must be soluble in the concentration necessary to prevent the lectin-glycoprotein reaction. When this requirement is fulfilled, this method is well suited for quantitative purposes as the amount of sugar necessary to inhibit the lectin reaction of each glycoprotein can be precisely calculated.

In the HALT-technique all purified lectins were positive with neuraminidase-treated erythrocytes. However, food items containing the respective lectins were not always positive. This was most probably due to coating competition from other proteins in the extracts. Moreover, the reactions by many food items were difficult to interpret due to sedimentation in the wells. Thirdly pseudohaemadsorption (false positive results) could not be immediately suspected or detected. In this volume, others have described the pseudohaemagglutination caused by polyphenols (12).

Screening for lectins in the diet has been performed by others by haemagglutination tests and by agglutination of streptococci (13, 14). The results from those and from the present study are in agreement for most items. We used a simple method of extraction with PBS or pressing in a juicer, whereby we may have failed to detect lectins needing more sophisticated methods for extraction.

In conclusion the Line-Dive immunoelectrophoresis is well suited for screening of food items for lectins. It is also very well suited for comparison of different lectin preparations. The drawback of using a serum protein/anti-serum protein system as reference, is the relative large number of false negative results. This is probably due to the carbohydrate structure of the serum proteins not being complex enough for a number of lectins. Similarly, others have found that relatively few serum proteins precipitate with lectins (15). It could be of interest to find an antigen/antibody system where the antigen caries more complex carbohydrates.

Acknowledgements. This study was supported by Eckbo's Dansk-Norske Legat and Kommunehospitalets Jubilæumsfond. We are grateful to Ms Jytte Lentz for typing the manuscript and Mr P. Henriksen for taking the photographs.

References

1. Begbi, R., and King, T.P. In: Lectins, Biology, Biochemistry, Clinical Biochemistry (Bøg-Hansen, T.C. and Breborowicz, J. eds.) vol. 4, pp. 15-27, Walter de Gruyter, Berlin-New York (1985).

2. Pusztai, A., King, T.P., and Clarke, E.M.W.: Toxicon 20, 195-197 (1982).

3. Sjölander, A., Magnusson, K-E, and Latkovic, S.: Int.Archs. Allergy appl. Immun. 75, 230-236 (1984).

4. Hara, T., Tsukamoto, I., and Miyoshi, M.: J.Nutr.Sci.Vitaminol. 29, 589-599 (1983).

5. Noah, N.D., Bender, A.E., Reaidi, G.B., and Gilbert, R.J.: Br.Med.J. 280, 236-237 (1980).

6. Köttgen, E., Kluge, F., Volk, B., and Gerok, W.: Klin. Wochenschr. 61, 111-112 (1983).

7. Krøll, J., and Andersen, M.M.: J.Immunol.Meth. 9, 141-146 (1975).

8. Kint, J.A.: Lectins, Biology, Biochemistry, Clinical Biochemistry (Bøg-Hansen, T.C., ed.) vol. 3, pp.415-424, Walter de Gruyter, Berlin-New York (1983).

9. Bøg-Hansen, T.C., and Grudzinskas, J.G.: Electrophoresis '83 (Hirai, H., ed.) Walter de Gruyter, Berlin-New York, pp. 595-602 (1984).

10. Pollack, L., and Atkinson, P.H.: J.Cell Biol. 97, 293-300 (1983).

11. Andersen, M.M., Hau, J., Bøg-Hansen, T.C.: Lectins, Biology, Biochemistry, Clinical Biochemistry (Bøg-Hansen, T.C., ed.) vol. 2, pp. 779-790, Walter de Gruyter, Berlin-New York (1982).

12. Blunden, G., Rogers, D.I., Loveless, R.W., and Patel, A.V.: These Proceedings, vol. 5.

13. Nachbar, M.S., and Oppenheim, J.D.: Am.J.Clin.Nutr. 33, 2338-2345 (1980).

14. Gibbons, R.J., and Dankers, I.: Appl. Environ. Microbiol. 41, 880-888 (1981).

15. Dörner, R. and Sachs, V.: These Proceedings, vol. 5.

THREE SPERM-AGGLUTINATING ISOAGGLUTININS FROM TUBERS OF TARO (*COLOCASIA ESCULENTA*)

PANUDDA PROMPLOOK and MONTRI CHULAVATNATOL

Department of Biochemistry, Faculty of Science, Mahidol University, Bangkok 10400, Thailand

Despite their great abundance, tropical plants have only been limitedly explored as sources for new lectins. For this reason, we have recently started a screening program to search for new lectins from Thai plants that can agglutinate sperm. Such lectins should be useful not only as tools for sperm surface study, but also as potential vaginal contraceptive agents. Among some 50 different plants tested, about 15 of them have been found to have sperm-agglutinating activity. Of these, a new sperm-agglutinating lectin from seeds of Jack fruit (*Artocarpus heterophyllus*) has been reported (1). Purification and characterization of another new sperm agglutinin from the tubers of a tropical plant, taro (*Colocasia esculenta*), will be presented here.

Materials and Methods

Purification of taro agglutinin: All steps were performed at 4°C. Tubers of taro were chopped with a knife and homogenized in 3 vol of 5 mM phosphate-buffered saline, pH 7.4 (PBS) in a Waring blender. The suspension was left stirring for 2 h. After removing the insoluble debris by centrifugation at 3,000 x g for 15 min, the extract was made 60% saturation by adding solid ammonium sulfate. The precipitate formed was collected by centrifugation at 27,000 x g for 15 min, resuspended in PBS and dialyzed exhaustively overnight against PBS. The fraction was then applied to a column (2 x 55 cm) of Sephadex G-100, pre-equilibrated and eluted with PBS. The protein elution profile was monitored at 280 nm. The peak containing the sperm-agglutinating activity was dialyzed overnight against 5 mM phosphate buffer, pH 7.4 and then applied to the first column (2.5 x 18 cm) of DEAE-cellulose, pre-equilibrated with 5 mM phosphate buffer, pH 7.4. The column was washed with the same buffer until the absorbance at 280 nm of the eluate was negligible. Then a gradient of 0 to 0.25 M NaCl in the buffer was applied. Two peaks containing the sperm-agglutinating activity were obtained. The pooled fraction from each peak was separately dialyzed overnight against 25 mM phosphate buffer, pH 6.5. Then it was applied to a second column (2.5 x 18 cm) of DEAE-cellulose, pre-equilibrated with 25 mM phosphate buffer, pH 6.5. The column was further washed with the buffer to remove unbound material containing the sperm-agglutinating activity. The bound fraction without activity was washed off the column by 1M NaCl in the buffer. From the first DEAE-cellulose column, the first active peak gave only one unbound but active peak (T_1) in the second DEAE-cellulose column chromatography. On the other hand, the second active peak from the first DEAE-cellulose column gave two unbound but active peaks (T_2 and T_3) in the second DEAE-cellulose column chromatography. Each of these active fractions (T_1, T_2 and T_3) was separately dialyzed overnight against 25 mM imidazole-HCl buffer, pH 7.4. Each dialyzed fraction was applied to a chromatofocusing column (1.2 x 12 cm) of Polybuffer exchanger (PBE 94) (Pharmacia), pre-equilibrated with 25 mM imidazole-HCl buffer, pH 7.4. The chromatofocusing column was developed using Polybuffer 74 according to the manufacturer's procedure. A single sharp peak with sperm-agglutinating activity was obtained from each fraction, T_1, T_2 or T_3. After the elution was completed, 1M NaCl was applied to remove tightly bound material. The absorbance of the eluate was monitored at 280 nm while its pH was measured using a pH electrode. To precipitate the protein from the active fraction in order to remove the Polybuffer, the fraction was dialyzed overnight against a saturated solution of ammonium sulfate. The precipitate was collected by centrifugation, redissolved in PBS and dialyzed. The purified agglutinin was stored at $-20^{\circ}C$.

Sperm preparations: Adult albino rats weighing 250 to 300 g were used. Epididymal sperm were collected in PBS by puncturing the caput or the caudal regions of the tubule with a hypodermic needle as described previously (2). Liquified human semen samples with no bacteria were pooled and overlayed on a solution of 10% Ficoll in PBS in a centrifuge tube. Sperm pellet was obtained by centrifuging the tube at 500 x g for 30 min at 4° C. After removing the top layer, the washed sperm pellet were resuspended in PBS. Sperm count was determined using a hemocytometer.

Sperm-agglutination test: Sperm agglutination test was performed as described previously (1) with following modifications. Sperm suspension (50 µl) containing 8×10^6 rat epididymal sperm/ml or 4×10^7 washed human sperm/ml was used for each test.

SDS-polyacrylamide gel electrophoresis (SDS-PAGE): SDS-PAGE was carried out as described by Laemmli (3) and the proteins were stained by Coomassie Brilliant Blue R-250.

Amino acid analysis: Amino acid composition of each purified preparation was determined by using a LKB 4101 Analyzer according to the previous report (1).

Protein determinations: Protein content was assayed by Lowry's method (4), except for samples from chromatofocusing column which were determined by Coomassie Blue G-250 (5).

Results

Purification of taro isoagglutinins: Most of the sperm-agglutinating activity in the crude taro extract was precipitable at 60% saturation of ammonium sulfate. When the precipitated fraction was passed through a Sephadex G-100 column, the activity was associated with one major protein peak. When this active peak was further purified by the first DEAE-cellulose column at pH 7.4, the activity was resolved into two peaks, eluting at 0.075 M and 0.107 M NaCl respectively. When the first active peak was applied to a second DEAE-cellulose column at pH 6.5, a single active peak (T_1) was obtained. In contrast, the second active peak from the first DEAE-cellulose column was resolved into two active peaks (T_2 and T_3) by the second DEAE-cellulose column at pH 6.5. Using chromatofocusing, T_1 was further purified and its pI value was estimated to be 6.45. Similarly, T_2 was purified by the chromatofocusing column and its pI value was 6.16 while T_3 purified by the same column showed pI value of 5.85. During the purification, the active fractions were subjected to SDS-PAGE to monitor the improvement of purity (Fig. 1). The active peaks (T_1 , T_2 and T_3) from the final chromatofocusing columns were found by the SDS-PAGE to be homogenous, containing the same single protein band. Because their pI values were different, the T_1 , T_2 and T_3 preparations from the final column were called "purified taro isoagglutinins".

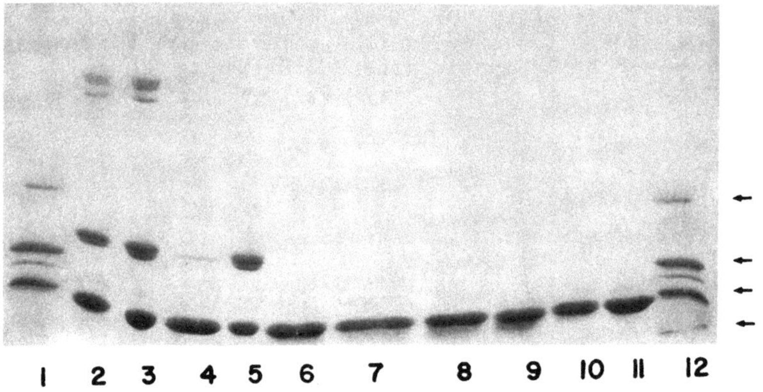

Fig. 1: Protein patterns on SDS-PAGE using a linear gradient of 7 to 15% acrylamide. Lanes 1 and 12, MW markers (arrows): con A (25,000), soybean trypsin inhibitor (20,500), cytochrome C (12,300) and aprotinin (6,700). Lane 2, ammonium sulfate fraction; lane 3, pooled active fraction from the Sephadex G-100 column; lanes 4 and 5, the first and second pooled active fractions from the first DEAE-cellulose column respectively; lanes 6,7 and 8, the pooled active fractions, T_1, T_2 and T_3, from the second DEAE-cellulose column respectively; lanes 9, 10 and 11, pooled active fractions of purified taro isoagglutinins, T_1, T_2 and T_3, from chromatofocusing respectively.

Table 1 Amino acid compositions of purified taro isoagglutinins

Amino acid	T_1 (mol/100 mol recovered)	T_2	T_3	Amino acid	T_1 (mol/100 mol recovered)	T_2	T_3
Asx	19.7	19.3	17.1	Ile	4.0	3.9	4.2
Thr	6.0	6.0	5.9	Leu	11.9	11.3	13.2
Ser	8.0	7.4	7.2	Phe	-	0.3	0.5
Glx	6.9	7.0	7.0	Lys	6.2	6.4	6.6
Pro	3.8	3.8	4.0	Arg	4.2	4.2	4.1
Gly	14.3	13.9	14.6	His	-	0.5	1.0
Ala	4.1	4.1	4.5	Cys	2.3	2.3	2.0
Val	7.6	7.7	7.7	Met	0.6	0.7	0.5

Properties of the purified taro isoagglutinins: Except for some minor differences, the amino acid compositions of the purified taro isoagglutinins (T_1, T_2 and T_3) were found to be very similar (Table 1). By gel filtration in a Sephadex G-100 column, the native MW values of T_1, T_2 and T_3 were found to be 29,000, 30,000 and 31,000 dalton respectively. By SDS-PAGE in a slab gel with a gradient of 12 to 15% acrylamide, the subunit MW of all purified taro isoagglutinins was determined to be 8,000 dalton. Thus, a tetrameric structure was proposed for each taro isoagglutinin.

The purified taro isoagglutinins were found to similarly agglutinate rat caput epididymal sperm and washed human sperm (Fig. 2) but not rat cauda epididymal sperm, unwashed human sperm or human red blood cells, with or without pre-treating the red blood cells with papain, trypsin or neuraminidase. The optimal pH for the activity of each isoagglutinin was 7.0. When each isoagglutinin was pre-heated at different temperatures for 20 min and cooled in ice, half of its activity was lost at 60° to $70^{\circ}C$. However, each purified isoagglutinin was insensitive to treatment with trypsin, papain or protease V (1 mg/ml) at $37^{\circ}C$ for 1 h. The presence of $CaCl_2$, $MgCl_2$, $MnCl_2$ or EGTA (5 mM) did not affect the agglutination activity. By light microscopic inspection, the agglutination induced by each purified isoagglutinin involved sperm midpieces (bodies) and tails but not sperm heads. Among 37 simple sugars (upto 0.2 M) and 4 glycoproteins (fetuin, orosomucoid, thyroglobulin and mucin; 2 mg/ml), none was found to be inhibitory to the agglutination activity.

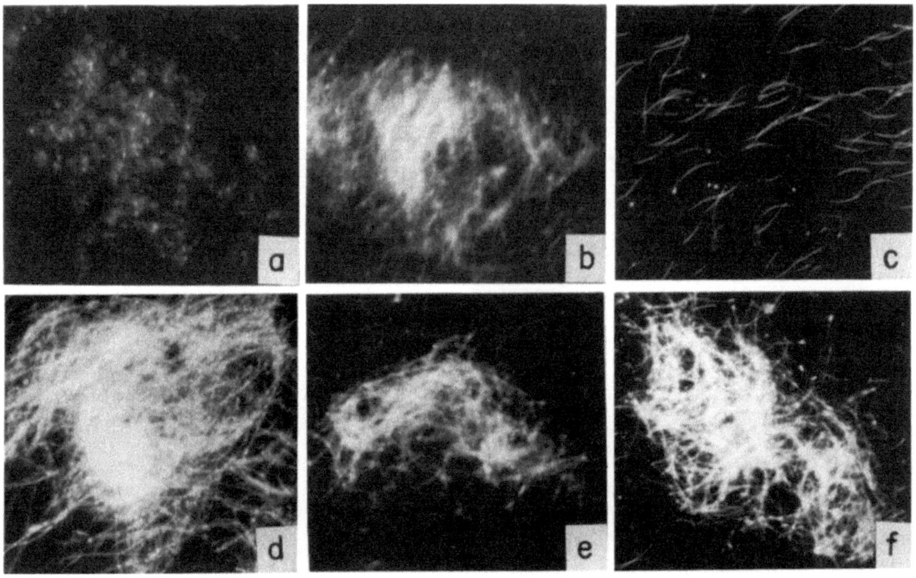

Fig. 2: Effects of ammonium sulfate fraction of taro on (a) human sperm, (b) rat caput and (c) rat cauda epididymal sperm. Effects of purified taro isoagglutinins, T_1(d), T_2(e) and T_3 (f), on rat caput epididymal sperm.

Discussion and Conclusion

To our knowledge, the three taro isoagglutinins with sperm-agglutinating activity have not been reported before. Since they do not agglutinate red blood cells, they would not have been discovered by the conventional hemagglutination test. This study has demonstrated that sperm agglutination may be another useful test to screen for new lectins or agglutinins. From the described properties, the three purified taro isoagglutinins are likely to be identical except for their charge differences as revealed by chromatofocusing. It remains to be shown if they contain different charged carbohydrates. As the taro isoagglutinins agglutinate the immature (caput epididymal) sperm but not the mature (cauda epididymal) sperm, they should be useful for studying sperm surface maturation. Other lectins useful in this regard include conA, WGA, RCA and UEA (6-8). Because the taro isoagglutinins do not agglutinate unwashed human sperm, they may be ineffective as a vaginal contraceptive agent. Since the sperm-agglutination inhibition test has so far failed to define the sugar specificity of the taro isoagglutinins, more work is needed to analyze their binding site(s) using the immature sperm. However, the data presented here support our contention that the three taro isoagglutinins are new and interesting sperm-agglutinins.

Acknowledgements

This work was supported by IDRC (3-P-82-0008) and PARFR-354. We thank Miss Panpaka Chalermisrachai for amino acid analysis and Mrs. Thitika Vajrodaya for secretarial assistance.

References

1 NAMJUNTRA, P., MUANWONGYATHI, P. and CHULAVATNATOL, M.: Biochem. Biophys. Res. Commun.: 128, 833-839 (1985)
2 CHULAVATNATOL, M., HASIBAUN, I., YINDEPIT, S. and EKSITTIKUL, T.: J. Reprod. Fert. 50, 137-139 (1977)
3 LAEMMLI, U.K.: Nature 227, 680-685 (1970)
4 LOWRY, O.H., ROSEBROUGH, N.J., FARR, A.L. and RANDALL, R.J.: J. Biol. Chem.: 193, 265-275 (1951)
5 BRADFORD, M.M.: Anal. Biochem.: 72, 248-254 (1976)
6 NICOLSON, G.L., USUI, N., YANAGIMACHI, R., YANAGIMACHI, H. and SMITH, J.R.: J. Cell Biol.: 74, 950-962 (1977)
7 OLSON, G.E. and DANZO, B.J.: Biol. Reprod.: 24, 431-443 (1981)
8 HAMMERSTEDT, R.H., HAY, S.R. and AMANN, R.P.: Biol. Reprod.: 27, 745-754 (1982)

A MELIBIOSE/MANNOSE-SPECIFIC LECTIN FROM SNOWDROP (*GALANTHUS NIVALIS*) BULBS

Catharina E.P. De Meirsman, Makuta Nsimba-Lubaki, Willy J. Peumans

Laboratorium voor Plantenbiochemie, Katholieke Universiteit Leuven,
Kardinaal Mercierlaan 92, B-3030 Leuven (Heverlee), Belgium

At present, over 100 plant lectins have been isolated and characterized in some detail with respect to their biochemical properties, molecular structure and carbohydrate-specificity (1). With the exception of the Gramineae lectins such as those from cereals and rice (2), all plant lectins which hitherto have been characterized were isolated from seeds or vegetative tissues of plant species belonging to the classis Magnoliatae (dicotyledons). To find out whether among species of the classis Liliatae (monocotyledons) which represents the second classis of the Angiosperms also lectins occur and if so whether they are related to or comparable with those isolated from dicotyledonous species, we looked for the presence of hemagglutinins in a number of Liliaceae and Amaryllidaceae species. Here, we describe the isolation and partial characterization of a lectin from bulbs of the snowdrop (*Galanthus nivalis*).

Materials and Methods

Snowdrop (*Galanthus nivalis*) is a typical representative of the plant family Amaryllidaceae. It is a perennial plant with surviving bulbs. Bulbs were collected locally just at the beginning of the growing season (just after the winter). They were thoroughly washed with tap water, decoated and cut into small pieces which were used immediately or stored at -20°C.

Extraction and purification of the lectin: Chopped bulbs were homogenized with a mixer in 5 volumes of phosphate-buffered saline (PBS; 1.5 mM KH_2PO_4/ 10 mM Na_2HPO_4 (pH 7.4)/ 3 mM KCl/ 140 mM NaCl) containing 5 mM thiourea (to inhibit the polyphenoloxydase activity in the extract). The homogenate was squeezed through cheesecloth and the filtrate centrifuged (10 000*g*; 10 min). The resulting supernatant was decanted and frozen overnight at -80°C. After thawing, the material that coagulated during freezing was spun down by centrifugation (10 000*g*; 10 min) and the resulting supernatant taken off and filtered through filter paper (Whatman 3MM) (to remove any particulate material). The clear extract was slowly applied to a column (5 ml bed volume) of mannose immobilized on cross-linked agarose (Selectin 10, from Pierce Chemical Company) equilibrated with PBS. The column was then washed with buffer and 1 M NaCl until the absorbance of the effluent had dropped below 0.01. Finally, the bound lectin was desorbed with 15 mM unbuffered DAP (DAP; 20 mM 1,3-diaminopropane, pH 10) and dialysed against the appropriate buffer solutions for further analysis. Purified *Galanthus nivalis* agglutinin will further be referred to as GNA.

Ion-exchange chromatography and gel filtration: Affinity purified snowdrop lectin was analyzed by ion-exchange chromatography using a Pharmacia Fast Protein Liquid Chromatography system (Type GP 250) equiped with a Mono Q (HR 5/5) anion-exchange column. Lectin samples were loaded on the column equilibrated with DAP-buffer. After washing the column with 4 ml of DAP-buffer, lectins were eluted with a linear gradient (32 ml) of increasing NaCl concentration (0 - 0.3 M). The position of the lectin was followed by monitoring the A_{280} of the eluate.

Purified lectin was analyzed by gel filtration using the same Fast Protein Liquid Chromatography system equiped with a Superose 12 gel filtration column (Type HR 10/30), with PBS as the running buffer.

Agglutination assays: Agglutination assays were carried out in small tubes in a final volume of 0.1 ml containing 80 µl of a 1 % suspension of trypsin-treated rabbit red blood cells and 20 µl of crude extracts or lectin solutions (each serially diluted with two-fold increments). Agglutination was monitored visually after the tubes had been left for 1 h at room temperature.

Analytical method: Protein estimation was carried out according to Lowry *et al.* (3), using bovine serum albumin as a standard. Total neutral sugar was determined by the phenol/H_2SO_4 method (4) with

D-glucose as a standard. Lectin preparations were analyzed by sodium dodecyl sulfate-polyacrylamide gel electrophoresis (SDS-PAGE) using a discontinuous system (5) on 12.5 - 25 % (w/v) acrylamide gradient gels.

Amino acid analysis: Lectin samples were hydrolysed (in sealed glass tubes under an N_2 atmosphere) for 24, 48 and 72 h at 110°C in 6 M HCl containing 0.02 % 2-mercaptoethanol. After hydrolysis, HCl was removed by evaporation, and the residue was dissolved in sodium citrate buffer, pH 1.9. Amino acid analysis was performed with a Biotronik LC 2000 amino acid analyser. This apparatus allows the separation of all 20 amino acids on a single column of Durrum resin, by using an elution program consisting of four different buffers. Amino acids were directly quantified using an Infotronics integrator.

Results

Occurrence of a lectin in snowdrop bulbs: Crude extracts from snowdrop bulbs were found to exhibit relatively high agglutinating activity. Higher titers were obtained with trypsin-treated than with untreated rabbit erythrocytes. For instance, when extracts were prepared by homogenizing bulb tissue in 5 volumes of PBS, titers of 25 and 200 were scored with untreated and trypsin-treated red blood cells respectively. Similar results were obtained in agglutination assays with the purified lectin. The minimal concentration of GNA to agglutinate untreated and trypsin-treated rabbit erythrocytes was 16 and 2 µg/ml respectively. Even at concentrations above 2 mg/ml, human erythrocytes were not agglutinated by GNA.

In order to determine the carbohydrate-binding specificity of GNA, hapten inhibition assays with a series of simple sugars were performed (Table 1). Since it was shown that GNA bound to Sephadex G-100 during gel filtration, some polymers of glucose were also tested for their possible inhibitory effect. As can be seen in Table 1, glycogen, soluble amylose and dextran effectively inhibited the agglutination of rabbit erythrocytes. The sugar specificity of the lectin was exploited to develop a purification scheme based on affinity-chromatography on immobilized mannose. Using the isolation procedure described in the Materials and Methods pure lectin was obtained in a single step. The yield usually varied around 1 mg lectin per gram bulb tissue.

Table 1

Carbohydrate-binding specificity of snowdrop bulb extracts and purified GNA. Agglutination assays were carried out with trypsin-treated rabbit erythrocytes. N-acetylgalactosamine, galactose, glucose, N-acetylglucosamine, galactosamine, glucosamine, lactose, fucose, arabinose, ribose, fructose, sorbose, xylose, sucrose, maltose and sorbitol did not inhibit the agglutination.

carbohydrate	concentration required for 50% inhibition of agglutination with	
	crude extract (titer=100)	purified GNA (100 µg/ml)
melibiose	1.5 mM	3 mM
mannose	6 mM	9 mM
trehalose	12.5 mM	25 mM
glycogen	1 mg/ml	1 mg/ml
amylose	1 mg/ml	1 mg/ml
dextran (MW 40 000)	0.5 mg/ml	0.5 mg/ml

Molecular structure of the Galanthus nivalis agglutinin (GNA): Affinity purified GNA was analyzed by SDS-polyacrylamide gel electrophoresis and gel filtration both in the presence and absence of 2% (v/v) 2-mercaptoethanol. As shown in Fig. 1, in both conditions, the protein migrates as a single polypeptide band of MW 10 000 dalton. The MW of the native GNA was estimated by gel filtration. As shown in Fig. 2, this method indicated a MW around 40 000 dalton. It appears, therefore, that GNA is a tetrameric protein build up of four identical subunits which are not held together by disulfide bounds. The amino acid composition of GNA is given in Table 2. GNA is particulary rich in aspartic acid (17.2 mole %)and glycine(13.9 mole %); no methionine, proline and phenylalanine could be detected. Using the phenol/H_2SO_4 method, no covalently bound sugar could be detected in purified GNA, indicating that the lectin is not glycosylated.

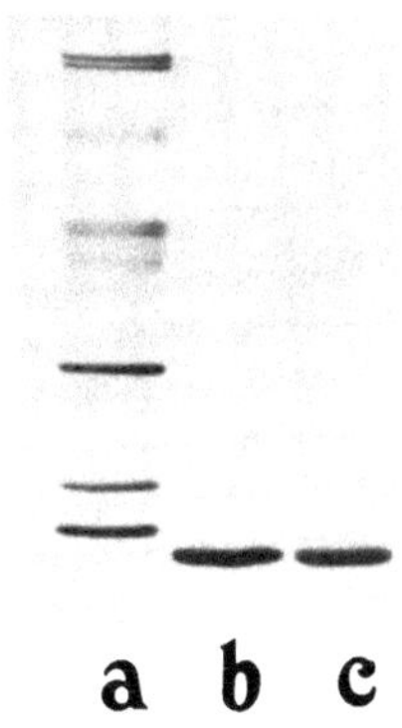

Fig. 1. *Sodium dodecyl sulfate-polyacrylamide gel electrophoresis of purified GNA in the presence (b) and absence (c) of 2% 2-mercaptoethanol. Reference proteins (a) were lysozyme (MW 14 300), soya-bean trypsin inhibitor (MW 21 000), carbonic anhydrase (MW 30 000), ovalbumin (MW 45 000), bovine serum albumin (MW 67 000) and phosphorylase b (MW 94 000). Gels were stained with Coomassie Brilliant Blue.*

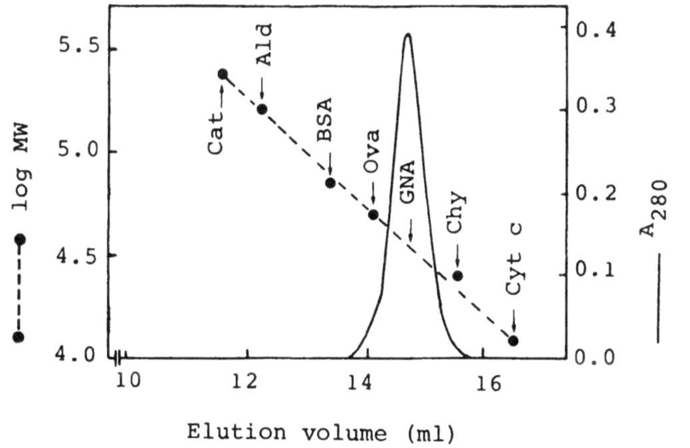

Fig. 2. *Gel filtration of GNA on Superose 12 (Pharmacia type HR 10/30). Conditions: flow rate: 2 ml/h; running buffer: PBS. The position of the MW reference proteins (cytochrome c (Cyt c; MW 12 500); chymotrypsin (Chy; MW 25 000); ovalbumin (Ova; MW 45 000); bovine serum albumin (BSA; MW 67 000); aldolase (Ald; MW 160 000) and catalase (Cat; MW 240 000)) was determined in a separate run.*

Table 2

Amino acid composition of GNA

Amino acid	mole %	Amino acid	mole %
Asp	17.2	Met	0
Thr	6.6	Ile	6.2
Ser	8.6	Leu	9.2
Glu	6.0	Tyr	8.1
Pro	0	Phe	0
Gly	13.9	His	3.8
Ala	5.8	Lys	2.8
Cys	Not determined	Arg	7.6
Val	4.2	Trp	Not determined

To find out whether GNA exists as a single molecular form or as a mixture of isolectins, the purified lectin was analyzed by high resolution ion-exchange chromatography. As shown in Fig. 3, the affinity purified lectin could be separated in at least 6 peaks indicating that several isolectins were present. To prove that each peak contained GNA, the individual isolectins were isolated and further analyzed by SDS-PAGE and agglutination assays. They all appeared to contain the same polypeptide of MW 10 000 dalton and in addition agglutinated rabbit red blood cells in a melibiose/mannose inhibitable manner (results not shown).

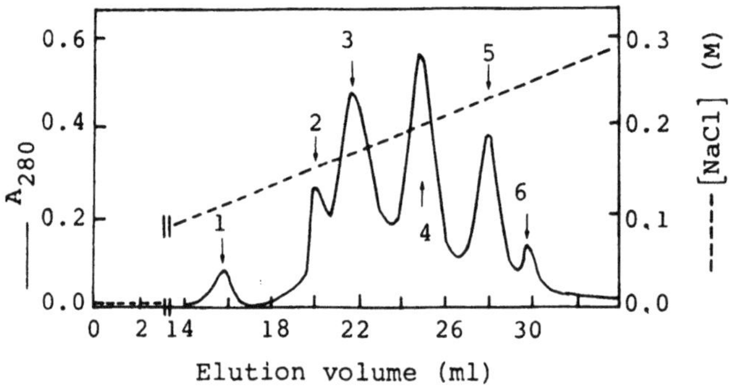

Fig. 3. *Ion-exchange chromatography of GNA by FPLC on a Pharmacia Mono Q (type HR 5/5) column. Individual isolectin peaks are numbered from 1 to 6.*

Discussion

Snowdrop bulbs contain reasonable amounts of a melibiose/mannose-specific lectin which can readily be isolated by affinity-chromatography on immobilized mannose. Purified GNA has been characterized and appears to be a protein composed of 4 identical subunits with a molecular weight of 10 000 dalton. By comparing GNA to previously described phytohemagglutinins, it can be concluded that on the basis of its molecular structure and sugar-specificity, the snowdrop lectin has to be considered as a new type of plant lectin. Further investigations are underway to find out whether GNA is a unique lectin or just represents an example of a more widespread class of lectins occurring in other Amaryllidaceae species too or even in members of other monocotyledonous plant families.

References

1. Goldstein, I.J., Hayes, C.E.: Adv. Carbohydr. Chem. Biochem. *35*, 127-340 (1978).
2. Stinissen, H.M., Peumans, W.J.: Biochem. Physiol. Pflanzen *180*, 85-106 (1985).
3. Lowry, O.H., Rosebrough, N.J., Farr, A.L., Randall, R.J.: J. Biol. Chem. *193*, 265-275 (1951).
4. Dubois, M., Gilles, K.A., Hamilton, J.K., Rebers, P.A., Smith, F.: Anal. Chem. *28*, 350-356 (1956).
5. Laemmli, U.K.: Nature (London) *227*, 680-685 (1970).

PURIFICATION AND CHARACTERIZATION OF AN α-D-GALACTOSYL-BINDING LECTIN FROM THE SEED OF JACKFRUIT, ANTOCARPUS INTEGRIFOLIA

Hafiz Ahmed and Bishnu Pada Chatterjee

Department of Marcromolecules, Indian Association for the Cultivation of Science, Jadavpur, Calcutta - 700 032, India.

Lectins are a group of proteins found in a wide variety of plants, animals and microorganisms that interact with glycoconjugates and polysaccharides by binding to specific carbohydrate residues. Some of these lectins have been purified and characterized in detail (1, 2). Among the lectins described in the literature showing α-D-galactosyl specificity are those from *Bandeiraea simplicifolia* and *Maclura pomifera*. Both the lectins have been purified by affinity chromatography employing α-D-galactosyl derivative as ligand to the matrix (3, 4). The haemagglutinin present in the seeds of Antocarpus integrifolia (jackfruit) has first been reported by us (5) and its specificity for the α-D-galactose sugar has been demonstrated by preciptin reactions with various galactomannans.

With a view to isolate biopolymers containing α-D-galactopyranosyl end groups using a jackfruit lectin immobilized affinity system, we have isolated this lectin with a melibiose-agarose column. In this communication, we describe the purification of the jackfruit lectin and its characterization with respect to some serological, chemical and physico-chemical properties.

Lectins, Vol. V

Materials and Methods

Jackfruit Antocarpus integrifolia family Moraceae grows as trees in the tropics. Seeds were obtained from inside the juicy and fleshly capsule of the fruit. After decapsulation the seeds were washed with 0.85% saline, dried under the sun and husked mannually.

Sugars: D-galactose was purchased from B.D.H., England. 2-acetamido-2-deoxy-D-galactose, O-nitrophenyl-α-D-galactopyranoside, p-nitrophenyl-α-D-galactopyranoside, Methyl-β-D-galactopyranoside were the product of Sigma. Methyl-α-D-galactopyranoside was obtained from Calbiochem.

Chemicals: Melibiose-agarose was purchased from Sigma, U.S.A. Sephadex G-100 was purchased from Pharmacia, Sweden. Acrylamide was a product of Sigma. N-N'-methylene bisacrylamide and ammonium persulfate were obtained from Serva, F.R.G. TEMED was purchased from B.D.H., England. All other chemicals used were of analytical grade from Merck and B.D.H.

Enzymes: Neuraminidase was obtained from Behringwerke, Marburg, F.R.G. Pronase was a product of Serva, Heidelberg, F.R.G.

Purification of A. integrifolia lectin: Dried and husked A. integrifolia seeds were soaked in 0.85% NaCl overnight and homogenised in a blender. The slurry was centrifuged in a Sorvall RC-5B refrigerated centrifuge at 15000 r.p.m. for 40 minutes. To the clear extract solid $(NH_4)_2SO_4$ was added to give 40%, 70% and 100% saturation. The precipitate obtained in each fraction after centrifugation was dissolved in a minimum quantity of water and dialyzed against water until freed from NH_4^+, centrifuged and lyophilized. The most active haemagglutinin fraction (40-70%) was applied to the melibiose-agarose column (1x3 cm) previously equilibrated with PBS (0.01 M phosphate containing 0.15 M NaCl, pH 7). The column was washed with the same buffer. The bound lectin was desorbed by eluting with the same buffer containing 0.2 M D-galactose. The fractions showing haemagglutinating activity were pooled, concentrated by membrane ultrafiltration (PM-10, Amicon) dialyzed against PBS till free of galactose, and stored at -20° until used.

Assay procedures: Haemagglutination assays were performed in Takatsy microtiter plate in PBS with two-fold serial dilution of the lectin as previously described (5). Titer was expressed as the reciprocal of the highest dilution showing positive agglutination; specific activity was expressed as titre per mg of protein per ml. Haemagglutination-inhibition tests were carried out as follows: to two-fold serial dilutions of test sugar solution (25 µl) was added an equal volume of two haemagglutinating doses of lectin. After incubation for 1h at room temperature 25 µl of 2% suspension of erythrocytes was added and the mixture was allowed to stand for 1h at room temperature. The degree of haemagglutination was examined and the maximum dilution of sugar solution showing haemagglutination-inhibition was recorded. The protein content of various fractions was measured by the method of Lowry et al (7) using bovine serum albumin as standard.

Disc electrophoresis: Disc electrophoresis in 7.5% polyacrylamide gel was performed according to Reisfeld et al (8) at pH 4.3 in β-alanineacetic acid buffer. Staining was performed with amido black in 7% acetic acid, and destained in an electric field in 7% acetic acid.

Analytical ultracentrifugation: Sedimentation velocity experiments were conducted according to Schahman (9) with a Beckman Spinco Model E analytical ultracentrifuge equiped with schlieren optics and using a A N-D-rotor with standard aluminium cell. Photographs were taken before and after the rotor attained a running speed of 56,000 at $4^{o}C$.

Immunological methods: Antiserum to a crude extract of A. integrifolia was produced in rabbits. Ouchternony double diffusion was performed in 1% agarose (Serva) in petri dishes as described previously (5).

Molecular weight estimation: Molecular weights were estimated by SDS-polyacrylamide gel electrophoresis and gel filtration. SDS-polyacrylamide gel electrophoresis was carried out in 10% polyacrylamide gels according to Laemmli (10). Dissociation and reduction of protein were performed by heating at $100^{o}C$ for 5 min in 0.1% SDS with or without 0.1% 2-mercaptoethanol. Protein markers employed were bovine serum albumin (M_r = 68,000), ovalbumin (M_r = 43,000), chymotripsinogen A (M_r = 25,000), myoglobin (M_r = 17,000) and cytochrome C (M_r = 12,400). The molecular weight of the lectin was estimated by comparing its mobility with those of markers of known molecular weight. Staining for proteins was performed with 0.05% Coomassie brilliant blue and destained in 7.5% acetic acid containing 5% methanol. Gel filtration on a Sephadex G-100 column (90 x 1.6 cm) was performed as described by Andrews (11). The molecular weight of the lectin was determined by comparison of its elution volume with those of molecular weight markers.

Carbohydrate analysis: The total neutral carbohydrate content of the lectin after Sephadex G-100 gel chromatography was determined by the phenol-H_2SO_4 method of Dubois et al (12). For identification and estimation of individual neutral sugars, the lectin sample (1.07 mg) was hydrolyzed with 0.1M HCl for 48h at $100^{o}C$ with inositol as standard. The hydrolyzate was neutralized with Ag_2CO_3 and the neutral sugars were analyzed by gas liquid chromatography (Hewlett Packard) as their alditol acetates using a 3% ECNSS-M column as previously described (13). Hexosamines were separated and quantitated during amino acid analysis.

Amino acid analysis: The purified lectin (1.24 mg) was hydrolyzed with 6N HCl in a sealed tube for 24h at 110°. The amino acid content of the hydrolyzate was determined in amino acid analyzer (Beckman) according to the method of Spackman et al. (14). Tryptophan was determined on unhydrolyzed protein samples by the spectrophotometric method of Goodwin and Morton (15).

Results

Purification of A. integrifolia lectin: Fractionation of the crude saline extract of A. integrifolia (jackfruit) seeds with solid ammonium sulfate gave three fractions of which the 40-70% fraction was found to be most active (Table 1). Further purification to homogeneity was achieved by a combination of affinity chromatography on a melbiose agarose column and gelfiltration on Sephadex G-100 (Table 1 and Fig. 1).

The homogeneity of jackfruit lectin was tested by analytical ultracentrifugation, immunodiffusion and polyacrylamide disc gel electrophoresis. Polyacrylamide disc gel electrophoresis at pH 4.3 produced a sharp single band (Fig. 1) while at pH 8.9 the lectin separated into a large diffuse band. The lectin sedimented as a symmetical peak at pH 7.0, and in Ouchterlony double diffusion, the JFL produced a sharp preciption band against JFL antiserum. A single major peak was obtained upon chromatography on a column of Sephadex G- 100.

Table 1

Purification of Antocarpus integrifolia lectin.

Fraction	Protein mg/ml	Titer[1]	*specific activity	Purification fold
Crude extract	3.1	128	41	1
Partially purified lectin				
0-40% $(NH_4)_2SO_4$ fraction	2.1	256	122	3
40-70% $(NH_4)_2SO_4$ fraction	1.65	1024	621	15
70-100% $(NH_4)_2SO_4$ fraction	1.50	256	171	4
Galactose eluate from affinity column	0.86	4096	4763	116
Sephadex G-100 eluate	0.80	16384	20480	500

**sp. Activity = titer per mg protein per ml.*

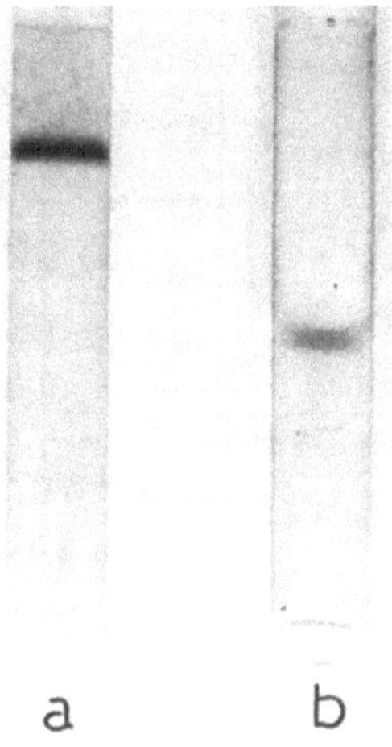

Fig. 1. (a) Disc gel electrophoresis of purified lectin at pH 4.3 in 7.5% polyacrylamide gel.
(b) SDS-gel electrophoresis in 10% polyacrylamide gel in presence of 0.1% SDS.

Haemagglutination assays: Jackfruit lectin agglutinated human O, A, B, and AB erythrocytes but its degree of agglutination with blood group B erythrocytes appears to be higher compared to other blood groups. The minimum amount of purified JFL required to agglutinate B erythrocytes was 48 ng/ml.

Molecular weight: The molecular weight of native jackfruit lectin was estimated to be 42,000 dalton (Fig. 2) by gel filtration on Sephadex G-100. However, SDS-polyacrylamide gel electrophoresis with or without 2-mercaptoethanol showed a single band corresponding to a molecular weight of 11,400 dalton (Figs. 1 & 3). When treated with 6 M guanidine, 1 M propionic acid (pH 4.6) and subjected to gel filtration on Sephadex G-100 the lectin gave a single protein peak with a molecular weight about 11,500 dalton (Fig. 2). Thus JFL is composed of four apparently identical subunits joined noncovalently in the same molecule.

Inhibition assay: The results of haemagglutination inhibition of the jackfruid lectin with simple haptenic sugars are shown in Table 2. Methyl-α-D-galactopyranoside was found to be most potent inhibitor. Neither D-glucose, D-mannose, D-fucose, D-arabinose, L-arabinose, methyl- β -D-galactopyranoside, lactose, nor galactosamine had any inhibitor effect even at a concentration of 0.4 M.

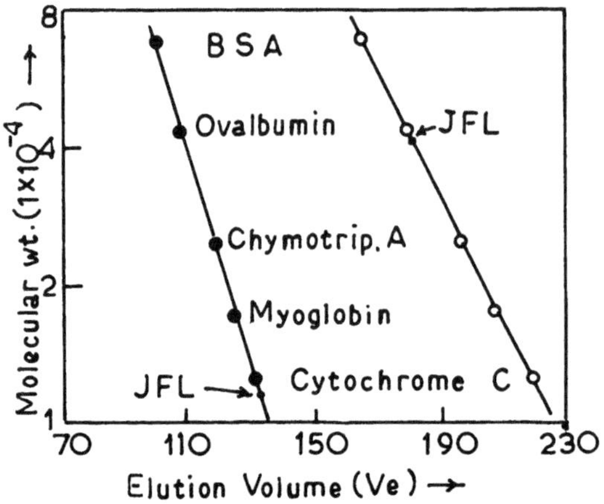

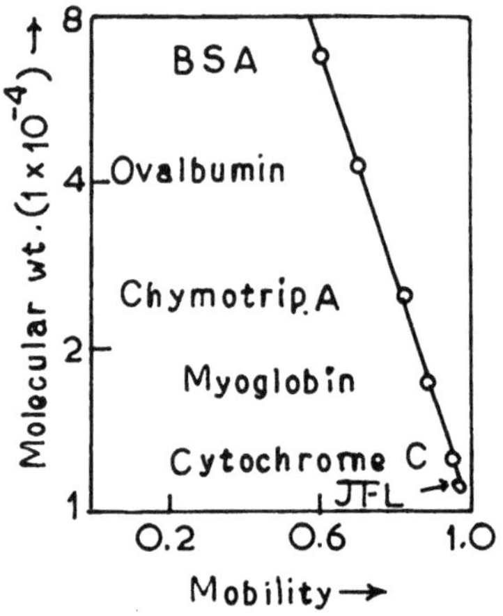

Figure. 2. Determination of the molecular weight of jackfruit lectin (JFL) by gel filtration. A column (1.6 x 90 cm) of Sephadex G-100 was used with PBS as eluent at 4°C (o-o), or with 6M Guanidine-1M propionic acid at 23°C (●-●). The protein solution (1 ml 10 mg/ml in eluting buffer) was applied to the column and 4 ml fractions were collected at a rate of 8 ml/h. Elution volumes were determined from the position of the maxima of the elution profiles.

Figure 3. Determination of the subunit molecular weight of JFL by polyacrylamide gel electrophoresis in presence of 0.1% SDS. Standard samples (50 ug each) were applied together with JFL and after staining with Coomassie brilliant blue R-250; the mobility was measured from the top of the gel to the leading edge of the protein band.

Carbohydrate and amino acid analyses: Table 3 summerizes the results of carbohydrate and amino acid analyses of jackfruit lectin. The carbohydrate analysis of the lectin revealed the presence of xylose, galactose, mannose, glucosamine and a trace amount of glucose and mannosamine. The lectin contains 6.5% neutral carbohydrate as determined by the phenol H_2SO_4 reaction and is in agreement with the amount of neutral sugar (6%) and aminosugar (2%) determined by gas liquid chromatography. The amino acid analysis is high in acidic amino acids and relatively low in basic amino acids with very small amounts of cysteic acid and absence of methionine. Remarkable is also the absence of proline in the hydrozylate. A high amount of ammonia indicates that most of the aspartic and glutamic acid residues are present in the amide form.

Discussion:

In the last decade lectins have gained tremendous importance as analytical tools for the detection and characterization of complex carbohydrates both in solution (16) and on cell surfaces (17). A few lectins have been chemically coupled to insoluble matrices and employed for the isolation of polysaccharides (18) and glycoproteins (19). Therefore, we have developed a practical method of purifying lectin from jackfruit seed by affinity chromatography on a melibiose-agarose column and we have studied its chemical and physiochemical properties. The isolated jackfruit lectin, JFL, is similar in its physical and chemical properties to many other purified lectins. The molecular weight of the lectin (42,000) and its subunits resemble those of the Maclura pomifera lectin (4). It is composed of four apparently identical polypeptide chains joined noncovalently in the molecule and thus it resembles B. simplicifolia (3) and peanut lectins (1). Like B. simplicifolia and M. pomifera lectins, JFL is specific for α-D-galactose. JFL precipitated polysaccharides (6) and blood group substances (5) thus showing a close similarity with the B. simplicifolia lectin. Its amino acid composition corresponds to that of numerous other lectins (the abundance of acidic and hydroxylic amino acids as well as the paucity of sulfur containing amino acids). It is a glycoprotein but unlike the B. simplicifolia lectin it does neither require any metal ion for saccharide binding nor has any blood group specificity. On prolonged storage at 4°C in the lyophilized form, the lectin tends to form an aggregate of82,000 dalton. Such a phenomenon was also observed for the Maclura pomifera lectin (4). The lectin activity was completely abolished at 75°.

During our study we have come across a report by Suresh Kumar et al. (20) who have isolated JFL by affinity chromatography on immobilized guar galactomannan. Their preparation showed the same molecular weight and polypeptide chains but contained only 3% carbohydrate, somewhat lower than our finding.

Acknowledgement: Ms Caroline Nielsen is cordially thanked for typing the manuscript.

Table 2

Hemagglutination inhibition of Antocarpus intergrifolia lectin by carbohydrates.

Carbohydrates	Mm
D-Galactose	300
Methyl- α-D-galactopyranoside	3.1
O-nitrophenyl- α-D-galactopyranoside	6.25
p-nitrophenyl- α-D-galactopyranoside	6.25
N-acetyl-D-galactosamine	150
Melibiose	12.50

Minimum concentration required for complete inhibition of two haemagglutinating doses of lectin

Table 3

Chemical composition of purified Antocarpus integrifolia lectin.

Amino acid	g/100g	Amino acid	g/100g	Carbohydrate	g/100g
Asp	8.48	Ile	4.96	Xyl	4.3
Thr	5.76	Leu	5.18	Man	0.8
Ser	5.72	Tyr	7.63	Gal	1.0
Glu	6.48	Phe	7.65	Glc	0.2
Pro	-	His	0.79	$GlcNH_2$	1.5
Gly	6.16	Lys	5.74	$ManNH_2$	0.5
Ala	1.52	NH_3	6.50		
Cys	0.11	Arg	1.61		
Val	6.40	Try	5.75		
Met	-				
Sum			86.44		8.3

References

1. Goldstein, I.J., Hayes, C.E. (1978), Adv. Carbohydr. Chem. Biochem., 35, 127-340.

2. Lis, H., Sharon, N. (1979). in: The Antigens. Vol. 4, p. 429-529, Sela, M. (ed.) Academic Press, New York

3. Hayes, C.E., Goldstein, I.L. (1974). J. Biol. Chem., 249, 1904-1914.

4. Bausch, J.N., Poretz, R.D. (1977), Biochemistry 16, 5790-5794.

5. Chatterjee, B.P., Vaith, P., Chatterjee, S., Karduck, D., Uhlenbruck, G. (1979) Int. J. Biochem., 10, 321-327.

6. Chatterjee, B.P., Sarkar, N., Rao, A.S. (1982), Carbohydr. Res., 104, 348-353.

7. Lowry, O.H., Rosebrough, N.J., Farr, A.L., Randall, R.J. (1951) J. Biol. Chem., 193, 265-275.

8. Reisfield, R.A., Lewis, U.J., Williams, D.E. (1962), Nature 195, 281-283.

9. Schachman, H.K. (1957)m, Methods Enzymol. 4, 32-103.

10. Laemmli, U.K. (1970), Nature 227, 680-685.

11. Andrews, P. (1965), Biochem, J. 96, 595-606.

12. Dubois, M., Gilles, K.A., Hamilton, J.K., Rebers, P.A., Smith, F. (1956) Anal. Chem., 28, 350-356.

13. Sarkar, N., Chatterjee, B.P. (1983), Carbohydr. Res., 112, 113-121.

14. Spackman, D.H., Stein, W.H., Moore, S., (1958), Anal Chem. 30, 1190-1206.

15. Goodwin, T.W., Morton, R.A. (1946), Biochem. J., 40, 628-632.

16. Goldstein, I.J. (1972), Methods Carbohydr. Chem., 6, 106-119.

17. Nicolson, G.L., (1974), Int. Rec. Cytol., 39, 89-120.

18. Ross, T.T., Hayes, C.E., Goldstein, I.J. (1976), Carbohydr. Res., 47, 91-97.

19. Asperg, K., Porath, J. (1970), Acta Chem. Scand., 24, 1839-1841.

20. Suresh Kumar, G., Appukuttan, P.S., Basu, D.K. (1982), J. Biosci. 4, 257-261.

PURIFICATION OF *ULEX EUROPEUS* HEMAGGLUTININ I BY AFFINITY CHROMATOGRAPHY.

Klaus G. Müller, C. Schäfer Nielsen and Allan Lihme
The Protein Laboratory, 34, Sigurdsgade, DK-2200 Copenhagen N, Denmark

Crude extracts of Ulex europeus seeds contain two lectins, designated hemagglutinin I and II. Both agglutinate human type O erythrocytes. The agglutination activity of purified hemagglutinin I is inhibited by L-fucose, while hemagglutinin II is inhibited by N-acetylchitobiose (1).

Previously, hemagglutinin I has been purified using affinity chromatography but only with relativly small yields, i.e. 2-10 mg/100 gram of seeds (1, 3-5). In an effort to improve the procedure for isolation of hemagglutinin I, we have developed a rapid method for the isolation of large quantities of pure hemagglutinin I. The average yield of hemagglutinin I by this method is 26 mg/100 gram of seeds.

Lectins, Vol. V

Materials and Methods

Unless otherwise stated all procedures were carried out at 4°C.

Extraction: Seeds from Ulex europeus were obtained from J.E. Ohlsens Enke, Copenhagen. The seeds were ground in a blender (Braun Multiquick) and extracted for 24 hours with continous stirring in a buffer solution of PBS containing 15 mM NaN_3 at a ratio of 100 grams of seeds/1 liter PBS. The suspension was centrifuged at 4000 r.p.m. for 15 minutes and then filtered twice on a Buchner funnel using two types of filter aid (Celite 545, Standard Supercel, Johns-Manville) on water suction.

Preparation of agarose 4B-fucose: Twenty grams of divinyl sulphone activated agarose 4B (Mini-Leak, Kem-En-Tec, Hellerup, Denmark was suspended in 20 ml of 0.5 M sodium carbonate solution and 2 grams of L-fucose was added. The mixture was adjusted to pH 11,0 with NaOH and incubated with gentle stirring at room temperature for 24 hours. The gel was then washed with distilled water and 4°C.

Affinity chromatography: Columns containing 10 grams (wet weight) of affinity gel equilibrated with PBS were used for affinity chromatography with flow rates of 0,85 ml. min^{-1}. cm^{-2} filtered extract the columns were washed with at least 50 bed volumes of PBS until the optical density at 280 nm of the effluent remained constant (less than 0.05). Hemagglutinin I was not detected in the effluent. The columns were finally eluted with 1% W/v L-fucose in PBS.

Immuno diffusion assays: These were carried out in 1% agarose gel by the Mancini technique using two fold serial dilutions of commercially obtained lectin (IBF, France) as standard. The wells contained 10 microliters of the liquid to be examined. The content of anti-UEA (Dakopatts, Danmark) in the gel was 0,2 μl/cm^2. The plates were incubated for 48 hours at room temperature in a moist chamber before measurement.

Electrophoresis: nalytical SDS-electrophoresis was performed according to Laemmli (6) in a 15% T, 5,7% C gel, and the gels were stained with Coomassie Brilliant Blue R.

Results

The differed extract from 100 grams of the seeds was found to contain an average of 0.07 mg of hemagglutinin/ml extract. The bright yellow colour of the crude extract turns into a dark brownish after a few days of storage.

The yield of hemagglutinin obtained from 100 grams of seeds was 27 mg. This could be increased by 1/3 by one reextraction giving a total yield of 36 mg of pure lectin/100 grams of seeds.

The capacity of the gel, defined as the amount of hemagglutinin bound when no lectin was detected in the effluent was estimated to be 4-5 mg of lectin/ml of gel. With larger loads the gel allows increasing amounts of lectin to pass through. After loading the gel with 10 mg hemagglutinin I/ml of gel about 10% of the applied lectin was detected in the effluent liquid.

Upon SDS electrophoresis, two closely migrating protein bands were observed. The molecular weight of these bands were estimated to be approximately 32, 00 D (Fig. 1).

A single coincident preciptin was observed upon double diffusion a.m. Ouchterlony of the purified lectin and the crude extract versus commercially obtained antibody against hemagglutinin I.

Discussion

A varity of methods have been reported for the isolation of Ulex europeus hemagglutinin I (1-5).

Matsumoto and Osawa (2) reported the isolation of hemagglutinin I with a yield of 20 mg/100 grams of seeds by use of ion-exchange chromatography and gelfiltration techniques.

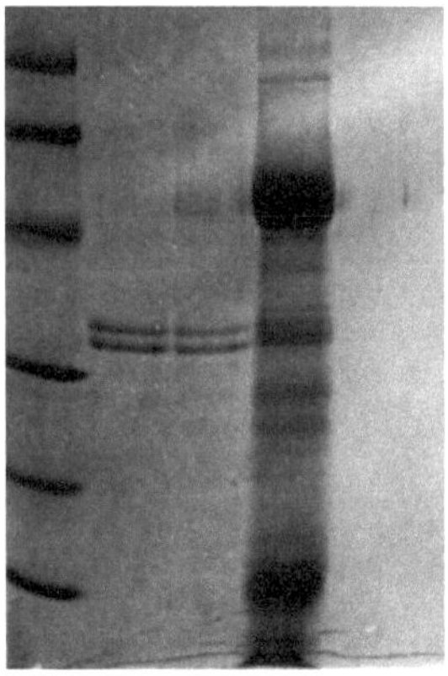

Fig. 1. SDS-PAGE. Lane A (from above): Phosphorylase, 92500 D; bovine serum albumin, 66200 D; ovalbumin, 45000 D; carbonic anhydrase, 31000 D; soybean trypsin inhibitor, 21500 D; lysozyme, 14400 D (Kem-En-Tec). Lane B: Eluate from affinity adsorbent. Lane C: Commercially obtained UEA I (IBF). Lane D: The crude extract

The use of affinity chromatography for the isolation of hemagglutinin I has been described previously but with smaller yields and lower capacity of the gel than reported here.

Horejsi, and Kocourek (3) using an O-α-fucosyl polyacrylamide adsorbent had a yield of 2.9 mg hemagglutinin I/100 grams of seeds. Frost et al. (1) reported a yield of 8.7-13.0 mg hemagglutinin I/100 grams of seeds. Their procedure requires the rather complex synthesis of 6-amino α-L-fucopyranoside and subsequent coupling to CNBr-activated Sepharose 4B.

The divinylsulphone-activated Sepharose with a capacity of 1.2 mg of hemagglutinin I/ml gel and a Yield of 4.5 mg/100 grams of seeds has been reported by Allen and Johnson (4). Smith et al. (5) using the same technique reported similar results.
The higher yield and capacity of the gel reported herein may partly be due to a more eddicient activation and coupling of L-fucose to the gel. Another crucial factor may be the steps before the affinity chromatography. Like Frost et al (1), who reported a relatively high yield, we have used a procedure with few steps compaired to the procedures reported by other authors, i.e. fractionation with ammonium sulfate, dialysis and lyophilization of the extract. Allen and Johnson mentioned that their yield of 4.5 mg/100 grams seeds could be doubled by use of an unfractionated seed extract.

The isolated lectin proved to be pure upon electrophoresis in the presence of sodium dodecyl sulphate. Two closely migrating bands with a molecular weight of 32,000 + 1000 D were observed in agreement with earlier published results (1, 4).

References

1. Russel, G.F. et al.: Anal. Biochem. 69, 170-179 (1975)
2. Matsumoto, I. and Osawa, T.: Biochim. Biophys. Acta. 194, 180-189 (1969)
3. Horejsi, V. and Kocourek, J.: Biochim. Biophys. Acta. 336, 329-337 (1974)
4. Allen, H.J. and Johnson, E.A.Z.: Carbohydr. Res. 58, 253-265 (1977)
5. Harris-Smith, P.W. and Fletcher, S.M.: Forensic Sci. Soc. 23, 225-231 (1983)
6. Laemmli, U.K.: Nature (Lond.): 227, 680-685 (1970)

HAEMAGGLUTININS IN MARINE ALGAE - LECTINS OR PHENOLS?

G. Blunden, D.J. Rogers, R.W. Loveless and A.V. Patel

School of Pharmacy, Portsmouth Polytechnic, King Henry I Street, Portsmouth, Hants., PO1 2DZ, U.K.

Over 100 species of British marine algae have been tested for the presence of agglutinins directed against human erythrocytes. Positive results were obtained for several species distributed in the Chlorophyceae, Charophyceae, Phaeophyceae and Florideophyceae (1,2). The activities reported were mainly weak and non-specific, an exception being that of the red alga *Ptilota plumosa*, which shows blood group B specificity; the agglutination is inhibited by α-(1$\rightarrow$3)-linked-D-galactose (3). Other marine algal lectins have been examined in detail and extracts of *Solieria chordalis* and *Codium fragile* subspecies *tomentosoides* were found to possess anti-sialic acid and anti-N-acetyl-D-galactosamine activities, respectively (4,5). Agglutinins from the red algae *Agardhiella tenera* (6) and *Cystoclonium purpureum* (7) have been characterised, but their specificities were not recorded. An agglutinin for sheep erythrocytes has also been reported from the brown alga *Fucus vesiculosus* (8).

Recently, Rogers and Loveless (9) suggested that the human erythrocyte agglutination produced by six species of brown algae was probably due to compounds which are polyphenolic in nature. In view of this, most of the British algal species previously reported to contain haemagglutinins (2) have been re-examined to elucidate the possible contribution of phenols to the reported haemagglutination.

Lectins, Vol. V

MATERIALS AND METHODS

Plant Materials: The algae tested and the places and dates of their collection are as follows: CHLOROPHYCEAE, *Enteromorpha flexuosa* + *E.intestinalis*, Farlington Marshes, Hampshire, October 1984, *Ulva lactuca*, Kimmeridge, Dorset, April 1984, *Codium fragile* subsp. *tomentosoides*, Kimmeridge, May 1983; CHAROPHYCEAE, *Chara vulgaris*, Cloudshill, the Fleet, Dorset, August 1984; PHAEOPHYCEAE, *Desmarestia aculeata*, Southsea, Hampshire, June 1985, *Laminaria saccharina*, Southsea, September 1984, *Ascophyllum nodosum*, Farlington Marshes, April 1984, *Fucus ceranoides*, Bell Harbour, Co. Clare, Ireland, June 1984, *F.serratus*, Southsea, October 1984, *F. spiralis*, Farlington Marshes, May 1985, *F.vesiculosus*, Farlington Marshes, May 1985, *Pelvetia canaliculata*, Farlington Marshes, May 1985, *Himanthalia elongata*, Seacombe, Dorset, May 1985, *Bifurcaria bifurcata*, near Black Head, Co. Clare, Ireland, June 1984, *Cystoseira baccata*, Kimmeridge, July 1985, *C.tamariscifolia*, Kimmeridge, June 1983, *Halidrys siliquosa*, Kimmeridge, May 1983, *Sargassum muticum*, Farlington Marshes, June 1984; FLORIDEOPHYCEAE, *Polyides rotundus*, Finavarra, Co. Clare, Ireland, June 1984, *Cystoclonium purpureum*, Langstone Harbour, Hampshire, June 1984, *Gracilaria bursa-pastoris*, the Fleet, June 1984, *G.verrucosa*, Langstone Harbour, June 1984, *Phyllophora crispa*, Fanore, Co. Clare, Ireland, June 1981, *Palmaria palmata*, Kimmeridge, Dorset, May 1984, *Chylocladia verticillata*, Finavarra, June 1984, *Callithamnion* species, Portsmouth Harbour, Hampshire, June 1984, *Griffithsia floscułosa*, Felpham, W. Sussex, September 1984, *Plumaria elegans*, near Black Head, June 1984, *Ptilota plumosa*, Keem Bay, Achill Island, Co. Mayo, September 1982, *Delesseria sanguinea*, near Black Head, June 1984, *Membranoptera alata*, Keem Bay, September 1982, *Phycodrys rubens*, Keem Bay, September 1982, and *Chondria coerulescens*, Langstone Harbour, June 1984.

Extraction: Each species, after collection, was carefully sorted to reduce contamination by extraneous material, such as epiphytes. The plant material was then stored at -20°C until extracted.

Each alga (10 g fresh weight) was ground to a fine powder in liquid nitrogen and the frozen material added to 40 ml phosphate buffered saline (PBS), pH 7.3, and stirred for 25 min at 18°C. Particulate matter was removed first by filtration through a bilayer of muslin cloth, followed by centrifugation at 2500 x g for 10 min. The supernatant liquid (PBS-extract) was stored at -20°C until tested. Extracts of each alga were also prepared in the same manner, but using 40 ml PBS containing 4 g polyvinylpolypyrrolidone (PVPP - Sigma Chemical Co., Poole, U.K.). The PVPP had been added to the PBS 24 hours before the addition of the alga. These extracts will be referred to as PVPP-extracts.

Tests for haemagglutinating activity: Serial doubling dilutions of both the PBS- and PVPP-extracts of each algal species were prepared in PBS, the dilutions ranging from 1:1 to 1:512. An equal volume (0.1 ml) of a 5% thrice-washed suspension of papainised erythrocytes of human blood group O R_1R_2 in PBS was added to each tube and, after gentle mixing of the contents, left at 18°C for 2 hours before examination for agglutination. The highest dilution showing macroscopic agglutination of erythrocytes was recorded as the titre of the extract. The erythrocytes were treated with papain according to the method referred to by Mollison (10). In the case of *Chara vulgaris*, erythrocytes of human blood group A_1rr were tested in addition to O R_1R_2 cells.

The procedure described above was repeated for both the PBS- and PVPP-extracts of each alga, but the PBS used for dilutions contained 1 g/l of polyoxyethylene sorbitan mono-oleate (Tween 80, Sigma Chemical Co.). The papainised erythrocytes used were also suspended in PBS containing 1 g/l Tween 80.

Both the PBS- and PVPP-extracts of each alga were dialysed separately against distilled water. The water was changed three times at 24 hour intervals. The non-dialysable fractions were freeze dried. The original PBS- and PVPP-extracts, after freeze drying, as well as the non-dialysable fractions of the extracts were examined, over 4 K data points, as solutions in deuterium oxide, by proton magnetic resonance spectroscopy ('H nmr) using a Brüker WH-270 MHz machine. All signals observed between δ5.50 and δ8.00 were noted. Chemical shift measurements are in ppm from DSS.

RESULTS AND DISCUSSION

The Phaeophyceae, in particular the Fucales, have been shown to contain large quantities of polyphenols (11), which are readily extracted by water. Both phenols and their oxidation products, quinones, bind strongly to proteins (12). The Fucales have been reported to be a rich source of "haemagglutinins" (2) and, recently, Rogers and Loveless (9) demonstrated that this "haemagglutinating" activity of six brown algal extracts is probably due to their content of polyphenols. To aid the investigation of marine algal lectins it was considered desirable to re-examine the algal species reported previously to contain haemagglutinins (2) in order to determine whether the activities were due to phenols or lectin-like molecules.

Of the algae tested all showed "haemagglutinating" activity, except the extracts of *Enteromorpha flexuosa* + *E.intestinalis*, *Desmarestia aculeata*, *Laminaria saccharina*, *Bifurcaria bifurcata*, *Cystoseira baccata*, *Sargassum muticum* and *Callithamnion* species. As previously reported (2), the extract of *Chara vulgaris* reacted only with blood group A_1 erythrocytes and not with O cells, although the activity against A_1 erythrocytes was very low.

Of the extracts of species in the Chlorophyceae, Charophyceae and Florideophyceae that produced "haemagglutination", the titres of both the PBS- and PVPP-extracts were not significantly different. PVPP binds polyphenols and, being insoluble, enables the polyphenols to be removed from the extract (13). Moreover, PVPP is a competitive inhibitor of phenoloxidase enzymes concerned with the formation of polyphenols and quinones during extraction processes (14). The similarity between the results obtained with the PBS- and PVPP-extracts suggests that phenols are not responsible for the "haemagglutination" produced. This conclusion is strengthened by the titres recorded for the extracts containing Tween 80, which are similar to those obtained for the extracts without this addition. Tween 80 reverses protein-polyphenol binding (15) and hence, if the "haemagglutination" had resulted from the reaction of polyphenols with either erythrocyte-surface proteins or glycoproteins Tween 80 would be expected to reduce or eliminate haemagglutination. This did not occur. Conversely, if any polyphenols present were bound to lectins, Tween 80 presence would be expected to increase agglutination by the release of lectin. This did not occur either. Heating the extracts at 100°C for 1 h abolished all haemagglutinating activity.

When the same PBS-extracts from the Chlorophyceae, Charophyceae and Florideophyceae were examined by proton magnetic resonance spectroscopy no resonance was observed in the region between $\delta 5.68$ and $\delta 6.48$, which is where signals produced by polyphenols are located. Some low level resonances were recorded between $\delta 6.57$ and $\delta 7.88$ in the spectra of the extracts of many of the species (Fig. 1B), and these were still observed in the spectra of both the PVPP-extracts and the extracts of the non-dialysable fractions of the PBS- and PVPP-extracts. The nature of the compounds producing these resonances was not determined, but they were not considered to be polyphenolic.

All the data available suggest that the "haemagglutination" produced by the species tested in the Chlorophyceae, Charophyceae and Florideophyceae was not due to polyphenols and could be the result of lectin-type molecules.

Unlike the extracts of the species of Chlorophyceae, Charophyceae and Florideophyceae, the activities of the PBS- and PVPP-extracts of the species of Phaeophyceae tested differed markedly. Of those species for which the PBS-extract produced "haemagglutination", the activity of the PVPP-extract was either zero (for example, *Fucus ceranoides* and *Cystoseira tamariscifolia*) or markedly reduced (for example, *Himanthalia elongata* 1:256 to 1:32 and *Halidrys siliquosa* 1:512 to 1:32). This indicated strongly that the "haemagglutination" was produced, at least partly, by phenols. This view was strengthened by comparison of the titres produced when the PBS- and PVPP-extracts were tested with and without the addition of Tween 80. The presence of Tween 80 reduced the levels of activity to either zero or very low levels. It is suggested that this is caused by Tween 80 either reducing or eliminating non-specific aggregation of erythrocytes by polyphenols binding to cell-surface proteins. However, the possibility exists that any lectins present may be destabilised by the Tween 80, although no evidence exists for the destabilisation of other algal lectins by this substance.

Examination of the proton magnetic resonance spectra of all the PBS-extracts of the active species of brown algae showed resonances in the region between δ5.68 and δ6.48, indicating the presence of phenols (Fig. 1A and 1C). The spectra of the PVPP-extracts showed either considerable reductions in the intensities of the phenol signals (for example, *Halidrys siliquosa* and *Fucus serratus*) or their absence (for example, *Ascophyllum nodosum* and *Cystoseira tamariscifolia*). After dialysis of both the PBS and PVPP-extracts, the spectra of all the non-dialysable fractions showed the signals observed in the spectra of the original extracts. This indicated that the resonances were produced by polyphenols which could not be removed by dialysis.

The information obtained indicates that the human erythrocyte "agglutination" produced by all the species of Phaeophyceae tested results from the presence of polyphenols. However, the presence of a lectin-type agglutinin for sheep erythrocytes in extracts of *Fucus vesiculosus* has been reported (8) and the possibility of similar lectins in other species of Phaeophyceae cannot be disregarded.

Sargassum muticum has been reported to contain phenols (16), yet the PBS-extract of this alga did not produce "haemagglutination" and was also reported as inactive in a previous survey (2). The proton magnetic resonance spectrum of the PBS-extract showed the presence of phenols, but, after dialysis, the non-dialysable fraction appeared to be phenol free. The phenols present in the PBS-extract, therefore, had relatively low molecular weights and it has

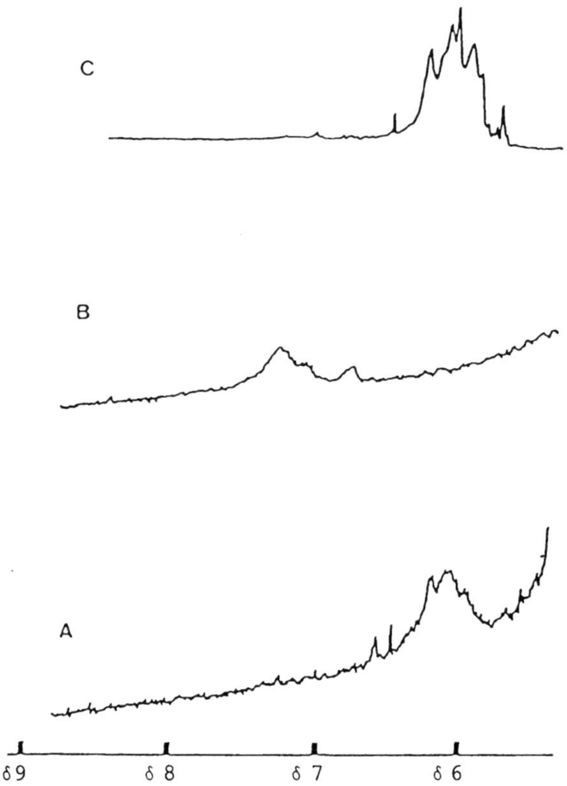

Fig.1 Proton magnetic resonance spectra of the non-dialysable fractions of the PBS-extracts of A, Ascophyllum nodosum; B, Plumaria elegans; and C, Fucus serratus.

been reported previously that protein-phenol interactions vary with the molecular weight of the phenol (17).

Extracts of *Bifurcaria bifurcata* were found previously to agglutinate erythrocytes (2), but the extracts prepared in this current study were inactive. Phenols were not detected in the proton magnetic resonance spectra of the PBS-extract either before or after dialysis. However, when a relatively large quantity of the alga (50 g) was extracted in a Soxhlet apparatus with methanol, the extract concentrated, dialysed, as before, and the non-dialysable fraction examined by proton magnetic resonance spectroscopy, polyphenols were detected, although in low quantity. It is possible, therefore, that the activity found in the earlier collection was produced by polyphenols which may have been present in larger amounts than in the present sample of alga.

ACKNOWLEDGEMENTS

We are grateful to Dr. W.F. Farnham for the collection and identification of many algae used in this study. We thank Mrs. A. Brightwell for her technical assistance.

REFERENCES

1. Blunden, G., D.J. Rogers and W.F. Farnham: Lloydia *38*, 162-168 (1975).
2. Blunden, G., D.J. Rogers and W.F. Farnham: In Modern Approaches to the Taxonomy of Red and Brown Algae (D.E.G. Irvine and J.H. Price, eds) Academic Press, London, pp. 21-45 (1978).
3. Rogers, D.J., G. Blunden and P.R. Evans: Med.Lab.Sci. *34*, 193-200 (1977).
4. Rogers, D.J. and J.A. Topliss: Bot.Mar. *26*, 301-305 (1983).
5. Rogers, D.J. and R.W. Loveless: Br.phycol.J. *20*, 190-191 (1985).
6. Shiomi, K., H. Kamiya and Y. Shimizu: Biochim.Biophys.Acta *576*, 118-127 (1979).
7. Kamiya, H., K. Shiomi and Y. Shimizu: J.Nat.Prod. *13*, 136-139 (1980).
8. Ferreiros, C.M. and M.T. Cirado: Revista Española de Fisiologia *39*, 51-60 (1983).
9. Rogers, D.J. and R.W. Loveless: Bot.Mar. *28*, 133-137 (1985).
10. Mollison, P.L.: In Blood Transfusion in Clinical Medicine, 6th ed., Blackwell, Oxford (1983).
11. Ragan, M.A.: Bot.Mar. *19*, 145-154 (1976).
12. Loomis, W.D.: Methods Enzymol. *16*, 528-544 (1974).
13. Anderson, J.W.: Phytochemistry *7*, 1973-1988 (1968).
14. Mayer, A.M. and E. Harel: Phytochemistry *18*, 193-215 (1979).
15. Goldstein, J.L. and T. Swain: Phytochemistry *4*, 185-192 (1965).
16. Glombitza, K.W., M. Forster and W.F. Farnham: Bot.Mar. *25*, 499-453 (1982).
17. Bate-Smith, E.C.: Phytochemistry *12*, 907-912 (1973).

THE SUGAR SPECIFICITY OF *MACHAEROCEREUS ERUCA* ISOLECTINS.

Edgar Zenteno, Henri Debray, Jean Montreuil and José-Luis Ochoa*.

Laboratoire de Chimie Biologique, Université des Sciences et Techniques de Lille I, Laboratoire Associé au CNRS n° 217, 59655 Villeneuve d'Ascq Cedex, France.

* Centro de Investigaciones Biologicas de Baja California, La Paz Baja California Sur, México.

It has been shown recently that the cactus *Machaerocereus eruca* contains 4 isolectins which could be isolated by hydrophobic interaction chromatography on phenyl- or octyl-Sepharose (1). They are glycoproteins of about 35 kDa immunochemically indistinguishable but having different physico-chemical properties and carbohydrate content. They agglutinate erythrocytes from different animal species and are not specific for any blood group in the ABO system (2). In order to define the specificity of the *M.eruca* isolectins, various glycoproteins and their related glycans and glycopeptides have been used for determining their ability to inhibit the agglutination of human erythrocytes.

MATERIALS AND METHODS

Isolation of M.eruca isolectins. The cactus *Machaerocereus eruca* was collected in Baja California Sur Mexico and isolated as described elsewhere (1). The isolectins I (the less hydrophobic) and IV (the more hydrophobic) obtained by chromatography on phenyl-Sepharose and further purified by affinity chromatography on stroma from desialylated rabbit erythrocytes have been studied in this work.

Hemagglutinating activity. Agglutination of native or desialylated human A erythrocytes (2×10^6 cells/ml in 0.02 M phosphate-0.9 % NaCl, pH 7.4) and the inhibition of this agglutination by various glycoproteins and sugars were performed in microtiter U-plates (Titer tek, Linbro Scientific Co., Hamden, USA) by a two-fold serial dilution procedure (3). The inhibition was expressed as the minimum concentration (µM) of either glycoproteins, glycopeptides or glycans required to inhibit 4 hemagglutinating units.

Neuraminidase treatment of cells. The erythrocytes were treated as in (4) by using *Vibrio cholerae* neuraminidase (Sigma Fine Chem., St Louis, MO, USA).

Glycoproteins. Pig stomach mucin grade II, fetuin grade III and all sugars were from Sigma. Human serotransferrin, human lactotransferrin, milk and serum human IgA as well as O-glycans from IgA were gifts from Prof. G. SPIK and Dr. A. PIERCE. Sheep submaxillary gland mucin, α-2,3 and α-2,6 sialyllactose from human milk were gifts from Drs. G. STRECKER and J.C. MICHALSKI. Hen ovomucoid was isolated by Dr. B. FOURNET. Human α_1 acid glycoprotein (orosomucoid) was isolated according to (5).

Isolation of oligosaccharides from glycoproteins. Oligosaccharides were released by β-elimination as follows : 100 mg of mucin or fetuin were incubated in 30 ml of 0.05 M NaOH with 1 M $NaBH_4$ for 24 h at 45°C. The reaction was stopped by addition of Dowex 50 W x 8 (H^+ ; 200-400 mesh) to pH 5. The solution was then passed on a Dowex column (1.6 x 60 cm) which was further washed with 5 vol. of distilled water. The filtrate and wash solutions were pooled and concentrated by flash evaporation. The mixture of oligosaccharides was fractionated on an Ultrogel AcA-202 (IBF-France) column (1.8 x 134 cm) equilibrated with 0.15 M NaCl-0.05 M Tris , pH 7.4. By this procedure, 4 fractions were obtained from pig mucin. Mucin fractions, as well as the N-glycosylpeptides and glycans from fetuin, were concentrated separately and desalted on a Bio-gel P-2 (Bio-Rad, Richmond, CA, USA) column (1.6 x 60 cm) before sugar analysis by gas liquid chromatography.

Chemical and enzymatic modification of fetuin N-glycosylpeptides. N-glycosylpeptides were desialylated by heating at 80°C for 1 h in 0.1 M trifluoroacetic acid. The latter was evaporated under a nitrogen stream and free sialic acid was eliminated by gel filtration on Bio-gel P-2. Galactose residues of desialylated N-glycosylpeptides from fetuin were removed by Jack bean β-galactosidase as described in (6).

Sugar analysis. Gas liquid chromatography was carried out according to (7) and sugar concentrations were determined by a phenol-sulphuric acid procedure (8), using lactose as standard.

Protein determination. Proteins were determined by the method of Lowry et al. (9) using bovine serum albumin (Sigma Fine Chem.) as standard.

RESULTS

The *M.eruca* isolectins isolated by hydrophobic chromatography and purified by affinity chromatography on a stroma column possess no hemagglutinating specificity toward human erythrocytes in the ABO system. 1 µg/ml of isolectin I was necessary to induce the agglutination of a suspension of 2 x 10^6 cells/ml and 16 µg/ml of isolectin IV to induce the same effect. Desialylated erythrocytes are more easily agglutinated by the lectins : 0.01 and 0.04 µg/ml of isolectins I and IV, respectively, was necessary to induce agglutination of desialylated cells.

Table 1. - *Minimal concentration (µM) of sugars necessary to inhibit 50 % of the hemagglutinating activity of M.eruca isolectins.*

Sugars	Isolectin I	Isolectin IV
L-Fucose	100	25
D-Fucose	N.I. (*)	N.I.
D-Galactose	25	12.5
D-Galactosamine	100	100
N-Acetyl-D-galactosamine	50	50
Nitrophenyl-D-galactose	50	50
D-Glucose	N.I.	200
D-Glucosamine	N.I.	200
N-Acetyl-D-glucosamine	N.I.	200
D-Mannose	N.I.	N.I.
D-Mannosamine	N.I.	N.I.

(*) *N.I. : not inhibitory at 200 mM*

Data from Table 1 show that both isolectins exhibit a weak sugar specificity for galactose and that isolectin IV can be inhibited by L-fucose. O-glycosylproteins as mucin and fetuin, as well as both kinds of immunoglobulin A were 100-fold more inhibitory than N-glycosylproteins as shown in Table 2. In Table 3 are reported the monosaccharide molar ratios of the glycans and glycopeptides liberated by β-elimination from O-glycosylproteins, the expressed values are relative to 1 galactosaminitol residue for released glycans or to 3 mannose residues for N-glycosylpeptides. Molar ratios of the N-glycosylpeptides from fetuin after ATFA desialylation and after β-galactosidase treatment are also shown.

Table 2. - *Minimal concentration (μM) of glycoproteins necessary to inhibit 4 hemagglutinating units of M.eruca isolectins*

Substrates	Sources	MW (in kDa)	Type of glycosidic linkages	Isolectins I	IV
Mucin	Pig stomach	1000	O-	0.04	0.02
Mucin	Sheep	400	O-	0.05	0.03
Fetuin	Fetal calf serum	35	O,N-	0.01	0.005
IgA	Human milk	400	O,N-	0.02	0.01
IgA	Human serum	400	O,N-	0.2	0.05
Serotransferrin	Human serum	78	N-	25	3.0
Lactotransferrin	Human milk	78	N-	6.0	1.5
Orosomucoid	Human serum	45	N-	110	55

Table 3. - *Monosaccharide molar ratios of glycans and glycopeptides used in the experiments*

Origin	Fuc	Gal	Man	GlcNAc	GalNAc	GalNAc-ol	NeuAc
Pig gastric mucin							
Glycan I	5.25	6.8	0	3.7	3.3	1	0
II	2.2	6.1	0	3.6	0	1	0
III	4.0	9.0	0	6.0	1	1	0
IV	2.6	7.4	0	6.3	0.7	1	0
Calf fetuin							
N-glycosylpeptides	0	4.0	3	3.6	0.8	0	2
Glycan-alditols	0	1.9	0	0	0	1	1.3
Asialoglycosylpeptides	0	4.0	3	4.1	0	0	1.0
Asialo-agalactoglycosylpeptides	0	1.0	3	4.1	0	0	0.8
Human milk							
IgA	0.05	0.9	0.05	0.14	1	0	0.2
Hen ovomucoid	0	0.8	3	5	0.7	0.2	0.1

Table 4. - *Minimal concentration (μM) of glycans liberated by β-elimination and of glycosylpeptides necessary to inhibit 4 hemagglutinating units of M.eruca isolectins.*

Substrates	Sources	Isolectins I	Isolectins IV
Mucin	Pig stomach		
Glycan I		2.0	1.0
II		0.6	1.8
III		3.0	12
IV		70	0
Fetuin N-glycosylpeptides	Fetal calf serum	58	20
Fetuin O-glycans	Fetal calf serum	10	5.0
IgA	Human serum and milk	70	70
Ovomucoid(*)	Hen	N.I.	N.I.
α-2,3 sialyl lactose(*)	Human milk	N.I.	N.I.
α-2,6 sialyl lactose(*)	Human milk	N.I.	N.I.

(*) *At concentrations of 2 mM, 1.4 mM and 4 mM respectively.*

The oligosaccharides liberated by β-elimination from O-glycosylproteins are good inhibitors of the red blood cells agglutination induced by both isolectins (Table 4) but fraction II from pig mucin is the most powerful inhibitor for both isolectins. The N-glycosylpeptidic fraction isolated from fetuin is also a good inhibitor but, after removal of galactose residues, the N-glycosylpeptides are unable to recognize the lectins (Table 5).

Table 5.- *Minimal concentration (μM) of chemically and enzymatically modified fetuin glycopeptides to inhibit 4 hemagglutinating units of M.eruca isolectins.*

Subtrates	Isolectin I	Isolectin IV
Native N-glycosylpeptides	58	29
Asialo-N-glycosylpeptides	12	15
Asialoagalacto-N-glycosylpeptides	N.I.	N.I.

N.I. : At concentrations of 50 μM

DISCUSSION

Machaerocereus eruca isolectins express little affinity for galactose and fucose, but both isolectins seem to recognize the same oligosaccharidic determinants of glycans released by β-elimination from O-glycosylproteins, such as pig stomach mucin or calf fetuin. The minimal structure recognized by the isolectins is constituted of the trisaccharide NeuAc(α 2-3)Gal(β 1-3)GalNAc (glycan released from fetuin by β-elimination). Lectins from *Vicia graminea* and *Agaricus bisporus* recognize also similar structures, but the presence of external sialic acid residues strongly inhibits saccharide recognition (10, 11).

Removal of galactose residues from fetuin N-glycosypeptides abolishes the lectin affinity proving that these residues are important determinants.

Native O-glycosylproteins are better inhibitors than their released glycans suggesting that hydrophobic interactions could also take place during the lectin-glycoprotein interaction as described for ConA (12, 13) or for *Lens*, *Vicia*, *Pisum*, *Solanum* and *Datura* lectins (14).

SUMMARY

The preliminary results we present in this paper show that isolectins I and IV from the cactus *M.eruca* constitute an additional example of lectins interacting with specific domains of complex oligosaccharidic chains belonging either to O- or N-glycosylproteins as demonstrated for other lectins (14-16).

ACKNOWLEDGEMENTS

Edgar ZENTENO is a fellowship recipient from the Mexican Government (VIII Program CONACYT-CEFI, Mexico-France). This work was supported in part by the Centre National de la Recherche Scientifique (LA n° 217 : Relations structure-fonction des constituants membranaires : Director : Pr. J. MONTREUIL). The authors are very gratefull to Drs. G. SPIK, G. STRECKER, A. PIERCE, J.C. MICHALSKI, B. FOURNET and S. BOUQUELET for providing the glycoproteins and enzymes used in this study and for helpful discussions.

REFERENCES

1. Zenteno, E., and Ochoa, J.L. : In : Lectins, Biology, Biochemistry and clinical Chemistry, T.C. Bøg-Hansen and J. Breborowics eds., Vol. 4, pp. 437-445, De Gruyter, Berlin, 1985.
2. Zenteno, E., and Cordoba, F. : In : Lectins, Biology, Biochemistry and Clinical Chemistry, T.C. Bøg-Hansen ed., Vol. 2, pp. 721-729, De Gruyter, Berlin, 1982.
3. Debray, H., and Rouge, P. : FEBS-Lett. 176 (1984) 120-124.
4. Monsigny, M., Roche, A.C., Sene, C., Maget-Dana, R., and Delmotte, F. : Eur. J. Biochem. 104 (1980) 147-153.
5. Weimer, H.E., Mehl, J.W., and Winzler, R.J. : J. Biol. Chem. 185 (1950) 561-568.
6. Spik, G., Bayard, B., Fournet, B., Strecker, G., Bouquelet, S. and Montreuil, J. : FEBS-Lett. 50 (1975) 296-299.
7. Zanetta, J.P., Breckendrige, W.C., and Vincendon, G. : J. Chromatog. 69 (1972) 291-304.
8. Dubois, N., Gilles, K.A., Hamilton,J.K., Rebers, P.A., and Smith, F. : Nature 168 (1951) 167-169.
9. Lowry, H.O., Rosenbrough, N.J., Lewis-Farr, A., and Randall, R. : J. Biol. Chem. 193 (1951) 265-268.
10. Duk, M., Lisowska, E., Kordowikz, M., and Wasniowska, K. : Eur. J. Biochem. 123 (1982) 105-111.
11. Presant, C.A., Kornfeld, S. : J. Biol. Chem. 247 (1972) 6937-6945.
12. Davey, M.W., Huang, J.W., Sulkowsky, E. and Carter, W.A. : Biochemistry 15 (1976) 704-713.
13. Davey, M.W., Sulkowsky, E. and Carter, W.A. : J. Biol. Chem. 249 (1974) 6354-6355.
14. Debray, H., Decout, D., Strecker, G., Spik, G., and Montreuil, J. : Eur. J. Biochem. 117 (1981) 41-55.
15. Kornfeld, R., and Ferris, C. : J. Biol. Chem. 250 (1975) 2614-2619.
16. Gallagher, J.T. : Biosci. Rep., 4 (1984) 621-632.

ISOLATION AND CHARACTERISATION OF THE LECTINS FROM SUB-SPECIES OF *Codium fragile*.

D.J. Rogers[+], R.W. Loveless[+] and P. Balding[*].

[+]Haematology and Immunology Laboratory, Park Building, Portsmouth Polytechnic, King Henry I Street, Portsmouth, Hants., PO1 2DZ, UK.

*Production Division, Sigma Chemical Company, Fancy Road, Poole, Dorset, BH17 7NH, UK.

Although it is well established that extracts of marine algae contain lectin-type agglutinins (1, 2) relatively few have been characterised biochemically. Most of those which have been investigated in detail belong to the division Rhodophyta (red algae) and Ingram has reviewed the relevant literature (3).

The situation with regard to lectins from the Chlorophyta (green algae) is quite different as no detailed reports of their isolation and characterisation have appeared.

Two groups of workers have described the occurrence of agglutinins for human-erythrocytes (4) and rabbit-erythrocytes (5) in extracts of the green marine alga *Codium fragile*. The agglutinins in the cell sap of the two British sub-species of this alga have now been isolated and characterised.

Lectins, Vol. V

MATERIALS AND METHODS

Plant Material: *Codium fragile* subs. *atlanticum* was collected at Lerwick, Shetland Isles, U.K. in June 1984 and subsp. *tomentosoides* at Kimmeridge, Dorset, U.K. in July 1984. After collection, the algae were cleaned of epiphytes, added to phosphate buffered saline (0.15 M pH 7.3) (PBS) in a ratio of 1:2 weight to volume and homogenised. Particulate matter was removed by centrifugation at 15,000 x g and the supernatants stored at -20°C until required.

Haemagglutination and haemagglutination inhibition tests were performed by conventional methods using papain-treated human A_1 erythrocytes as described previously (6).

Purification: Affinity resin was prepared by a two-stage process. Washed Sepharose 6B, 10 g, was mixed for 8 h at 18°C with 10 ml 1-4 butanediol diglycidyl ether and 10 ml 0.6 M sodium hydroxide containing 2 mg ml^{-1} sodium borohydride. This step was stopped by washing with distilled water. The epoxy activated Sepharose was washed with 0.1 M sodium hydroxide, resuspended in an equal volume of 0.1 M sodium hydroxide containing 50 mg ml^{-1} N-acetyl galactosamine (GalNAc) and mixed for 15 h at 45°C. The affinity resin was then washed sequentially with distilled water, 0.05 M Tris buffer, 0.05 M sodium chloride pH 8.0 and 0.05 M sodium formate pH 4.0. The prepared resin was stored at 4°C in 0.02% sodium azide in distilled water.

Crude extracts of the sub-species of *C.fragile* were mixed for 3 h at 4°C with the affinity resin in a ratio of 4 ml:1 g. The resin was then recovered by centrifugation and washed with distilled water. The absorbed extract was retained and tested for residual haemagglutinating activity. Material adsorbed to the affinity resin was eluted for 3 h at 4°C with 0.05 M GalNAc in a ratio of 1 ml GalNAc : 1 g resin. Eluates were dialysed against distilled water and then freeze dried.

Protein concentrations were measured by the modified Folin-Lowry method (7).

Assessment of purity and characterisation: Conventional flat-bed polyacrylamide gel electrophoresis (PAGE) and sodium dodecyl sulphate PAGE were carried out using LKB Multiphor electrophoretic apparatus and power pack. Protein bands were stained with Brilliant Blue G-250. Bovine albumin, ovalbumin, α-chymotrypsinogen A and horse myoglobin were used as standards.

Broad range isoelectric focusing was carried out between pH 3.5-9.5 using Ampholine PAG plates (LKB) and electrofocusing over a pH range of 2.5-4.5 was conducted in granulated gels using an Ampholine electrofocusing kit (LKB).

The amino acid composition of the purified lectins from both subspecies was assayed on a Rank-Hilger Chromaspex analyser following digestion of the proteins in 10 mM hydrochloric acid. Cysteine and tryptophan are not assayable in this system and asparagine is assayed as aspartic acid.

Carbohydrate content of the lectins was determined using the method of Dubois *et al* (8).

RESULTS

Crude extracts of both subsp. of *Codium fragile* agglutinated native human erythrocytes suspended in PBS (Table 1). No ABO blood-group specificity was demonstrated. When papain-treated erythrocytes were tested, the extract from subsp. *atlanticum* agglutinated A_1 cells strongly, but failed to react with B and O cells. The extract from subsp. *tomentosoides* agglutinated papain-treated A_1 cells strongly, but also agglutinated B and O cells, albeit to lower titration values.

Haemagglutination inhibition studies using a wide range of mono-, di- and trisaccharides revealed that both lectins were preferentially inhibited by N-acetyl-α-D-galactosamine (GalNAc). The minimal inhibitory concentration of this substance was 6.25 mM. N-acetyl-α-D-glucosamine failed to inhibit the lectin from subsp.

Table 1

Haemagglutination titres of extracts of sub-species of Codium fragile

	Titration value with human erythrocytes					
	Phosphate buffered saline (PBS)			Papain-treated erythrocytes		
	A_1	B	O	A_1	B	O
Codium fragile subsp. *atlanticum*	1:8	1:4	1:8	1:256	0	0
Codium fragile subsp. *tomentosoides*	1:8	1:16	1:16	1:512	1:32	1:16

Table 2

Increase in specific activity and yield of product following affinity purification of the lectins from sub-species of Codium fragile

Codium fragile subsp.	Specific activity (haemagglutinating units μg protein^{-1} ml^{-1})		Increase in specific haemagglutinating activity	Yield of purified product μg g alga^{-1}
	Crude extract	Affinity purified extract		
atlanticum	3.1	136.53	44.0	35.0
tomentosoides	0.7	28.1	40.2	16.0

tomentosoides, but showed a minimal inhibitory concentration of 50 mM with the extract from subsp. *atlanticum*.

The results of the affinity purification of both lectins are shown in Table 2.

Although the increase in specific haemagglutinating activity was similar with the lectins from both sub-species the yield of freeze-dried product from subsp. *atlanticum* was slightly more than double the yield from subsp. *tomentosoides*.

Conventional polyacrylamide gel electrophoresis (PAGE) demonstrated that each of the purified lectins consisted of a single band of protein and no minor contaminants. Gel eluates from the position of the band produced strong haemagglutination.
When PAGE was conducted in the presence of sodium dodecyl sulphate (SDS) each lectin produced a single band with a calculated molecular weight of 60,000 daltons. Treatment of the lectins with 2-mercaptoethanol prior to SDS PAGE produced gels showing a single band of approximately 15,000 daltons.

Isoelectric focusing showed that both lectins consisted of two iso-lectins with a pI of 3.8 or 3.9.

Haemagglutination inhibition studies were repeated using the purified lectins and identical results to those observed with the crude extracts were obtained (Table 1). Titration values of the purified lectins with papain-treated A_1 erythrocytes were unaffected by variations in incubation temperature between 4°C and 45°C or by variation in pH between 4.5 and 9.5. The chelation of divalent cations by the incorporation of 0.01M ethylenediamine tetra acetic acid (EDTA) in the PBS used as diluent did not affect the haemagglutinin titres.
Table 3 shows that the amino acid composition of the purified lectins from both sub-species were very similar. Total carbohydrate was calculated to be 13.1% for the lectin from subsp. *atlanticum* and 11.9% for that from subsp. *tomentosoides*.

Table 3
The amino acid composition of the lectins from Codium fragile

Amino acid	Amino acid composition (% mol) *C.fragile* subsp. *atlanticum*	*C.fragile* subsp. *tomentosoides*
Basic		
Arg	4.98	4.63
His	1.05	1.03
Lys	3.39	3.37
Acidic		
Asp	7.31	8.60
Glu	13.17	13.36
Neutral		
Ala	8.01	8.53
Gly	11.79	12.32
Ile	3.94	3.56
Leu	5.38	6.12
Met	0.73	0.72
Phe	6.97	6.86
Pro	5.66	5.53
Ser	3.19	2.41
Thr	5.70	5.60
Tyr	6.30	7.78
Val	12.39	9.60

DISCUSSION

This report represents the first description of a biochemical study on lectins from green macroalgae.

The lectins from both sub-species of *C.fragile* investigated agglutinate native human erythrocytes regardless of blood group, but react preferentially with papain-treated A_1 cells. This preference for papainised A_1 cells is particularly marked with the lectin from subsp. *atlanticum*. In view of this, the inhibition of haemagglutination of both lectins by GalNAc, preferentially, is not surprising since GalNAc is known to be the immunodominant sugar of the human erythrocyte A antigen (9).

Electrophoretic studies indicate that both lectins have a molecular weight of 60,000 daltons and are composed of 4 sub-units of 15,000 daltons joined by disulphide bonds. This molecular weight is higher than that recorded for the lectins from red algae using similar methods (10 12). Unlike the lectin from the red alga *Ptilota plumosa* (13) the lectins from *C.fragile* react in the presence of EDTA indicating no requirement for divalent cations. Quite clearly the lectins from macroalgae are as structurally heterogenous as those from angiosperm plants.

However, these studies show that the lectins from the two subspecies of *C.fragile* are biochemically very similar. The amino acid and carbohydrate content are comparable although the degree of human blood-group specificity appears to differ. This could be due to the relative proportions of the two isolectins present in each plant demonstrable by isoelectric focusing. This aspect of the work and studies on the reactions of human A subgroups with lectins is under investigation.

REFERENCES

1. Boyd, W.C., Almodovar, L.R., Boyd, L.C.: Transfusion *6*, 82-83 (1966).
2. Blunden, G., Rogers, D.J., Farnham, W.F.: In: Modern Approaches to the Taxonomy of Red and Brown Algae. (Irvine, D.E.G., Price, J.H., Eds) Academic Press London. p. 21-45 (1978).
3. Ingram, G.A.: Developmental and Com.Imm. *9*, 1-10 (1985).
4. Blunden, G., Rogers, D.J., Farnham, W.F.: Lloydia *38*, 162-168 (1975).
5. Kobayshi, A., Shimizu, Y.: Lloydia *39*, 474 (1976).
6. Rogers, D.J., Blunden, G., Evans, P.R.: Med.Lab.Sci. *34*, 193-200 (1977).
7. Daughaday, W.H., Lowry, O.H., Rosebrough, N.J., Fields, W.S.: J.Lab.Clin.Med. *39*, 663-665 (1952).
8. Dubois, M., Gilles, K.A., Hamilton, J.K., Rebers, P.A., Smith, F.: Anal.Chem. *28*, 350-356 (1956).
9. Morgan, W.T.J., Watkins, W.M.: Brit.med.Bull. *25*, 30 (1969).
10. Shiomi, K., Kamiya, H., Shimizu, Y.: Biochim.Biophys.Acta *576*, 118-127 (1979).
11. Kamiya, H., Shiomi, K., Shimizu, Y.: J.Nat.Prod. *43*, 136-139 (1980).
12. Kamiya, H., Ogata, K., Hori, K.: Bot.Mar. *25*, 537-540 (1982).
13. Rogers, D.J., Blunden, G.: Bot.Mar. *23*, 459-462 (1980).

VICIA FABA α-GALACTOSIDASES WITH LECTIN ACTIVITIES : AN OVERVIEW

Prakash M. Dey and John B. Pridham

Department of Biochemistry, Royal Holloway and Bedford New College, University of London, Egham Hill, Egham, Surrey, TW20 0EX, U.K.

Kocourek and Hořejší (1) in their definition of a lectin state that it should not covalently alter the carbohydrate ligand. Some enzymes (2-6) will, under certain conditions, cause haemagglutination because they possess multiple substrate binding sites and these enzymes are categorised as lectins according to an alternative definition (7) advanced by the International Union of Biochemistry which ignores covalent modification of the bound carbohydrate which may occur in such cases. One of us has pointed out (8) that the definition of Kocourek and Hořejší is more appropriate in this respect and this was prompted by an earlier observation (9) that a tetrameric α-galactosidase from *Vigna radiata* (mung bean) seeds possessed D-galactose-specific haemagglutinating properties. This enzyme was able to agglutinate rabbit erythrocytes but the agglutinated cells subsequently re-dispersed at a rate which was dependent on the pH, temperature and enzyme concentration. A monomeric α-galactosidase from the same source possessed no haemagglutinating properties. It was later concluded (8) that the tetrameric enzyme was not a true lectin and that the 'lectin-like' activity was due to the initial formation of a relatively stable enzyme-substrate complex which was subsequently hydrolysed to galactose and degalactosylated erythro-

Lectins, Vol. V

cytes as depicted below:

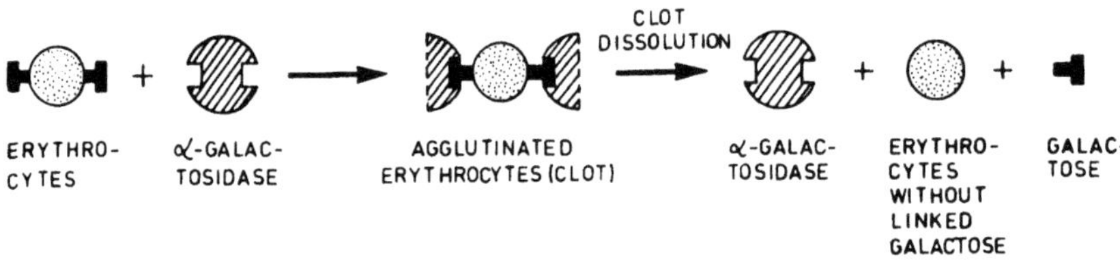

Tetrameric α-galactosidases (but no monomeric forms) from seeds of a number of different legume species possess this galactose-specific 'lectin-like' activity (2, 9) and in all cases the haemagglutinating properties are probably due to the presence of multiple catalytic sites. In contrast, α-galactosidases (tetrameric and monomeric forms) from *Vicia faba* (broad bean) are unique examples of enzymes that display true lectin activity (10, 11). In this paper we review the properties of these enzymes.

Results and Discussion

α-Galactosidase (EC 3.2.1.22) activity from extracts of resting *V. faba* seeds can be resolved into two fractions, I and II by Sephadex G-100 gel-filtration (12). Fraction I elutes as a single peak on further chromatography using either DEAE-cellulose or CM-cellulose whereas II can be resolved into two active fractions, II^1 and II^2, with the cation-exchange column (13). Enzymes II^1 and II^2 have been purified to homogeneity via a multi-step protocol involving acid precipitation, $(NH_4)_2SO_4$ fractionation, Sephadex gel-filtration and CM-cellulose chromatography (13). Recently, affinity chromatographic methods using Con A-Sepharose and melibiose-Sepharose columns have been included in the purification procedure which have resulted in improved recovery of the enzymes (10, 11). Molecular weights (MW) determined by sedimentation equilibrium were 160400 ± 2850, 45730 ± 3073 and 43390 ± 1409 dalton for α-galactosidases I, II^1 and II^2, respectively (13).

These galactosidases bind strongly to Con A-Sepharose suggesting that they are glycoproteins with accessible D-glucose/D-mannose residues. Carbohydrate analyses of the three enzymes demonstrated the presence of combined D-xylose, D-mannose, D-glucose and D-glucosamine in all forms (10,14). The significance of D-glucose and D-xylose in the glycoproteins is not entirely clear; D-glucose could have arisen from starch contamination and/or from nascent, intermediary forms of the glycoproteins (15). A fraction obtained by pronase digestion of α-galactosidase I, was shown to contain 'high-mannose' type glycopeptides by acetolysis followed by fast atom bombardment mass spectrometry; D-xylose could not be detected in these fractions (16).

When α-galactosidase I was subjected to SDS-polyacrylamide-gel electrophoresis, a single protein band (MW ~ 44500 dalton) with evidence of microheterogeneity was observed; the same band pattern was obtained with enzyme II^2. The MW of enzyme II^1 obtained by this method was ~ 49500 dalton (7). Thus it appears that I is a tetramer of II^2 and serological tests which show a close structural relationship between I and II^2 (and less similarly between I and II^1) support this concept (10, 14). In view of the observed microheterogeneity of enzyme I it is likely that it consists of a family of closely related tetramers composed of near identical monomers. This would be expected if enzyme I is assembled from monomers with different degrees of glycosylation (18) or monomers with structural differences produced by minor post-translational changes in the polypeptide chains. The micro-heterogeneity of enzymes I and II[2] was also reflected in their chromatofocusing patterns (Fig. 1). In the presence of 2.5M urea a small proportion of enzyme I was dissociated to an active form with MW ~ 80,888 dalton (a dimer?); at higher urea concentrations (3.5M), most of I was converted to lower MW forms, mainly dimer and possibly some trimer (MW ~ 120,000 dalton). However, if enzyme I was treated with 1.5 M urea, no dissociation was observed unless 1M methyl α-D-mannoside was present and in this case both dimer and monomer were formed (19). Similar dissociation results were obtained by incubation with mixture of 2.5 M urea and 0.5 M methyl α-D-mannoside. Methyl α-D-mannoside alone produced no apparent dissociation of the tetrameric enzyme. Do these observations indicate that lectin-glycan bonds in part stabilize the quaternary structure of enzyme I?

All forms of the *V. faba* α-galactosidase displayed lectin activity (10, 11). The carbohydrate specificity of these lectin activities, determined by hapten inhibition of haemagglutination, are shown in Table 1. The data clearly indicate that all three forms are D-glucose/D-mannose specific with a probable preference for α-D-anomers as is generally the case with other lectins with the same hexose specificity (20). The observed hapten inhibition by certain oligo- and poly-saccharides is presumably due to the presence of D-glucose or D-mannose residues in these structures. D-Glucose and D-mannose and their derivatives produced no significant inhibition of the catalytic activity of the enzymes, whereas D-galactose was a potent competitive inhibitor (~ 70%

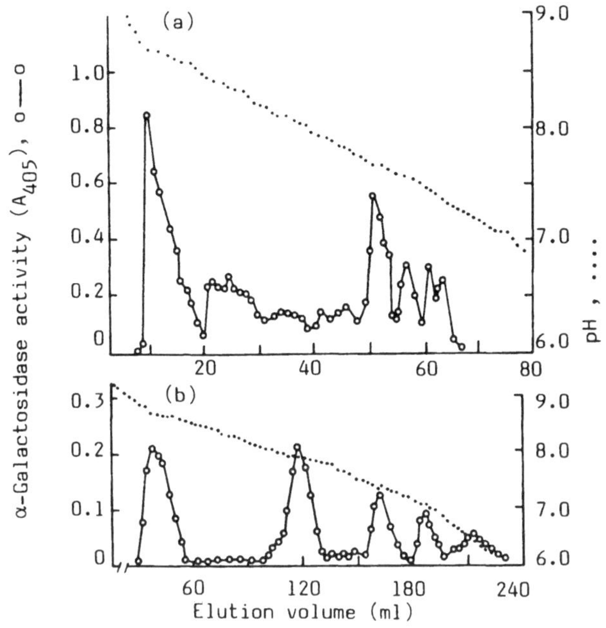

Fig. 1. Chromatofocusing profiles of Vicia faba α-galactosidase I (a) and II^2 (b) (see ref. 19).

inhibition at 5mM) (21). α-Galactosidases I, II^1 and II^2 were able to precipitate yeast mannan, glycogen and soluble starch from their solutions. With yeast mannan, a bell-shaped curve, typical of a lectin-polysaccharide interaction (22), was obtained when turbidity was plotted against mannan concentration. D-Mannose inhibited turbidity formation. Enzyme I was observed to bind to starch granules leaving catalytic sites accessible to a galactoside substrate (Fig. 2). Further evidence for the existence of separate catalytic and lectin sites on the enzyme proteins came from investigations of the effects of pH and heat on two activities (11). All three forms of the enzyme exhibited two

Table 1

Carbohydrate inhibition of the lectin activities of α-galactosidases I, II^1 and II^2 from V. faba seeds (17).

Sugar	Minimum concentration (mM) causing 50% inhibition of 6-haemagglutinin units		
	I	II^1	II^2
Methyl-α-D-mannopyranoside	3.1	2.00	1.00
D-Mannose	15.6	4.68	9.40
D-Glucose	31.3	8.36	19.60
3-O-Methyl-D-glucopyranose	31.3	8.36	19.60
N-Acetyl-D-glucosamine	31.3	8.36	19.60
Methyl-α-D-glucopyranoside	62.5	16.72	39.20
Methyl-β-D-glucopyranoside	250	100	125
D-Xylose	250	100	125
L-Arabinose	250	125	250
D-Galactose	1000	1000	1000
Methyl-α-D-galactopyranoside	1000	1000	1000
Melibiose	100	62.5	125
Stachyose	100	62.5	125
Raffinose	100	62.5	125
Sucrose	125	N.D.	N.D.
Starch (soluble)	0.03 mg/ml	0.02 mg/ml	0.03 mg/ml
Glycogen	0.03 mg/ml	0.02 mg/ml	0.03 mg/ml
α-Mannan (Baker's yeast)	N.D.	0.03 mg/ml	0.03 mg/ml
Galactomannan (guar gum)	0.09 mg/ml	0.05 mg/ml	0.10 mg/ml
Galactomannan (clover)	0.12 mg/ml	0.06 mg/ml	0.10 mg/ml

N.D., not determined

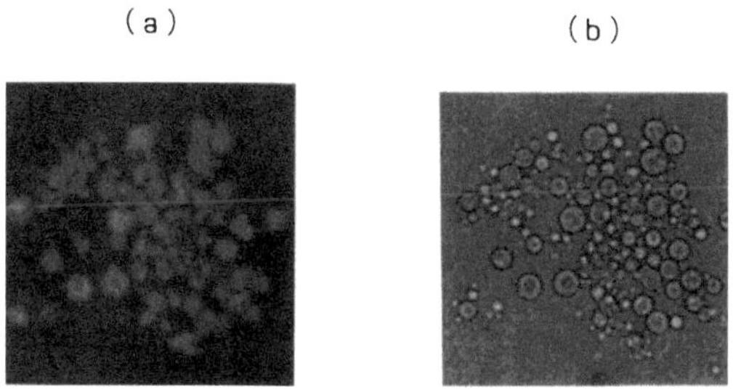

Fig. 2. Ultraviolet microscopy showing fluorescence by starch granules after incubating with Vicia faba α-galactosidase I followed by the addition of 4-methylumbelliferyl α-D-galactoside and leaving for (a), 2 min and (b), 15 min. Under the latter condition the light background is caused by diffusion of the liberated 4-methylumbelliferone away from the starch granules (23).

optima (at pH ~ 3.5 and pH ~ 5.5 at 30°C) for the hydrolysis of p-nitrophenyl α-D-galactoside, but a single optimum (pH ~ 7.2) for haemagglutination. After heating at 65°C for 20 min the residual catalytic activities of enzymes I, II^1 and II^2 were 75%, 20% and 12%, respectively whilst the corresponding lectin activities were, 100%, 50% and 48%. The decay rates of the two activities were faster and indistinguishable at 75°C (Fig. 3). However, at this temperature they responded differently to the presence of monosaccharides; inclusion of D-galactose during incubation protected the catalytic functions but not the lectin activities of the three enzymes, whereas D-mannose had the opposite effect (Fig. 3).

In addition demetallization of enzyme I with EDTA resulted in a 60% increase in the catalytic activity but an 87% decrease in lectin activity (19) providing further support for the two-site theory. Scatchard analysis (24) of the data obtained from equilibrium dialysis experiments for the binding of [U-^{14}C] mannose revealed that 8, 4 and 2 molecules of D-mannose were bound per

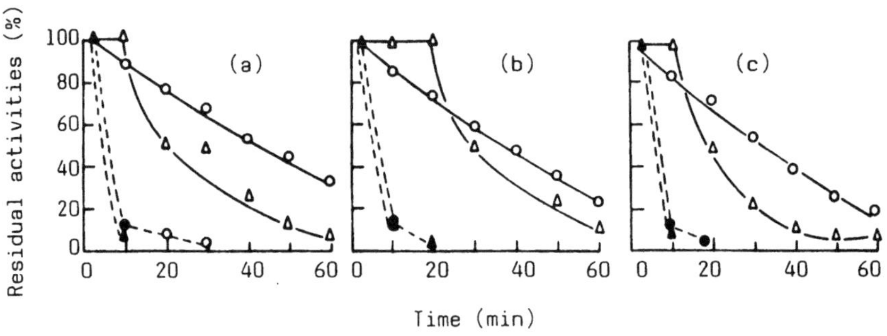

Fig. 3. Protection of catalytic and lectin activities at 75°C of Vicia faba α*-galactosidases (a), I (b), II^1 and (c), II^2 in the presence of D-mannose and D-galactose (both at 0.1M).* o———o, *catalytic activity in the presence of D-galactose;* ●———●, *catalytic activity in the absence of D-galactose or, in the presence of D-mannose;* Δ———Δ, *lectin activity in the presence of D-mannose;* ▲———▲, *lectin activity in the absence of D-mannose or, in the presence of D-galactose.*

molecule of enzyme I, II^1 and II^2, respectively. The corresponding Ka values were, $1.37 \times 10^4 M^{-1}$, $2.18 \times 10^3 M^{-1}$ and $3.14 \times 10^3 M^{-1}$. The addition of D-galactose (0.1M) to the dialysis mixtures did not alter either the number of bound D-mannose molecules or the Ka values. Further, D-mannose but not D-galactose induced changes in the U.V. absorption by the three enzymes resulting in differenc spectra with peaks between 282.5 and 279nm (presumably due to participation of tyrosine residues in the interactions). These changes allowed Ka values to be determined (25) [$1.48 \times 10^4 M^{-1}$, $2.51 \times 10^3 M^{-1}$ and $3.05 \times 10^3 M^{-1}$ for enzymes I, II^1 and II^2, respectively] which were close to those found by the equilibrium dialysis method.

The three forms of *V. faba* α-galactosidase-glycoprotein-lectin therefore, represent an interesting system of considerable potential physiological importance. α-Galactosidase activity appears to be compartmentalized in resting seeds. In *Vigna radiata* for example, its location is reported to be the protein bodies (26) and in *V. faba* a proportion of the activity, at least, is believed to be closely associated with the cell wall (unpublished) as is the case with 'acid' α-galactosidase in squash (*Cucurbita pepo*) leaf tissues (27). In seeds compartmentalization of the enzyme may allow galactose-containing oligosaccharide reserves to be laid down without loss through enzymic hydrolysis during seed maturation. Both the lectin sites and the glycoprotein glycan chains of the *V. faba* α-galactosidases may be important for binding of the enzymes to cell components via lectin interactions. Such bonds could conceivably regulate the activity of the catalytic sites. With the tetrameric α-galactosidase I from *V. faba* it is possible that internal lectin bonds involving the glycan chains of one monomer with the lectin sites of another may exist and that these might also modulate enzyme activity and contribute to the quaternary structure.

It would be interesting to know how many, so-called, soluble enzymes possess lectin sites which would allow them to bind to cell components.

References

1. Kocourek, J., and Hořejší, V.: Nature (Lond.) 290, 188 (1981).
2. Shannon, L.M., and Hankins, C.N.: Recent Adv. Phytochem. 15, 83-113 (1981).
3. Levitzki, A., Heller, J., and Schramm, M.: Biochim. Biophys. Acta 81, 101-107 (1964).
4. Hořejší, V.: Biochim. Biophys. Acta 577, 383-388 (1979).
5. Hořejší, V., and Kocourek, J.: Experientia 30, 1348-1349 (1974).

6. Selinger, Z., and Schramm, M.: Biochem. Biophys. Res. Commun. 12, 208-214 (1963).

7. Nomenclature Committee of the International Union of Biochemistry: Arch. Biochem. Biophys. 206, 458-462 (1981).

8. Dey, P.M.: Eur. J. Biochem. 140, 385-390 (1984).

9. Hankins, C.N., Kindinger, J.I., and Shannon, L.M.: Plant Physiol. 65, 618-622 (1980).

10. Dey, P.M., Pridham, J.B. and Sumar, N.: Phytochemistry 21, 2195-2199 (1982).

11. Dey, P.M., Naik, S., and Pridham, J.B.: FEBS Lett 150, 233-237 (1982).

12. Dey, P.M., and Pridham, J.B.: Phytochemistry 7, 1737-1739 (1968).

13. Khaleque, A.: Ph.D. thesis. University of London (1973).

14. Sumar, N.: Ph.D. thesis. University of London (1983).

15. Hughes, R.C.: Glycoproteins, Chapman and Hall Ltd., London and New York (1983).

16. Naik, S., Oates, J.E., Dell, A., Taylor, G.W., Dey, P.M., and Pridham, J.B.: Biochem. Biophys. Res. Commun. 132, 1-7 (1985).

17. Naik, S.: Ph.D. thesis. University of London (1984).

18. Hatton, M., März, L., and Regoeczi, E.: Trends Biochem. Sci. 8, 287-291 (1983).

19. Dey, P.M., Naik, S. and Pridham, J.B.: Phytochemistry, in press.

20. Goldstein, I.J., and Hayes, C.E.: Adv. Carbohydr. Chem. Biochem. 35, 127-340 (1978).

21. Dey, P.M., and Pridham, J.B.: Biochem. J. 113, 49-55 (1969).

22. Smith, E.E., Gunja-Smith, Z.H., and Goldstein, I.J.: Biochem. J. 107, 715-725 (1968).

23. Dey, P.M., Jones, B.M.G., Naik, S. and Pridham, J.B.: Phytochemistry 23, 2169-2171 (1984).

24. Scatchard, G.: Ann. N.Y. Acad. Sci. 51, 660-672 (1949).

25. Matsumoto, I., Jinbo, A., Kitagaki, H., Golovchenko-Matsumoto, A.M., and Seno, N.: J. Biochem. (Tokyo) 88, 1093-1096 (1980).

26. Dey, P.M.: Phytochemistry 23, 257-260 (1984).

27. Thomas, B., and Webb, J.A.: Can. J. Bot. 57, 1904-1911 (1979).

THE DERIVED AMINO ACID SEQUENCE OF THE SEED LECTIN PRESENT IN THE PINTO UI 111 CULTIVAR OF *PHASEOLUS VULGARIS* AND A COMPARISON WITH PHA-E AND PHA-L

Toni Voelker, Paul Staswick, Brian Tague, and Maarten J. Chrispeels

Department of Biology, University of California, San Diego, La Jolla, CA 92093

Phytohemagglutinin (PHA) is an abundant protein which accounts for 5-10% of the seed protein in most cultivars of the common bean, *Phaseolus vulgaris*. This lectin is a tetrameric glycoprotein with two distinct biological activities. Erythroagglutinating activity resides in the E subunits (Mr = 36000) while leucoagglutinating and mitogenic activities are associated with the L subunits (Mr = 34000) (for review, see ref. 1). The two subunits form 5 different oligomers with the compositions L_4, L_3E, L_2E_2, LE_3 and E_4 (2). The two polypeptides are encoded by two linked genes, dlecl and dlec2, whose nucleotide sequence has recently been determined (3). The gene dlecl encodes the E subunit and the gene dlec2 encodes the L subunit.

Based on the inability of seed extracts to agglutinate red blood cells, some bean cultivars have been classified as having very low levels or no PHA (4,5). A detailed analysis of one such variety, Pinto UI 111, shows that it contains small amounts of a protein which is closely related to PHA (6-9). This protein has immunological cross-reactivity with PHA (7-9), is similar in size to the L-subunit of PHA (7,9), has 2 oligosaccharide side chains on each polypeptide in the same way as PHA (9), and forms dimers (6). The protein has no erythroagglutinating activity, but is as potent a mitogen as PHA (9). The N-terminal amino acid sequence of this protein is similar to the N-terminal sequences of both E and L subunits of PHA (10).

The genetic basis for the low level of expression of the lectin genes in Pinto UI 111 is not yet understood. Brown et al. (5) postulated that the absence of lectin might be due to a lectin gene deletion, or a defective structural gene. Determinations of the levels of mRNA show that the cotyledons have very low levels of mRNA for PHA (11). Southern analysis of genomic DNA cut with several restriction endonucleases, and using cDNA for PHA as a probe, shows that PHA genes (or parts of genes) are present in Pinto UI 111 (Vitale and Chrispeels, unpublished).

We have determined the nucleotide sequence of one of these putative PHA genes and report here the derived amino acid sequence. The protein encoded by this gene is probably the same one which was found at very low levels in the Pinto UI 111 seeds (10).

Materials and Methods

The methods used for this work are all described in Maniatis et al. (12). The probes pSC1 and pSC2 represent cDNA clones for PHA as described in ref. 11. The genomic library of *Phaseolus vulgaris* cv. Pinto was prepared by Dr. J. Horowitz in the laboratory of Dr. R.B. Goldberg. Using pSC1 as a probe, we screened the genomic library and obtained a number of positive isolates. Several isolates were mapped with restriction enzymes and fragments subcloned into the *E. coli* plasmid pUC8. Promising fragments were subcloned into bacteriophage M13 and the nucleotide sequence determined. The hydropathy plot of the derived amino acid sequence was obtained with the computer program of Kyte and Doolittle (13).

Results and Discussion

Southern blot analysis of Pinto UI 111 genomic DNA cut with different restriction enzymes and probed with cDNA of PHA mRNA indicates that this cultivar contains nucleotide sequences which are highly homologous to PHA genes. The intensity of the hybridizing bands is similar to that obtained with genomic DNA from cultivars (e.g. Greensleeves) which have normal levels of PHA in their seeds (Vitale, unpublished results). The bands have different mobilities indicating alterations in the restriction enzyme sites.

Our first complete set of nucleotide sequence data showed a gene with an open reading frame coding for a polypeptide of 273 amino acids. It has a consensus TATA box and a polyA addition signal at the appropriate positions. The exact nucleotide sequence will be published when we have determined the nucleotide sequence of the second lectin gene in the Pinto UI 111 cultivar.

The predicted amino acid sequence coded for by the open reading frame of this gene (Figure 1) has greater than 85% homology to the sequences of PHA-E and PHA-L from the cultivar Tendergreen recently reported by Hoffman and Donaldson (3). The first 21 amino acids are identical to the signal peptide of PHA-E, and constitute a putative signal sequence with 8 hydrophobic amino acids in the mid-portion of the sequence.

A PHA-like protein present in small amounts in Pinto UI 111 seeds has been isolated and characterized, and its amino acid composi-

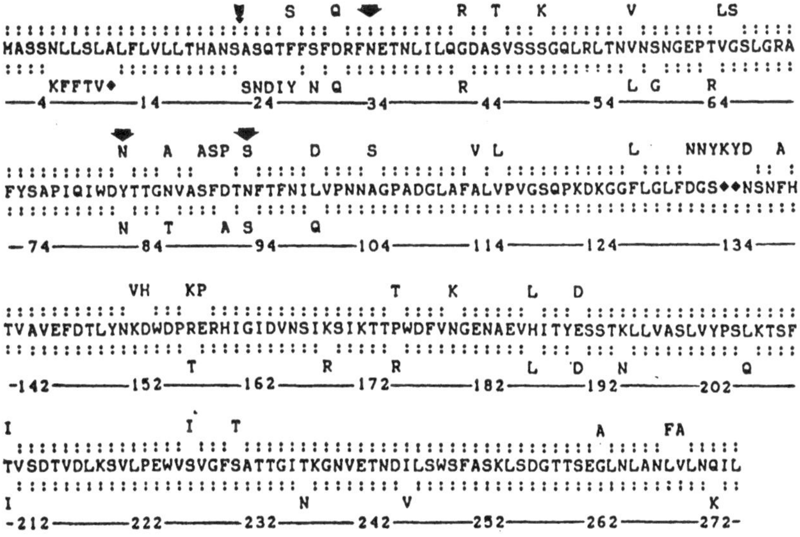

Figure 1. The derived amino acid sequence of the seed lectin of Phaseolus vulgaris cv. Pinto UI 111, and a comparison with PHA-E and PHA-L sequences from the cv. Tendergreen (3). Top line: PHA-E; middle line: Seed lectin from Pinto UI 111; bottom line: PHA-L; narrow arrow: cleavage site of the signal peptidase; peptidase broad arrows: glycosylation sequons.

```
1       5         10        15        20
A S Q T F F S F D R F N E T N L I L Q G D A S
| | | | | | | |   | | | | | | | | | | | | | |
A S Q T F F S F N R F N E T N L I L Q G D A S
```

Figure 2. A comparison of the N-terminal amino acid sequence derived from the nucleotide sequence (top line) and obtained by direct protein sequencing (bottom line) as reported by Strosberg et al., (Ref. 10), of the seed lectin from Pinto UI 111.

tion determined (6). A comparison of the published amino acid composition of the Pinto UI 111 lectin with the composition of our derived amino acid sequence (minus the putative signal peptide) shows a close correlation of the abundance of all amino acids, the highest aberration being 15%. Strosberg et al. (10) reported the amino acid sequence of the 23 N-terminal residues of this protein. A comparison of the amino acid sequences of the Pinto UI 111 lectin and our derived sequence (Figure 2) shows that both sequences are identical with the exception of residue 9. Our data predict an aspartic acid residue while the amino acid sequencing showed an asparagine residue in position 9. This difference could be due to an error in the amino acid sequencing or a single base substitution related to a difference in cultivars. We suggest that the open reading frame sequenced by us may be the gene which encodes the protein isolated by Pusztai et al. (6,7), reported by Strosberg et al. (10), and studied by Vitale et al. (9).

A detailed comparison of the amino acid sequence of the Pinto UI 111 lectin with PHA-E and PHA-L illustrates the evolutionary relationships between these proteins. The Pinto UI 111 lectin has one amino acid insertion (residue 10) and 33 amino acid substitutions relative to PHA-L, and two amino acid deletions (between residues 133 and 134) and 40 amino acid substitutions relative to PHA-E. A closer look at the differences shows that the Pinto UI 111 amino acid sequence closely follows the sequence of the PHA-E polypeptide up to residue 50, but this portion of the protein contains 2 clusters of differences with the PHA-L polypeptide (residues 5-10 and 22-26 in Figure 1). Beyond residue 50, there is greater homology between the Pinto UI 111 lectin and PHA-L, but a significant difference with PHA-E (residues

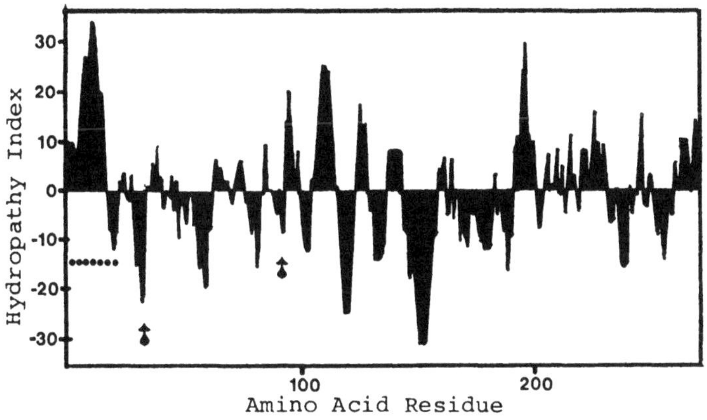

Figure 3. Hydropathy plot of PHA-L (Pinto UI 111). Hydrophobic areas are positive, hydrophilic areas are negative. The short line of dots shows the position of the putative signal sequence, and the vertical arrows show the positions of the glycosylation sequons.

131-134). Thus, the Pinto UI 111 lectin may be a hybrid molecule resembling PHA-E from residues 1-50 and PHA-L until the C-terminal end. Such a gene may have arisen by an unequal recombination between dlec2 and dlec1, the genes which code for PHA-L and PHA-E respectively, and are linked on the chromosome (3). Our data on the nucleotide sequences at the 3' and 5' ends of the gene give additional support for this interpretation.

Overall, the sequence of the mature lectin from Pinto UI 111 is much more similar to PHA-L than to PHA-E. This is in agreement with the findings that this lectin has a mobility on denaturing gels closest to PHA-L (6,9) shows very little erythroagglutinating (7,9) but strong mitogenic activity (9) similar to L_4 isolated from beans (2).

A hydropathy plot obtained with the computer program of Kyte and Doolittle (12) is shown in Figure 3. It shows first of all that the signal peptide is, as expected, a strongly hydrophobic domain (dotted line). There is no significant difference between the hydropathy plots of the 3 lectin sequences. The glycosylation sequons, indicated by vertical arrows, are both found in short hydrophilic domains. In this respect, the Pinto UI 111 lectin resembles the PHA-E and PHA-L polypeptides. The potential glycosylation sites for PHA-E are at residues 12, 82, and 101, for

PHA-L at residues 12 and 82, and for PHA-L (Pinto UI 111) at residues 12 and 92 (see Figure 1). These findings are consistent with the observations of Vitale et al. (9) that the Pinto UI 111 lectin is synthesized on polysomes attached to the rough endoplasmic reticulum, and co-translationally glycosylated with two oligosaccharide chains. It is not at all clear why this protein, which is so similar to PHA-E and PHA-L, should be a dimer rather than a tetramer (7).

Acknowledgements

Supported by grants from the United States Department of Agriculture and the National Science Foundation and by a fellowship to Toni Voelker from the Max Kade Foundation.

References

1. Goldstein, I.J. and Hayes, C.E.: Progress in carbohydrate chemistry and biochemistry *35*, 127-340 (1978).
2. Miller, J.B., Hsu, R., Heinrikson, R., and Yachnin, S.: Proc. Nat. Acad. Sci. (USA) *72*, 1388-1391 (1975).
3. Hoffman, L.M. and Donaldson, D.D.: EMBO Journal *4*, 883-889 (1985).
4. Brücher, O.: Proc. Tropical. Region. Amer. Soc. Hort. Sci. *12*, 68-85 (1968).
5. Brown, J.W.S., Osborn, T.C., Bliss, F.A., and Hall, T.C.: Theor. Appl. Genet. *62*, 263-271 (1982).
6. Pusztai, A., Grant, G., and Stewart, J.C.: Biochim. Biophys. Acta *671*, 146-154 (1981).
7. Pusztai, A., Grant, G., and Stewart, J.C.: In: Lectins-Biology, Biochemistry, Clinical Biochemistry, Vol. II, pp. 743-758, Ed. T.C. Bog-Hansen, Walter de Gruyter, Berlin-New York (1982).
8. Horowitz, J.: Mol. Gen. Genet. *198*, 482-485 (1985).
9. Vitale, A., Ceriotti, A., and Bollini, R.: Planta, In Press (1985).
10. Strosberg, A.D., Lauwereys, M., and Foriers, A.: In: Chemical Taxonomy, Molecular Biology and Function of Plant Lectins, Eds. Goldstein, I.J. and Etzler, M.E., pp. 7-20, Alan R. Liss, Inc. New York (1983).
11. Staswick, P. and Chrispeels, M.J.: J. Mol. Appl. Genet. *2*, 525-535 (1984).
12. Maniatis, T., Fritsch, E.F., and Sambrook, J.: Molecular Cloning, Cold Spring Harbor Laboratory, Cold Spring Harbor, New York (1982).
13. Kyte, J. and Doolittle, R.F.: J. Mol. Biol. *157*, 105-132 (1982).

TYROSINE RESIDUES AND THE BINDING-SITES OF LECTINS

N. Martin Young and Gail E.D. Jackson

Division of Biological Sciences, National Research Council of Canada, Ottawa, Canada K1A 0R6

Crystallographic studies of concanavalin A have shown that two tyrosine residues, Y12 and Y100, are key features of the sugar-binding site (1). Tyrosine residues have also been implicated in the sites of other legume lectins. For example, peanut agglutinin showed a UV difference spectrum indicative of tyrosine, on interaction with galactosides (2). When a group of lectins specific for D-galactose and/or N-acetyl-D-galactosamine was studied by CD difference spectroscopy (3), each gave a unique difference spectrum on sugar binding, most with features in the tyrosine region. Hence tyrosine residues often participate in sugar binding, but there is generally low structural homology among the binding-sites.

This is substantiated by the presently available amino acid sequence data. It has been proposed that four loops of peptide form the binding-site in concanavalin A (4). Comparison of the sequences of other lectins in these regions (Fig. 1) indicates that a) the four peptide loops are not well conserved among the lectins and b) these loops frequently contain tyrosine residues. The formation of the site by loops of peptide between β-sheet strands and the low site homology between lectins of similar specificity are features analogous to the six hypervariable

Lectins, Vol. V

	11	97	204	225
Con A	TYPNTDIGD	TGLYKET	S HPAD	STGRLL
SBA	TFRNS WD	TGLDIPG	TKRLAD	AGY L
Favin	TFYNAA WD	TGAEYAT	GYNVAD	GGY L
Sainfoin	TFSNR WD	TGDLVEQ	INRGGD	GGY L

Figure 1. Sequence comparison of con A (5) SBA (6) favin (7) and sainfoin (8) in the regions proposed by Carver et al. (4) to form the con A binding site. The site-forming peptides are indicated by arrows, and the various tyrosines are underlined. The tyrosine in the SBA sequence is Y107.

regions of immunoglobulins. It is important to determine whether the region proposed for the con A site is also the location of sugar-binding in other lectins. The frequent requirement for metal atoms (9) suggests that it is, but further evidence is desirable and may be provided by identifying site residues such as tyrosines.

To investigate tyrosine residues, we are developing methods based on the modifying reagent 1-fluoro-2,4-dinitrobenzene (FDNB). This can be used to modify histidine and tyrosine side-chains and coupled with high-performance liquid chromatography (HPLC) gives a selective mapping procedure for peptides with these residues. In the experiments described here [^{3}H]- and [^{14}C]FDNB were used to look for changes in the chemical properties of tyrosines that might accompany sugar binding. The systems studied were con A with methyl α-D-glucoside or with mixed ovalbumin glycopeptides and soybean agglutinin (SBA) with N-acetyl-D-galactosamine. In the peptide segments of SBA homologous to the site-forming ones of con A (Fig. 1) only one tyrosine occurs, Y107, and it is in a different segment from the tyrosines of con A. Yet relatively large CD changes indicative of tyrosine occur on interaction with carbohydrate (3, 10). SBA should be a good test of the site proposal.

Materials and Methods

Con A was purified on Sephadex G75 (11) and SBA on a p-aminophenyl-N-acetyl-D-galactosamine derivative of AffiGel 10 (BioRad Laboratories) as previously described (4). Ovalbumin glycopeptides were mixtures prepared by pronase digestion and gel-filtration on BioGel P-2. [^{3}H]- and [^{14}C]FDNB were obtained from Amersham Corporation.

To identify peptides containing histidine or tyrosine residues, samples of con A and SBA (5 mg) were treated with maleic anhydride to modify amino groups followed by FDNB to modify histidine and tyrosine side-chains. The modified proteins were digested with thermolysin and the resulting peptide mixtures were separated by HPLC on 5 μm Spherisorb ODS-2 columns, using a 0.01 M Na phosphate:acetonitrile gradient system and a Beckman model 165 dual wavelength detector. The selected peptides were recycled in a 0.01 M ammonium acetate pH 5.3:acetonitrile gradient system, and amino acid compositions were determined after hydrolysis in 6 N HCl, 20 hr at 110°.

The effect of sugar-binding on the reactivities of binding-site tyrosines was studied in three systems:- con A plus methyl α-D-glucoside (10 mM), con A plus ovalbumin glycopeptides (5:1 by weight) and SBA plus N-acetyl-D-galactosamine (10 mM). For each system, aliquots of native lectin (2.5 mg in 1 ml of N-methyl morpholine buffer, pH 7.6, containing 375 nmoles of Gln-Gly, as an internal standard) were reacted with [^{14}C]FDNB (5 μCi) in the dark at room temperature protected from light. To further duplicate aliquots were added the appropriate sugars, and these mixtures were reacted with [^{3}H]FDNB (5 μCi) as above. Each [^{3}H] sample was then combined with an appropriate [^{14}C] sample. The radiolabelled internal standards were partially removed by a brief dialysis against water (20 ml) and purified by HPLC. They were then hydrolysed to give Dnp-glutamic acid which was purified by HPLC as described previously (12) and the ^{3}H/^{14}C ratios were determined by liquid scintillation counting. The radiolabelled lectins were then treated as described above for the native proteins, and the resulting peptides were separated by HPLC. The desired peptides were hydrolysed to give O-Dnp-tyrosine and imidazolyl-Dnp-histidine which were also purified by HPLC (12) and their ^{3}H/^{14}C ratios were determined.

Results

The peptide maps obtained for Dnp-labelled con A and SBA are shown in Figs. 2 and 3. Detection at two wavelengths allows peptides containing O-Dnp-tyrosine (A320/A270 = 1.1) to be distinguished from peptides containing imidazolyl-Dnp-histidine (A320/A270 = 2.0). The major tyrosine peptides from the two lectins could be

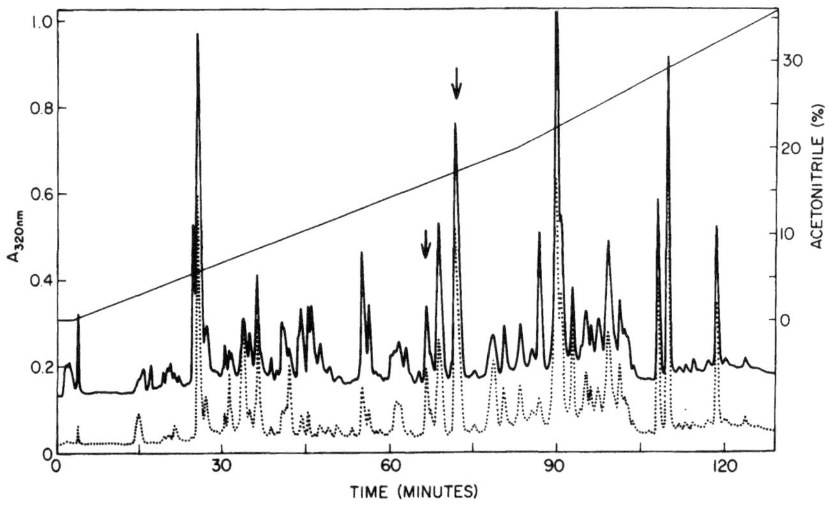

Figure 2. Peptide map of maleyl-Dnp-con A, digested with thermolysin. The peptides marked contained residues Y12 and Y100. Upper line, A320 nm, lower line, A270 nm.

placed in the amino acid sequences (5,6) from their compositions. The peptides with the residues of interest, Y12 and Y100 for con A and Y107 for SBA, are indicated.

The effect of sugar binding on the reactivities of these residues was assessed by a double-labelling procedure in which [^{14}C]FDNB was used to trace label the native lectin, and [^{3}H]FDNB to trace label lectin-carbohydrate mixtures. The samples were combined then subjected to the same modification and HPLC procedures used for the mapping above. The ratios of the two isotopes in the tyrosine peptides compared to the ratio of the internal standard should indicate whether carbohydrate binding causes any significant change in the chemical properties of the tyrosine side-chains. The results obtained for con A with methyl α-D-glucoside or ovalbumin glycopeptides and SBA with N-acetyl-D-galactosamine are summarised in Table 1. There was no significant change in the

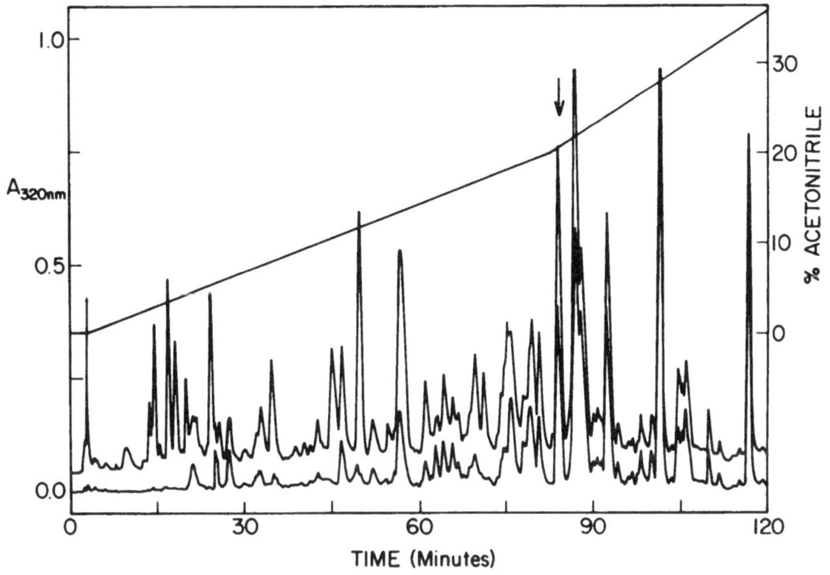

Figure 3. Peptide map of maleyl-Dnp-SBA, digested with thermolysin. The peptide marked contained residue Y107. Upper line, A320 nm, lower line, A270 nm.

TABLE 1

Chemical reactivity ratios for tyrosine residues.

Residue	Ligand	$^{3}H/^{14}C$ residue / $^{3}H/^{14}C$ internal standard
Con A Y12	methyl α-D-	1.00
Con A Y100	glucoside	1.09
Con A Yavg[a]		1.10
Con A Y12	ovalbumin	0.74
Con A Y100	glycopeptides	0.73
Con A Yavg[a]		0.70
SBA Y107	N-acetyl-D-	0.92
SBA Yavg[b]	galactosamine	0.96

[a] Average of Y22, Y54, Y67, Y77 and Y176

[b] Average of Y53, Y79, Y158 and Y168

chemical properties of the key tyrosine residues.

Discussion

Modification of proteins with FDNB coupled with separation by HPLC allows selective peptide mapping for histidine and tyrosine residues as exemplified by the maps in Figs. 2 and 3. A fuller account of this procedure and its application to study the chemical properties of individual histidine and tyrosine residues in con A will be given elsewhere (G.E.D. Jackson and N.M. Young, submitted for publication). Two particular merits are the different spectral properties of the side-chain derivatives permit assignment of a labelled peptide to histidine or tyrosine, and the hydrophobicity of the Dnp group leads to later elution of the labelled peptides.

The use of a double trace labelling procedure to study tyrosine residues was originally developed with radio-iodine reagents several years ago and was applied in studies of the exposure of tyrosine residues in a Bence-Jones protein (13). In addition to the chromatographic merits noted above, dinitrophenylation has advantages over iodination in that the isotopes are more stable and only one unique derivative is formed with tyrosine.

The two tyrosine residues that are at the carbohydrate-binding site of con A, Y12 and Y100, have significantly lower pKs and reactivities than the other tyrosines (G.E.D. Jackson and N.M. Young, submitted for publication). This may be due to charge transfer effects between the two residues. The data obtained in the present experiments show that the reactivities are not significantly changed upon binding either methyl α-D-glucoside or ovalbumin glycopeptides. CD studies have shown that the binding-site tyrosines have different conformations when binding the simple monosaccharide compared to the more complex glycopeptide (14) yet the chemical properties of the tyrosines in these or the native conformation do not differ. This suggests that the movements of residues in the site are relatively small on binding, which would validate experiments conducted by molecular graphics involving "docking" glycopeptides to the native con A structure (4).

For SBA, it was hoped that the postulated participation of Y107 in the site could be confirmed by the double-labelling approach, but no change in Y107 properties could be seen. No other tyrosine appears to occur at all near the site region on the basis of a rough structural model for SBA derived by superimposing its amino acid sequence on the backbone structure of con A. Also, the other tyrosines did not change in properties upon sugar binding. Hence, if the site is located in the postulated region, the environment of the residue when interacting with a sugar molecule is not significantly different to that in the native protein, as was the case with the con A tyrosines. Alternatively, the spectroscopic changes might arise indirectly from conformational changes caused by interactions of neighbouring residues with the carbohydrate.

Acknowledgements

We thank Dr. Harvey Kaplan for helpful discussions, Dr. Makato Yaguchi for amino acid analyses and Carol Ward for technical assistance.

References

1. Hardman, K.D.: Carbohydrate-Protein Interaction (Goldstein, I.J., ed.) pp. 12-26, American Chemical Society, Washington, D.C. (1979).

2. Neurohr, K.J., Young, N.M., and Mantsch, H.H.: J. Biol. Chem. 255, 9205-9209 (1980).

3. Young, N.M., and Williams, R.E.: Can. J. Biochem. Cell Biol. 63, 268-271 (1985).

4. Carver, J.P., Mackenzie, A.E., and Hardman, K.D.: Biopolymers 24, 49-63 (1985).

5. Cunningham, B.A., Wang, J.L., Waxdal, M.J., and Edelman, G.M.: J. Biol. Chem. 250, 1503-1512 (1975).

6. Vodkin, L.O., Rhodes, P.R., and Goldberg, R.B.: Cell 34, 1023-1031 (1983).

7. Hopp, T.P., Hemperly, J.J., and Cunningham, B.A.: J. Biol. Chem. 257, 4473-4483 (1982).

8. Kouchalakos, R.M., Bates, O.J., Bradshaw, R.A., and Hapner, K.D.: Biochemistry 23, 1824-1830 (1984).

9. Lonnerdal, B., Borrebaeck, C.A.K., Etzler, J.E., and Ersson, B.: Biochem. Biophys. Res. Comm. 115, 1069-1074 (1983).

10. Thomas, M.W., Rudzki, J.E., Walberg, E.F., and Jirgensons, B.: Carbohydrate-Protein Interaction (Goldstein, I.J., ed.) pp. 67-75, American Chemical Society, Washington, D.C. (1979).

11. Agrawal, B.B.L., and Goldstein, I.J.: Biochim. Biophys. Acta 147, 262-271 (1967).

12. Cockle, S.A., Kaplan, H., Hefford, M.A., and Young, N.M.: Analyt. Biochem. 125, 210-216 (1982).

13. Seon, B.-K., Roholt, D.A., and Pressman, D.: Biochim. Biophys. Acta 200, 81-88 (1970).

14. Young, N.M., Neurohr, K.J., and Williams, R.E.: Biochim. Biophys. Acta 701, 142-145 (1982).

MOLECULAR STRUCTURE OF *LATHYRUS* LECTINS AND ISOLECTINS

Pierre Rougé [1], Michael Richardson [2], Christian Chatelain [1], Alan Yarwood [2], Benildo Sousa-Cavada [3] and Daniel Père [1]

(1) Laboratoire de Biologie cellulaire, Faculté des Sciences Pharmaceutiques, Université Paul Sabatier, 35 chemin des Maraîchers, 31062 Toulouse, France

(2) Department of Botany, University of Durham, Durham DH1 3LE, England

(3) Departamento de Bioquimica, Universidade Federal de Pelotas, Pelotas RS, Brasil

Like other *Vicieae* lectins, lectins from the genus *Lathyrus* are tetramers built up of two light (α) and two heavy (β) subunits with respective MW of 6 000 and 20 000 daltons (1). Thus, the *Lathyrus* lectins so far studied, including lectins from *L. odoratus* (2,3), *L. silvestris* (2), *L. sativus* (4), *L. tingitanus* (5), *L. ochrus* (6), *L. cicera* (7), *L. aphaca* and *L. articulatus* (8) posses such a tetrameric structure, although Gupta et al. (9) have found a *L. sativus* lectin composed of two identical subunits with a MW of 21 000 daltons. In addition, many *Lathyrus* species contain isolectins (5-8), like the lentil (10) and pea (11,12) lectins which also belong to the same tribe of *Vicieae*. It is of particular importance to study and compare these isolectins in order to obtain a better understanding of the lectin biosynthesis and its regulation in *Vicieae* seeds. Previous studies (13,14) have shown that in pea, a lectin precursor is synthsized prior to the lectin which subsequently accumulates following a limited proteolysis of the precursor (15). Other very similar examples have been described in *Vicia faba* (16) and *Phaseolus vulgaris* (17) seeds. Moreover, a better knowledge of the isolectin molecular structures could give an insight into the changes that have occurred at the gene level during the speciation process within the genus *Lathyrus*. In this respect, comparisons of the amino acid sequences of the legume lectins have shown (18-21) that a high degree of homology exists among these very conserved proteins. This is especially true for the *Lathyrus* lectins (22) whose subunit amino acid sequences appear to be particularly well conserved. We present here some preliminary results concerning the molecular structure of various *Lathyrus* lectins and isolectins and suggestions about the possible origin of such a lectin heterogeneity, associated with their biosynthetic pathway within the plant.

Lectins, Vol. V

MATERIALS AND METHODS

1. Plant material: Seeds from *Lathyrus aphaca* L., *L. articulatus* L., *L. cicera* L., *L. hirsutus* L. and *L. ochrus* (L.) DC., were harvested from plants cultivated under field conditions.

2. Isolation of lectins and isolectins: As previously described (5), lectins were isolated from crude seed meal extracts with 50 mM Tris, 0.15M NaCl (pH 7.6) followed by fractional precipitation between 30-60% ammonium sulfate and subsequent affinity chromatography on a Sephadex G-100 column equilibrated with the same buffer. The retained lectins were eluted by adding 0.1M glucose to the eluting buffer, precipitated with 90% ammonium sulfate and extensively dialysed against Tris buffered saline. They were stored at -30°C until used.
Isolectins from *L. cicera* and *L. ochrus*, were separated by chromatofocusing on a PBE 94 (Pharmacia) column equilibrated with 25 mM Tris-acetic acid (pH 8.4) and elution was performed with a 10-fold dilution mixture of Polybuffers 96 and 74 (Pharmacia) (30/70, v/v) adjusted to pH 5.0 with 1.0M acetic acid. The separation of the two *L. hirsutus* isolectins was achieved under similar conditions but with a pH 11.0 starting buffer (25 mM triethylamine-HCl) and a pH 7.0 eluent (45-fold diluted Pharmalyte pH 10.5-8.0 adjusted to pH 7.0 with HCl). The isolated isolectins were precipitated with 90% ammonium sulfate, extensively dialysed against distilled water, and stored frozen at -30°C until used.

3. Isolation of light and heavy subunits: The light and heavy subunits from both the isolated lectins and isolectins were separated by gel filtration on a Bio-Gel P60 (Bio-Rad) column equilibrated with 6.0M guanidine-HCl. Two peaks corresponding respectively to the heavy (1st peak) and the light (2nd peak) subunits were collected. Guanidine-HCl was eliminated by extensive dialysis against distilled water and the precipitated subunits were lyophilised and stored frozen at -30°C.

4. Electrophoresis and isoelectrofocusing: Polyacrylamide gel electrophoresis of lectins was performed either at basic (23) or acidic (24) pH.
The lectin and isolectin subunits were separated by sodium dodecyl sulfate-polyacrylamide gel electrophoresis (SDS-PAGE) on gradient polyacrylamide (4-30% acrylamide) gel slabs using a 40 mM Tris-acetate (pH 7.4) buffer containing 0.1% ethylene diamine tetraacetic acid disodium salt and 0.2% SDS, for electrophoresis. Electrophoresis of samples, previously made 1% SDS and 5% mercaptoethanol and heated at 100°C for 10 min, was conducted at 150 V until the tracking dye (bromophenol blue) had migrated to the bottom of the gel slabs.
Analytical isoelectrofocusing was performed with LKB Ampholine PAGE-plates in the pH range between 9.5 and 3.5 and isoelectrofocusing in the presence of 8.0M urea was carried out over pH 8.0-5.0 according to (25).
All of these techniques were carried out in order to verify the purity of our lectin, isolectin and subunit preparations.

5. Chemical analyses: Samples for amino acid analysis were hydrolysed in twice glass-distilled 6.0M HCl at 110°C for 18 h and analyses were performed on a Beckman model 119 BL amino acid analyzer.
Protein was estimated colorimetrically according to Goa (26), using bovine serum albumin (Sigma) as standard.

6. Sequence determination: Samples of the light subunits were digested separately with various proteases (trypsin, chymotrypsin and *Staphylococcus aureus* V8 protease) as previously described (27). The resulting mixtures of peptides were purified by reverse-phase high performance liquid chromatography (HPLC) on a Micropac MCH-10 column (30 x 0.4 cm, Varian) in a Varian model 5 000 HPLC appa-

ratus, using a linear gradient of 0-70% acetonitrile in 0.1% trifluoroacetic acid. Peptides were detected by measuring the absorbance at 214 nm.
Similarly, samples of the heavy subunits were digested with trypsin, chymotrypsin, pepsin and *S. aureus* V8 protease and the resulting mixtures of peptides were fractionated on a Bio-Gel P6 (Bio-Rad) column (1 x 200 cm) in 50 mM ammonium bicarbonate pH 8.1. Further purification of peak fractions was achieved by reverse-phase HPLC as previously described for light subunits. In addition, samples of the heavy subunits were cleaved at Trp-X peptide bonds by iodosobenzoic acid in 80% acetic acid in 4.0M guanidine-HCl for 21 h at 20°C as reported in (28). The resulting peptides were fractionated on a Bio-Gel P30 (Bio-Rad) column in 70% formic acid and subsequently purified by reverse-phase HPLC on a Vydac C18 column (0.5 x 25 cm, Technicol. Ltd, Stockport) using a linear gradient of 0-70% acetonitrile in 0.1% trifluoroacetic acid.
The intact light and heavy subunits and their derived peptides were subjected to micro-sequence analysis using the 4-N,N-dimethylaminobenzene-4'-isothiocyanate (DABITC)/phenylisothiocyanate (PITC) double coupling method of Chang et al. (29), together with the dansyl-Edman procedure and digestion with carboxypeptidase A, all as in (28). The presence of tryptophan in peptides was detected by staining on paper with p-dimethylaminobenzaldehyde.

RESULTS

The *Lathyrus aphaca* and *L. articulatus* seed lectins gave a single protein band by PAGE, either at basic or acidic pH. By isoelectrofocusing in the presence of 8.0M urea, both lectins exhibited two distinct subunits (α and β) whose amino acid compositions (Table 1) and amino acid sequences (Fig. 1 and 2) have been determined. All these results confirm the presence in both species of a single lectin having an $(\alpha\beta)_2$ molecular structure.
A quite different situation was observed with the *L. cicera* lectin which could be resolved into three distinct protein fractions by PAGE at basic pH, although a single broad diffuse band was obtained by PAGE at acidic pH. The lectin showed two distinct α subunits but a single β subunit by isoelectrofocusing under denaturating conditions. Two peaks were obtained by chromatofocusing over pH 8.4-5.0, corresponding respectively to two distinct isolectins whose subunit amino acid compositions (Table 1) clearly showed that these two isolectins possess different α subunits whereas their β subunits are identical. Amino acid sequence determinations gave confirmation to such a composition (Fig. 1 and 2). Thus, the *L. cicera* lectin appears to be composed of two isolectins having respectively the $(\alpha 1\beta)_2$ and $(\alpha 2\beta)_2$ molecular structures. The additional protein fraction observed by PAGE at basic pH could presumably correspond to a hybrid combination exhibiting an $\alpha 1\alpha 2(\beta)_2$ molecular structure, even though it remains difficult to predict if such an isolectin was readily present in *L. cicera* seeds or occurred as an artifact of the extraction or separation procedures. A very similar finding of a heterogeneous light subunit composition has been pre-

Table 1

Amino acid composition of light (α,α1,α2) and heavy (β,β1,β2) subunits from L. aphaca, L. articulatus, L. cicera, L. hirsutus and L. ochrus lectins and isolectins, expressed as residue/100 residues (values for Cys, ranging from 0.1 to 0.5, have not been included)

Amino	*L. aphaca*		*L. articulatus*		*L. cicera*				*L. hirsutus*				*L. ochrus*			
acids	α	β	α	β	α1	α2	β1	β2	α1	α2	β1	β2	α1	α2	β1	β2
Asp	5.4	15.9	4.3	15.4	4.7	5.4	16.4	15.9	5.4	5.1	13.0	14.5	4.0	5.5	13.4	13.0
Thr	9.7	11.4	9.7	11.8	8.9	8.8	11.4	11.9	8.9	9.6	10.1	11.6	9.3	9.6	12.2	12.9
Ser	13.8	9.0	13.9	7.5	13.1	14.0	7.5	7.2	11.6	11.8	10.4	8.6	10.3	13.3	8.1	7.7
Glu	12.0	5.8	11.7	5.7	12.4	13.0	6.3	6.8	11.0	11.0	8.7	6.9	9.3	11.0	4.5	4.0
Pro	1.9	2.6	3.8	3.1	4.9	4.2	3.8	3.8	4.4	4.1	4.6	4.4	4.9	4.0	3.9	4.7
Gly	8.5	6.2	6.5	7.3	6.5	7.1	6.8	6.7	11.2	9.4	10.6	8.3	7.4	4.4	7.7	8.4
Ala	12.6	7.0	12.3	7.1	9.5	10.4	7.4	7.6	11.5	11.3	8.9	8.4	11.4	8.3	8.0	7.3
Val	12.4	8.2	8.8	8.4	11.8	12.2	9.3	9.4	11.0	12.3	7.3	8.2	11.0	14.0	8.9	9.2
Ile	0.2	6.3	2.2	6.3	2.2	2.1	5.8	5.6	2.1	2.2	4.6	5.6	2.5	2.1	6.2	6.2
Leu	8.5	5.8	8.4	5.4	8.0	7.7	6.0	5.8	6.8	7.6	4.5	5.0	7.2	8.4	5.2	4.9
Tyr	1.9	3.7	2.1	3.5	1.7	1.4	3.6	3.6	1.2	1.6	2.6	3.2	1.4	2.4	3.1	2.9
Phe	6.3	8.5	8.2	8.0	6.0	5.7	7.3	7.4	5.0	5.5	5.9	6.7	10.7	9.1	8.0	7.7
Lys	2.6	7.8	2.7	6.6	4.1	2.3	5.6	5.4	4.3	2.7	5.9	6.4	3.2	2.6	6.7	6.6
His	1.9	1.2	3.5	1.4	4.1	3.8	1.2	1.3	3.7	3.9	1.5	1.2	3.9	2.1	1.1	1.5
Arg	1.9	1.1	1.9	2.4	2.1	1.9	1.6	1.6	2.2	2.0	1.5	1.3	2.3	1.9	2.1	2.2

```
                          10         20         30         40         50
L. aphaca        : ATSYTLNEV VALKDVVPEW VRVGFSATTG AEFAAHEVLS WSFQSELSGT SGSN
L. articulatus   : --------- -P-EF----- --I------- ---------- ---H---A-- -S--
                                                                           K
L. cicera α1     : V-------- -P-------- --I------- ---------- ---H---GE- -A-VQ
L. cicera α2     : V-------- -P-------- --I------- ---------- ---H---GE- -A-Q
L. hirsutus α1   : V-------- -P-------- --I------- ---------- ---H---G-- -A-KQ
L. hirsutus α2   : V-------- -P-------- --I------- ---------- ---H---G-- -A-Q
L. ochrus α1     : E-------- -P-EF----- --I------- ---------- --FH---A-- -S--
L. ochrus α2     : E-------- -P-EF----- --I------- ---------- --YN----V- -S--
```

Figure 1. Amino acid sequences of light subunits (α, α1, α2) from L. aphaca, L. articulatus, L. cicera, L. hirsutus and L. ochrus lectins and isolectins (data in part from Ref. 30,31)

viously reported for the pea lectin by Entlicher et al. (32), that was apparently related to pH changes depending upon the selected electrophoretic running buffers.

Two distinct isolectins were obtained by PAGE at basic pH for the *L. hirsutus* and *L. ochrus* lectins and, in addition, these lectins showed two α and two β subunits when subjected to isoelectrofocusing in the presence of 8.0M urea. Amino acid compositions (Table 1) of the α and β subunits of the isolectins previously separated by chromatofocusing, clearly indicated that these isolectins are built up of different light and heavy subunits. This was subsequently corroborated by the determination of their amino acid sequences (Fig. 1 and 2). The two *L. hirsutus* and *L. ochrus* isolectins obviously correspond to the $(\alpha 1 \beta 1)_2$ and $(\alpha 2 \beta 2)_2$ molecular structures. In both cases, no hybrid combination, resulting from the possible association of different light and/or heavy subunits, could be detected.

The *Lathyrus tingitanus* lectin corresponds to a quite different and more complex situation since more than three distinct protein fractions (three major and two minor bands) could be detected by PAGE at basic pH. In addition, chromatofocusing gave different results depending on how the lectin was stored following its isolation. Thus, chromatofocusing over pH 7.4-4.0 of samples of freshly isolated *L. tingitanus* lectin gave three protein peaks, the first of which (pI 6.3) was predominant whereas, under similar conditions, samples of the lectin stored for a few months at -30°C showed higher amounts of the more acidic protein peak (pI 5.5) associated with a concomitant decrease of the 1st peak. These changes could be explained on the basis of a hybridization process occurring between different but related subunits during the storage of lectin solutions even at -30°C.

```
                                10         20         30         40         50
(1) L. aphaca     : TETTSFLIT KFSPDQQNLI FQGDGYTTSE KLLLTKAVKN TVGRALYSSP IHIWD
(2) L. cicera β1  : --------- -G-------- --------K- R-T-----R- ---------- -----
(3) L. cicera β2  : --------- -G-------- --------K- R-T-----R- ---------- -----
(4) L. ochrus β1  : ------S-- -G-------- --------K- R-T-----R- ---------- -----
(5) L. ochrus β2  : ------S-- ------P--- --------K- R-T-----R- ---------- -----

         60         70         80         90        100        110
                                      F
(1) : SKTGN VANFITSFTF VINAPNSYNV ADGFTFYIAP VDTKPQTGGG YLGVFNSKDY DKTSKTVAVE
(2) : ----- ---V------ -D-------- ------F--- ---------- -------V-- ----Q-----
                        N
(3) : ----- ---V------ -D-------- ------F--- ---------- -------V-- ----Q-----
                        N
(4) : ----- ---V------ -D-------- ------F--- ---------- ---------- ----Q-----
(5) : ----- ---V-A---- -D-------- ------F--- ---------- ---------- ----Q-----

           120        130        140        150        160        170   180
(1) : FDTFYNAAWD PSNKDRHIGI DVNSIKSVNT KSWKLQNGKE ANVVIAFNAE TNVLTVSLTY SN
(2) : ---------- --RE------ -------I-- --------V- --------A- ---------- P-
(3) : ---------- --RE------ -------I-- --------V- --------A- ---------- P-
(4) : -----T---- --G------- -------I-- ---------- --------A- ---------- P-
(5) : -----T---- --G------- -------I-- ---------- -------GA- ---------- P-
```

Figure 2. Amino acid sequences of heavy subunits (β, β1, β2) from L. aphaca, L. cicera and L. ochrus lectins and isolectins (data in part from Ref. 33,34)

DISCUSSION

According to their lectin, isolectin and subunit composition, the *Lathyrus* species so far studied can be divided into three distinct groups :

- species from the first group (e.g. *L. aphaca* or *L. articulatus*) contain only one lectin made of two identical light and heavy subunits. Such lectins have an $(\alpha\beta)_2$ molecular structure.
- species belonging to the second group possess two different isolectins built up of similar or distinct light and/or heavy subunits. Thus, *L. cicera* contains two isolectins $(\alpha 1\beta)_2$ and $(\alpha 2\beta)_2$ which only differ in their α subunits while *L. ochrus* exhibits two isolectins $(\alpha 1\beta 1)_2$ and $(\alpha 2\beta 2)_2$ having different light and heavy subunits. Isolectins with similar α subunits but distinct β subunits have not yet been found, although the presence in some molecules of favin (the fava bean (*Vicia faba* L.) lectin) of a non-glycosylated β' heavy subunit, slightly shorter than the β one, could correspond to such a situation (35). As in the case of the *L. cicera* isolectins, the two pea isolectins also differ

in their light subunits (12,32,36).

- species from the third group correspond to a quite different and more complex situation since they have more than two isolectins for which an *in vitro* hybridization could occur between their light and heavy subunits. *Lathyrus tingitanus* offers an interesting example of such a situation, but other species, e.g. *L. luteus* and perhaps *L. odoratus*, could exhibit a very similar lectin behaviour. However, such a group has to be considered with caution since the amino acid sequences of the *L. tingitanus* isolectin subunits have not been determined. On the other hand, examples of subunit exchanges between lectins from different cereal species occurring either *in vitro* or *in vivo* conditions, have been already reported (37-39). It seems therefore reasonable to consider that isolectins from *L. tingitanus* could result at least in part by such a subunit exchange that had occurred during the lectin storage.

It has been clearly demonstrated that the biosynthesis of various legume lectins (40) starts with a lectin precursor, which undergoes a co-translational removal of a signal peptide when sequestered into the endoplasmic reticulum, associated with a co-translational glycosylation for those lectins that are glycoproteins. For two-chain lectins like the pea (13) and fava bean (16) lectins, the resulting oligomers are subsequently post-translationally cleaved to yield the light and heavy subunits of the mature lectin. When different isolectins occur simultaneously within the same seed, the question arises whether they are derived from a single or different lectin precursor(s). Slightly different α or β subunits could result from a further post-translational cleavage occurring at the C-terminal end of some of the α and β subunits derived from a single precursor. The slightly different $\alpha 1$ and $\alpha 2$ chains from *L. cicera* (see Fig. 1), together with the β and β' chains from favin (35), which all differ at their very C-termini, could have such an origin. In the case of *L. ochrus* however, whose $\alpha 1$ and $\beta 1$ subunit sequences differ from the $\alpha 2$ and $\beta 2$ ones in such a way (see Fig. 1 and 2) that a C-terminal cleavage could not explain the observed differences, two distinct precursors are likely to occur. In addition, their occurrence could satisfactorily explain the fact that combinations other than $(\alpha 1 \beta 1)_2$ and $(\alpha 2 \beta 2)_2$ could not exist in this species. Preliminary results (C. Boisseau and P. Rougé, unpublished) indicate that two lectin precursors could be isolated from immature seeds. Whatever the origin of such a lectin heterogeneity, it involves well regulated processing steps since all the above mentioned *Lathyrus* isolectins always occur in well defined proportions (e.g. 1/3 and 2/3 for the *L. ochrus* isolectins I and II, respectively (6)). In this respect, we need more work on lectin heterogeneity at their subunit or gene levels, in order to get a clear picture of the regulation processes involved in lectin/isolectin biosynthesis.

Another interesting point concerns the mechanism by which lectin changes have occurred throughout evolution, during the speciation process. From the comparison of the light and heavy subunit amino acid sequences of various *Lathyrus* species (Fig. 1 and 2), it appears that the simplest type of mutations, i.e. point mutations, are sufficient to explain the main observed differences. These observations are in full agreement with the previously noted close phylogenetical relatedness among the *Vicieae* and Leguminosae lectins (18-21).

REFERENCES

1. Rougé, P.: in Chemical Taxonomy, Molecular Biology, and Function of Plant Lectins, (Goldstein, I.J. and Etzler, M.E. ed.), Alan R. Liss Inc., New York, 265-267 (1983).
2. Ticha, M., Zeineddine, I., and Kocourek, J.: Acta Biol. Med. Ger. 39, 649-655 (1980).
3. Kolberg, J., Michaelsen, T.E., and Sletten, K.: FEBS Lett. 117, 281-283 (1980).
4. Kolberg, J., and Sletten, K.: Biochim. Biophys. Acta 704, 26-30 (1982).
5. Rougé, P., and Chabert, P.: FEBS Lett. 157, 257-260 (1983).
6. Rougé, P., and Sousa-Cavada, B.: Plant Sci. Lett. 37, 21-27 (1984).
7. Sousa-Cavada, B., and Rougé, P.: Arch. Biol. Technol. 28, 421-430 (1985).
8. Rougé, P.: Biochem. Syst. Ecol. 12, 47-51 (1984).
9. Gupta, B.K.D., Chatterjee-Ghose, R., and Sen, A.: Arch. Biochem. Biophys. 201, 137-146 (1980).
10. Howard, I.K., Sage, H.J., Stein, M.D., Young, N.M., Leon, M.A., and Dyckes, D.F.: J. Biol. Chem. 246, 1590-1595 (1971).
11. Marik, T., Entlicher, G., and Kocourek, J.: Biochim. Biophys. Acta 336, 53-61 (1974).
12. Trowbridge, I.S.: J. Biol. Chem. 249, 6004-6012 (1974).
13. Higgins, T.J.V., Chrispeels, M.J., Chandler, P.M., and Spencer, D.: J. Biol. Chem. 258, 9550-9553 (1983).
14. Higgins, T.J.V., Chandler, P.M., Zurawski, G., Button, S.C., and Spencer, D.: J. Biol. Chem. 258, 9544-9549 (1983).
15. Lauwereys, M., Van Driessche, E., Strosberg, A., Dejaegere, R., and Kanarek, L.: in Lectins, Biology, Biochemistry, Clinical Biochemistry, (Bøg-Hansen, T.C., and Spengler, G.A. ed.), De Gruyter, Berlin, 3, 603-610 (1983).
16. Hemperly, J.J., Mostov, K.E., and Cunningham, B.A.: J. Biol. Chem. 257, 7903-7909 (1982).
17. Hoffman, L., Ma, Y., and Berker, R.F.: Nucl. Acid Res. 10, 7819-7828 (1982).
18. Foriers, A., De Neve, R., and Strosberg, A.D.: Physiol. Vég. 17, 597-606 (1979).

19. Cunningham, B.A., Hemperly, J.J., Hopp, T.P., and Edelman, G.M.: Proc. Natl. Acad. Sci. U.S.A. 76, 3218-3222 (1979).

20. Strosberg, A.D., Lauwereys, L., and Foriers, A.: in Chemical Taxonomy, Molecular Biology, and Function of Plant Lectins, (Goldstein, I.J., and Etzler, M.E. ed.), Alan R. Liss Inc., New York, 7-20 (1983).

21. Becker, J.W., Cunningham, B.A., and Hemperly, J.J.: in Chemical Taxonomy, Molecular Biolohy, and Function of Plant Lectins, (Goldstein, I.J., and Etzler, M.E. ed.), Alan R. Liss Inc., New York, 31-45 (1983).

22. Rougé, P., Richardson, M., Yarwood, A., Ranfaing, P., Sousa-Cavada, B., and Père, D. (in preparation).

23. Davis, B.J.: Ann. N. Y. Acad. Sci. 121, 404-427 (1964).

24. Reisfeld, R.A., Lewis, U.J., and Williams, D.E.: Nature 195, 281-283 (1962).

25. Wrigley, C.: Sci. Tools 15, 17-23 (1968).

26. Goa, J.: Scand. J. Clin. Lab. Invest. 5, 218-222 (1953).

27. Campos, F.D.A.P., and Richardson, M.: FEBS Lett. 152, 300-304 (1983).

28. Richardson, M., Campos, F.D.A.P., Moreira, R.A., Ainouz, I.L., Begbie, R., Watt, W.B., and Pusztai, A.: Eur. J. Biochem. 144, 101-111 (1984).

29. Chang, T.Y., Brauer, D., and Wittman-Liebold, B.: FEBS Lett. 93, 205-214, (1978).

30. Richardson, M., Rougé, P., Sousa-Cavada, B., and Yarwood, A.: FEBS Lett. 175, 76-81 (1984).

31. Sousa-Cavada, B., Richardson, M., Yarwood, A., Père, D., and Rougé, P.: Phytochemistry (in press).

32. Entlicher, G., and Kocourek, J.: Biochim. Biophys. Acta 393, 165-169 (1975).

33. Yarwood, A., Richardson, M., Sousa-Cavada, B., and Rougé, P.: FEBS Lett. 184, 104-109 (1985).

34. Sousa-Cavada, B., Richardson, M., Yarwood, A., Père, D., and Rougé, P. (in preparation).

35. Hopp, T.P., Hemperly, J.J., and Cunningham, B.A.: J. Biol. Chem. 257, 4473-4483 (1982).

36. Van Driessche, E., Foriers, A., Strosberg, A.D., and Kanarek, L.: FEBS Lett. 71, 220-222 (1976).

37. Rice, R.H., and Etzler, M.E.: Biochemistry 14, 4093-4099 (1975).

38. Peumans, W.J., Stinissen, H.M., and Carlier, A.R.: Planta 154, 562-567 (1982).

39. Peumans, W.J., Stinissen, H.M., and Carlier, A.R.: Planta 154, 568-572 (1982).

40. Chrispeels, M.J.: Phil. Trans. Roy. Soc. Lond. B 304, 309-322 (1984).

PART 3

ANIMAL LECTINS

MAMMALIAN LECTINS AND THEIR FUNCTION - A REVIEW

Victoria Kolb-Bachofen

Institute for Biophysics and Electron Microscopy, University of Düsseldorf, Moorenstr. 5, D-4000 Düsseldorf 1, FRG

The scope of this paper will not be to give you an updated list of animal lectins, this is a rapidly increasing list where many saccharide inhibitable phenomena have not been substantiated so far. So I will deal mainly with those lectins that are well characterized and this list is still surprisingly small.

The emphasis of this paper will be placed on the function of lectins in animals, especially in mammals. As far as can be seen to date, recognition by lectins plays a role in developmental processes, in receptor mediated uptake of various compounds and in defense mechanisms.

In the following I will discuss the lectin activities known so far according to these functional groups.

A Lectins involved in developmental processes

A group of lactose inhibitable lectins has been described to occur in various organs of various animals (1,2). These lectins - often named galaptins - share a common saccharide specificity: they bind to thiodigalactoside and to lactose with no ion requirement for binding. They all are dimers of proteins with a M_r of around 13 - 16 kd. They are soluble lectins, not integrated into biological membranes. The isolated lectins need a high concentration of reducing agents to maintain their lectin activity. A lectin activity falling into this group has been found in the electric organ of the electric eel and is termed electrolectin. Nothing is known about the function of this lectin nor about its natural ligand. Antibodies raised against this electrolectin however are found to react with galaptins for instance in chicken, mouse, rat, cow and man. These galaptins are localized in muscle, lung, liver, pancreas, intestine, thymus and bone marrow. They are seen to occur intra- and extra-cellularly. One striking feature about these lectins is, that their expression within the organ is developmentally regulated.

In rat lung, for instance, no galaptin is present in the embryo. Around the 12th day post partum a peak in lectin activity is found in this organ. The lectin reaches high concentrations, so that it comprises 0.1 % of total lung protein during this stage. The peak lectin expression coincides with the peak alveolarization of rat lung, the lectin level in adult lung being much smaller (3). In addition lectin can be readily isolated from organ homogenate at day 12 p.p. but isolation of the same lectin from adult lung needs the presence of a competing saccharide (4). Both results as well as the observation, that this lung lectin is localized intracellularly in the alveolar cells as well as extracellularly as a component of the elastic fibers (5) leads to the general assumption, that this lectin is functioning in organ development. Though we do not know what developmental function the lectin mediates, a recent work has shown, that lung membrane protein(s) can be isolated binding with high affinity to the endogenous lectin (6). This finding makes it probable, that the galaptin contributes to cellular adhesion and/or organization of extracellular matrix.

More is known about the function in cellular differentiation of a galaptin found in mouse thymus. Here it has been shown, that this mouse thymic lectin (MTL) binds and agglutinates immature T-lymphocytes, whereas it expresses no affinity to mature, circulating T-lymphocytes (7). Injection of electrolectin (sharing many characteristics with MTL) prevents onset of experimentally induced autoimmune myasthenia gravis (8) in mice. Function of MTL in retaining immature thymocytes within the thymus ensuring their proper maturation is discussed.

A similar functional characteristic is described for the erythroid developmental agglutinin localized in bone marrow. This lectin agglutinates erythroblasts making it probable, that it functions in retention of erythroblasts in the bone marrow and their attachment to the "nurse" macrophages, prior to enucleation (9).

There is a preliminary report hinting to the probability that in bone a lectin activity, eventually similar to a galaptin, might serve the attachment of macrophages to bone for either formation of osteoclasts or bone resorption or both (10).

A lectin, not belonging to the group of galaptins, appears to play a role in organ organisation,too. Laminin, a component of basement membranes, was reported to exhibit lectin activity. It was speculated, that this binding activity may contribute to cell adhesion on basement membranes (11).

B Lectins functioning in receptor-mediated endocytosis

To this group belongs a lectin, which is by far the best known, which in fact was the first lectin found in mammals. The galactose-specific receptor of liver parenchymal cells, first decribed by Ashwell and Morell (12), also termed asialoglycoprotein-receptor (ASGPR) or hepatic binding protein (HBP). This lectin is a glycoprotein (13), it is a transmembrane protein the peptide moiety spanning the bilayer (14). It is localized on the plasma membrane as well as on membranes of the cell's interior (15). It appears to have the same binding specificity in all mammals investigated so far, including man and expresses a strict ion dependency, needing Ca^{++}-ions for binding (16). Around 10^6 binding sites per hepatocyte can be found (13). This receptor mediates binding and uptake of glycoproteins exhibiting terminal galactosyl residues, i.e. asialoglycoproteins. The receptor has a half-life of around 80 days (15) and can mediate the uptake of several hundred ligands during this time using the phenomenon described as receptor recycling (17). In fact, this is one out of 4 receptor systems where uptake routes and receptor recycling have been studied with. Favoured by the high number of binding sites per cell and the simple inhibition characteristics of the hepatocyte binding protein (GalNAc and/or ion chelators) some aspects of receptor-mediated endocytosis have been elucidated using this receptor system. The experiments of several groups have shown how hepatocytes take up asialoglycoproteins:

The ligand is bound to the distributed receptors on the sinusoidal surface of the hepatocyte plasma membrane, ligand-receptor complexes microaggregate and are caught within the coated pits (18), specialized plasma membrane areas. Coated vesicles transport both, receptor and ligand, to the endosome, an organelle also termed receptosome or CURL. This is the place where receptor and ligand are uncoupled (19). It is achieved by maintenance of an acidic pH ($pH \sim 5$) within this organelle, the binding of ligand and receptor being unstable at low pH, ligands and receptors are subsequently sorted into different organelles, the asialoglycoprotein is ultimately degraded in lysosomes and/or secreted into the bile, the receptor recycles back to the plasma membrane, where it can start another round of ligand uptake (20).

The presence of the hepatic binding proteins on hepatocytes is dependent on the state of cell differentiation. Dividing hepatocytes do express little lectin activity. Accordingly, fetal liver is devoid of receptor activity. Its expression starts around 20 days after gestation in mice, adult levels of activity being reached around the end of the suckling period (21). In regenerating liver binding activity is low (22), and during growth of liver tumors receptor activity is also low. It has been reported that the titre of asialoorosomucoid or asialotransferrin is high in patients with primary tumors of the liver, with hepatitis or liver cirrhosis (16). This may be of future clinical relevance.

Now, it is long known that not only glycoproteins expressing terminal galactosyl groups rapidly home to the liver but large compounds as well: Both, desialylated erythrocytes as well as desiaalylated lymphocytes are both retained in the liver (23,24), although ligands as large as these blood cells do not have access to the hepatocyte membrane, prevented by the endothelial layer. We were able to show that large, particulate ligands exposing terminal galactosyl groups cannot be taken up by hepatocytes, and are found to be endocytosed by the resident liver macrophages, the Kupffer cells (25). These bear a receptor, having the same saccharide specificity (26). The receptor can also mediate endocytosis of particles but is strikingly ineffective in mediating uptake of molecular compounds (27). Since it appears to function in host defence it will be dealt with in the next chapter.

A receptor sharing many molecular features including cellular localization with the Ashwell-receptor described above is found in chicken liver, but there it has a different binding specificity. In accord with the normal occurrence of unsialylated glycoproteins, the chicken hepatic binding protein recognizes agalactoglycoproteins (16).

Another very well understood receptor system, is the mannose-6 phosphate receptor which functions in recognition, transport and proper package of lysosomal enzymes (28,29). It is found in many different cell types, it has been studied excessively in fibroblasts and hepatocytes. With this lectin researchers have been extremly lucky since there are two human diseases that are consequences of either lack of proper recognition signals on the ligands (I-cell disease) or lack of synthesis of one of the hydrolases (Hurler's Syndrome). With these two mutant fibroblasts at hand the function of the Man-6P receptor could be elucidated (29) and the clinical relevance of such knowledge demonstrated impressively.

Two other membrane-bound lectin activities mediating endocytosis have been described, both occurring in mammalian liver: hepatocytes are able to clear lactoferrin from the circulation, a process that can be completely inhibited by fucoidin (31), but the lectin nature of the receptor mediating the uptake of lactoferrin has recently been questioned (32).

A mannose-specific receptor is present on macrophages residing in several organs: Glycoproteins exhibiting mannosyl or N-acetyl-D-glucosaminyl residues are avidly taken up by macrophages of the lung, liver and peritoneum. The uptake is inhibited by mannan (33). Since these cells are also capable to recognize and bind yeast via this receptor system it functions probably also mainly in host defense (34).

Another mannan-binding protein has been isolated from rat liver hepatocytes that is localized intracellularly. Although it was initially described to occur within hepatocytes exclusively (35), recent data make it probable that a minor percentage of this lectin activity is also found on the cell surface and functions in receptor-mediated endocytosis (36).

C Lectin involved in host defense mechanisms

It is an intriguing concept, that some of the lectins found in primitive multicellular organisms serve the recognition of "non-self" and hence function as a primitive immune system, and that furthermore these primitive defense mechanisms also exist in mammals as a simple means to deal with "foreign", the highly specialized and variable immune system serving the more complicated defense problems, where i.e. structural mimicry has out-maneouvered the simple systems (37).

Lectins do play an important role in immunology. Lymphocytes can be stimulated to divide and differentiate by the addition of various lectins (38) including the endogenous thymic lectin mentioned earlier.

On the other hand lymphocytes express lectin activities themselves (39,40). And a lectin activity specific for fucose (41) and/or for mannose-6P (42) on either lymphocytes or endothelial cells of the high endothelial venules is discussed to mediate lymph node homing of lymphocytes.

Lectin activity has been discussed to be associated with MIF, a soluble factor that serves the retention of macrophages at an injured site (43).

Two well characterized lectin activities are associated with host defense mechanisms. It is probable though not proven, that the man/GlcNAc-specific receptor on macrophages of lung, liver and peritoneum serves this function because - as mentioned earlier - it can mediate recognition of yeast cells (44).

Finally as indicated earlier a GalNAc/Gal-specific lectin on liver macrophages has been described by us (45) and on peritoneal macrophages by the group of Schauer. We have extensively studied this lectin in rat livers. Its binding specificity is very similar - if not identical - to the hepatic binding protein (26). But this is the only common characteristic. The macrophage receptor is a different protein, sharing no common antigenic sites, having a different M_r of 30 kd and exhibiting different solubility (46). Recently we could show, that in contrast to the hepatic binding protein the macrophage receptor is not an intrinsic membrane protein. It can be removed from the macrophage plasma membrane by washing the cells in the presence of divalent ion chelators (47). So, although it is a soluble protein, loosely attached to the membrane it nonetheless functions in recognition, binding and uptake of ligands. However, ligands for the macrophage receptor have to fullfill more conditions than just exposing terminal galactosyl groups: As mentioned earlier *in vivo* the uptake of soluble glycoproteins as for instance asialofetuin by liver macrophages is low (48,27). This is also one out of several reasons, why not all investigators of galactose-specific recognition in liver were able to demonstrate a galactose-specific binding activity associated with the macrophages. If asialoglycoproteins or neoglycoproteins bearing galactosyl-groups (Gal-BSA) are adsorbed onto particles as for instance onto colloidal gold 17 nm of diameter, hepatocytes normally are unable to effectively bind and take up this particulate ligand but liver macrophages are found to ingest these particles *in vitro* and *in vivo* (45). The uptake of the particulate ligand is completely inhibitable by the addition of GalNAc/Gal and related saccharides but is unaffected by the presence of unrelated saccharides or BSA (49). The adherence of desialylated autologous erythrocytes or lymphocytes to the liver macrophages shows the same binding specificity (50,51).

The macrophages have been found to take up any particulate ligands exposing terminal galactosyl groups irrespective of their size. We found that such particles are endocytosed via the coated pit/vesicle pathway, when they do not exceed 35 nm, whereas particles bigger than this are phagocytosed via a vacuole that does not show any coated membrane areas (52).

The difference in handling of particulate ligands found between hepatocytes and macrophages is best explained by the difference seen in initial receptor distribution on the plasma membranes. Whereas binding sites on hepatocytes are initially statistically distributed (18), the binding sites on liver macrophages are always arranged in a preclustered pattern (49). In fact two lines of evidence support this hypothesis: if the binding sites on hepatocytes are clustered - achieved by prolonged incubation in the cold - these cells are capable to bind and take up particulate ligands (45). In addition we found that liver macrophages of rats 2 days of age have less binding activity than from adults, the binding sites not being preclustered and particulate ligands injected into the tail vein is not taken up by the liver macrophages of rats at this age (53). This shows, that not any cell bearing a sugar specific receptor will take up any ligand with the proper recognition signal, but that receptor arrangement plays an essential role in effective binding of particulate or molecular ligands.

During the last years we were able to show, that recognition by the galactose-specific macrophage receptor mediates recognition of desialylated erythrocytes (54), old and aged erythrocytes (55) and several tumor cells (56,57). We could show, that the retainment of asialo-lymphocytes within the liver is due primarily to recognition by the liver macrophages (51,58) and we were able to demonstrate that a subpopulation of these lymphocytes is then activated and infiltrates the liver parenchyma destroying hepatocytes (58).

References

1. Barondes, S.H.: Ann. Rev. Biochem. *50*, 207-231 (1981)
2. Harrison, F.L. and Chesterton, C.J.: FEBS Lett. *122*, 157-165 (1980)
3. Powell, J.T. and Whitney, P.L.: Biochem. J. *188*, 1-8 (1980)
4. Sanford, G.L., Davis, D.L. and Powell, J.T.: Biochem. J. *204*, 97-102 (1982)
5. Cerra, R.F., Haywood-Reid, P.L. and Barondes, S.H.: J. Cell Biol. *98*, 1580-µ589 (1984)
6. Powell, J.T. and Whitney, P.L.: Biochem. J. *223*, 769-774 (1984)
7. Levi, G. and Teichberg, V.I.: Immunol. Lett. *7*, 35-39 (1983)
8. Levi, G., Tarrab-Hazdai, R. and Teichberg, V.I.: Eur. J. Immunol. *13*, 500-507 (1983)
9. Harrison, F.L. and Chesterton, C.J.: Nature *286*, 502-504 (1980)
10. Bar-Shavit, Z., Teitelbaum, S.L. and Kahn, A.J.: J. Clin. Invest. *72*, 516-525 (1983)
11. Ozawa, M., Sato, M. and Muramatsu, T.: J. Biochem. *94*, 479-485 (1983)
12. Ashwell, G. and Morell, A.: Adv. Enzymol. *41*, 99-128 (1974)
13. Schwartz, A.L., Marshak-Rothstein, A., Rup, D. and Lodish, H.F.: Proc. Natl. Acad. Sci. USA *78*, 3348-3352 (1981)
14. Harford, J. and Ashwell, G.: Proc. Natl. Acad. Sci. USA *78*, 1557-1561 (1961)
15. Tanabe, T., Pricer, W.E. and Ashwell, G.: J. Biol. Chem. *254*, 1038-1043 (1979)
16. Ashwell, G. and Harford, J.: Ann. Rev. Biochem. *51*, 531-554 (1982)
17. Steer, C.J. and Ashwell, G.: J. Biol. Chem. *255*, 3008-3013 (1980)
18. Kolb-Bachofen, V.: Biochim. Biophys. Acta *645*, 293-299 (1981)
19. Geuze, H.J., Slot, J.W., Strous, G.J.A.M., Lodish, H.J. and Schwartz, A.L.: Cell *32*, 277-287 (1983)
20. Harford, J., Klausner, R.D. and Ashwell, G.: Biol. Cell *51*, 173-180 (1984)
21. Collins, J., Stockert, R.J. and Morell, A.: Hepatology *4*, 80-83 (1984)

22. Howard, D.J., Stockert, R.J. and Morell, A.: J. Biol. Chem. *257*, 2856-2858 (1982)

23. Gattegno, L., Bladier, D. and Cornillet, P.: Carbohydrate Res. *34*, 361-369 (1971)

24. Woodruff, J.J. and Woodruff, J.F.: J. Immunol. *117*, 852-858 (1976)

25. Kolb, H., Kolb-Bachofen, V. and Schlepper-Schäfer, J.: Biol. Cell *36*, 301-306 (1979)

26. Kolb, H., Vogt, D., Herbertz, L., Corfield, A., Schauer, R.and Schlepper-Schäfer, J.: Hoppe-Seyler's Z. Physiol. Chem. *361*, 1747-1750 (1980)

27. Kolb-Bachofen, V., Schlepper-Schäfer, J., Teradaira, R., Vogt, D. and Kolb, H.: In "Structural Carbohydrates of the Liver", Popper, H., Gudat, F., Kötggen, E. and Reutter, W. (Eds.), 277-286 (1983)

28. Reitman, M.L. and Kornfeld, S.: J. Biol. Chem. *256*, 11977-11980 (1981)

29. von Figura, K. and Weber, E.: Biochem. J. *176*, 943-950 (1978)

30. Reitman, M.L., Varki, A. and Kornfeld, S.: J. Clin. Invest. *67*, 1574-1579 (1981)

31. Prieels, J.P., Pizzo, S.V., Glasgow, L.R., Paulson, J.C. and Hill, R.L.: Proc. Natl. Acad. Sci. USA *75*, 2215-2219 (1978)

32. Imber, M.J. and Pizzo, S.V.: Biochem. J. *212*, 249-257 (1983)

33. Stahl, P.D., Wileman, T.E., Dimet, S. and Shepherd, V.L.: Biol. Cell *51*, 215-218 (1984)

34. Warr, G.A.: Biochem. Biophys. Res. Commun. *93*, 737-745 (1980)

35. Maynard, Y. and Baenziger, J.U.: J. Biol. Chem. *257*, 3788-3794 (1982)

36. Tolleshaug, H., Berg, T. and Blomhoff, R.: Biochem. J. *223*, 151-160 (1984)

37. Kolb, H.: Dev. Comp. Immunol. *1*, 193-206 (1977)

38. Miller, K.: Immunobiol. *165*, 132-146 (1983)

39. Kieda, C., Roche, A.C., Delmotte, F. and Monsigny, M.: FEBS Lett. *99*, 329-332 (1979)

40. Teodorescu, M.: Biol. Cell *51*, 251-258 (1984)

41. Stoolman, L.M. and Rosen, S.D.: J. Cell Biol. *96*, 722-729 (1983)

42. Butcher, E.C.: Biol. Cell *51*, 247-250 (1984)

43. Higgins, T.J., Liu, D.Y., Remold, H.G. and David, J.R.: Biochem. Biophys. Res. Commun.*93*,1259-1265 (1980)

44. Sung, J.S., Nelson, R.S. and Silverstein, S.C.: J. Cell Biol. *96*, 160-166 (1983)

45. Kolb-Bachofen, V., Schlepper-Schäfer, J., Roos, P. Hülsmann, D. and Kolb, H.: Biol. Cell *51*, 219-226 (1984)

46. Roos, P., Hartmann, H.J., Schlepper-Schäfer, J., Kolb, H. and Kolb-Bachofen, V.: Biochim. Biophys. Acta, in press.

47. Kempka,G. and Kolb-Bachofen,V.: Biochim. Biophys. Acta,in press

48. Steer, C. and Clarenburg, R.: J. Biol. Chem. *254*, 4457-4461 (1979)

49. Kolb-Bachofen, V., Schlepper-Schäfer, J., Vogell, W. and Kolb, H.: Cell *29*, 859-866 (2982)

50. Schlepper-Schäfer, J., Kolb-Bachofen, V. and Kolb, H.: Biochem. J. *186*, 827-831 (1980)

51. Kolb, H., Kriese, A., Kolb-Bachofen, V. and Kolb, H.A.: Cell Immunol. *40*, 457-462 (1978)

52. Kolb-Bachofen, V., Schlepper-Schäfer, J. and Kolb, H.: Exptl. Cell Res. *148*, 173-182 (1983)

53. Dini, L. and Kolb-Bachofen, V., in preparation

54. Kolb, H., Friedrich, E. and Süss, R.: Hoppe-Seyler's Z.Physiol. Chem. *362*, 1609-1614 (1981)

55. Schlepper-Schäfer, J., Kolb-Bachofen, V. and Kolb, H.: Biochem. Biophys. Res. Commun. *115*, 551-559 (1983)

56. Schlepper-Schäfer, J., Friedrich, E. and Kolb, H.: Europ. J. Cell Biol. *25*, 95-102 (1981)

57. Schlepper-Schäfer, J., Holl, N., Kolb-Bachofen, V., Friedrich, E. and Kolb, H., Biol. Cell *52*, 253-258 (1984)

58. Kolb-Bachofen, V. and Kolb, H.: J. Immunol. *123*, 2830-2834 (1979)

A GALACTOSE-SPECIFIC LECTIN ALSO ON ENDOTHELIAL CELLS OF RAT LIVERS

J. Schlepper-Schäfer, D. Hülsmann, H. Kolb* and V. Kolb-Bachofen
Institute for Biophysics and Electron Microscopy and
*Diabetes Research Institute, University of Düsseldorf,
D-4000 Düsseldorf, FRG

Receptors specific for terminal galactose-residues have been described on plasma membranes of hepatocytes of mammalian livers (1) and on macrophages of rat livers (2). Most studies have been performed with isolated liver cell fractions and the results often depend on the purity of the cell fraction. While studying receptors on the cell surface of different cell populations of an organ under identical conditions the electron microscope provides a useful technique for *in vivo* and *in situ* (isolated perfused organ) experiments. Here we describe a galactose-specific lectin on the cell surface of endothelial cells in rat liver.

Lectins, Vol. V

Materials and Methods

Electron-dense ligands: Gold particles 5 nm of diameter were prepared according to Horisberger and Rosset (3). Bovine serum albumin (BSA) derivatized with lactose [ß-Gal(1→4)Glc] was adsorbed to the gold particles (LacBSA-Au_5).

In situ experiments: Rat livers were perfused with Eagle's medium at room temperature to remove blood. Livers were prefixed with icecold 0.1 % glutaraldehyde in s-collidine buffer (2) and subsequently perfused with BSA containing Eagle's medium. Prefixed livers were perfused with icecold Eagle's medium to remove BSA. Ligands (about 3 x 10^{13} LacBSA-Au_5 particles) were added to the perfusion medium (Eagle's medium). Unbound ligand was washed out by further perfusion with Eagle's medium. Livers were fixed with 0.5 % glutaraldehyde in s-collidine buffer, cut into small pieces, postfixed with 2 % OsO_4 and processed for electron microscopy (2). In inhibition experiments N-acetylgalactosamine (GalNAc) or N-acetylglucosamine (GlcNAc) at a final concentration of 80 mM were added to the perfusion medium prior to ligand addition.

Electron-dense ligands: Gold particles 5 nm of diameter were prepared according to Horisberger and Rosset (3). Bovine serum albumin (BSA) derivatized with lactose [ß-Gal(1-4)Glc] was adsorbed to the gold particles (LacBSA-Au_5). LacBSA was prepared by spontaneous glycation of BSA in the presence of lactose. A borate buffered mixture of BSA (essentially fatty acid free, Sigma,Munich, FRG) and lactose at a molar ratio of 1 : 50 was incubated at 37°C for 4 days. The glycated protein was purified by passing through a Sephadex G50 column (Pharmacia, Upsala, Sweden).

In situ experiments: Rat livers were perfused with Eagle's medium at room temperature to remove blood. Livers were prefixed with icecold 0.1 % glutaraldehyde in s-collidin buffer (2) and subsequently perfused with BSA containing Eagle's medium and then with icecold Eagle's medium to remove BSA. Ligands (about 3 x 10^{13} LacBSA-Au_5 particles) were added to the perfusion medium (Eagle's medium). Unbound ligand was washed out by further perfusion with Eagle's medium. Livers were fixed with 0.5 % glutaraldehyde in s-collidine buffer, cut into small pieces, postfixed with 2 % OsO_4 and processed for electron microscopy (2). In inhibition experiments N-acetylgalactosamine (GalNAc) or N-acetylglucosamine (GlcNAc) at a final concentration of 80 mM were added to the perfusion medium prior to ligand addition.

Results

Examination of prefixed rat livers which had been perfused with LacBSA-Au_5 revealed binding of the ligand to liver macrophages and hepatocytes as expected and to endothelial cells. No other liver cell type was found to bind the galactose-exposing ligand, nor normal blood cells eventually seen within the sinusoids. Binding of ligands to the surface of liver cells was found in clusters on

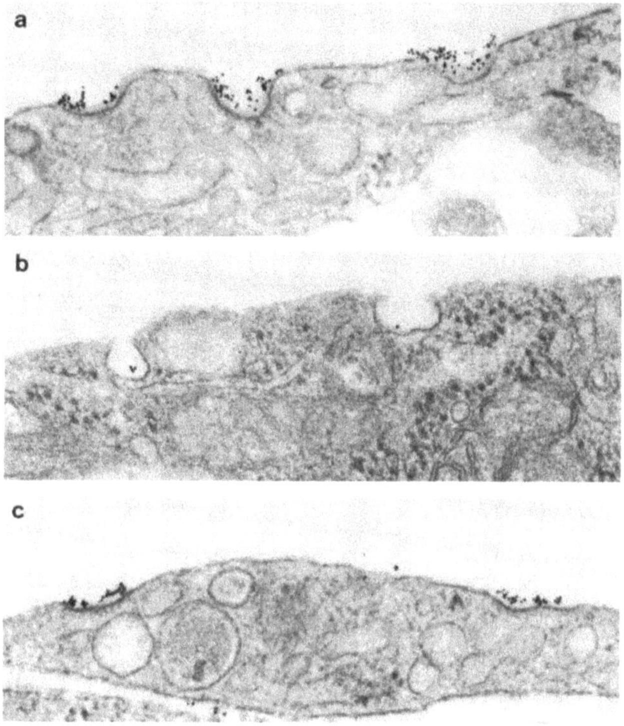

Fig. 1. Binding of LacBSA-Au_5 to endothelial cells in a prefixed rat liver. Binding is restricted to coated pits. (a) without inhibitor, (b) in the presence of 80 mM GalNAc, (c) in the presence of 80 mM GlcNAc. Transmission electron microscopy, 84 000 x.

liver macrophages and statistically distributed on hepatocytes as described previously (4). Ligand binding to endothelial cells was observed in coated pits only (Fig. 1a). To confirm specificity of ligand binding we performed binding experiments in the presence of 80 mM GalNAc and 80 mM GlcNAc, respectively. In the presence of GalNAc binding of LacBSA-Au_5 is strongly reduced (Fig. 1b). Inhibition exceeds 80 %. Experiments in the presence of GlcNAc reveal no difference in ligand binding (Fig. 1c) as compared to controls.

Discussion

We describe a receptor on the cell surface of endothelial cells specific for terminal galactose residues. Evidence that binding of LacBSA-Au_5 is mediated by a galactose-specific lectin and not due to unspecific adsorption comes from four different results.

I. Ligand binding is not distributed statistically all over the cell surface but restricted to coated pits. Less than 25 % of total binding occurs outside coated membrane areas. These specialized membrane areas are known to be involved in receptor-mediated endocytosis (5).

II. Ligand binding is reduced to < 20 % of control in the presence of GalNAc, a monosaccharide which blocks binding activity of the asialoglycoprotein receptor on hepatocytes as well as the galactose-particle receptor on liver macrophages. GlcNAc does not interfer with ligand binding.

III. Other ligands known to be recognized by the galactose-specific receptors on liver macrophages and hepatocytes are bound (and endocytosed) by endothelial cells in coated pits, i.e. asialofetuin (ASF) bound to gold particles 17 nm of diameter (ASF-Au_{17}) (6) and GalBSA-Au_{17} (BSA derivatized with D-galactose bound to 17 nm gold particles) (data not shown). Both of these ligands show the same inhibition profile described.

IV. As demonstrated in the transmission electron microscope endothelial cells participate in homing of neuraminidase treated erythrocytes (4), a process mediated by a galactose-specific recognition mechanism (7).

Acknowledgments. This study was supported by grants from the Deutsche Forschungsgemeinschaft (Ko 801-2) and Minister für Wissenschaft und Forschung des Landes Nordhein-Westfalden.

References

1. Ashwell, G. and Harford, J.: Ann. Rev. Biochem. *51*, 531-554 (1982)
2. Kolb-Bachofen, V., Schlepper-Schäfer, J., Vogell, W. and Kolb, H.: Cell *29*, 859-866 (1982)
3. Horisberger, M. and Rosset, J.: J. Histochem. Cytochem. *25*, 295-305 (1977)
4. Kolb-Bachofen, V., Schlepper-Schäfer, J., Roos, P., Hülsmann, D. and Kolb, H.: Biol. Cell. *51*, 219-226 (1984)
5. Goldstein, J., Anderson, R. and Brown, N.: Nature *279*, 679-685 (1979)
6. Kolb-Bachofen, V., Schlepper-Schäfer, J., Teradaira, R., Vogt, D. and Kolb, H.: In "Structural carbohydrates in the liver", eds. H. Popper, W. Reutter, E. Köttgen and F. Gudat, MTP Press, Boston, p. 277-286, 1983.
7. Kolb, H., Friedrich, E. and Süss, R.: Hoppe-Seyler's Z. Physiol. Chem. *362*, 1609-1614 (1981)

ADULT RAT BRAIN EXTRACT INDUCES LACTOSE-INHIBITABLE AGGREGATION OF DISSOCIATED EMBRYONIC MOUSE BRAIN CELLS

R. Joubert, M. Caron, M.A. Deugnier, J.C. Bisconte

Laboratoires de Biochimie Fondamentale et de Neurobiologie Quantitative
U.F.R. Santé, Médecine, Biologie Humaine, F-93012 Bobigny Cedex, France

Rat brain extracts have been reported to induce growth, differentiation and maturation of cultured nerve cells (1-3). Attempts to identify the active components of these extracts pointed out the importance of polypeptides and proteins (2), but the mechanisms underlying the *in vitro* activities of such fractions still remain unknown. On one hand, earlier works showed the presence of endogenous lactose-inhibitable lectin(s) in extracts of neonatal (4) and adult (5) rat brain. On the other hand, it has been shown that binding sites for lectins, especially sites for lectins implying a recognition of galactose moieties, are exposed on the surface of embryonic mouse nerve cells (6,7). Such sites are part of glycoconjugate molecules known to be involved in various cellular functions (8,9) as well as in cell-cell recognition (10). Thus we have postulated that these sites may act as specific receptors of the lectin fraction present in the adult rat brain extract. In testing this assumption, we have searched for the ability of this extract to induce specifically the aggregation of dissociated embryonic brain cells in suspension at low density that strongly limits their spontaneous reassociation. We present here studies on aggregation mediated by extracts and its carbohydrate specificity.

Lectins, Vol. V

Experimental

1) Rat brain extracts. A total brain extract was prepared from brains of adult rats by homogeneization of tissues in tyrode solution (200 mg fresh tissue weight/mL Tyrode solution) and centrifugation at 105 000 g for 1 h. (1).
A depleted rat brain extract was obtained by adsorption on D-lactose-O-amino-phenyl-agarose beads (Lactogel, E.Y. Laboratories) ; 5 mL of adult rat brain extract was filtered on a gel column (0.8 cm x 4 cm) in Eagle's minimum essential medium (Flow Laboratory). The column eluate was monitored for absorbance at 280 nm, and the fractions containing unadsorbed molecules were pooled. The solution obtained (5 mL) had a protein concentration of 6 mg/mL and showed no ability to agglutinate fixed rabbit erythrocytes. Controls were performed with mannose-substituted agarose beads and with unsubstituted agarose beads to test the specificity of adsorption.
2) Mouse brain cells. Whole brains from 14-day-old embryos were dissected and the meningeal membranes carrefully removed. No spinal cord was taken. The brains were completely dissociated into single cells by a standardized mechanical procedure (11). The viability of the cells was tested by trypan blue exclusion test and was at least 90 %. The initial cell suspension was divided into 2 parts. One part was kept alive in minimal essential medium. The other part was fixed in 4 % (v/v) glutaraldehyde (Sigma) in buffered saline. After 1 h the cells were washed in PBS, rinsed twice in 0.2 M glycine and dissociated again in appropriate medium. Isolated cells (95-97 %) were suspended at low density (5.10^5 cells/mL).
3) Cell aggregation assays. These assays measure the decrease in particle number that accompanies cell aggregation after their gyration in different experimental conditions : a) 100 µL of total brain extract or depleted brain extract was added to tubes containing 5.10^5 isolated living or fixed cells in 0.9 mL of medium supplemented with fetal calf serum (Gibco) at a final concentration of 10 % ; b) inhibition tests were conducted in presence of lactose or mannose (Sigma) at a final concentration of 3-100 mM. Total brain extract was preincubated 30 min with sugar and the cell suspension (5.10^5 cells/0.8 mL) was added thereafter ; c) Total brain extract-mediated aggregation was tested in different media : minimal essential medium unsupplemented or supplemented with bovine serum albumin (Sigma ; 3.5 mg/mL) and 0.15 M NaCl instead of minimal essential medium with fetal calf serum. Controls were performed without extract ; 1 µL of anti-protease (Iniprol, Choay) was added to all living samples whose total volume was 1 mL. The tubes containing living cells were rotated at 70 rpm at 20°C. After 0, 20, 40 and 60 minutes of gyration small aliquots were gently removed and the number of particles per unit volume was determined using a Malassez hemocytometer. The aggregation was expressed as the percentage decrease in particle number.

Results and discussion

The effects of total brain extracts and depleted brain extracts were studied on living as well as fixed embryonic mouse brain cells suspended at low density. The percent decrease in particle number with or without addition of any extract is shown in figure 1. Only a discrete aggregation was observed on

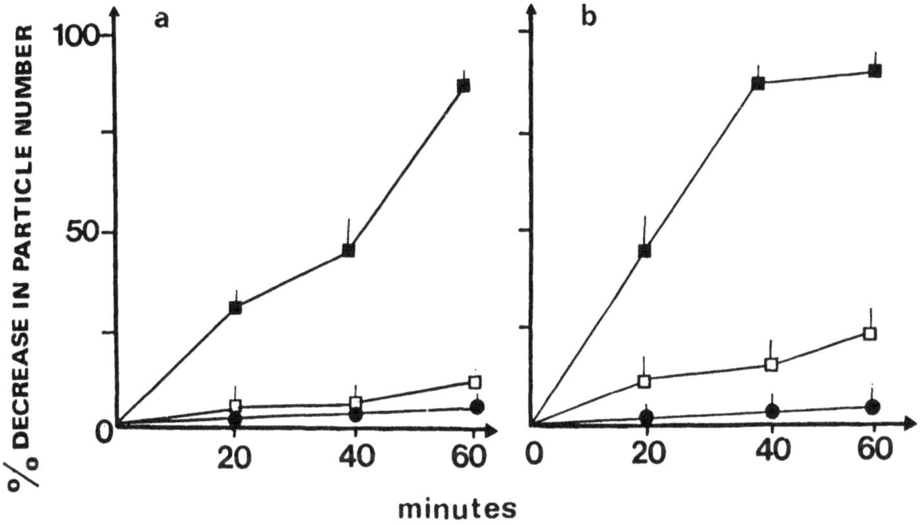

Fig. 1 : Compared effects of total brain extract and depleted brain extract on the mouse brain cell reaggregation.
(a) aggregation of living cells (b) aggregation of fixed cells. The aggregation is expressed as the percent decrease in particle number after 60 min of gyration at 70 rpm.
(●——●) Controls without extract
(□——□) Depleted brain extract-treated suspension
(■——■) Total brain extract-treated suspension
All points indicate mean +/- S.D. of 3 samples from at least 3 independent experiments.

living cells in controls without extract ; conversely, it appeared that total brain extract strongly promoted the cell reaggregation. This phenomenon depended on the duration of contact with the total brain extract and was complete after about 40 min with fixed cells and 60 min with living cells. Furthermore, we found that specifically depleted extract obtained by adsorption of lactose-binding molecule(s) on lactogel was a poor inducer of cell-cell adhesion. Thus promoting effects of the total brain extract in the reaggregation of dissociated nerve cells seem to be correlated to the presence of a lactose-binding fraction, which is lost in the depleted brain extract. The assays performed with fixed cells support the idea that total brain extract-mediated cell-cell adhesion *in vitro* is not secretion-dependent and only concerns the cell surface. It presumably acts by interacting with carbohydrate structures of the embryonic brain cell membrane, but this has not been directly demonstrated by the preceding results. To determine the contribution of such structures to the interaction with the total brain extract, attempts were made to inhibit cell reassociation with simple sugars. We previously reported (5) that the total brain extract possesses an hemagglutinating activity i) highly inhibitable by lactose and non-inhibitable by mannose, ii) adsorbable on lactose-agarose beads but non-adsorbable on mannose-agarose beads. Thus we studied the effects of these two sugars as potent inhibitors of total brain extract aggregation of suspended fixed cells. The presence of sugars in the reaction medium do not lead to interferring perturbations of the cellular reactions. The results of the inhibition tests are shown in figure 2 a and b. The aggregation effect of total brain extract was inhibited in presence of 25 mM lactose. It should be noted that its inhibitory activity is dose-dependent (Fig. 3). Mannose was not an inhibitor in the concentration range tested (0-100 mM mannose). The inhibitory effect of lactose as well as the consequence of the adsorption of lactose-binding molecules on lactose-agarose beads are in favor of the presence of a carbohydrate binding (lectin) fraction in rat brain extract able to interact with glycoconjugates of dissociated embryonic brain cells. However this does not prove that total brain extract induced reagregation is the direct consequence of a single lectin-carbohydrate interaction. It remains possible that one or more medium components may be implicated in the phenomenon. In order to elucidate this problem, the ability of the total brain extract to induce cell aggregation was studied in different media. Dissociated cells were suspended in 0.15 M NaCl, unsupplementel minimal essential medium and in BSA (at a protein concentration

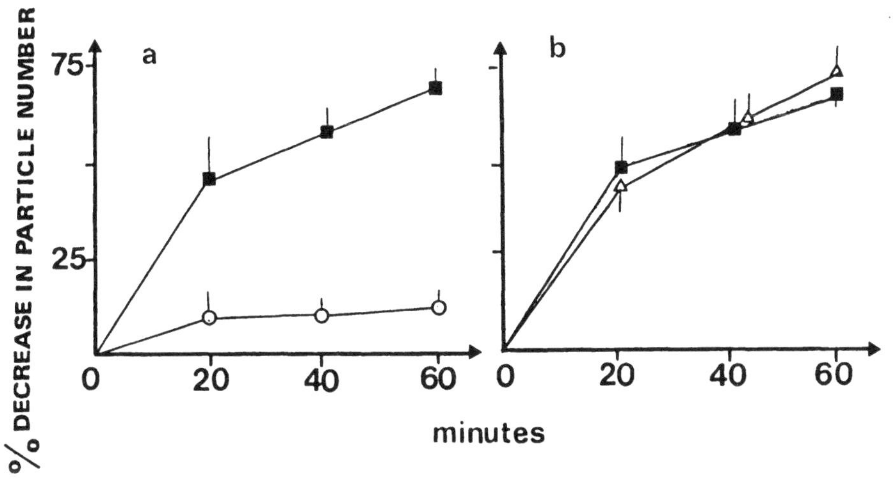

Fig. 2 : Saccharide inhibition of total brain extract-mediated cell-cell aggregation
a) in presence of 25 mM lactose ; b) in presence of 100 mM Mannose
(■——■) total brain extract-treated control without saccharide
(○——○) total brain extract-treated suspension in presence of lactose
(△——△) total brain extract-treated suspension in presence of mannoes
All points indicate mean ± S.D. of at least 2 samples from 3 or 4 independent experiments.

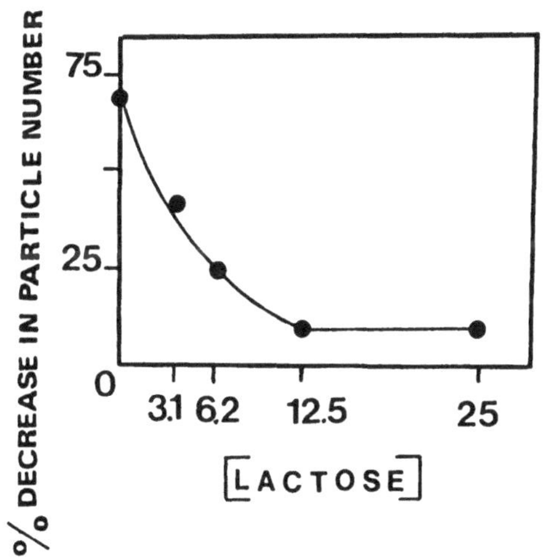

Fig. 3 : Total brain extract-mediated cell aggregation as a function of the lactose concentration.

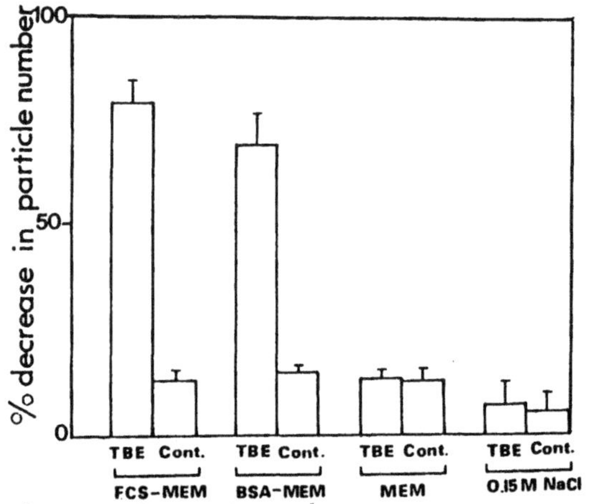

Fig. 4 : Percent decrease in particle number in different media. Controls were performed without total brain extract . The total protein concentration was 3.5 mg/mL. Results are expressed as the mean ± S.D. of 4 independent samples Abbreviations : TBE : total brain extract Cont : Control sample MEM : Minimum essential medium BSA : bovine serum albumin

equal to that found in minimum essential medium supplemented with 10 % fetal serum). As shown in figure 4, total brain extract cannot induce aggregation of mouse brain cells in unsupplemented minimal essential medium or in NaCl. From the experiments with BSA and calf serum supplement, we conclude that the protein concentration of the reaction medium is probably more important than any component of the fetal calf serum or of the minimal essential medium. It is possible that this dependence is related to the electric charge of the cell membranes, as in the case of agglutination by so called "uncomplete-antibodies" or "uncomplete-lectins" (12, 13).

Specific attachment of lectins to glycolipids and/or glycoproteins has been demonstrated to be a critical initial process in a number of important phenomena including cell differentiation and cell adhesion (10,14,15). Our results show that adult rat brain contains molecules that bind-lactose and other β-galactosides, and aggregate dissociated embryonic brain cells in suspension. As several works have shown that nerve cell surfaces possess terminal galactosyl residues (6,7), a reasonable extrapolation is that one or more soluble lectins of brain tissues play a role in developmentally regulated nerve cell interactions.

Acknowledgments

The authors are grateful to Dr. M. Sensenbrenner for the gift of adult rat brain extracts and for helpful discussion. We thank S. Margules for her assistance in the preparation of the manuscript.

References

1. Pettmann, B., Delaunoy, J.P., Courageot, S., Devilliers, G., and Sensenbrenner, M. : Develop. Biol. 75, 278-287 (1985).

2. Barakat, I., Sensenbrenner, M., and Labourdette, G. : J. Neurosc. Res. 8, 303-314 (1982).

3. Sensenbrenner, M., Barrakat, I., Delaunoy, J.P., Labourdette, G. and Pettmann, B. : Neurosciences approached through cell culture, (Pfeiffer, S.E., ed.) Vol. 1, pp. 87-105, CRC. Press, Boca Raton (1981).

4. Simpson, D., Thorne, D., and Loh, H. : Nature 266, 367-369 (1977).

5. Joubert, R., Caron, M., Deugnier, M.A., Rioux, F., Sensenbrenner, M., and Bisconte, J.C. : Cell. Mol. Biol. 31, 131-138 (1985).

6. Hatten, M., and François, A. : Develop. Biol. 87, 102-113 (1981).

7. Sacks, H.J., Stohr, M., and Schachner, M. : Cell Tissue Res. 228, 183-204 (1983).

8. Gartner, T.K., and Poldeski, T.R. : Biochem. Biophys. Res. Commun. 70, 1142-1149 (1976).

9. Poldeski, T.R., and Greenberg, I. : Proc. Natl. Acad. Sci. USA 77, 1054-1058 (1980).

10. Monsigny, M., Kieda, C., and Roche, A.C. : Biol. Cell 47, 95-110 (1983).

11. Joubert, R., Buys, B., Moschetto, Y., and Bisconte, J.C. : ITBM 2, 563-579 (1981).

12. Sharon, N., and Lis, H. : Sciences 177, 949-959 (1972).

13. Ochoa, J.L. : Pathol. Biol. 27, 103-113 (1979).

14. Barondes, S. : Science 223, 1259-1264 (1984).

15. Grant, C.W.M., and Peters, M.W. : Biochim. Biophys. Acta 779, 403-422 (1984)

THE GALACTOSE RECEPTOR OF RAT PERITONEAL MACROPHAGES
BINDING OF SIALIDASE-TREATED BLOOD CELLS AND GLYCOPROTEINS

Sørge Kelm, Saif ed-Din Jibril, Haekyung Lee, Teruo Yoshino* and Roland Schauer

Biochemisches Institut, Universität Kiel, D-2300 Kiel, FRG
*Department of Chemistry, International Christian University, Tokyo, Japan

One of the first hints at the galactose-specific receptor of macrophages was the observation that sialidase treatment of erythrocytes or lymphocytes leads to their rapid sequestration from the blood stream (1,2). However, in contrast to erythrocytes, lymphocytes reappear in the blood after 24 to 48 hours (2). Further investigations showed that sialidase-treated erythrocytes are bound by macrophages in liver and spleen (3) and that this interaction can be inhibited with galactose-containing substances (4). Since the first experiments in vitro sialidase-treated erythrocytes are well established as ligands for the galactose-specific receptor on peritoneal macrophages (5). Binding of erythrocytes was estimated either by the use of ^{51}Cr-labelled erythrocytes or, more recently, by light microscopy (6). We used the latter system to study the specificity of the receptor. Furthermore, it is shown that sialidase-treated lymphocytes and radio-iodinated glycoproteins are bound in a similar way. Finally, we give evidence for the receptor-mediated internalisation of ^{125}I-labelled asialo-orosomucoid (ASOR).

Lectins, Vol. V

Materials and Methods

Substances: Lactose derivatives were synthesized by procedures that will be described elsewhere. Antifreeze glycoprotein 8 was a gift from R. Feeney, Davis (USA), and Galα(1-4)Glc was a present from M. Mori, Nagoya (Japan). Asialo-orosomucoid (ASOR) was prepared from orosomucoid (Sigma, Munich, FRG) by treatment with Vibrio cholerae sialidase (Behringwerke, Marburg, FRG), purified by ion-exchange chromatography on DEAE-Sephacell (Pharmacia, Freiburg, FRG) and radio-iodinated with ^{125}I by the chloramine T method (7).

Cells: Peritoneal macrophages and erythrocytes were isolated from male Wistar rats as described before (5). Lymphocytes were isolated from rat EDTA-blood, thymus, and spleen on a Ficoll-Pacque gradient (8). Residual erythrocytes were removed by lysis with ammonium chloride, monocytes by adherence to glass and thrombocytes on a gradient of fetal calf serum.

Sialidase treatment: $4x10^9$ erythrocytes were incubated with 50 mU Vibrio cholerae sialidase in 500 µl phosphate-buffered saline (PBS) for 1 h at 37°C. 10^8 lymphocytes were incubated with 50 mU sialidase in 1 ml PBS for 1 h at 37°C, followed by washing of the cells with PBS.

Binding assays: a) Sialidase-treated cells were incubated with peritoneal macrophages in the presence or absence of inhibitors (6). Unbound cells were removed by washing, and after staining the number of bound cells was estimated by light microscopy.
b) For receptor assays with radio-iodinated ASOR peritoneal macrophages were seeded into microtiter plates and after 1 h incubation at 37°C in medium 199E (5% CO_2 in humidified air) non-adherent cells were removed by washing with incubation medium (10 mM HEPES, 137 mM NaCl, 5.36 mM KCl, 1.17 mM $MgSO_4$, 1 mM $CaCl_2$, 5.56 mM glucose, 0.15% BSA, pH 7.5) and incubated with ligands for the times and temperatures indicated. Unbound ligand was removed by 3 washes with incubation medium and bound radioactivity was estimated in a Philips gammacounter PW 4800.Unspecific binding was estimated by incubation in the presence of 30 mM lactose. For binding assays with fixed macrophages the adherent peritoneal macrophages were incubated with 0.025% glutardialdehyde in PBS for 10 min. at room temperature. The reaction was stopped by aspiration of the buffer and subsequent addition of ice-cold 0.3% BSA in PBS. After 5 min. the cells were washed, treated for 15 min. at 4°C with binding buffer (10 mM HEPES, 130 mM NaCl, 1 mM $CaCl_2$, 0.05% Triton X-100) which was used for assays with fixed macrophages instead of incubation medium. After washing with binding buffer these microtiter plates were used as immobilized receptors immediately or stored at 4°C in binding buffer or at -18°C without buffer.

Results and Discussion

It was unknown whether sialidase-treated lymphocytes are sequestrated from the blood stream by the same mechanism as sialidase-treated erythrocytes (2). We found that similarly to erythrocytes, sialidase-treatment of lymphocytes leads to binding of these cells to peritoneal macrophages in a way dependent on the degree of sialic acid removal (Fig. 1). Binding can be inhibited with galactose, lactose, or ASOR, but not glucose, mannose, or N-acetylglucosamine. As with sialidase-treated erythrocytes, ASOR is an inhibitor 10^5 fold more potent than lactose, 5 nM ASOR leading to 50% inhibition. An even 100-fold higher concentration of native orosomucoid does not inhibit binding. These observations exclude the participation of the well known receptor specific for mannose/N-acetylglucosamine on macrophages (9) and clearly demonstrates the galactose-specificity of the binding of sialidase-treated lymphocytes by peritoneal macrophages.

Whereas sialidase treatment of erythrocytes leads to receptor-mediated phagocytosis of these cells by macrophages (6), this does not occur with lymphocytes. In contrast to erythrocytes, the increase of phagocytosis of lymphocytes observed after exposure to sialidase cannot be inhibited by lactose. This may indicate a second mechanism following binding by the receptor, which leads to

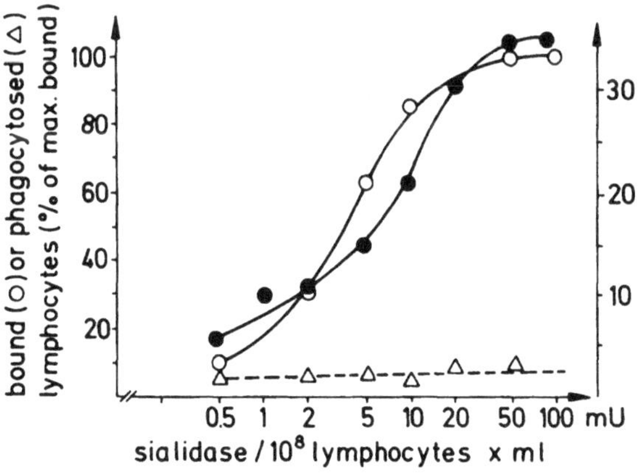

Fig. 1. Binding and phagocytosis of lymphocytes by peritoneal macrophages depending on the amount (% of total) of enzymatically removed sialic acid. Lymphocytes were incubated for 60 min. at 37°C with the amount of sialidase indicated. Binding and phagocytosis of these cells was estimated as described (n = 3).

the phagocytosis of erythrocytes but not of lymphocytes. Another possibility is that lymphocytes themselves inhibit phagocytosis mediated by the galactose-specific receptor, for example by resialylation.

Besides erythrocytes and lymphocytes, gold particles coated with neoglycoproteins or asialofetuin were used as ligands for study of the galactose receptor on Kupffer cells (10). However, binding of soluble molecules to macrophages (from peritoneum or liver) was demonstrated so far only in an indirect way via their potency to inhibit cell or particle binding. Using a radio-receptorassay we studied the binding of radio-iodinated ASOR to native and glutardialdehyde-fixed peritoneal macrophages. This ligand is bound by peritoneal macrophages in a way similar to that found with cells. The same results were obtained with ^{125}I-labelled antifreeze glycoprotein 8. Binding of the radioactive glycoproteins is mediated by recognition of the structure of galactose, as is shown in Fig. 2 for ASOR, the attachment of which is inhibited by galactose, N-acetylgalactosamine, and interestingly also by the structural analogue D-fucose. Binding is also inhibited by unlabelled ASOR; this occurs at a 5×10^3-fold lower concentration than by native orosomucoid. The amount of bound ligand correlates with the number of macrophages seeded into the microtiter plate without changing the inhibition curves of specific ligands like lactose.

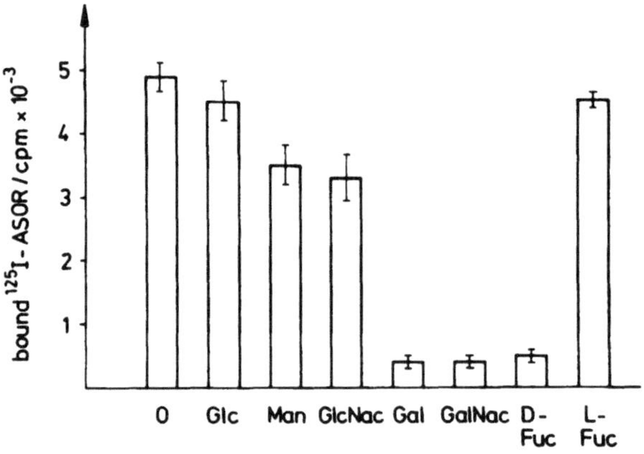

Fig. 2. Influence of monosaccharides (30 mM) on ASOR binding to peritoneal macrophages. Fixed macrophages were incubated with 3×10^5 cpm ^{125}I-ASOR in the presence or absence (O) of sugars at 4 °C for 20 h before estimation of bound ^{125}I-ASOR (n = 4).

TABLE

Influence of galactose substituents on the potency of this sugar to inhibit binding of ASOR or sialidase-treated erythrocytes to peritoneal macrophages

position	substituent	influence on inhibitory potency
1	α- or ß-CH_3, α- or ß(1-4)Glc, α(1-4)- or α(1-6)Glc, α(1-4)Gal, ß(1-4)Fru, ß(1-4)sorbitol, ß(1-4)Glc-1-P	0
2	$-NH_2$, -NHAc, $-O-CH_3$	0
3	$-O-CH_3$	-
4	$-O-CH_3$	--
6	$-O-CH_3$, $-CH_3$ (D-Fuc)	0
6	$-COO^-$	---
2,6	$(-O-CH_3)_2$	--
4,6	$(-O-CH_3)_2$	--
3,4	isopropyl	--

(0) no influence, (-) little decrease, (--) modest decrease, and (---) strong decrease in inhibitory activity.

In contrast to previous methods using native cells, the binding activity of the galactose-specific receptor can be assayed in the presence of Triton X-100 using the radio-receptorassay with glutardialdehyde-fixed macrophages. In the presence of detergent unspecific binding is lower and the receptor activity is higher, probably due to the presence of an internal receptor pool. A similar assay was recently described by Ray et al. using glutardialdehyde-fixed hepatocytes as a tool for the immobilized asialoglycoprotein receptor (11).

The structural features of the galactose molecule itself required for recognition by the receptor were studied with ^{125}I-ASOR as ligand and synthetic lactose derivatives as possible inhibitors. The specificity with regard to the "environment" of the galactose on the ligand molecule was studied with sialidase-treated erythrocytes as ligand and a series of galactose-containing molecules as inhibitors. The potency to inhibit the binding of either ASOR or sialidase-treated erythrocytes was estimated from inhibition curves using different concentrations of inhibitor for the determination of the concentration leading to 50% inhibition.

As summarized in the Table all galactose derivatives investigated are inhibitors except lacturonic acid. The sugar is recognized, although at reduced degree, even if two hydroxyl groups of galactose in lactose are substituted by O-methyl ethers. However, if the mono-methyl lactoses are compared, there are differences in the inhibitory potency. From these results it can be concluded that the OH-group at position 4 is most important for recognition of the galactose moiety, whereas at position 6 the hydroxyl function does not play any role. Remarkably, however, oxidation of this group to a carboxyl group abolishes the binding capacity.

The inhibitory potency of galactose-containing substances does not depend on the anomeric configuration of galactose, as there is no difference between the inhibition curves of methyl-α- and -ß-glycosides of galactose on the one hand or lactose and Galα(1-4)Glc on the other hand. Furthermore, the kind of aglycon or penultimate sugar has no influence on binding, since all disaccharides investigated containing terminal galactose have the same inhibitory potency (Table). Even the negative charge on glucose in lactose-1-phosphate cannot alter binding to the receptor.

However, if galactose residues are clustered in oligoantennary systems of glycoproteins, their inhibitory potency is increased depending on the grade of clustering. Most potent inhibitors on the basis of galactose concentration are ASOR and antifreeze glycoprotein 8. This was found with erythrocytes, lymphocytes, and ^{125}I-ASOR as ligand.

Besides binding, the dissociation of ^{125}I-ASOR was studied. Like the corresponding hepatocyte receptor the galactose-specific receptor of macrophages requires Ca^{2+} for binding activity. Whereas removal of unbound ligand does not lead to the dissociation of bound ASOR within 1 h, EDTA treatment (10 mM, 5 min, 0°C) releases all the specifically bound radioactivity, similar to 30 mM lactose.

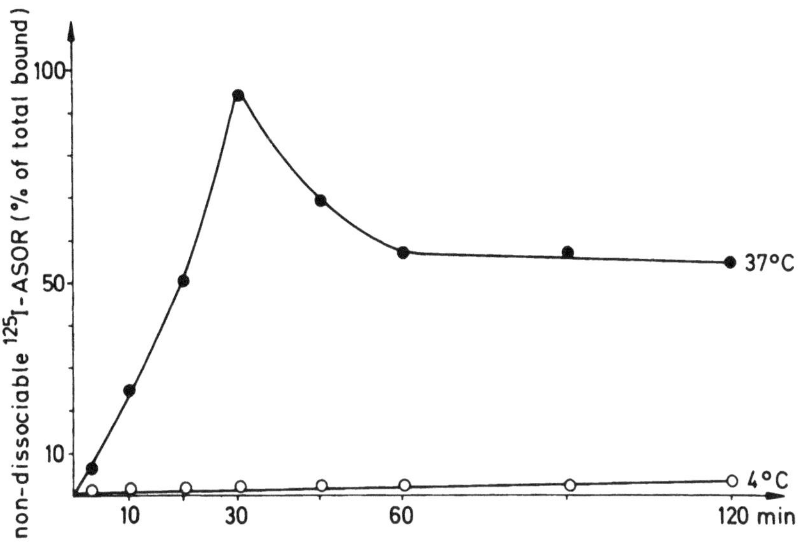

Fig. 3. Uptake of ASOR by peritoneal macrophages. Native cells were incubated with $3x10^5$ cpm ^{125}I-ASOR at 37 °C (●) or 4°C (O) for times indicated. Non-dissociable radioactivity was estimated by treatment with EDTA (10 mM, 5 min., 0°C).

This demonstrates that the binding of ^{125}I-ASOR is reversible.

With vital macrophages the use of a ligand in solution allows dynamic studies. If macrophages were incubated with ^{125}I-ASOR at 37°C a significant portion of the bound radioactivity could not be released with EDTA already after 3 min (Fig. 3). The amount of non-releasable ligand increased and reached a maximum (90% of the total bound ^{125}I) after 30 min., before decreasing to 40-50%. This points to the possibility of an internalisation of the ligand, especially because at 4°C, where no internalisation is known to take place (12), all specifically bound radioactivity could be released with EDTA. Similar kinetics were observed with isolated hepatocytes (13) or macrophages for the mannose/N-acetylglucosamine receptor (9). However, for the explanation of this phenomenon other mechanisms like covalent binding by enzymatic reactions cannot be ruled out completely.

Conclusion

These studies demonstrate that peritoneal macrophages bind sialidase-treated erythrocytes, lymphocytes and glycoproteins via a receptor specific for galactose. Differences can be found in post-binding events, since erythrocytes are phagocytosed and glycoproteins probably internalized. In contrast, lymphocytes are not taken up by the macrophages as a receptor-mediated event.

For recognition of the galactose residue it is not important how and to which molecule it is linked. Clustering of the ligand leads to a strong increase in binding, which may be due to simultaneous binding of more than one galactose residue. No single hydroxyl group of the sugar seems to be responsible for recognition by the receptor, leading to the suggestion that the whole galactose residue is "seen" by the receptor. Only a negative charge at position 6 can abolish the binding. The radio-receptorassay described will be useful for studies of post-binding events as well as further characterisation of the galactose-specific receptor on macrophages.

Acknowledgements

The authors thank Juliane Frank for expert technical assistance. Financial support by the Deutsche Forschungsgemeinschaft (Scha 202/11-2) and the Fonds der Chemischen Industrie is acknowledged.

References

1. Jancik, J., and Schauer, R.: Hoppe-Seyler's Z. Physiol. Chem. 355, 395-400 (1974).
2. Kaufmann, S.H.E., Schauer, R., and Hahn, H.: Immunobiol. 160, 184-195 (1981).
3. Jancik, J.M., Schauer, R., Andres, K.H., and von Düring, M.: Cell Tiss. Res. 186, 209-226 (1978).
4. Kolb, H., and Kolb-Bachofen, V.: Biochem. Biophys. Res. Commun. 85, 678-683 (1978).
5. Küster, J.M., and Schauer, R.: Hoppe-Seyler's Z. Physiol. Chem. 362, 1507-1514 (1981).
6. Schauer, R., Schröder, C., Müller, E., and von Gaudecker, B.: Biomed. Biochim. Acta 43, 711-717 (1984).
7. Hunter, W.M., and Greenwood, F.C.: Nature 194, 495-496 (1962).
8. Bøyum, A.: Nature 204, 793 (1964).
9. Stahl, P.D., Wileman, T.E., Diment, S., and Shepherd, V.L.: Biol. Cell 51, 215-218 (1984).
10. Kolb-Bachofen, V., Schlepper-Schäfer, J., Roos, P., Hülsmann, D., and Kolb, H.: Biol. Cell 51, 219-226 (1984).
11. Ray, D.A., and Weigel, P.H.: Anal. Biochem. 145, 37-46 (1985).
12. Michl, J.: Am. J. Clin. Nutr. 33, 2462-2471 (1980).
13. Baenziger, J.U., and Fiete, D.: Cell 22, 611-620 (1980).

LIVER LECTINS AS RECEPTORS FOR TUMOR CELLS IN METASTASIS

J. Beuth[1], K. Oette[1] and G. Uhlenbruck[2]

Department of Clinical Chemistry[1] and Medical Clinic I[2], University Clinics of Cologne, Joseph-Stelzmann-Straße 9, D-5000 Cologne 41, Federal Republic of Germany

The metastasis of malignant tumor cells from a primary site to near and distant secondary sites is probably the most important event in the pathogenesis of cancer and it accounts for most cancer deaths (1,2). Whereas advances in the treatment of primary cancer have led to increased patient survival, metastatic cancer is still the most difficult group of diseases to treat successfully (3). Provided that organ-specific lectins do play an important role in the organ-manifestation of metastatic islets (5-7), then there may be a possibility, for instance during surgery, to block organ-specific lectins with the appropriate receptor-bearing glycoconjugates.

Earlier experiments from our laboratories have shown that cryptic or incomplete carbohydrate chains may act as receptors on the surface of various tumor cells, because these precursor substances accumulate and cannot be utilized in a normal way. One of these cryptic carbohydrate receptors is, for instance, the Thomsen-Friedenreich antigen (Galß(1→3)GalNAc), lectinologically detected by the peanut agglutinin. After the discovery of vertebrate lectins by Ashwell and his group (4), it has been postulated by us that the organ-characteristic lectins may act as acceptors in the metastatic process of tumors by interacting with the cryptic carbohydrate receptors of the circulating tumor cells (5,6,9). Accordingly, we studied the cell contact formation between different human tumor cells and human hepatocytes and found evidence for such a carbohydrate-dependent recognition mechanism. An evident conclusion is based on competitive carbohydrate inhibition by which glycoconjugates with terminal D-galactose residues inhibit the tumor cell adhesion. In accord with our hypothesis it could be demonstrated that the tumor cells bind to the asialoglycoprotein receptor on hepatocytes which again indicates the presence of complementary D-galactose-like residues on the tumor cell membranes.

Lectins, Vol. V

MATERIALS AND METHODS

Human tumor cells (HT-1080, fibrosarcoma; HeLa, cervix; MDA-231, mamma; Wi 38 VA 13, lung) were obtained from Flow Laboratories, West Germany.
All preparations were done under sterile conditions in a laminar flow hood.

Preparation of human hepatocytes

Pieces of liver biopsies were obtained from the Surgical University Clinic of Cologne and washed several times with Earle's Balanced Salt Solution (without Calcium and Magnesium) (Flow Laboratories). After slicing the biopsies with scalpels, the liver pieces were incubated with hyaluronidase (0.15 mg/ml) (Sigma Chemicals Company, USA) for 15 min at 37° C with moderate shaking (90 oscillations/min). After incubation the mixture was filtered through a 150 µm nylon mesh to remove undigested material and the resulting filtrate was centrifuged at 200 rpm for 10 min. The supernatant was decanted and the cellular pellet was suspended in Minimun Essential Medium, Eagle (modified) (Flow Laboratories), fortified with 10% Ultroser G (LKB, Sweden) plus glutamine (2 mM, final concentration) and protected against bacterial growth by addition of Penicillin (10 U/µl) and Streptomycin (10 µg/ml). Yield and viability of the cells were determined in a Neubauer Chamber using the Trypan Blue exclusion technique.

Treatment of tumor cells and the "reversed" Rosette Assay

Tumor cells (1×10^7/ml) were suspended in Minimum Essential Medium, Ultroser G (10%), glutamine (2 mM) and incubated in Petri dishes at 37° C in 5% CO_2 for 12 h.
20 µl of neuraminidase test (1 U/ml from Vibrio cholerae) (Behring Werke AG, Marburg, FRG) was added to 2×10^7 tumor cells in 1 ml and incubated at 37° C for 30 min. After three washes of the neuraminidase-treated tumor cells with Earle's Balanced Salt Solution a hepatocyte suspension (1×10^7/ml) was added to the tumor cell culture and incubated at 37° C in 5% CO_2 for 1 h. The supernatant was decanted and after careful washing with phosphate-buffered saline (Dulbeccos Formula, modified) (Flow Laboratories) the number of the rosette-forming cells was determined.
In other assays glycoconjugates with terminal D-galactose residues (5 mg/ml arabinogalactan, 5 mg/ml edible bird's nest glycoprotein, 5 mg/ml Pullulan as control), which inhibit the tumor cell-hepatocyte rosette formation, were added to the growth medium of the hepatocyte suspension to block the D-galactose-specific receptors (lectins) on the surface of the hepatocytes. The reversed rosette forming assay was performed in the above described manner.

RESULTS

Various tumor cells (fibrosarcoma, cervix carcinoma, lung carcinoma, mammary carcinoma) were tested in a binding assay with freshly isolated human hepatocytes to demonstrate the role of the D-galactose-specific lectin for cell contacts and adhesion. Figures 1 - 3 show that the hepatocytes spontaneously bind to tumor cells in vitro at 37° C and 5% CO_2, and that addition of 5 mg/ml D-glucose does not inhibit the rosette-forming process, whereas treatment of the tumor cells with neuraminidase (1 U/ml) leads to an increased adhesion of hepatocytes. In accordance with the hypothesis that blood group antigens of carbohydrate character and their precursors (exhibiting terminal galactosyl residues) seem to be important signals during malignant transformation of normal cells and that D-galactosyl-specific lectins represent the most important receptors for adhesion and binding of glycoproteins and cells (5), we could demonstrate: 1) that there is a spontaneous rosette formation of tumor cells and hepatocytes in carbohydrate-free culture medium, 2) that addition of 5 mg/ml D-glucose or other glycoconjugates without terminal D-galactose residues do not inhibit rosette formation, 3) that neuraminidase treatment of the tumor cells leads to increased rosette formation, and 4) that galactose oxidase (Sigma Chemicals Company, USA) treatment of the tumor cells completely abolishes rosette formation. These experiments demonstrate the importance of special glycoconjugates (e.g. terminal Gal or GalNAc, the Thomsen-Friedenreich antigen) for adhesion and binding of tumor cells and hepatocytes and confirm the discovery by Ashwell and his group (4) of ß-galactosido-specific lectins, which may be responsible for the organotropy of metastasis (5), in vertebrates and in various organs of the human body.

Based on these experiments and in accordance with our hypothesis that blocking the receptors (lectins) with ß-galactosido-glycoconjugates would diminish or inhibit contact and adhesion of the cells, we incubated hepatocytes with different glycoconjugates (5 mg/ml arabinogalactan, 5 mg/ml D-galactose, 5 mg/ml edible bird's nest glycoprotein) and repeated the reversed rosette forming assays with various tumor cells. The results of our carbohydrate inhibition experiments for the adhesion of hepatocytes confirmed our hypothesis. D-Galactose and related carbohydrates inhibit cell contacts strongly (edible bird's nest glycoprotein, Fig. 4) or completely (D-galactose, arabinogalactan, Fig. 5). Our inhibition studies with different glycoconjugates show that cell contacts between tumor cells and hepatocytes involve D-galactose-specific lectin-carbohydrate interactions and that blocking the receptors leads to complete loss of adhesive properties (7,8). Our experiments, however, do not rule out the other possibility, which has also been described (14), that lectins of the tumor cell may interact with organ-characteristic carbohydrate structures. This would be analogous to the organotropic infection modus of bacteria, which is also mediated by lectins, either on the membrane of the invading microorganism or on the cell surface of the infected organ.

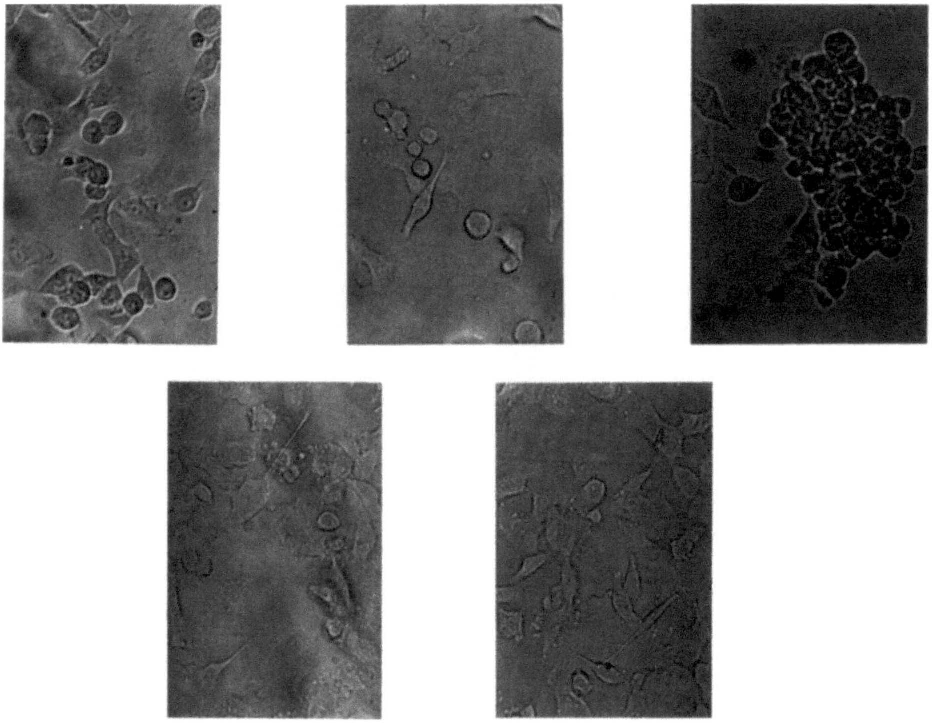

Figures 1 - 5. Reversed rosette forming assay of human tumor cells (fibrosarcoma) and freshly isolated human hepatocytes without any carbohydrates in the growth medium (Fig. 1) and with D-glucose (5 mg/ml) (Fig. 2) as control; after neuraminidase treatment (1 U/ml) of the tumor cells (Fig. 3) and after inhibition of the rosette formation by addition of edible bird's nest glycoprotein (5 mg/ml) (Fig. 4) and arabinogalactan (5 mg/ml) (Fig. 5).

DISCUSSION

Reversed rosette forming assays demonstrate that in vitro human hepatocytes spontaneously adhere to various tumor cells at 37° C,

5% CO_2, independent of the metabolic activities of the cells. Probably they form a privileged target for metastasis in vivo. Binding of hepatocytes by tumor cells is not a non-specific aggregation of cells, but is the result of a specific recognition process due to D-galactose-specific lectin-carbohydrate interactions (10). Inhibition studies with various glycoconjugates exhibiting terminal D-galactose residues confirmed that cell contacts between tumor cells and hepatocytes involve lectin-carbohydrate interactions. This lectin mediating the D-galactose-specific adhesion phenomenon of hepatocytes (5,11) seems to be indistinguishable from the hepatic receptor for asialoglycoproteins described first by Ashwell and Morell (4). Lectin-like receptors of hepatocytes recognize membrane-bound glycoconjugates with terminal D-galactose residues which are characteristic for tumor cells and often belong to membrane-integrated tumor markers (12,13). These ß-galactosido structures, which appear de novo after loss of neuraminic acid, may be a signal for the removal of their carriers from the circulation. This may be the case for all "glycophorin"-bearing cell membranes or for serum asialoglycoproteins which may have different "uncovered" receptors for lectin-like acceptors in the liver (5).

The discovery by Ashwell et al. (4) of the occurrence of ß-galactosido-specific lectins in vertebrates and various organs of the human body created new aspects about the mechanism of metastasis. The organotropic distribution of metastases might be the result of interactions of organ-characteristic lectins with anti-galactosyl specificity as acceptor and tumor cells with cryptic or exposed precursor ß-galactosido structures as receptor (5). Our in vivo experiments with human hepatocytes and human tumor cells offer strong evidence for this hypothesis. Moreover, the reversed rosette-forming assays show that blocking these receptors with ß-galactosido-glycoconjugates (arabinogalactan and other galactans) inhibits cellular adhesion. The significance of this finding with special emphasis on the inhibition of the metastatic process (e.g. during surgical operations) by blocking organ lectins with the appropriate carbohydrates, for instance neoglycoproteins or galactans has to be confirmed by further experiments with labeled arabinogalactans and their remaining in the circulation.

However, sometimes things do not seem to be so simple as printed out here, because our own studies on the organotropic bacterial adherence of certain microorganisms with a high content of sialic acid, for instance streptococci of type B, and the investigations of the Schirrmacher group (15) have shown that we must take into consideration also the possibility that the human organism or certain human cells may have neuraminic acid-specific lectins too. Accordingly, both masking by sialic acid and unmasking (loss of sialic acid) of certain lectin receptor sites may be of great importance for the metastatic potential of the invasive tumor cells.

ACKNOWLEDGEMENT

The work has been supported by Deutsche Forschungsgemeinschaft (Uh 8/14-2) and has been generously supplied by Medac GmbH, Hamburg, FRG.

References

1. Fidler, I.J., and Nicolson, G.L. (1981) Cancer Biol. Rev. 2, 171.

2. Nicolson, G.L. (1984) Exp. Cell Res. 150, 3.

3. Nicolson, G.L., and Poste, G. (1982) Curr. Prob. Cancer 7(6), 4.

4. Ashwell, G., and Morell, A.G. (1974) Adv. Enzymol. 41, 99.

5. Uhlenbruck, G. (1981) Immunol. Commun. 10, 251.

6. Uhlenbruck, G., Beuth, J., and Weidtman, V. (1983) Experientia 39, 1314.

7. Uhlenbruck, G. (1982) Ärztliche Praxis 43, 1723.

8. Roos, E. (1984) Biochim. Biophys. Acta 738, 263.

9. Hakomori, S.-I. (1985) Cancer Res. 45, 2405.

10. Schirrmacher, V., Altevogt, P., Fogel, M., Dennis, J., Waller, C.A., Barz, D., Schwartz, R., Cheingsong-Popov, R., Springer, G., Robinson, P.J., Nebe, T., Brossmer, W., Vlodavsky, J., Pawletz, N., Zimmermann, H.P., and Uhlenbruck, G. (1982) Invasion Metastasis 2, 2.

11. Maury, C.P.J. (1983) Scand. J. Gastroenterol. 18, 321.

12. Springer, G.F., Desai, P.R., and Banatwala, J. (1975) J. Natl. Cancer Inst. 54, 335.

13. Springer, G.F., Desai, P.R., Tegtmeyer, H., Schirrmacher, V., and Cheingsong-Popov, R. (1983) Naturwiss. 70, 98.

14. Gabius, H.J., Vehmeyer, K., Engelhardt, R., Nagel, G.A., and Cramer, F. (1985) Cell and Tissue Research 241, 9.

15. Fogel, M., Altevogt, P., and Schirrmacher, V. (1983) J. Exp. Med. 157, 371.

LECTIN IN CARCINOMA CELLS: LEVEL REDUCTION AS POSSIBLE REGULATORY EVENT IN TUMOR GROWTH AND COLONIZATION

Hans-Joachim Gabius, Reinhild Engelhardt, Gerhart Graupner and Friedrich Cramer

Max-Planck-Institut für experimentelle Medizin, Abteilung Chemie, Hermann-Rein-Str. 3, D-3400 Goettingen, FRG

Glycoconjugates (glycoproteins and glycolipids) are known to exert various biological functions and are as part of membranes intimately involved in cognitive cellular interactions (1-3). Endogenous lectins that bind to the carbohydrate moieties of glycoconjugates have been shown to be located within a wide variety of cells and membranes (4,5). Whereas alterations of glycoconjugates upon malignant transformation are comparatively well documented (6-8), possible changes in the concentration of lectins in tumor cells are thus far only studied for the asialoglycoprotein receptor of liver (9,10). Furthermore the presence of lectin activities have been detected for a limited number of tumor cell lines (11) and have been biochemically characterized for some solid tumors (12,13). The evidence of the presence of endogenous lectins in tumors indicates the possibility that the interplay between lectins and their carbohydrate receptors may contribute to the process of tumor growth at the primary site, to cohesiveness of tumor cells and embolization and organotropic distribution of metastasis. Therefore we intended to compare the abundant ß-galactoside binding lectin from spontaneously occurring uterus adenocarcinoma of DA rats and lung carcinoma of Han:NMRI mice with that of the corresponding normal cells. This ß-galactoside binding lectin differs from the hepatic binding glycoprotein in molecular weight, in specific requirement of thiol-reducing groups and in independence from divalent cations for activity (9,10). We found that the levels of this lectin from both types of tumor cells were significantly reduced and that the purified lectin from tumor cells appeared to

Lectins, Vol. V

display characteristics nearly identical to those of the lectin from normal, non-malignant cells in all functional tests.

Materials and Methods

The ß-galactoside specific lectins from tumor and normal tissue of the same rodent strain, obtained from Central Institute for Animal Breeding (Hannover, FRG), were purified using affinity chromatography on asialofetuin-Sepharose 4B and subsequently on lactose-Sepharose 4B by elution with 0.3 M lactose in 75 mM phosphate buffer (pH 7.2), 75 mM NaCl, 2 mM ethylenediaminetetraacetic acid and 2 mM ß-mercaptoethanol (MEPBS) essentially as described (14,15). The tumors were trimmed of necrotic parts. Haemagglutination activity using trypsinized, glutaraldehyde-fixed rabbit erytrocytes was assayed and quantitated defining the specific activity as the titer^{-1} divided by milligrams of protein per ml as described (16,17). Orosomucoid-fragment binding inhibition is performed as described (15). Aggregation of cerebroside micelles and mediation of adhesion of trypsinized baby hamster kidney (BHK) cells were carried out using established methods (18-20). Gel electrophoresis in the presence of 0.1 % sodium dodecyl sulfate was performed on 10 % running gel with a 4 % stacking gel (21). Fluoresceinyl-lactosylated bovine serum albumin was prepared essentially as described (22).

Results and Discussion

Optimal extraction for both tumor and normal cells was performed in the presence of 0.3 M lactose for uterus and 0.1 M lactose for lung, suggesting in situ-binding of lectin and receptor. Repeated extractions with 0.6 M NaCl and lactose resulted only in less than 4 % of the activity excluding inhibition of extraction by conjugative ligands in tumor cells. After centrifugation the supernatant was dialyzed four times against 20 volumes of MEPBS to remove lactose and pellets were further treated with MEPBS containing 2 % Triton X-100 and 0.4 M KCl instead of 75 mM NaCl with nearly no further extraction of lectin activity. Therefore the extractability by sugar and salt suggests that the lectin is not an integral membrane, but a soluble cytoplasmic and/or extracellular protein. Relatively to the corresponding control the specific activity in extracts for the uterus tumor was 15 % (27 % per initial cell protein due to relatively increased amount of extractable protein by a factor of 1.8) and for the lung tumor 40 % without significant changes in extractable proteins (Fig. 1). Only for mouse carcinoma cells a low, but clearly detectable amount of lectin activity was extracted by detergent. For all cases, subsequent chromatographic

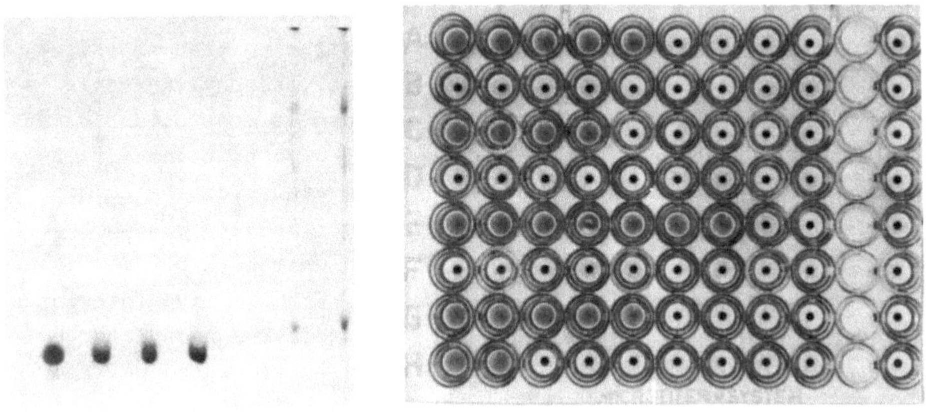

Fig. 1. Lectin activity determination and purity characterization. Left: Haemagglutination assay in microtiter wells using serial 2-fold dilutions of extracts from normal (A,B) and malignant (C,D) rat uterus cells, from normal (E,F) and malignant (G,H) mouse lung cells. The plate shows a typical result of 8 independent determinations with standard error of the mean of 12 %. The first row for each type of cells with dilutions from 1:2 (position 1) to 1:512 (position 9) refers to extraction in the presence of lactose, the second to subsequent extraction in the presence of detergent with undiluted extracts (position 1) to a dilution of 1:256 (position 9) after removal of detergent by chloroform extraction. The protein concentrations and resulting specific activities for lectins were: (A) 2.5 mg/ml, 12.8 U/mg; (B) 29 mg/ml, -; (C) 8.0 mg/ml, 2 U/mg; (D) 23 mg/ml, -; (E) 12 mg/ml, 10.6 U/mg; (F) 40 mg/ml, -, (G) 7.5 mg/ml, 4.3 U/mg; (H) 40 mg/ml; 0.125 U/mg. Control wells without extract are shown in position 11. Right: Disc gel electrophoresis of rat uterus lectins from normal and malignant (first and second lane) cells and of mouse lung lectins from normal and malignant (third and fourth lane) cells using the following molecular weight standards (fifth and sixth lane): bovine serum albumin (66 000), egg ovalbumin (45 000) and ß-lactoglobin (18 400).

steps on asialofetuin-Sepharose 4B and lactose-Sepharose 4B resulted in purification of the ß-galactoside specific lectin to homogeneity, as judged by gel electrophoresis under denaturing conditions (MW 14 000) (Fig. 1). Using the highly sensitive silver stain, a faint band at an apparent molecular weight of 34 000 was visible, excluding any major interference of the assay, described later, by other lactose binding proteins. A reduced level of lectin in tumor cells as reason for diminished specific activity in extracts is suggested by nearly identical specific activity for all preparations and a decreased yield of the soluble protein from 0.07 % and 0.045 % for normal uterus and lung, respectively, to 0.02 % for malignant cells.

The saccharide specificity of haemagglutination and orosomucoid-fragment binding inhibition is closely similar to that of calf heart (14), human and rat lung (15) and embryonic chick muscle lectins (23) with thiodigalactoside and lactose as most potent sugar inhibitors and strong inhibition caused by glycoproteins with ß-galactose terminus like asialofetuin. Periodate oxidation eliminated the inhibitory activity. Both types of lectins also caused aggregation of cerebroside micelles (12 mol % N-palmitoyl DL-dihydrolactocerebroside) indicating the possibility of an in vivo-interaction with this type of glycoconjugate. Fluoresceinyl-lactosylated bovine serum albumin demonstrated for frozen sections and unfixed culture cells of uterus that the lectin is located predominantly intracellularly with a measurable amount located on the cell surface in accordance to other reports (16,24-27). In this respect, it is noteworthy that externalization of an endogenous lectin has been shown to occur during development (28).

When the different purified and iodinated lectins were coupled by glutaraldehyde to inert, aminopropyltriethoxysilane-treated glass surface to test for mediation of adhesion of trypsinized baby hamster kidney (BHK) cells (19,20), they all caused cell attachment to the glass surface and cell spreading with similar efficiency at a threshold concentration of 45 - 55 ng lectin cm^{-2} (Fig. 2). This

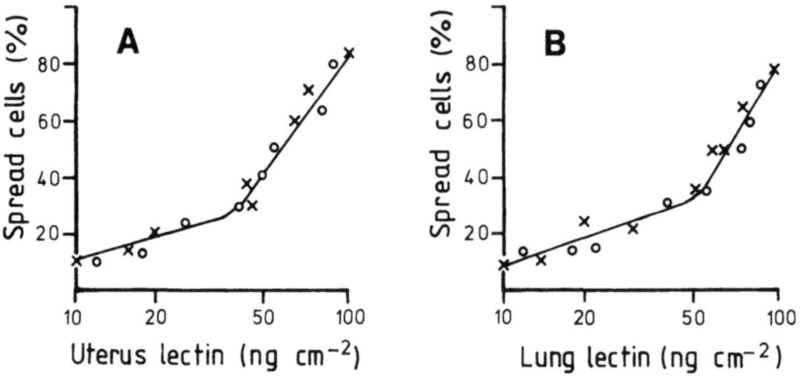

Fig. 2. Cell spreading of trypsizized baby hamster kidney cells on glass coated with different concentrations of rat uterus (A) and mouse lung lectins (B) from normal (X) and malignant (O) cells. Results are means of 5 - 9 repeated experiments with a standard error of the mean from 8 - 15 %.

value is in aggreement with 40 ng lectin cm^{-2}, reported earlier for the endogenous lectin from bovine heart (20). Thus the lectin from both types of cell pairs appeared to mediate attachment and spreading as a prerequisite for a possible role in cognitive intercellular lectin-receptor interaction. No functional test could detect any difference between the lectin of the described normal and malignant cells besides the reduced level in the malignant phenotype. Since ß-galactoside binding lectins from various normal tissues of the same animal showed different subcellular locations including the cell surface (16,24-27) and furthermore induction of homotypic aggregation of B16 melanoma by asialofetuin indicated the presence of a lectin of this type on the cell surface (29), a possible function for this lectin type in cellular recognition for at least some tissues appears to be a reasonable proposal. In terms of this suggestion a decrease of the lectin level in malignant cells can influence cell cohesiveness in the primary tumor and certain subsequent steps up to arrest at another site, leading to metastasis with typical non-random pattern (30,31). Uterus adenocarcinoma cells for example primarily lead to tumor colonization at pulmonary sites. It is interesting to note that among rat organs uterus and lung showed a particular abundance for this lectin (15).

The idea of selective level alteration of lectins at a definite point in cellular development is further supported by occurrence of developmentally regulated lectins in different animals (for review see 4,32) and activation-dependent expression of the surface receptor for mannose-containing glycoproteins by murine macrophages (33). It now becomes important to test critically the possible functional involvement of lectins in the pathway of metastasis formation by analysis of type and concentration of receptor glycoconjugates in different tissues. Studies by antibodies directed against lectins and/or the receptor molecules will be helpful in this respect.

References

1. Hughes, R.C. Membrane Glycoproteins Butterworth, London (1976).
2. Olden, K., Parent, J.P., White, S.L. BIochim. Biophys. Acta 650, 209-232. (1982).
3. Sharon, N. Biol. Cell 51, 239-246. (1984).
4. Barondes, S.H. Ann. Rev. Biochem. 50, 207-231. 1981).
5. Uhlenbruck, G.G. Funkt. Biol. Med. 2, 40-48. (1983).

6. Hynes, R.O., Fox, C.F. Tumor Cell Surfaces and Malignancy. Alan R. Liss, New York, (1980).

7. Yogeeswaran, G. Adv. Cancer REs. 38, 289-350. (1983).

8. Hakomori, S. Cancer Res. 45, 2405-2415. (1985).

9. Stockert, R.J., Becker, F.F. Cancer Res. 40, 3632-3632. (1980).

10. Schwartz, A.L., Fridovich, S.E., Knowles, B.B., Lodish, H.F. J. Biol. Chem. 256, 8878-8881. (1981).

11. Monsigny, M., Kieda, C., Roche, A.-C. Biol. Cell. 47, 95-110. (1983).

12. Gabius, H.-J., Engelhardt, R., Rehm, S., Cramer, F. J. Natl. Cancer Inst. 73, 1349-1357. (1984).

13. Gabius, H.-J-. Engelhardt, R., Cramer, F., Bätge, r., Nagel, G.A. Cancer Res. 45, 253-257. (1985).

14. DeWaard, A., Hickman, S., Kornfeld, S. J. Biol. Chem. 251, 7581-7587. (1976).

15. Powell, J.T. Biochem. J. 187, 123-129. (1980).

16. Nowak, T.P., Haywood, P.L., Barondes, S.H. Biochem. Biophys. Res. Commun. 68, 650-657. (1977).

17. Roberson, M.M., Barondes, S.H. J. Biol. Chem. 257, 7520-7524. (1982).

18. Curatolo, W., Yau, A.O., Small, D.M., Sears, B. Biochem. 17, 5740-5744. (1978).

19. Aplin, J.D., Hughes, R.C. Anal. Biochem. 113, 144-148. (1981).

20. Stojanovic, D., Hughes, R.C., Feizi, T., Childs, R.A. J. Cell. Biochem. 21, 119-127. (1983).

21. Laemmli, U.K. Nature 227, 680-685. (1970).

22. Kieda, C., Roche, A.-C-. Delmotte, F., MOnsigny, M. FEBS Lett. 99, 329-332. (1979).

23. Nowak, T.P., Kobiler, D., Roel, L.E., Barondes, S.H. J. Biol. Chem. 252, 6026-6030. (1977).

24. Kobiler, D., Beyer, E.C., Barondes, S.H. Develop. Biol. 64, 265-272. (1978).

25. Briles, E.B., Gregory, W., Fletcher, P., Kornfeld, S. J. Cell Biol. 81, 528-537. (1979).

26. Beyer, E.C., Tokuyasu, K.T., Barondes, S.H. J. Cell Biol. 82, 565-571. (1979).

27. Childs, R.A., Feizi, T. Cell Biol. Int. Rep. 4, 755. (1980).

28. Barondes, S.H., Haywood-Reid, P.L. J. Cell Biol. 91, 568-627. (1981).

29. Raz, A., Lotan, R. Cancer Res. 41, 3642-3647. (1981).

30. Fidler, I.J., Gersten, D.M., Hart, I.R. Adv. Cancer Res. 28, 149-250. (1978).

31. Poste, G., Fidler, I.J. Nature 283, 139-146. (1980).

32. Barondes, S.H. Science 223, 1259-1264. (1984).

33. Imber, M.J., Pizzo, S.V., Johnson, W.J., Adams, D.O. J. Biol. Chem. 257, 5129-5135. (1982).

ISOLATION AND PARTIAL CHARACTERIZATION OF THE N-ACETYL HEXOSAMINE SPECIFIC LECTIN IN THE MUCUS OF *ARCHACHATINA (CALACHATINA) MARGINATA*

R.O. Okotore and E.I. Nwakanma.
Department of Biochemistry, College of Medicine, University of Lagos, P.M.B. 12003, Lagos, Nigeria.

The three major biological compounds which have been identified and isolated in snails are:

(a) polysaccharides of the galactan type (1),

(b) several proteinase inhibitors (2), and

(c) anti-N acetyl-D-galactosamine specific lectins and other lectins, (3)

They have been found predominantly in the albumin glands, eggs and body fluids of different snails (3,4). The snail agglutinins are species-specific. They do exhibit diverse blood group specificity but most of them are of anti-A type (3). Suprisingly, erythroagglutinin has neither been detected in the body fluids nor in the albumin glands of African land snail species.

This paper reports the investigation on the occurence of erythroagglutinins in the body-surface mucus secretion of an African giant land snail, Archachatina marginata (Calachatina) which has not been examined so far. Such information may lend credence to the concept of protective functional role attributed to snail agglutinins (3). Also, it may be useful for establishing taxonomic and ecological differences within snail species (4).

Lectins, Vol. V

MATERIALS AND METHODS

SNAILS: The African giant land snails, *Archachatina marginata* (approx. body wt, 200-300g) were purchased from the local markets in Lagos, Nigeria. Each was deshelled to gain access to the visceral tissues. The body-surface mucus material was scrapped carefully from the foot and the visceral tissue with a scarpel into clean containers.

CHEMICALS: Agarose, Highly purified agglutinins from peanut, *Arachis hypogaea*, PNA, were purchased from E-Y Laboratories, San Mateo, U.S.A. Sepharose 4B was a product of Sigma Chemical Company, Missouri, U.S.A. whilst Cyanogen bromide was purchased from Koch-Light. Colnbrook U.K. Acrylamide, bis-acrylamide, ammonium persulfate, TEMED, and Coomassie blue were bought from Bio-Rad Laboratories, U.S.A.

EXTRACTION OF AGGLUTININ: The pooled material was diluted with twice the volume of saline, thereby reducing the viscosity. Solid crystals of ammonium sulphate was added in small portions to the solutions with continuous stirring up to the 50% saturation level (313g of salt/1000ml). The mixture was allowed to settle for two hours and thereafter centrifuged at 10,000 r.p.m. for 30 minutes. The supernatant was saved whilst the precipitate was discarded. Additional quantities of the salt were added to the supernatant to reach 70% saturation level. After standing for 2 hr at $4^{o}C$, the precipitate was collected by centrifugation at 8,000 r.p.m. for 15 min. Thereafter, it was dialysed against several changes of distilled water overnight at $4^{o}C$. The dialysate constituted the crude lectin preparation.

SYNTHESIS OF AFFINITY GEL: The affinity gel, N-acetylglucosamine-Sepharose 4B was synthesized as follows: Sepharose-4B (5ml suspension of swollen gel) was activated by cyanogen bromide (5) and washed with water on a glass filter for 30 min, then washed with 50ml of 0.2M sodium bicarbonate solution followed by water and then Tris-saline buffer, pH 7.5. A solution of N-acetylglucosamine (25mg) in 50mM Tris-saline buffer pH 7.5 (5ml) was added and the mixture was incubated at $20^{o}C$ for 20 h on a shaker. The product was washed extensively with water, 0.05M Tris-HCl (0.15M NaCl) pH 7.5. It was finally transfered to a column and equilibrated with the same buffer solution.

Affinity Chromatography: The dialyzed crude agglutinin preparation was applied to a column (20x4mm) containing N-acetylglucosamine-Sepharose 4B gel at a flow rate of 6ml/h. The effluent was continually recycled at $4^{o}C$ for 16h. Unbound material was then removed by washing with buffer, 0.05M Tris-HCl (0.15M NaCl) pH 7.5 after which the same buffer containing 0.2M N-acetylglucosamine was applied. Fractions (1ml) were collected and absorbance read at 450nM. The fractions under the peak eluted after the application of N-acetylglucosamine were pooled (approx. 5ml) and extensively dialysed against a large volume of saline overnight, and then centrifuged. The supernatant was analyzed by haemagglutination and electrophoresis.

ANALYTICAL PROCEDURES: Protein was determined by using the methods of Lowry *et al* (6) using crystalline bovine serum albumin as a standard. Carbohydrate content was measured by the phenol-sulfuric acid reaction (7) using D-galactose as a standard. Ouchterlony double-diffusion experiments were performed in a Tris /Ca^{2+} buffer system (10mM Tris, 1mM $CaCl_2$, 0.9% (w/v) NaCl, pH 7.3) and were allowed to proceed at 4°C for 24 hours. Analytical disc SDS-polyacrylamide gel electrophoresis was performed according to (8) using 10% separating gel at pH 8.0. The gels were fixed and stained with Coomasie blue.

Haemagglutination assay: This was performed in microtitre V-plates (Cooke, Coy U.K). Two-fold serial dilutions of agglutinin were made in Tris-buffered saline and 25ul of the 2% erythrocyte suspension. For sugar inhibition studies 25ul of Tris-buffered saline containing 10mM $CaCl_2$ were placed in control wells and 25ul of Tris-buffered saline was added to the experimental wells before addition of serial dilutions of agglutinin and the erythrocyte suspension. The plates were kept at room temperature and agglutination titers were read at 1 h and again at 16 h. The titers recorded at these two times agreed closely in all experiments. The red blood cells were treated with neuraminidase as follows: The red blood cells were washed several times with 0.01M Tris-HCl buffer, pH 7.2, containing 0.15M NaCl. A 0.5ml sample of the washed red blood cells was treated for 30 min. at 37°C with 0.002 units of neuraminidase dissolved in 0.1ml of the same Tris-HCl buffer. The neuraminidase-treated cells were washed twice with the same buffer before they were used in the haemagglutination assay.

RESULTS:

The analytical data of the product during the main stages of isolation procedure of the agglutinin is presented in Table 1. The purified agglutinin contained a relatively high proportion of carbohydrate which suggests that it is a glycoprotein.

The elution of the agglutinin from the affinity gel, N-acetyl-D-glucosamine sepharose was effected by the same amino sugar, N-acetyl-D-glucosamine but added at considerably high concentration, 0.5M. On SDS-polyacrylamide electrophoresis of the purified agglutinin, a single major protein band and two faint bands were observed with indicates presence of minor impurities in the preparation (fig.1). From the calibrated run of standard proteins (product of Kem-En-Tec, Copenhagen, Denmark) on similar gels of identical porosities, the estimated molecular weitht of the major band was found to be approx. 67.6 kD. The purified agglutinin agglutinated

TABLE 1

ANALYTICAL DATA OF AGGLUTININ FROM MUCUS OF ARCHACHATINA MARGINATA

PROPERTY	CRUDE MUCUS	CRUDE LECTIN	PURIFIED LECTIN
TOTAL CARBOHYDRATE (mg/ml.)	21.0	10.80	68.0
TOTAL PROTEIN (mg/ml.)	268.8	166.5	1600.0
RATIO OF PROTEIN CONTENT/TOTAL CARBOHYDRATE	12.80	15.42	23.53

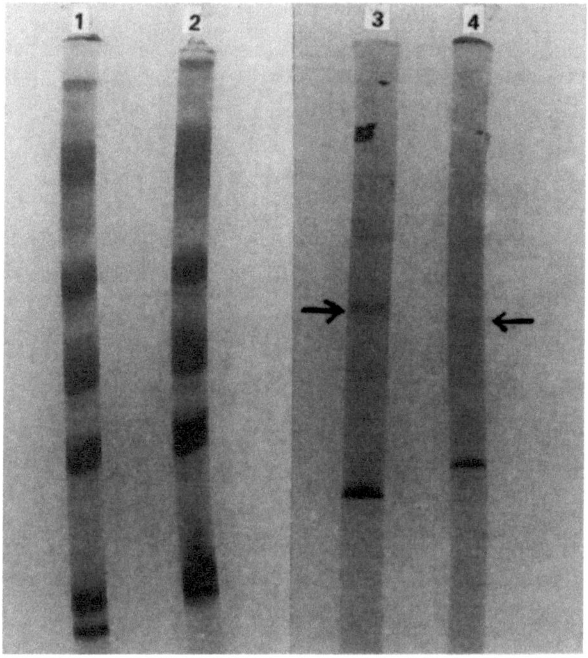

Fig.1

Sodium dodecyl suphate (SDS) polyacrylamide electrophoresis of erythroagglutinin in the body-surface mucus of the land snail, Archachatina marginata.
Lanes (1 and 2): SDS-PAGE M.Wt Markers-14/20/29/45/66/97 kd.
Lane 3: Unbound material from the affinity column, Lane 4: Bound agglutinin eluted from the affinity column.

TABLE 2

COMPARATIVE AGGLUTINATION OF ARCHACHATINA MARGINATA AGGLUTININ BY VARIOUS ERTHYROCYTES

ERTHYROCYTE SOURCE	TITRE-1 UNTREATED CELLS	TITRE-1 NEURAMINIDASE-TREATED
PIG	512	1024
RAT	128	256
HUMAN BLOOD GROUP A	2048	2048
HUMAN BLOOD GROUP B	Ø	32
HUMAN BLOOD GROUP O	Ø	16
SHEEP	256	1024
RABBIT	128	512

TABLE 3

HAEMAGGLUTINATION - INHIBITION OF ARCHACHATINA MARGINATA AGGLUTININ BY DIFFERENT CONCENTRATIONS OF N-ACETYL-D-GLUCOSAMINE

SUGAR CONCENTRATION (mM)	INHIBITION TITRE^{-1}
25	8
50	64
100	256
200	512
400	Ø
600	Ø

Cell - Human blood group A erythrocytes. 4 Hemagglutinating doses of agglutinin. The values represent average of 3 experiments with two different lectin preparations.

erythrocytes from various origin pig, rat, sheep, rabbit, and human blood group A but not human B and human O erythrocytes (Table 2). Neuraminidase-treated red cells were agglutinated with correspondingly slight increase in titre values (Table 2). The inhibition of haemaggluting activity was tested with various suggars: D-galactose, D-glucose, D-mannose, L-fucose, and N-acetyl-D-glucosamine at the concentration of 25mM. Of all tested sugars, only N-acetyl-D-glucosamine inhibited the haemaglutination reaction with an inhibition titre of 32. Subsequently, this amino sugar derivative was tested at various concentrations as high as 600mM. Inhibition was observed at low concentrations (25mM-200mM) with increasing titre values, but high concentration (400mM-600mM) caused no detectable inhibitory effect (Table 3).

DISCUSSION:

Mucus has been described as a unique protective secretion of complexity that occurs as a viscous sticky water-insoluble glycoprotein gel (9). Mucus is not confined to the higher vertebrates, i.e it is secreted externally by fish and invertebrates, snails and earthworms as slime trails.

This investigation has probed the occurence of agglutinin in the mucus of the African giant snail which was extracted at 70% ammonium sulphate saturation. The lectin was precipitated and further purified by affinity chromatography on a synthetic biospecific absorbent, N-acetyl-D-glucosamine-Sepharose which has been employed for isolation of wheat-germ lectin (10). This approach was similar to that employed by Iguchi and co-workers who reported the isolation of a novel sialic acid-specific lectin (Achatinin) from the mucus of a different African giant snail species, Achatina fulica (11). The elution of the agglutinin by N-acetyl-D-glucosamine suggests that the binding of the agglutinin to the matrix is biospecific. However, non-specific interactions between the agglutinin and the ligand may not be ruled out since higher concentration of N-acetyl-D-glucosamine was employed for the elution. Experiments using other sugars were not carried out.

The estimated molecular weight of the major protein band observed for the agglutinin in the SDS-polyacrylamide gel was 67,600 daltons. The preparation from the mucus of Achatina fulica had molecular weight of 78,000, (11). Generally, mammalian mucus glycoproteins possess inherent polydispersity. Whilst in most cases undegraded mucus glycoproteins have molecular weight of $2x10^6$ the proteolytically degraded counterparts have molecular weights of $2\text{-}7x10^5$ da/ton

Noteworthy is the specific agglutinating activity of the lectin towards human A erythrocytes. Agglutinins with anti-A-specificity have been well characterised from many snails, notably Helix snails for example Helix hortensis, Helix pomatia by Prokop and co-workers (3). Similar observation on Otala lactea was reported by Boyd and Brown (12). The immuno-determinant of the human A erythrocyte is ascribed to the terminal non-reducing N-acetylgalactosamine. The agglutinating activity against red cells from various origin and the specific inhibition of the agglutination reaction by N-acetylglucosamine indicates that it is most likely heterophile agglutinin with a much broader specificity. Moreover, the results also suggest that N-acetyl groups are essential to the binding to the agglutinin. In summary, the results of this investigation have brought to light the presence of another human red cell agglutinin of anti-A specificity from the body-surface mucus of African giant land snail. To our knowledge, this finding has not been reported in the literature for this pulmonate snail Archachatina marginata (Calachatina). The lectin shows some immunological resemblances to the "Achatinin", a sialic acid-specific lectin isolated from the body-surface mucus of a different African giant snail Achatina fulica (11). Despite the limited experimental data on the inhibition studies, it is apparent that N-acetyl groups are essential to the binding to the lectin. Thus, the novel feature of this lectin relates to the sugar specificity that could be ascribed to N-acetyl hexosamine residues.

The future prospect of this investigation lies in detailed chemical and immunological characterization of this agglutinin. It may serve as a useful tool to verify the immunobiological protective function in snails that has been advanced by Prokop and co-workers (13). Moreover, the snail agglutinins have been shown to be excellent tools for bacterial agglutination and therefore also for bacterial classification (13).

ACKNOWLEDGEMENTS: This work was supported by the University of Lagos Central Research Grant Award. The authors thank Mrs. A Solabi for the expert technical assistance and Mr. C. Owunna for typing the manuscript.

REFERENCES:

1. Uhlenbruck, G. Steinhausen, G & Baldo, B.A. Forschung and Lehre, Scriftenlehre Medizin, Verlag Stippak, Aachen, (1975).
2. Uhlenbruck, G. Sprenger, I, & Ishiyama, I.Z. Klin Chem. Klin Biochem. 9 36 (1971).
3. Prokop, O. Uhlenbruck, G, and Kohler, W. Vox. Sanquinis 14 321-333 (1968).
4. Uhlenbruck. G. Steinhausen, G, Palatnik, M. Comp. Biochem Physiol. 57B 335-339 (1977).
5. Cuatrecasas, P. J.Biol. Chem. 245, 3059-3065 (1970).
6. Lowry, O.H. Rosebrough, N.J. Farr, A.L. Randall, R.J. - J. Biol. Chem. 193 265-275 (1951).
7. Dubois, M. Gilles, K.A., Hamilton, J.K. Rebers, P.A. and Smith F. Analyt. Chem. 28 350-356 (1956).
8. Weber, K. and Osborn, M. J.Biol. Chem. 244 4406-4412 (1969).
9. Allen, A. Trends Biochem. Sci. 8 169-173 (1983).
10. Vretblad, P. B.B. Acta 434, 169-176 (1976).
11. Iguchi, S.M.M. Egawa, K. and Momoi, T. Proc. 7th Intern. Symp. on Glycoconjugates, Lund, Sweden 304 (1983).
12. Boyd, W.C. and Brown, R. Nature 208, 593 (1965).
13. Prokop, O. Uhlenbruck, G. Rothe, A and Cohen, E. Ann. N.Y. Acad. Sci. 234, 228-231 (1974).

COLD AGGLUTININ FROM *ACHATINA FULICA* SNAILS HAVING SPECIFICITY TOWARDS N-ACETYLLACTOSAMINE.

M. Sarkar, D. Mitra, B.K. Bachhawat, C. Mandal.
Indian Institute of chemical biology,
Jadavpur, Calcutta, India.

Invertebrate agglutinins, mainly from gastropoda (Mollusca) agglutinate human erythrocytes of specific blood group types (1). An anti-A agglutinin was first isolated from the albumin gland of the snail *Helix pomatia* (2). The presence of agglutinin in *Achatina fulica* and *Achatina granulata* snails has been reported earlier, but these were not purified and characterized (3-5). In our previous communication (6), we have reported the presence of a cold agglutinin in the albumin gland of *Achatina fulica* snail, which agglutinates rabbit and human umbilical cord erythrocytes only at low temperature (optimum 13°C), though there is a report of absence of any agglutinin in the albumin gland of *Achatina fulica* snail. This inconsistancy is probably due to the fact that this agglutinin is temperature sensitive. The cold agglutinin was purified to homogenity using asialofetuin-Sepharose 4B as an affinity column followed by ion-exchange chromatography. Inhibition data showed that it was specific for N-acetyllactosamine (Galβ1→4 GlcNAc). An agglutinin of such specificity has also been described from calf heart (7, 8) and from the seeds of *Erythrina crystagallis* (9), though these agglutinins were not reported to be cold agglutinins. The cold agglutinins so far reported are of human origin: monoclonal IgM antibodies (10) showing specificity to blood group I, i and Pr antigens (11, 12). Some of the cold agglutinins specific for blood group Ii antigens, can differentiate human adult and cord cells. The common core carbohydrate moiety of the Ii antigen is N-acetyllactosamine (13-15). Thus, the specificity toward N-acetyllactosamine and the agglutinating activity of the present cold aggllutinin bear some similarity with the cold agglutinin of human origin. This communication deals with further characterization of the N-acetyllactosamine specific cold agglutinin.

Lectins, Vol. V

Materials and Methods:

Achatina fulica snails were purchased from the local market. The albumin glands (165 g) from 180 *Achatina fulica* snails were dissected, homogenized and extracted with 0.01M phosphate-buffered saline (PBS), pH7.2 at 10° C. The agglutinin was purified to homogenity as described in our earlier Communication (6).

The molecular weight of the native agglutinin was estimated by gel filtration. The Sephadex G-200 column was calibrated using thyroglobulin, catalase, aldolase, as markers. Blue dextran was used to determine the void volume of the column.

The molecular weight of the subunits was measured by SDS-polyacrylamide gel electrophoresis acording to the method of Laemmli et al (16) in 10% polyacrylamide gel (pH8.8). The samples were treated with 1% SDS and 1% 2-mercaptoethanol for 10 minutes. Electrophoretic mobilities were calculated after staining with Coomassie blue.

Analytical isoelectrophocusing was carried out as described by Cisar et al (17) in 5% polyacrylamide gel containing 2% carrier pharmalyte (pH 3-10).

An antisera to crude extract of the albumin glands was prepared in rabbits. Each rabbit was injected with 0.2 ml diluted crude extract (1:10 dilution) intracutaneously without adjuvants three times during one week and then given one injection per week into the foot pads each with about 4 mg of crude extract in Freund's complete adjuvant (Difco Laboratories, Detroit, Mich.) for 3 weeks. They were bled after one week by cardiac puncture. The sera were tested by immunodiffusion according to Ouchterlony (18) using 1.5% agar in 0.05 M sodium barbital buffer (pH 8.3) and immunoelectrophoresis (19).

The carbohydrate composition was determined by gas-liquid chromatography. The glycoprotein was hydrolysed by acid and then derivatised to alditol acetates. The aminosugars were determined by modified Elson and Morgan method (20).

The agglutinin was treated with a variety of protein-modifying reagents in order to investigate the role of different amino acids in its hemagglutinating activity. All the treatments were performed under mild conditions and with reagents that had a high specificity for particular amino acid. A bulk of purified agglutinin were dialysed against saline to remove phosphate ions and used for chemical modifications (21). The results of chemical modifications are summerized in Table-1.

The agglutinin was treated with 0.1M NaOH and 0.05M $NaBH_4$ at 25°C for 24 hours followed by acidification by acetic acid and dialysis. The carbohydrate modified agglutinin was tested for agglutination both at 10°C and 30°C.

Hemagglutination assays were carried out in Takatsy microtitrator (Cooke Engineering Co., Alexandria, Va) using 25µl loops and 25µl of a 2% suspension of rabbit erythrocytes. Equal volumes of the red blood cell suspension and the agglutinin solution or its dilutions were mixed, placed at 10°C and 30°C temperature for 1 hour and scored.

Circular dichroism spectra of the agglutinin were measured on JASCO-J-20A recording spectropolarimeter at two different temperatures. Cell path length was 2 mm. Ellipticity was calculated as mean residue ellipticity,

$$[\theta] = \frac{M\theta}{10cl},$$

where M is the mean residue weight taken 110, θ is the observed ellipticity in degrees, c is the concentration of the agglutinin in g/ml and l is the path length in cm. No Lorentz corrections were made. The base line was determined in 0.01M sodium phosphate buffer. The errors due to noise was ± 50 to 80 degree cm ^{2} d mole^{-1} at 200-240nm. The concentration of the agglutinin was (0.4 mg/ml) as determined from known extinction coefficient ($E^{1\%}_{280nm}$ — 13.4). The change CD spectra was also checked by the addition of 3.2 mM and 6.4 mM of lactose (an inhibitor of the agglutinin). All readings were repeated three or four times both at 10°C and 30°C.

RESULTS

The cold agglutinin was purified to homogenity as checked by alkaline gel electrophoresis and further confirmed by gel filtration on Sephadex G 200 with a single peak corresponding to a molecular weight 200,00. It is composed of three non-identical subunits with a molecular weight of 84,000, 74,000 and 62,000 dalton as revealed by SDS-polyacrylamide gel electrophoresis. The isoelectric point of the agglutinin is 4.5. It is a glycoprotein containing 13.56% as D-galactose and 2.2% aminosugar. The purified agglutinin showed a single band by immunodiffusion and immunoelectrophoresis. The antiserum gave three bands with the crude extract.

The effect of chemical modification of some of the amino acid residues on the hemagglutination activity are summerized in Table 1. Hemagglutination was checked both at 10°C and 30°C. The agglutinin is highly sensitive to 1-ethyl-3-(3-dimethylaminoporpyl) carbodiimide followed by α-aminobutyric acid methyl ester which activate the carboxy group of glutamic acid, aspartic acid and the end terminal. Treatment with acetic anhydride which modifies the amino group of lysine and the N-terminal group as well as the hydroxyl group of tyrosine reduces agglutination. Treatment with citraconic anhydride which modifies the amino group of lysine and the N-terminal has no effect. The agglutinin is relatively insensitive to 2-hydroxy-5-nitrobenzylbromide, a reagent to modify tryptophan residues.

Table-1

Effect of chemical modification on hemagglutinating property.

The values for relative agglutination represent the ratio of the amount of native protein required for hemagglutination to the amount of chemically modified protein for achieving the same degree of agglutinability.

Modification	**Type of group modified**	**Relative Agglutination**
Native		1(at 10°C) -(at 30°C)
Acetylation	Lys ε-amino, N-Terminal α-amino, Tyr hydroxy	0.44(at 10°C) -(at 30°C)
Citraconylation	Lys ε-amino, N-Terminal α-amino	1(at 10°C) -(at 30°C)
2-Hydroxy -5-nitro-benzyl bromide.	Trp indole ring	1(at 10°C) -(at 30°C)
Carbodi-imide, then α-aminobutyric acid methyl ester	Glu γ-carboxy, Asp β-carboxy, C-Terminal α-carboxy	0.02(at 10°C) -(at 30°C)
Carbohydrate modified	Removed 33% Carbohydrate	1(at 30°C) 6.9 (at 10°C)

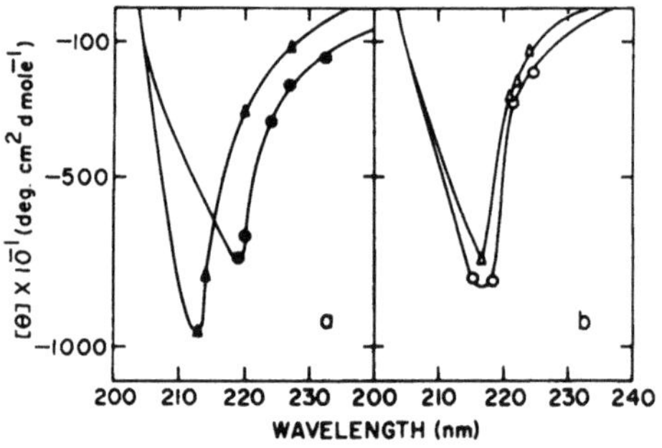

Fig 1(a): CD spectra of purified cold agglutinin at two temperatures at 10°C (—▲—) and 30°C (—●—).
(b): CD spectra of purified cold agglutinin in presence of 3.2 mM(—○—) and 6.4 mM (—△—) lactose respectively.

Removal of carbohydrate residues (33%) from the cold agglutinin by β-elimination using alkaline $NaBH_4$ changes the agglutinin to agglutinate both at 10°C and 30°C. From Table 1, it is evident that the relative agglutination of the modified agglutinin increases 6.9 times at 10°C.

The circular dichroism spectrum of the cold agglutinin at two different temperatures are shown in Fig. 1. A negative band in the far UV region at 213nm was noticed. An increase in the temperature from 10°C to 30°C caused a change in the intensity of the band and the maximum shifted to 219nm. Addition of lactose changed the $[\theta]$ value.

DISCUSSION

The cold agglutinin from the albumin gland of *Achatina fulica* snail is different from agglutinins from plants and animals as it agglutinates only at low temperature. The cold agglutinins known so far are of human origin. Therefore, purification and characterization of the *A*chatina cold agglutinin seems to be important. The cold agglutinin was demonstrated to be homogeneous by a variety of techniques. Gel filtration studies indicate that the purified agglutinin has a molecular weight of 220,000 consisting of three subunits of molecular weight 84,000, 74,000 and 62,000 as determined by SDS-polyacrylamide gel electrophoresis. The purified agglutinin is a glycoprotein containing mainly galactose and some aminosugar. Some of the galactose residues are probably linked to protein through aminosugars by O-glycosidic linkages as it undergoes β-elemination reactions.

The change of the cold agglutination behavior by removing 33% carbohydrate is not yet clear. Chemical modification of amino acids confirms involvement of the carboxyl group of glutamic acid, aspartic acid, the terminal end and also the hydroxyl group of tyrosine.

The circular dichroism data of the purified cold agglutinin suggest that the cold agglutinin undergoes a considerable conformational alteration due to changes in temperature and addition of lactose. The alternation involves the peptide backbone.

Finally from these results it appears that the cold agglutinating property of the agglutinin is modulated by its carbohydrate chain and also the tertiary structure of the molecule.

ACKNOWLEDGEMENTS

This work was supported in part by grants from the Department of Science and Technology Government of India. D.M. is a junior research fellow of the Council of Scientific and Industrial Research, India. The authors wish to thank Dr. A. Sen, Indian Institute of Chemical Biology for his help in carbohydrate analysis by glc. The helpful suggestions of Dr. M. Choudhuri is also acknowledged. Mr. Hjalmar Johansen is thanked for typing the manuscript.

References:

1. Prokop, O., Uhlenbruck, G., and Kohler, W. (1968) Vox. Sang. **14**, 321-333.
2. Hammarstrom, S., and Kabat, E.A. (1971) Biochemistry **10**, 1684-1692.
3. Gold, E.R., Cann, G.B., and Thompson, T.E. (1967) Vox. Sang **12**, 461-464.
4. Brain, P. and Grace, H.J. (1968) Vox. Sang **15**, 297-299.
5. Khalap, S., Thompson, J.E., and Gold, E.R. (1971) Vox. Sang **20**, 150-173.
6. Sarkar, M., Bachhawat, B.K. and Mandal, C. (1984) Arch. Biochem. Biophys. **233**, 286-289.
7. De Waard, A., Hickman, S., and Kornfeld, S. (1976) J. Biol. Chem. **25**, 7581-7587.
8. Childs, R.A., and Feizi, T. (1979) FEBS Lett. **99**, 175-179.
9. Iglesias, J.L., Lis, H., and Sharon, N. (1982) Eur. J. Biochem. **123**, 247-252.
10. Tsai, C.M., Zopf, D.A., Yu, R.K., Wister, R. Jr. and Ginsburg, V. (1977) Proc. Natl. Acad. Sci. USA **74**, 4591-4594.
11. Roelcke, D. (1974) Clin. Immunol. Immunopathol. **2**, 266-280.
12. Feizi, T., Kabat, E.A., Vicari, G., Anderson, B., and Marsh, W.L. (1971) J. Immunol. **106**, 1578-1592.
13. Kemler, R., Baninet, C., Eisen, H. and Jacob, F. (1977) Proc. Natl. Acad. Sci. USA **74**, 4449-4452.
14. Niemann, H., Watenabe, K., Hakomori, S., Childs, R. and Feizi, T. (1978) Biochem. Biophys. Res. Commun. **81**, 1286-1293.
15. Watanabe, K., Hamomori, S., Childs, R.A. and Feizi, T. (1979) J. Biol. Chem. **254**, 3221-3228.
16. Laemmli, U.K., and Favre, M. (1973) J. Mol. Biol. **80**, 575-599.
17. Cisar, J., Kabat, E.A., Dorner, M.M. and Liao, J. (1975) J. Exp. Med. **142**, 435-458.
18. Ouchterlony, O. (1948) Acta. Pathol. Microbiol. Scand. **25**, 186-191.
19. Grabar, P. and Williams, C.A. (1983) Biochem. Biophys. Acta. **10**, 193-194.
20. Kabat, E.A., Kabat and Mayer's Experimental Immunochemistry, 2nd edn., Thomas Springfield, Ill. (1961), pp 499-502, 504-511.
21. Ashford, D., Menon, R., Allen, A.K. and Neuberger, A. (1981) Biochem. J., **199**, 399-408.

FISCH CORTICAL VESICLE LECTINS - A NEW GROUP OF CARBOHYDRATE-BINDING PROTEINS

A Review

A. Krajhanzl and J. Kocourek
Department of Biochemistry, Faculty of Science, Charles University, Albertov 2o3o, CS-1284o Praha, Czechoslovakia.

One of the new and very interesting groups of sugar-binding proteins are the lectins of sexual cells. There are basically two types of lectin present in animal and perhaps also in plant gametes. These lectin systems differ mainly in their localization which itself bears witness to their different functions in living organisms. Lectins of the first group are predominantly hydrophobic proteins and occur as components of the outer gamete membranes (both sperm and oocytes). Reports of these lectins, designated more often as lectin-like proteins, are sporadic (e.g. papers 1-5). The evidence on their existence is more indirect (6-10). Sexual aggregation factors of lectin nature can be classified into the first group also. These have been demonstrated in the green algae of the genus Chlamydomonas (11-14). A little more information is available on the lectins of the second group, the hydrophilic proteins, which are localized intracellularly in the oocytes of both invertebrate and vertebrate animals. A number of these lectins can already be isolated and characterized from the physicochemical point of view (15-41). At present the most extensive is our knowledge of the lectins obtained from oocytes of bony fishes, the most numerous subgroup of intracellular lectins of animal sexual cells (21-34).

Papers on lectins from fish oocytes were first published in 1967 (42-5o). They dealt especially with the remarkable erythroagglutinating properties of extracts from ovaries of some bony-fish species. Previously Ginsburg had found that the perivitelline fluid of activated trout eggs agglutinates homologous spermatozoa

Lectins, Vol. V

(51), but he did not study the basis of this phenomenon. A successful mastering of isolation procedures for these lectins rendered it possible to start the study of their function in vivo. In this review we briefly summarize our present knowledge of the physicochemical and biological properties of these lectins and, on the basis of previous studies, we suggest their function in fish oocytes.

Isolation of lectins from fish oocytes

Lectins from oocytes of bony fishes were first obtained in pure form by Krajhanzl et al. (21-23) and Voss et al. (24). Affinity chromatography on carriers with immobilized D-galactose, D-glucose or L-fucose residues was successfully used, resp.. Most often O-glycosyl polyacrylamide gels poly(glycosyloxyalken-acrylamide) copolymers (52) or glycosyl derivatives of Separon (Spheron) (53) were used (21-23, 25-3o),

In some cases unmodified Sepharose (24, 31, 35), Sephadex (21, 32) or direct specific adsorption on human B or O red blood cells (33, 34) were also successfully used. Up to now lectins have been isolated from oocytes of sixteen fish species which belong taxonomically to the families Clupeidae, Osmeridae, Salmonidae, Cyprinidae, Cobitidae and Percidae. The simplest and, at the same time, most effective procedure proved to be direct batchwise affinity adsorption of the crude ovary extracts in mild acid buffer (acetate buffer pH 5.5) of higher ionic strength; after that the lectin is eluted by a solution of inhibiting sugar in the mentioned buffer (21-23, 25-3o). Fractionation of crude extracts with ammonium sulfate or with with acetone caused in some lectins a loss of their activity or significantly decreased yield (25, 28-3o). Of great importance for the yield of all hitherto studied lectins has been the stage of oocyte maturity (25, 28-3o). For final purity and maximal yields of lectin the most advantageous starting material is ovaries containing previtellogenic oocytes. Binding interactions with components of the eggs which are being formed during the later stages of the oocyte development (see below), exert negative influence in the homogenity, purity and the yield of lectins in isolations starting from vitellogenic oocytes (25, 27-3o).

In spite of the undisputable advantages of the simple isolation procedure (isolation in one step), the work with crude oocyte extracts evidently leads to partial degradation of lectins by proteinases. This modification is revealed most distinctly in the heterogeneity of preparations obtained from late vitellogenic oocytes. We do not know whether this degradation is physiological, i.e. whether it occurs also in situ in cortical granules.

Physicochemical properties

From the physicochemical point of view the cortical vesicle lectins form a very interesting group of proteins with a number analogous properties (Table 1). Common to most of the oocyte lectins is their ability to bind D-galactose or its derivatives and oligomers. It appears that two or three types of cortical lectins have been developed during phylogenesis, with different sugar-binding specifities. Besides the affinity for D-galactose which is common in the great majority of animal egg lectins, lectins isolated from oocytes of fishes belonging taxonomically to the families Clupeîdae, Osmeridae, Salmonidae, Cyprinidae, Cobitidae and Sparidae possess a very pronounced and otherwise very uncommon affinity for L-rhamnose. An analogous specifity was hitherto found only in a lectin produced by Streptomyces sp. (69). In addion to L-rhamnose and D-galactose these lectins bind also further C-2 and C-4 carbohydrate analogs of L-rhamnose, e.g. L-arabinose, D-fucose, L-mannose and their oligomers, and also D-galacturonic and galactaric acids (25, 3o). Quantitative interaction studies of some of these lectins show that the corresponding association constants are of the order of magnitude of 10^4 M^{-1} (25, 26, 29, 3o). Thus, they are one order higher than those of the majority of known plant lectins for their inhibiting monosaccharides.

The presence of lectins with binding specificity for L-rhamnose has been shown by erythroagglutination studies in species besides these quoted above (Table 1), e.g. in extracts from ovaries of further fish-species of the family Salmonidae: Salmo salar (43,

44, 5o, 54, 57, 59, 6o, 7o, 71), S. eriox (51, 71), S. trutta (54, 57, 58, 71), Oncorhynchus keta, O. gorbuscha, O. kisutch, O. nerka (24, 58, 59, 72), O. rhodurus (73), and in the species Silurus glanis (family Siluridae) (5o, 74) and Phoxinus percnurus (Cyprinidae) (58).

A specificity quite different from most of fish lectins was shown in lectins from oocytes of fish family Percidae, and in lectin I from oocytes of the tench, Tinca tinca and of Sparus berda (later Spondyliosoma cantharus) (see Table 1). They bind specifically D-glucose and/or L-fucose, and their C-2 and C-4 analogues, e.g. D-arabinose, D-mannose, D-xylose etc. The binding sites of all presently described fish cortical-vesicle lectins thus recognize in their interaction with saccharides preferentially the configuration of hydroxyl groups on C-2 and C-4 of their pyranose ring. For a very strong interaction the trans-position of substituents on C-4 and C-5 is usually necessary. In some cases the position of the anomeric hydroxyl group is also relevant. Thus the lectin from the roach oocytes binds more strongly methyl oc-L-rhamnoside and methyl β-D-galactoside than their anomeric counterparts, whereas the reverse is true with the oocyte lectins of the fish-genus Coregonus and the silver carp Hypophtalmichthys molitrix (21, 22, 25, 26, 3o).

Somewhat unexpected was a lectin specific for D-glucose found in ovary-extracts of the carp Cyprinus carpio, taxonomically a typical representative of the family Cyprinidae (75). In oocytes of all other fishes of this genus only lectins specific for L-rhamnose were detected. Therefore, it appears probable that in addition to the serologically-detected lectin agglutinating streptococci of the group E (75) the eggs of carp contain another lectin specific for L-rhamnose. Two lectins with different specifities were already isolated from oocytes of a related fish, the tench Tinca tinca (25). According to some serological papers a lectin specific for L-fucose and/or D-glucose is contained also in oocytes of fishes Blicca bjoerkna, Lepidopus caudatus, Phoxinus percnurus, and Dicentrarchus labrax (34, 58, 59, 65).

Pronounced analogies can be found in amino acid composition of the fish oocyte lectins. As in other lectins of animal eggs studied so far, strongly represented are aspartic and glutamic acids, serine, cysteine, glycine and threonine, i.e. acidic or polar amino acids. On the other hand the content of aromatic acids, especially tryptophane is usually relatively low. A further common feature is a relatively high content of calcium (usually one atom of Ca^{2+} per main subunit) and a low content of covalently bound neutral and amino sugars.

As shown by circular dichroism studies extensive domains of polypeptide chains of fish cortical lectins studied so far have a pleated sheet conformation, with surface-exposed aromatic amino acid residues (21, 25, 26, 28, 29). This predominantly β-conformation is considerably rigid; it withstands both pH and temperature changes, thanks evidently to the numerous intra- and inter-chain disulfidic cystin groups. There are indications that in the case of perch lectin I the native β-conformation, except for the binding-site region, is stabilized by calcium ions (21, 28, 32). The results of the circular dichroism studies cannot be at present reliably interpreted because of the heterogeneity of most isolated lectins. However, it appears that the oc-helical conformation is relatively uncommon. It corresponds roughly to the values obtained in lectins from other biological sources.

The quarternary structure of most fish cortical-vesicle lectins is relatively complicated. The isolated lectin preparations are considerably heterogeneous mixtures or oligomers of several (often three) main subunits accompanied with several minor subunits of similar molecular mass. There are, however, differences in the degree of aggregation. Thus, the lectin from oocytes of fishes belonging to the genus Coregonus are present in solution predominantly in form of monomers and noncovalently linked dimers (25, 26). In the case of the lectin from oocytes of the white fish, Coregonus peled, the main subunits A, B and C can be separated by affinity re-chromatography of the lectin with stepwise elution. All of them bind specifically L-rhamnose. The differ, however, in their affinity for other monosaccharides. There are differences also in their electrophoretic mobility, amino acid composi-

Table 1

Some properties of lectins isolated from bony-fish oocytes

Family Species	M_r (X10^3) (subunit types	Specificity	Sugar content (W/W%)	Metal content	Agglutinating properties	Biological properties	References
Clupeidae							
Clupea harengus	(1)	L-Rha			B P a		21,54
Osmeridae							
Hypomesus transpacificus nipponensis	68	L-Rha	yes		B c		35
Salmonidae							
Coregonus lavaretus maraena	b (3)	L-Rha	3-10 i	Ca,Mg	B d e	j	22,25,61
Coregonus peled	b (3)	L-Rha	4-6 i	Ca,Mg	B d e		25,26
Oncorhynchus tschawytscha	122	L-Rha			pB d f	k l	24,58,59
Salmo gairdneri irideus	200 (1)	L-Rha	5.4		pB P d g h	l	21,24,45,46 50,54-56
Cyprinidae							
Abramis brama	120 (3)	L-Rha	4.0	Ca,Mg	B		21,23,28,30
Hypophthalmichthys molitrix	170 (1)	L-Rha	3.7	Ca,Mg	B d e	m	25,26,64
Rutilus rutilus	180 (3)	L-Rha	2.2-4.6 i	Ca,Mg	B P^k d h	j l m	23,25,29,30 47,48,54-69-62
Scardinius erythrophthalmus	(3)	L-Rha	3.2-4.0	Ca,Mg	pB	j m	21,23,30,62,63

Tinca Tinca	I	(1)	L-Rha			B d	m	28,62,64
	II	(1)	D-Glc			d		
Vimba vimba		(3)	L-Rha	8.4		pB	j	21,23
Cobitidae								
Misgurnus anguillicaudatus		50	L-Rha	yes		B c		31
Percidae								
Perca fluviatilis	I	140 (1)	D-Glc	1.3-3.3 i	Ca,Mg	pO n	m q	23,25,27,28,32 44,58,61-63, 65,68
	II	(1)	L-Fuc		Ca	o		
Stizostedion lucioperca	I	(1)	L-Rha			B h		21,23,44,46 58,60
	II	(1)	L-Fuc			nonspec		
Percichtyidae								
Dicentrarchus labrax			L-Fuc			pO		34
Sparidae								
Spondyliosoma cantharus	I		L-Rha			B		33
	II					O		

Abbreviations: A,B,O,P,P^k = the lectin agglutinates human erythrocytes of quoted groups; pB, pO = predominantly human erythrocytes as specified; a = agglutinates streptococcus type A; b = the lectin is composed mainly of subunit monomers; c = the lectin agglutinates Ehrlich carcinoma and AH 109A cells; d = rabbit erythrocytes; e = sheep erythrocytes; f = bacterial fish pathogens (Vibrio anguillarum, Aeromonas salmonicida, Yersinia ruckeri, Aeromonas hydrophila); g = dog, rat and pig erythrocytes; h = various types of streptococci; i = sugar content depends on oocyte development; j = the lectin is a weak mitogen, but blocks protein synthesis in higher concentration; k = the lectin inhibits growth of some bacterial fish pathogens; l = proposed protective functions; m = the lectin is localized in cortical vesicles; n = agglutinates homospecific spermatozoa; o = does not agglutinate human erythrocytes; p = predominantly; q = the lectin is a potent mitogen but in higher concentration blocks protein synthesis. L-Rha = L-rhamnose; D-Glc = D-glucose; L-Fuc = L-fucose.

tion and erythroagglutinating properties (25, 26).

Other isolated lectins are mostly high-molecular-mass aggregates (M_r 12o - 2oo kD) composed most frequently of disulfide-bridges-linked subunits. It appears, however, that in some cases, in addition to the covalent linkages noncavalent binding interactions of the subunits also participate (22-25, 3o). The isolated main subunits of the roach lectin have different affinities for monosaccharides, including L-rhamnose. There is also a more evident difference in their molecular mass. Their amino acid composition, their electrophoretic, sedimentation and erythroagglutinating properties are also somewhat different (25, 26, 29),

Of great significance is the finding of different properties of lectins isolated from vitellogenic and previtellogenic oocytes, resp., of the roach and the perch. Lectins from vitellogenic oocytes are much less soluble,have a higher content of neutral sugar, contain also sialic acids, have also a partially different amino acid composition as well as sedimentation and erythroagglutinating properties (25, 27-3o),

These changes are due to interaction of part of the lectin binding sites with carbohydrate chains of glucoproteins and polysaccharides (see below) and a more intensive activity of endogenous proteinases during the vitellogenic phase of oocyte maturation.

Biological properties of the fish cortical lectins

Though the biological properties of fish lectins have not been hitherto studied systematically, the results obtained so far indicate that they represent a very interesting group of proteins also from this point of view. Thanks to numerous studies dealing with the serological properties of ovarian extracts, the best examined biological property of fish oocyte lectins is agglutination.
Most of these lectins agglutinate specifically human erythrocytes of the group B, some of them recognize the P^k or P antigen. For routine application in blood-group-typing the cortical lectins of Salmonidae and of some Cyrpinidae fishes seem to be useful (22-25, 43, 59). The importance of this finding is somewhat im-

paired by the decrease of the expressed serological specificity during storage. Lectin preparations stored at 4° C for longer than one year are almost unapplicable for routine use in blood--typing (25). During storage evidently conformation alteration of the lectin binding sites take place, analogous to cross-linking with dimethylsuberimidate (25). The observed changes can be explained by formation of new disulfide bridges between the free -SH groups, or by rearrangements of some intra- and inter-chain disulfide bonds (25). With regard to the potential practical application of these lectins it will be necessary to investigate this problem in more detail. Other lectins, those specific for D-glucose or L-fucose agglutinate preferentially human erythrocytes of group O, but their specificity is less clean-cut by far (23,25-27, 34). Besides human erythrocytes all the lectins tested agglutinate also rabbit erythrocytes and some of them erythrocytes of other vertebrates, as well. A number of fish lectins agglutinates specifically some species of streptococci (76); the component of their capsular polysaccharide is L-rhamnose. Not less interesting is the agglutination of some kinds of malignant cells, Ehrlich ascites tumor and the AH 109A cells (31, 35). Here there is a quite evident analogy with lectins isolated from eggs of some frog species of the genus *Rana* and also with lectins from clawed toad *Xenopus laevis* oocytes (36-42, 77, 78). This ability could perhaps be used for preparation of some specific cytostatics, in which lectins would play the role of affinity carrier, haptomer. The homospecific sperm-agglutination in cases of *Salmonidae* fishes (51) and the perch (66, 68) and agglutination of some bacterial pathogens by chinook salmon ova lectin (24) might be connected with the postulated function of these lectins *in vivo*.

Only one paper has hitherto been published on interaction of fish cortical vesicle lectins with serum glycoproteins and polysaccharides (79). It shows that some of the lectins precipitate some less common serum glycoproteins and polysaccharides.

The study of mitogenic activity of fish cortical lectins (21-23, 8o) has yielded interesting results, too. Experiments in cell--free systems have shown that the mitogenic activity of the powan (*Coregonus lavaretus maraena*), the roach (*Rutilus rutilus*) and the perch (Perca fluviatilis) ova lectins is evidently dependent on their concentration (61, 8o, F. Stirpe, private communication 1979, 1980, 1983). These lectins have toxic effects in certain concentrations (most conspicuously perch lectin I and the roach lectin) and prevent incorporation of radioactively labeled thymidine. Stirpe and co-workers who have done these experiments assume that the blocking of protein synthesis by the lectins has a mechanism entirely different from the action of the Ricinus lectin. With regard to the finding of inhibition of proteinsynthesis, of special interest are the results obtained by Voss et al. (24). In their study of the chinook salmon (O. tschawytscha) ova lectin activity against salmonids' bacterial pathogens, they have shown that in addition to the above mentioned agglutination, this lectin inhibits strongly growth of some fish pathogens (see Table 1).

Probable functions of fish cortical-vesicle lectins

The conspicuous presence of sugar-binding proteins and carbohydrate receptors in gametes has lead a number of authors to the hypothesis that the endogenous lectins take part in the recognizing mechanisms during sperm-egg interaction. This function appears really very likely, but can be attributed only to those lectins which are components of outer membranes of sexual cells. According to some recent reports this interaction is mediated by carbohydrate receptors containing fucose (8, 81, 82) and sialic acids (1o).

A role in the mediation of the sperm-egg contact was tentatively suggested for the fish oocyte lectins (66, 68). However, experimental evidence for this hypothesis has been missing. The key moment in this study is the localization of lectins in fish ovaries. We have demonstrated that the hydrophilic oocyte lectins of all studied fish species are present during the whole oocyte development in cortical vesicles (hence the term cortical lectins) (25-27, 62-64, 67).

The biosynthesis of the lectins is associated with the formation of these organelles at the beginning of oogenesis. In the later stages of oocyte maturation lectins have been localized in morfologically different organelles - cortical granules which originate from cortical vesicles (83). As found already earlier by Greve and Hedrick cortical vesicles of clawed toad oocytes also contain an endogenous lectin, specific for β-D-galactosyl-(1-1)-β-D-thiogalactose (42, 84-86). In the eggs of the sea urchin Anthocidaris crassispina (Echinodermata) a lectin with the same carbohydrate specificity was localized in formations which could have been identical with cortical granules (18). In another invertebrate, the horseshoe-crab Tachypleus gigas (Arthropoda) a lectin was isolated directly from the perivitelline fluid of the fertilized eggs (19, 2o).

Cortical vesicles are the organelles of animal eggs of almost all taxonomic groups from invertebrates to mammals. During fertilization the individual components of cortical vesicles join the so--called cortical reaction, which prevents penetration of supernumerary sperm into already fertilized egg (blocking of polyspermy), for reviews see (87-9o). Except for some animals, oocyte of which lack these organelles (e.g. insects, birds), fertilization of the great majority of animals is monospermic. An artificially initiated polyspermy leads in these animals to an abnormal development of embryo or to its abortion (91). In general, both in higher and lower animals, two processes preventing the penetration of excessive sperm take place after gamete fusion. The first mechanism connected with the cortical reaction mentioned above, leads to elevation of the vitelline membrane and to the formation of a perivitelline space and fertilization membrane. The second process is so-called fast blocking of polyspermy which is based on depolarization of the egg plasmamembrane. This process prevents the penetration of supernumerary sperm up to formation of the fertilization envelope, e.g. in sea urchin eggs up to two minutes after the first-sperm-penetration. However, in the fertilized fish eggs the depolarization does not take place (92), but, in fact the cortical reaction in these lower vertebrates occurs substantially faster than in other animals. The fertilization membrane is formed about 10 seconds after penetration of the first sperm through a small opening (micropyle) (52, 93). The first distinct change after the gamete fusion is the Ca^{2+}-dependent exocytosis of cortical granules. Their content is set free into the perivitelline space, where together with the egg-jelly envelope and the remnants of cortical-granule membranes a new fertilization membrane is formed. The results of localization studies of the fertilized or artificially activated fish oocytes have shown that cortical lectins, as cross-linking agents, become a part of the arising fertilization membrane (25, 27, 63, 67), similarly as in the case of Xenopus laevis egg lectin (84, 86). A part of the total lectin content likely remains after exocytosis in the perivitelline space. A complementary mechanism of the block to polyspermy could be sperm-agglutination by the free cortical lectin (51, 66, 68, 94). As already mentioned above, the perivitelline fluid of the fertilized egg of the horseshoe crab Tachypleus gigas contains a lectin (19, 2o). Of particular interest is the finding that even after interaction with carbohydrate components of the perch-egg--jelly-envelope a part of the binding sites of the cortical lectin remains free, as they bind specifically fluoresceinylthiocarbamyl derivatives of glycosylated bovine serum albumine (63). The lectin preparation from the late vitellogenic oocytes of the perch binds to the D-glucosyl residues of the affinity carrier, but it loses its agglutinating properties (25, 28).

The participation of lectins in the cortical reaction by no means excludes these specific proteins from sharing completely different events. Some papers indicate that in the initial stage of the egg cleavage, and presumably also in the mature oocytes themselves, at least some of the cortical lectins have a protective function, which (as implied by the term "protectins") was intuitively suggested already earlier (45-47, 73, 76). As bound components with partially free binding sites, or as free components of the perivitelline fluid, the lectins can, in principle, bind L-rhamnosyl residues of capsular polysaccharides of fish pathogens (24, 45-47, 73, 76).

The study of properties of cortical vesicle lectins obtained from oocytes in various stages of development have shown some differences in the isolated preparations (see above). Conspicuous is the uniformly higher content of neutral sugar and sialic acids, and the relation of these changes approximately to the commencement of vitellogenesis (25, 27-3o). These changes suggest that the lectins in vivo interact with carbohydrate components of the contents and envelopes of the eggs. We have, unfortunately, no detailed information on the nature of these carbohydrate components. They could be acid glycosaminoglycans (the presence of which could be histochemically demonstrated directly in cortical vesicles - (hence their alternative designation "carbohydrate globules") of fresh-water fishes (87, 94-98), and sea urchins (88). These components are being synthezised in the Golgi complex and then transported into the growing vesicles in the form of electronvisible vesicules (99). Their intensive accumulation in cortical vesicles is associated with the beginning of vitellogenesis. With the growing content of these components, which form a complex with a protein rich in cysteine and tyrosine, the contents of the vesicles become denser and less soluble (87). The interaction with glycosaminoglycans could be in agreement with the transport function of cortical lectins. According to the recently presented hypothesis of Olden et al. (loo) these lectins could transport glycosaminoglycans and glycoproteins from the site of their biosynthesis to cortical vesicles. However, it has to be admitted that at the present, reliable experimental evidence for the transport function of cortical-vesicle lectins is almost missing. Sialic acid residues in lectins isolated from vitellogenic oocytes could originate from their interaction with sialoglycoproteins. Polysialoglycoproteins containing N-glycolylneuraminic acid and (among other components) D-galactose were isolated from some salmonids' oocytes (lol-lo5). As these components are not contained in cortical vesicles (lo2), cortical lectins of Salmonidae fishes can interact with them only when the vesicles (or granules) are broken during the cortical reaction, or artificially, during homogenization of oocytes.

From the above it follows that the cortical lectins could have several functions in oogenesis and embryogenesis. All of them are due to the sugar-binding capacity of lectins. In addition to presenting reliable evidence of the suggested complex function of fish cortical lectins, it will also be undoubtedly interesting to investigate their fate during the further development of the embryo. Some of the results obtained in the study of frog embryogenesis indicate that cortical lectins participate in the adhesion of cells in the differentiating embryo (41, 42, lo6, lo7).

References

1. Glabe, C.G., and Vacquier, V.D.: Nature 267, 836-837 (1977).

2. Glabe, C.G., and Vacquier, V.D.: Proc. Natl. Acad. Sci. U.S.A. 75, 881-885 (1978).

3. Glabe, C.G., Grabel, L.B., Vacquier, V.C., and Rosen, S.D.: J. Cell Biol. 91, 178a (1981).

4. Aketa, K., Miyazaki, S., Yoshida, M., and Tsuzuki, H.: Biochem. Biophys. Res. Commun. 8o, 917-922 (1978).

5. Heller, E., and Raftery, M.A.: Biochemistry 12, 41o6-4113 (1973).

6. Aketa, K., Yoshida, M., Miyazaki, S., and Ohta, T.: Exp. Cell Res. 123, 281-284 (1979).

7. Glabe, C.G., and Lennarz, W.J.: J. Cell Biol. 83, 595-6o4 (1979).

8. Huang Jr., T.T.F., Ohzu, E., and Yanagimachi, R.Y.: Gamete Res. 5, 355-363 (1982).

9. Bleil, J.D., and Wasserman, P.M.: Cell 2o, 873-883 (198o).

1o. Lambert, H., and van Le, A.: Gamete Res. 1o, 153-163 (1984).

11. Wiese, L.: J. Phycol. 1, 46-54 (1965).

12. Musgrave, A., van Eijk, E., Te Welscher, R., Broekman, R., Lens, P., Homan, W., and van den Ende, H.: Planta 153, 362-369 (1981).

13. Adair, W.S., Monk, B.C., Cohen, R., Hwang, C., and Goodenough, U.W.: J. Biol. Chem. 257, 4593-46o3 (1982).

14. Saito, T., and Matsuda, Y.: Arch. Microbiol. 139, 95-99 (1984).

15. Renwrantz, L., and Berliner, U.: J. Invertebr. Pathol. 31, 171-179 (1978).

16. Kamiya, H., and Shimizu, Y.: Nippon Suisan Gakkaishi 47, 255-259 (1981); Chem. Abstr. 94 (21): 171278 s.

17. Bretting, H., Stanislawski, E., Jacobs, G., and Becker, W.: Biochim. Biophys. Acta 749, 143-152 (1983).

18. Sasaki, H., and Aketa, K.: Exp. Cell Res. 135, 15-19 (1981).

19. Shishikura, F., and Sekiguchi, K.: J. Biochem. (Tokyo) 96, 621-628 (1984).

2o. Shishikura, F., and Sekiguchi, K.: J. Biochem. (Tokyo) 96, 629-636 (1984).

21. Krajhanzl, A.: Some Lectins from Fish Roe, RNDr. Thesis, Department of Biochemistry, Charles University, Praha 1977 (in Czech).

22. Krajhanzl, A., Horejsi, V., and Kocourek, J.: Biochim. Biophys. Acta 532, 2o9-214 (1978).

23. Krajhanzl, A., Horejsi, V., and Kocourek, J.: Biochim. Biophys. Acta 532, 215-224 (1978).

24. Voss Jr., E.W., Fryer, J.L., and Banowetz, G.M.: Arch. Biochem. Biophys. 186, 25-34 (1978).

25. Krajhanzl, A.: Properties of Lectins from Oocytes of Teleost Fishes, their Interaction with Saccharides and Localization, PhD. Thesis, Department of Biochemistry, Charles University, Praha 1984 (in Czech).

26. Krajhanzl, A., Nosek, J., Delmotte, F., Monsigny, M., Danisová, A., Smolek, P., and Kocourek, J.: 16th Meeting of the Federation of European Biochemical Societes, Abstr. Vol., Poster XIV-o71, Moscow 1984.

27. Nosek, J., Krajhanzl, A., and Kocourek, J.: These Proceedings Vol. 3 (Bøg-Hansen, T., and Spengler, G.A., eds.), W. de Gruyter, Berlin-New York 1983, pp. 453-459.

28. Krajhanzl, A., Kocourek, J., Pancoska, P., and Nosek, J.: These Proceedings Vol. 4 (Bøg-Hansen, T.C., and Breborowicz, J., eds.), W. de Gruyter, Berlin-New York 1985, pp. 377-396.

29. Krajhanzl, A., Danisová, A., Kocourek, J., and Pancoska, P.: These Proceedings Vol. 4 (Bøg-Hansen, T.C., and Breborowicz, J., eds.), W. de Gruyter, Berlin-New York 1985, pp. 397-4o8.

3o. Danisová, A.: Isolation and Characterization of Lectins from Vitellogenic Oocytes of Some Cyprinid Fishes, Thesis, Department of Biochemistry, Charles University, Praha 1983 (in Czech).

31. Sakakibara, F., Takayanagi, G., and Kawauchi, H.: Yakugaku Zasshi lol, 918-925 (1981); Chem Abstr. 96 (o5): 33156 r.

32. Bazil, V.: Isolation and Characterization of the Lectin from the Roe of the Perch, Thesis, Department of Biochemistry, Charles University, Praha 1983 (in Czech).

33. Rogers, D.J.: Med. Lab. Sci. 35, 239-245 (1978).

34. Topliss, J.A., and Rogers, D.J.: Med. Lab. Sci. 42, 199-200 (1985).

35. Sakakibara, F., Takayanagi, G., Mukai, C., Sue, H., Terasaki, Y., Nitta, K., and Kawauchi, H.: Annu. Rep. Tohoku Coll. Pharm. 29, 115-124 (1982); Chem. Abstr. 99 (15) 12o549b.

36. Sakakibara, F., Takayanagi, G., Kawauchi, H., Watanabe, K., and Hakamori, S.: Biochim. Biophys. Acta 444, 386-395 (1976).

37. Sakakibara, F., Takayanagi, G., Ise, H., and Kawauchi, H.: Yakugaku Zasshi 97, 855-862 (1977).

38. Sakakibara, F., Watanabe, K., Takayanagi, G., Ise, H., and Kawauchi, H.: Tohoku Yakka Daigaku Kenkyu Nempo 23, 125-13o (1976); Chem. Abstr. 89 (15): 127524 s.

39. Sue, H., Takayanagi, G., Koseki, T., Nitta, K., Sakakibara, F., and Kawauchi, H.: Yakugaku Zasshi 1oo, 7o6-712 (1981).

4o. Yokota, M., Sakakibara, F., and Kawauchi, H.: Yakugaku Zasshi 95, 5o-55 (1975); Chem Abstr. 82, 123o88 k.

41. Roberson, M.M., and Armstrong, P.B.: Proc. Natl. Acad. Sci. U.S.A. 77, 346o-3463 (198o).

42. Roberson, M.M., and Barondes, S.H.: J. Biol. Chem. 257, 752o-7524 (1982).

43. Prokop, O., Schlesinger, D., and Geserick, G.: Z. Immun.-Forsch. Allergie Klin. Immunol. 132, 491-494 (1967).

44. Prokop, O., Schnitzler, S., and Uhlenbruck, G.: Acta Biol. Med. Germ. 18, K7-K9 (1967).

45. Jarosch, K., Schnitzler, S., Prokop, O., and Uhlenbruck, G.: Z. Arztl. Fortbild 61, 758-759 (1967).

46. Köhler, W., and Prokop, O.: Z. Immun.-Forsch. Allergie Klin. Immunol. 133, 491-5ol (1967).

47. Köhler, W., Prokop, O., and Kühnemund, O.: Z. Immun.-Forsch. Allergie Klin. Immunol. 134, 462-466 (1967).

48. Schnitzler, S., Müller, G., and Prokop, O.: Z. Immun.-Forsch. Allergie Klin. Immunol. 134, 45-53 (1967).

49. Uhlenbruck, G., and Prokop, O.: Vox Sang. 12, 465-466 (1967).

5o. Dürwald, W., and Leopold, D.: Dtsch. Gesundh.-Wes. 22, 2392-2393 (1967).

51. Ginsburg, A.S.: J. Embryol. Exp. Morphol. 9, 173-19o (1961).

52. Horejsí, V., and Kocourek, J.: Methods Enzymol. 34, 361-367 (1974).

53. Filka, K., Coupek, J., and Kocourek, J.: Biochim. Biophys. Acta 539, 518-529 (1978).

54. Anstee, D.J., Holt, P.D.J., and Pardoe, G.I.: Vox Sang. 25, 347-36o (1973).

55. Voak, D., Todd, G.M., and Pardoe, G.I.: Vox Sang. 26, 176-188 (1974).

56. Leopold, D., and Dürwald, W.: Dtsch. Gesundh.-Wes. 25, 1757-1759 (197o).

57. Todd, G.M.: Vox Sang. 21, 451-454 (1971).

58. Kolokolova, G.P., and Potapov, M.I.: Zh. obschej biologii 35, 14o-143 (1974).

59. Potapov, M.I., and Kolokolova, G.P.: Problemi Gematol. Pereliv. Krovi 22, 43-46 (1977).

6o. Kuznetsova, G.P.: Sud. Med. Ekspertiza 14, 35-4o (1971).

61. Licastro, F., Chiricolo, M., Barbieri, L., Falasca, A., Rossi, C.A., Franceschi, C., and Stirpe, F.: These Proceedings, Vol. 3 (Bøg-Hansen, T.C., and Spengler, G.A., eds.), W. de Gruyter, Berlin-New York 1983, pp. 293-3o2.

62. Nosek, J., Krajhanzl, A., and Kocourek, J.: Histochemistry 79, 131-139 (1983).

63. Krajhanzl, A., Nosek, J., Monsigny, M., and Kocourek, J.: Histochem. J. 16, 426-428 (1984).

64. Krajhanzl, A., Nosek, J., Habrová, V., and Kocourek, J.: Histochem. J. 16, 432-434 (1984).

65. Schenkel-Brunner, H., Splechtna, H., and Kothbauer, H.: Experientia 28, 1479 (1972).

66. Schenkel-Brunner, H., and Kothbauer, H.: Z. Immunitätsforsch. 154, 155-161 (1978).

67. Nosek, J., Krajhanzl, A., and Kocourek, J.: Histochem. J. 16, 429-431 (1984).

68. Kothbauer, H., and Schenkel-Brunner, H.: Z. Immunitätsforsch. 147, 87-9o (1974).

69. Fujita, Y., Oishi, K., Suzuki, K., and Imahori, K.: Biochemistry 14, 4465-447o (1975).

7o. Prokop, O.,Uhlenbruck, G., and Köhler, W.: Vox Sang. 14, 321-333 (1968).

71. Kothbauer, H., and Schenkel-Brunner, H.: Comp. Biochem. Physiol. A5o, 27-29 (1975).

72. Kolokolova, G.P.: Sud. Med. Ekspert 17, 19-21 (1974).

73. Ozaki, H., Ohwaki, H., and Fukada, T.: Dev. Comp. Immunol. 7, 77-89 (1983).

74. Schnitzler, S.: Acta Biol. Med. Germ. 2o, 259-262 (1968).

75. Köhler, W., Kühnemund, O., and Prokop. O.: Z. Immunitäts-forsch. 135, 324-33o (1968).

76. Köhler, W.: Fortschritte der Hämatologie (Perlick, E., Pleuert, W., and Prokop, O., eds.), J.A. Barth, Leipzig 197o, pp. 41-7o.

77. Nitta, K., Takayanagi, G., and Kawauchi, H.: Chem. Pharm.Bull. 32, 2325-2332 (1984).

78. Yokota, M. Sakakibara, H., and Kawauchi, H.: Yakugaku Zasshi 95, 5o-55 (1975); Chem. Abstr. 82, 123o88 k.

79. Uhlenbruck, G., Baldo, B.A., Steinhausen, G., Schwick, H.G., Chatterjee, B.P., Horejsí, V., Krajhanzl, A., and Kocourek, J.: J. Clin. Chem. Clin. Biochem. 16, 19-23 (1978).

8o. Barbieri, L., Lorenzoni, E., and Stirpe, F.: Biochem. J. 182, 633-635 (1979).

81. Bolwell, G.P., Callow, J.A., Callow, M.E., and Evans, L.V.: J. Cell Sci. 36, 19-3o (1979).

82. de Santis, R., Pinto, M.R., Cotelli, F., Rosati, F., Monroy, A.,and D'Álessic, G.: Exp. Cell Res. 148, 5o8-513 (1983).

83. Nosek, J.: Histochem. J. 16, 435-437 (1984).

84. Greve, L.C., and Hedrick, J.L.: Gamete Res. 1, 13-18 (1978).

85. Nishihara, T., and Hedrick, J.L.: Fed. Proc. 36, 811 (1977).

86. Wyrick, R.E., Nishihara, T., and Hedrick, J.L.: Proc. Natl. Acad. Sci. U.S.A. 71, 2o67-2o71 (1974).

87. Guraya, S.S.: Int. Rev. Cytol. 78, 257-36o (1982).

88. Rossignol, D.P., Decker, G.L., and Lennarz, W.J.: Cell Fusion: Gene Transfer and Transformation (Beers Jr., R.F., and Bassett, E.G., eds.), Raven Press, New York 1984, pp. 5-26.

89. Epel, D., Schmidt, T., and Sasaki, H.: Cell Fusion: Gene Transfer and Transformation (Beers Jr., R.F., and Bassett, E.G., eds.), Raven Press, New York 1984, pp. 39-48.

9o. Nagahama, Y.: Fish Physiology, Vol. 9, Pt. A (Hoar, W.S., Randall, D.J., and Donaldson, E.M., eds.), Acad. Press, Orlando 1983, pp. 223-277.

91. Boldt, J., Schuel, H., Schuel, R., Dandekar, P.V., and Troll, W.: Gamete Res. 4, 365-377 (1981).

92. Hagiwara, S., and Jaffe, L.A.: Annu. Rev. Biophys. Bioeng. 8, 385-416 (1979).

93. Kudo, S.: Develop., Growth and Differ. 22, 4o3-414 (198o).

94. Kudo, S.: Develop., Growth and Differ. 18, 167-176 (1976).

95. Kudo, S.: Develop., Growth and Differ. 24, 327-339 (1982).

96. Yamamoto, K.: J. Fac. Sci. Hokkaido, Univ. Ser. IV, 12, 391-4oo (1956).

97. Arndt, E.A.: Z. Zellforsch. 51, 2o9-224 (196o).

98. Anderson, E.: J. Morphology 125, 23-6o (1968).

99. Droller, M.J., and Roth, T.F.: J. Cell Biol. 28, 2o9-232 (1966).

1oo. Olden, K., Parent, J.B., and White, S.L.: Biochim. Biophys. Acta 65o, 2o9-232 (1982).

1o1. Inoue, S., and Iwasaki, M.: Biochem. Biophys. Res. Commun. 83, 1o18-1o23 (1978).

1o2. Inoue, S., and Matsamura, G.: Carb. Res. 74, 361-368 (1979).

1o3. Inoue, S., and Iwasaki, M.: Biochem. Biophys. Res. Commun. 93, 162-165 (198o).

1o4. Shimamura, M., Endo, T., Inoue, Y., Inoue, S., and Kambara, H.: Biochemistry 23, 317-322 (1984).

1o5. Iwasaki, M., Inoue, S., Kitajima, K., Nomoto, H., and Inoue, Y.: Biochemistry 23, 3o5-31o (1984).

1o6. Roberson, M.M., and Barondes, S.H.: J. Cell Sci. 97, 1875-1881 (1983).

1o7. Harris, H.L., and Zalik, S.E.: Wilhelm Roux. Arch. Develop. Biol. 191, 2o8-21o (1982).

AFFINITY OF SOLUBLE AND IMMOBILIZED PLASMA FIBRONECTIN FOR ARGINYL SULFAMIDE POLYSTYRENE BEADS

M. Caron

U.F.R. Santé, Médecine, Biologie Humaine, 74, rue Marcel Cachin, F-93012 Bobigny

A.Faure

C.N.T.S, F-91943 Les Ulis Cedex

D. Gulino, C. Boisson, J. Jozefonwicz

Lab. Rech. Macromolécules, F-93430 Villetaneuse

Plasma fibronectin is a high-molecular-mass glycoprotein which has been shown to have several lectin-like activities, often inhibited by non-acetylated aminosugars and also arginine (1). We have previously shown that arginine derivatives bound to modified polystyrenes interact with human plasma fibronectin (2). However, biological "adhesive" activities of fibronectin appear to be controlled by alterations in its molecular form, for example by its immobilization on insoluble supports (3).

In this communication, we provide evidence that both soluble and immobilized plasma fibronectin bind to particles of polystyrene modified by arginyl groups, and that this binding is inhibitable by saccharides.

Lectins, Vol. V

Material and methods

Plasma fibronectin was isolated and purified from cryoprecipitate by selective precipitation using glycin and heparin (4). Arginyl sulfamide polystyrene was obtained by sulfonation of the aromatic residues and binding of L-arginine. Sugars and arginine were from Sigma.
A solution of plasma fibronectin in 0.1 M NaCl, 0.05 M Tris-HCl (pH 7.6) containing 1 mM EDTA was applied on a column of arginyl sulfamide polystyrene particles (2 ml gel bed) at 20°C, washed with the previous buffer, and eluted with various concentrations of tested compound solutions. The retained fibronectin was eluted using a solution of 4 M urea. The flow rate in the column was about 20 ml per h and 2 ml fractions were collected. Absorbance was read at 280 nm and the concentration of plasma fibronectin was determined assuming an extinction coefficient of 1.28 for 1 mg/ml.
Fibronectin was immobilized by coating microtest V plates with 100 µl of plasma fibronectin solution. Some wells were incubated with bovine serum albumin, cryoprecipitate or human plasma instead of plasma fibronectin. The wells were finally reincubated with bovine serum albumin 1 % and washed three times with distilled water. To observe binding of arginyl polystyrene beads onto the surface of coated wells, each well was filled with 50 µl of beads in suspension (30-40 particles per mm^3) and with 50 µl of 0.15 M NaCl. The effects of inhibitors on interactions were studied by adding 50 µl of inhibitor solution in 0.15 M NaCl, replacing the equivalent volume of saline. The microtest plates were scored after 1 h at room temperature. The adsorption of particles was evaluated by direct reading of the reaction, as for hemagglutination test, or using a densitometer (2).

Results

Interaction of arginyl sulfamide polystyrene with soluble plasma fibronectin.

The results of these experiments are summarized in Fig. 1 and Table 1. Essentially 95 % of the protein was bound and could be subsequently eluted with 4 M urea. Washing with 0.5 M NaCl released 10 % of bound molecules and 1 M NaCl only 30 %, indicating that electrostatic forces do not dominate this interaction. No significant dissociation was brought about with 0.5 M arginine. By contrast, various saccharidic compounds eluted fibronectin.

Interaction of arginyl sulfamide polystyrene with immobilized plasma fibronectin.

Arginyl sulfamide polystyrene beads bound to wells coated with purified plasma fibronectin and gave no reaction with serum albumin-coated or with uncoated wells. Several saccharidic substances were potent inhibitors of the adhesion (Table 2). A significant interaction was also noted with cryoprecipitate which

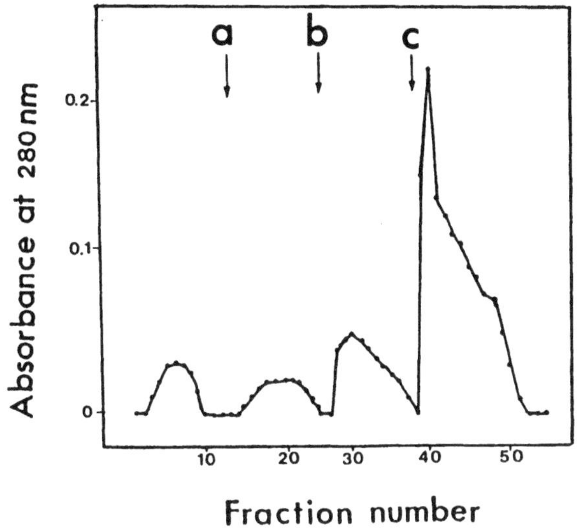

Fig. 1 : Affinity chromatography of 1.5 mg of human plasma fibronectin on an arginyl sulfamide polystyrene column. Elutions were performed with buffers containing : (a) 0.5 M NaCl, (b) 1 M NaCl, (c) 1 M NaCl, 4 M urea.

Table 1

Summary of affinity chromatography results

Applied Fibronectin (mg)	Bound fibronectin (%)	Displacer	Eluted fibronectin (%)
1.5	94	0.5 M NaCl	10
1.5	94	1 M NaCl	30
1.5	94	4 M Urea	94
1	96	0.1 M GalNAc	76
1	95	0.1 M Arabinose	71
1	95	0.1 M Lactose	57
1	95	0.1 M Fructose	33
1	96	0.1 M Glucosamine	< 10
1	96	0.1 M Glucose	< 10
1	96	0.1 M Arginine	< 10

Column volume, 2 ml ; Buffer : 0.1 M NaCl, 0.05 M Tris HCl, 1 mM EDTA

Table 2

Inhibition of adhesion of arginyl sulfamide polystyrene to immobilized human plasma fibronectin.

Inhibitor	Minimal inhibitory concentration
Heparin	50 μM
N-acetyl-D-galactosamine	6 mM
Lactose	25 mM
D-mannose	50 mM
D-galactose	> 50 mM
D-glucose	> 50 mM
D-glucosamine	> 50 mM
L-fucose	> 50 mM
β-nitrophenyl-D-galactose	> 50 mM
β-methyl-D-galactose	> 50 mM
N-acetyl-D-glucosamine	> 50 mM
Dextran sulfate	> 50 mM

contains fibronectin and a weak but detectable interaction with total plasma. The incubation of microtest plates with various concentrations of plasma fibronectin (1 μg to 10 mg/ml) demonstrated that a maximal adsorption was obtained for 1 mg/ml.

Discussion

It is well established that different forms of fibronectin show an affinity for arginine. Twenty years ago, Yamada demonstrated that arginine was an inhibitor of the hemagglutination by chick embryo fribroblasts cellular fibronectin (5). Since several studies have shown critical alterations in the molecular form of fibronectin occuring when it attaches to a solid phase (3,6,7), we studied the interaction of polystyrene-arginine with soluble as well as immobilized plasma fibronectin. The results of this investigation indicate that an interaction can successfully be obtained with the two forms. It is also inferred that a "lectin-like" phenomenon could play a role in this binding since some saccharides inhibit it. This observation is in accordance with several studies showing the ability of arginine to inhibit activities of several human lectins (8,9,10) and with the affinity of several phytolectins for arginyl sulfamide polystyrene particles (2).

References

1. Caron, M. and Joubert, R. : Marker Proteins in Inflammation (Laurent, P., Grimaud, J.A. and Bienvenu, J., eds) Vol. 3, Walter de Gruyter, Berlin. New York, in press.

2. Caron, M., Faure, A., Gulino, D., Boisson, C., Jozefonwicz, J. : Marker Proteins in Inflammation,(Arnaud, P., Bienvenu, J. and Laurent, P., eds) Vol. 2, pp. 627-630, Walter de Gruyter, Berlin.New York 1984.

3. Colvin, R.B. and Kradin, R.L. : Surv. Synth. Path. Res. 2, 627-630 (1984).

4. Faure, A., Caron, M. and Tepenier, D. : 1er Colloque National sur les Technologies de Purification des Protéines, Paris 1984.

5. Yamada, K.M., Yamada, S.S. and Pastan, J. : Proc. Natl. Acad. Sci. USA 72, 3158-3162 (1975).

6. Grinell, D. and Phan, T.V. : J. Cell. Physiol. 116, 289-296 (1983).

7. Tooney, N.M., Mosesson, M.W., Amrani, D.L., Hain, J.F. and Wall, J.S. : J. Cell. Biol. 97, 1686-1692 (1983).

8. Gartner, T.K. and Agin, P.P. : Biochem. Biophys. Res. Comm. 96, 1747-1754 (1980).

9. Gartner, T.K., Williams, D.C. and Phillips, D.R. : Biochem. Biophys. Res. Comm. 79, 592-599 (1977).

10. Lung, L.L.K. and Nachman, R.L. : J. Clin. Invest. 70, 542-549 (1982).

PART 4

MICROBIAL LECTINS

THE FIMBRIAL LECTINS OF *ESCHERICHIA COLI*

Frits K. de Graaf

Department of Microbiology, Biological Laboratory, Vrije Universiteit, de Boelelaan 1087, 1081 HV Amsterdam, The Netherlands.

Adherence of bacteria to the mucosal surfaces to rhe respiratory, the urinary, and the alimentary tract is generally recognized as an essential initial step in the pathogenesis of many bacterial infections. Depending on the type of the infecting organism, adherence is followed by colonization of the host epithelium, invasion of epithelial cells, and/or the eventual penetration of bacteria into the underlying tissues. Lectin-specific antibodies may prevent the establishment of infectious diseases through a failure of adherence or interference with the adherence.

The interaction of a particular bacterium with the mucosal or the epithelial cell surface is a complicated physico-chemical process which is rather difficult to analyze. Not only specific bacterial lectins (or adhesins) and their receptors, but other cell-surface components, like lipopolysaccharides or capsular antigens, and environmental or growth conditions that contribute to surface charge and hydrophobicity, play a more or less important role in this interaction.

Most of our present knowledge about bacterial lectins has been gained from studies on the role of fimbriae (or pili) in bacterial adherence. Fimbriae are characterized as long filamentous surface appendages consisting of a single repeating protein subunit. Fimbriated cells have been shown to adhere more readily than non-fimbriated cells, and specific antifimbrial antibodies appear to block the bacterial adherence. Generally, the fimbriae are supposed to have a hydrophobic surface (2, 33). Fimbriated cells adsorb readily to hydrophobic gels and a general positive correlation exists between the presence of fimbriae, hydrophobicity and adherence. The hydrophobic fimbriae counteract the repulsive forces between the negatively charged surface of the bacterium and the eukaryotic cell. The fimbrial tip is supposed to be the point of initial contact and to penetrate the electrostatic barrier because of its lower charge density, which is a consequence of its small surface area and radius of curvature. Subsequently, the bacterial ligands interact in a stereospecific binding with complementary receptor molecules on the eukaryotic cell surface. In addition, hydrophobic sites on the bacterial lectin may become embedded in the lipid interior of the eukaryotic cell membrane (39). Such non-specific hydrophobic interactions may facilitate the specific binding of the lectin to its receptor molecule.

The specificity of the interaction between fimbrial lectins and their receptors is similar to plant lectin-sugar interactions since, (i) the interaction can be inhibited by haptens which are identical or closely resemble (part of) the native ligand or its receptor (11,17,18,24,25,32), (ii) chemical modification or enzymatic treatment abolishes or alters the interaction (3,10,29), and (iii) the interaction can be blocked by specific antibodies directed against epitopes on either the lectin or its receptor (4,16,41). With a few exceptions, the fimbrial lectin receptors have been identified as glycolipids or glycoproteins. The specificity of the binding is controlled by the oligosaccharide chain present on these glycoconjugates.

The occurrence of fimbrial lectins seems to be restricted to Gram-negative bacteria, but within this group of organisms the presence of fimbriae may be considered as a common character of both pathogenic and non-pathogenic strains (41). Most studies on the molecular biology of the fimbrial lectins concern *Escherichia coli* and *Neisseria gonorrhoea*. Since the general features of all types of fimbria are largely comparable, this paper is focussed on the fimbrial lectins of *E. coli* with reference to *N. gonorrhoea* where appropriate.

THE FIMBRIAL LECTINS OF *ESCHERICHIA COLI*.

E. coli is a normal inhabitant of the intestinal tract of man and various animal species, but also wellknown as a multipotent pathogen that has the ability to cause disease in several organs or tissues, in particular in the small bowel and the urinary tract.

Most of the commensal and pathogenic strains possess a common fimbrial lectin, designated as type 1 fimbriae. The synthesis of this lectin enables *E. coli* to adhere to a large variety of eukaryotic cells inducing epithelial cells, erythrocytes, leucocytes,

TABLE 1. Some characteristics of fimbrial adhesins of pathogenic E. coli strains.

Adhesin	Morphology	Diameter (nm)	Mol.wt subunit	Genetic location	Epithelia colonized
Type I	rigid	7	15.700	chromosome	
K88	flexible	2.1	27.500	plasmid	pig intestine
K99	flexible	4.8	17.500	plasmid	claf, lamb, piglet intestine
F41	flexible	3.2	29.500	?	calf, lamb, piglet intestine
987P	rigid	7	20,000	chromosome	pig intestine
CFA/I	rigid	7	15.000	plasmid	human intestine
CFA/II-CS1	rigid	7	16.300	plasmid	human intestine
CFA/II-CS2	rigdi	7	15.300	plasmid	human intestine
CFA/II-CS3	flexible	2	14.800	plasmid	human intestine
P(Pap)	rigid	7	18.200	chromosome	human urinary tract
P($F7_2$)	rigid	7	18.400	chromosome	human urinary tract
M	rigid	7		chromosome	human urinary tract
S	rigid	7	17.000	chromosome	
Type 1C	rigid	7	17.300	chromosome	

sperm cells, fungi, yeasts, and others. Type 1 fimbriae have a diameter of about 7 nm, an axial hole, and vary in length roughly between 0.3 and 2 μm. They are peritrichously arranged and usually number up to 200-300 fimbriae per bacterium. Other fimbrial lectins, known to be involved in the pathogenesis of *E. coli* infections, occur on a very limited number of *E. coli* serotypes, associated with a particular type of disease in man or in domestic animals (12,26). In contrast to type 1-fimbriated cells, strains carrying other fimbria appear to be much more restricted in their ability to adhere to different eukaryotic cells. The general features of some wellknown *E. coli* fimbrial lectins are summarized in Table 1.

Chemical analysis of fimbriae has shown that they are composed of a single major polypeptide subunit. Some types of fimbria have been analyzed for the presence of prosthetic groups added to the fimbrial subunit by post-translational modification. Type 1 fimbriae have shown to contain approximately one residue of reducing sugar per subunit (21). Aminosugar residues have been detected in 987P fimbriae and in the P fimbriae of *E. coli* strain 7714 (14,30). Another species of P fimbriae as well as gonococcal fimbriae have shown to contain phosphate residues (31,42). The function of these post-translational modifications is unknown but one may envisize a role in stability or solubility of the fimbriae, enhanced protection against proteolytic degradation or a role in receptor recognition.

Most fimbrial subunits contain 140-200 amino acid residues with the exception of K88 and F41 fimbriae which are composed of subunits containing 250-300 residues. Except for K99 fimbriae, all other fimbriae investigated have shown to be acidic proteins with isoelectric points of 4-5, and therefore possess a negative charge at neutral pH. This negative charge may hamper the approach of fimbriated cells to the epithelial cell surface unless the fimbriae possess a low charge density or because the overall hydrophobic character of the fimbriae plays a more important role. Furthermore, some indications have been obtained that divalent cations enhance adhesion, probably by the formation of ionic bridges between the fimbriae and the eukaryotic cells (34).

Amino acid analysis of fimbrial subunits has revealed that most subunits, with the exception of K88, CFA/I and probably F41, con-

Fim A	M K I K T L A I V V L S A L S L S S T A A L A A A T T V N
1 C	M K L K F I S M A V F S A L T L G V A T N A S A V T T V N
Pap A	M I K S V I A G A V A M A V V S F G V N N A A P T I P Q G
$F7_2$	M I K S V I A G A V A M A V V S F G A Y - A A P T I P Q G

G G T V H F K G E V V N A A C A V D A G S V D Q T V Q L G Q V R T
G G T V H F K G E V V N A A C A V N T N S F D Q T V N L G Q V R S
Q G K V T F N G T V V D A P C S I S Q K S A D Q S I D F G Q L S K
Q G K V T F N G T V V D A P C G I D A Q S A D Q S I D F G Q V S K

A S L A Q E G A T S S A V G F N I Q L N D C D T N V A S - - - K A
E R L K V D G A K S N P V G F T I E L N D C D S Q V S A - - - G A
S F L E A G G V S K P M D - L D I E L V N C D I T A F K - - - G G
L F L E N D G E S Q P K S - F D I K L I N C D I T N F K K A A G G

A V A F L G T A I D A G H T N V L A L Q S S A A G S A T N V G V Q
G I V F S G P A V - T G K T D V L A L Q S S A A G S A T N V G V Q
N G A K K G T V K L A F T G P I V N G H S D E L D T N G G T G T A
G G A K T G T V S L T F S G V P S G P Q S D M L Q T V G A T N T A

I L D R - - T G A A L T L D G A T F S S E T T L N N G T N T I P F
I T D H - - T G K V V P L D G - T A S S T F T L T D G T N K I P F
I V V Q - G A G K N V V F D G - S E G D A N T L K D G E N V L H Y
I V V T D P H G K R V K F D G A T A T G V S Y L V D G D N T I H F

Q A R Y F A - - - - G A A - T P G A A N A D A T F K V Q Y Q
Q A V Y Y A T - - - G Q A - T A G I A N A D A T F K V Q Y Q
T A V V K K S S A V G A A V T E G A F S A V A N F N L T Y Q
T A A V R K D G S - G N P V T E G A F S A V A N F N L T Y Q

Fig. 1. Comparison of the amino acid sequence of the fimbrial subunits of two different type 1 fimbriae (Fim A, 18, and 1C, 37) and two different P fimbriae (Pap A, (1) and $F7_2$, (36)). The arrows indicate the predicted cleavage site for the signal peptidase. Homologies between Fim A and 1 C, and between Pap A and $F7_2$ only, are not indicated.

tain two cysteine residues that form an intrachain disulphide bond. Furthermore, the fimbrial subunits constituting type 1 or P fimbriae, isolated from different strains, appear to be very similar in amino acid composition suggesting a close evolutionary relationship. In contrast, fimbrial lectins associated with diarrhoeal disease are more heterogeneous with respect to their amino acid composition (6,15,23).

The primary structure of a number of fimbrial subunits has been partially or completely resolved (23). Comparison of the amino acid allows one to discern between conserved and variable sequences within a group of functionally and structurally related fimbriae. This comparison may facilitate the assignment of particular sequences to various functions such as subunit polymerization, antigenic variation or receptor-binding activity. Representatives of two different classes of bacterial lectins (type 1 and P fimbriae) are compared with respect to their primary structure (Fig. 1). It appears that within each class, the fimbrial subunits are highly homologous in particular in their amino- and carboxyl-terminal region. Comparison between the two classes shows a more restricted but still evident homology between type 1 and P fimbriae. No homology between these two morphologically similar types of fimbriae and the more thin and flexible fimbriae associated with diarrhoeal disease was observed, except for some positions in the amino- and carboxyl terminal region (6,23). A relatively large number of fimbrial subunits of other bacterial species has been analyzed for their amino-terminal amino acid sequence. It appears that the stricking homology in this region is not restricted to *E. coli* fimbriae but also exist between type 1 fimbriae of *E. coli* and related bacterial species as *Klebsiella pneumoniae* and *Salmonella typhimurium* (6). The amino-terminal amino acid sequences of fimbrial subunits from *Neisseria*, *Moraxella* and *Pseudomonas* species are also shown to be highly homologous indicating that this group of proteins, like the *E. coli* fimbrial subunits, may be derived from a common ancestral gene (15). Comparison of the carboxyl-terminal regions of various fimbrial subunits shows an even more stricking homology at particular positions, primarily occupied by aromatic amino acid residues (6,23). Indications have been obtained that these aromatic residues are located at the subunit-subunit interface (21,40). In general, the high degree of homology in the amino- and carboxyl-terminal amino acid sequences of different fimbrial subunits strongly suggests a more or less common function for these parts of the molecules. Since in particular the hydrophobic residues in these regions are highly conserved, and because hydrophobic residues play an important role in maintaining tertiary and quaternary structures, it is likely to suppose that these regions are involved in subunit polymerization. In contrast, the highly variable amino acid sequences observed in other regions of for instance gonococcal fimbriae constitute (part of) the variable antigenic determinants of these fimbriae (27).

RECEPTORS FOR FIMBRIAL LECTINS.

Haemagglutination was the first observed manifestation of the adhesive properties of *E. coli* (13). The majority of all strains agglutinate red blood cells of a variety of animal species (9). The specificity of this interaction as well as the binding of *E. coli* to a variety of other eukaryotic cells has been studied in *in vitro* experiments where whole bacteria or isolated lectins are incubated with erythrocytes, yeasts or isolated epithelial cells. It appears that the binding reactions can be blocked by the addition of D-mannose or α-methyl-D-mannoside, but not with a large variety of other sugars (5). This mannose-specific lectin has been identified as type 1 fimbriae (2,8). The widespread occurrence of mannose-containing glycoconjugates in mucus gels and on eukaryotic cells may explain why *E. coli* cells carrying mannose-specific lectins adhere to a very large variety of host cells.

Other fimbrial lectins have been studied for their receptor specificity (38). With a possible few exceptions all lectins have been shown to interact with glycoconjugates which are generally abundant on eukaryotic cell surfaces. A classification of the different types of *E. coli* lectins according to their receptor is given in Table 2.

Only a very few receptor species have been studied in sufficient detail to provide information about the conformation of the carbohydrate that interacts with the fimbrial lectin. The binding specificity for type 1 fimbriae has been investigated in experiments on the inhibitory activity of various linear and branched oligosaccharides containing mannose-residues (11). It appears that the

Table 2. Classes of fimbrial adhesins of *Escherichia coli*

Class	Host tissue specificity in vivo	Receptor specificity	Syndromes associated with
Type 1	none?	mannose-containing glycoconjugates	none
K88	pig intestinal epithelium	galactose-containing glycoconjugates	diarrhoea
K99	piglet, calf and lamb intestinal epithelium	N-glycolyl-sialyl-lactosyl ceramide	diarrhoea
987P	pig intestinal epithelium	glycoprotein	diarrhoea
CFA	human intestinal epithelium		diarrhoea
P	human urinary tract epithelium	globotriaosyl and globotetraosyl ceramide*	cystitis, pyelonephritis
M	human urinary tract epithelium	blood group M glycophorin A	cystitis, pyelonephritis
S		sialyl glactosides	neonatal sepsis meningitis

*Globotriaosyl and globotetraosyl represent the p^k and P blood group antigen, respectively.

mannose-specific lectin preferentially binds to relative short α-D-oligomannoside units present in the eukaryotic cell surface glycoconjugates. The P fimbriae recognize the globoseries of glycosphingolipids known as the P blood group antigens, and all containing the α-Dgal(1→4)-β-Dgal carbohydrate unit (17,19). Purified P fimbriae bind to the synthetic disaccharide α-Dgal(1→4)-β-Dgal (17). The K99 fimbriae recognize a glycolipid containing Neu5Gc-α(2→3)gal-β-(1→4)Glc as carbohydrate (32). Treatment of the sialic acid containing glycolipid with neuraminidase destroys the receptor activity (10). The experimental data that have been gathered so far indicate that the minimal carbohydrate moiety that determines receptor specificity probably consists of an (internally placed) disaccharide sequence which interacts in a highly stereospecific reaction with the bacterial lectin.

Very little information is available about the nature of the amino acids or the amino acid sequences that interact with the carbohydrate receptor. The binding site on the K88 fimbrial lectin has recently been identified as a small stretch of mainly hydrophobic residues positioned in between two β-turns that are predicted to constitute an antigenic epitope in the conserved sequence of the K88 fimbrial subunits. Synthetic peptides with a β-turn potential and corresponding to conserved sequences in the gonococcal fimbrial subunit, known to be (part of) the receptor binding domain, elicit antibodies that inhibit the binding of heterologous gonococcal strains to eukaryotic cells (28).

FIMBRIAL SUBUNITS AND LECTIN-ACTIVITY

A very interesting question to address is whether fimbrial lectins adhere only with their tip or also with lateral binding sites along the fimbrial structure. Experimental data supporting both possibilities have been reported (2,29,35) but conclusive evidence has never been obtained. If each of the identical fimbrial subunits possess its own receptor-binding domain one may expect that fimbrial lectins interact at multiple sites with the host epithelial cell membrane. Tip-wise attachment seems likely when only the terminal fimbrial subunit exposes its binding site at the fimbrial surface, or when bacterial fimbriae carry a specific lectin molecule at their tip. Such a separate lectin molecule as a minor component of the fimbriae may be structurally related to the fimbrial subunits and could easily remain undetected in purified preparation of fimbriae.

Recent genetic analyses of fimbriae production has revealed that at least the chromosomally encoded P fimbriae contain minor components which are responsible for the digalactoside-specific adhesive properties of these fimbriae (20). The P-specific lectin and the P fimbrial subunit are encoded by separate genes. Mutations in either one of these genes completely abolish either fimbriae or lectin synthesis. A comparable situation may exist in the chromosomal determinant for the synthesis of type 1 fimbriae. One may presume that both P and type 1 fimbriae carry a lectin molecule at their tip and that the hydrophobic fimbrial structure itself is necessary to carry the lectin at some distance from the bacterial cell surface, presumably to facilitate the interaction between two cells of like surface charge.

A quite different organization has been found for the very thin and flexible, plasmid-encoded, K88 and K99 fimbriae (6,23). The K88 fimbriae are composed of structural subunits which also possess receptor-binding activity. A second fimbria-like subunit encoded by another gene of the K88 operon is produced in very low amounts and is supposed to anchor the polymerized K88 lectin molecules to the bacterial outer membrane (22). The K99 fimbriae seem to be organized in an intermediate fashion compared to K88 and either P or type 1 fimbriae. Only one product of the K99 gene cluster is transported to the bacterial surface. This gene product is identified as the K99 fimbrial subunit (7). Mutations in this gene completely abolish the production of fimbriae as well as the lectin activity. A comparable situation has been found for gonococcal fimbriae since the receptor-binding domain is part of the gonococcal fimbrial subunit.

It appears that the lectin activity of bacterial fimbriae is generally determined by highly conserved amino acid sequences, no matter whether those sequences are part of the structural fimbrial subunit or encoded by separate genes. Antigenic variation occurs mainly in fimbrial subunits without lectin activity or in parts of the (structural) lectin molecules that are not involved in receptor recognition or polymerization (23,27). Antibodies produced by the host are primarily directed against these variable, but immunodominant, components (28).

REFERENCES

1. Båga, M., Normark, S., Hardy, J., O'Hanley, P., Lark, D., Olsson, O., Schoolnik, G., Falkow, S., J. Bacteriol. 157 (1984) 330.
2. Brinton, C-C., Jr., Trans. N.Y. Acad. Sci. 27 (1965) 1003.
3. Buchanan, T.M., Pearce, W.A., Chen, K.C.S., *In* Immunobiology of Neisseria gonorrhoea, (Brook, G.F., Gotschlich, E.C., Holmes, K.N., Sawyer, W.D., Young, F.E., eds), American Soc. for Microbiology, Washington, D.C. p. 242.
4. Burrows, M.R., Sellwood, R., Gibbons, R.A., J. gen. Microbiol. 96 (1976) 269.
5. Collier, W.A., de Miranda, J.C., Antonie van Leeuwenhoek, J. Microbiol. Serol. 21 (1955) 133.
6. De Graaf, F.K., Mooi, F.R. Adv. Microb. Physiol. 28 (1986). in press.
7. De Graaf, F.K., Krenn, B.E., Klaasen, P., Infect. Immun. 43 (1984) 508.
8. Duguid, J.P., Anderson, E.S., Campbell, I., J. Path. Bact. 92 (1966) 107.
9. Duguid, J.P., Old, D.C., *In* Bacterial Adherence (Beachey, E.H., ed.), Chapman and Hall, London and New York, series B, Vol. 6, (1980) p. 185.
10. Faris, A., Lindahl, M., Wadström, T. FEMS Microbiol. Lett. 7 (1980) 265.
11. Firon, N., Ofek, I., Sharon, N., Biochem. Biophys. Res. Comm. 105 (1982) 1426.
12. Gaastra, W., De Graaf, F.K. Microbiol. Rev. 46 (1982) 129.
13. Guyot, G., Zbt. Bakt. Abt. I. Orig. 47 (1908) 640.
14. Isaacson, R.E., Richter, P., J. Bacteriol. 146 (1981) 784.
15. Jones, G.W., Isaacson, R.E., CRC Crit. Rev. Microbiol. 10 (1983) 229.
16. Jones, G.W., Rutter, J.M., Infect. Immun. 6 (1972) 918.
17. Källenius, G., Möllby, R., Svenson, S.B., Winberg, J., Hultberg, H., Infection 8 (1980) S288.
18. Klemm, P., Eur. J. Biochem. 143 (1984) 395.
19. Leffler, H., Svanborg Edén, C., Infect. Immun. 34 (1981) 920.
20. Lindberg, F.P., Lund, B., Normark, S., The EMBO J. 3 (1984) 1167.
21. McMichael, J.C., Ou, J.T., J. Bacteriol. 138 (1979) 969.
22. Mooi, F.R., van Buuren, M., Koopman, G., Roosendaal, R., De Graaf, F.K., J. Bacteriol., 159 (1984) 482.
23. Mooi, F.R., De Graaf, F.K., Curr. Topics Microbiol. Immunol., 118 (1985) 119.
24. Ofek, I., Beachey, E.H., Infect. Immun. 22 (1978) 247.
25. Ofek, I., Mirelman, D., Sharon, N., Nature 265 (1977) 623.
26. Parry, S.H., Rooke, D.M., *In* Sussman, M. (ed.), The Virulence of *Escherichia coli*. Soc. Gen. Microbiol. 13, Academic Press, (1985) p. 79.
27. Rothbard, J.B., Fernandez, R., Schoolnik, G.K., J. Exp. Med., 160 (1984) 208.
28. Rothbard, J.B., Fernandez, R., Wang, L., Teng, N.N.H., Schoolnik, G.K., Proc. Natl. Acad. Sci. USA, 82 (1985) 915.
29. Salit, I.E., Gotschlich, E.C., J. exp. Med. 146 (1977) 1182.

30. Salit, I.E., Vavougics, J., Hofmann, T., Infect. Immun. 42 (1983) 755.
31. Schoolnik, G.K., Fernandez, R., Tai, J.Y., Rothbard, J., Gotschlich, E.C., J. Exp. Med. 159 (1984) 1351.
32. Smit, H., Gaastra, W., Kamerling, J.P., Vliegenthart, J.F.G., De Graaf, F.K., Infect. Immun. 46 (1984) 578.
33. Smyth, C.J., Jonsson, P., Olsson, E., Söderlind, O., Rosengren, J., Hjertén, S., Wadström, T., Infect. Immun. 22 (1978) 462.
34. Sugarman, B., Epps, L-R., Stenback, W.A., Infect. Immun. 37 (1982) 1191.
35. Sweeney, G., Freer, J.H., J. gen. Microbiol. 112 (1979) 321.
36. Van Die, I., Bergmans, H., Gene 32 (1984) 83.
37. Van Die, I., Van Geffen, B., Hoekstra, W., Bergmans, H., Gene 34 (1984) 187.
38. Wadström, T., Trust, T.J., Brooks, D.E., *In* Lectins, Biology, Biochemistry, Clinical Biochemistry, (Bøg-Hansen, T.C., Spengler, G.A., eds.), W. de Gruyter, Berlin, Vol. 3 (1983), p. 479.
39. Watt, P.J., Ward, M.E., *In* The Gonococcus (Roberts, R.B., ed.), John Wiley and Sons, New York, p. 355.
40. Watts, T.H., Kay, C-M, Paranchynch, W., Biochem. 22 (1983) 3640.
41. Wilson, M.R., Hohmann, A.W., Infect. Immun. 10 (1974) 776.
42. Wevers, P., Picken, R., Schmidt, G., Jann, B., Jann, K., Golecki, J.R., Kist, M., Infect. Immun. 29(1980) 685.

LECTIN-MEDIATED ADHERENCE OF ACTINOMYCES VISCOSUS IN THE ORAL CAVITY

Othmar Gabriel and Marybeth Hinrichs

Department of Biochemistry, Georgetown University,
Schools of Medicine and Dentistry, Washington, D.C. 20007

The process of "cellular recognition" is significant for many biological processes such as growth, differentiation and host-parasite interactions. We wish to examine the basis for the specific and localized population of bacteria in the human oral cavity and have chosen to study this process with Actinomyces viscosus. This microorganism has been implicated in the etiology of gingivitis and periodontitis and was shown to lead in germfree animals to plaque formation followed by root surface caries and bone loss characteristic of periodontal disease (1). The understanding of the molecular events involved in the specific interaction between bacterial cells and human tissues is a requisit to find ways to prevent bacterial adherence and colonization followed by pathological sequelae. Recently, considerable progress was made when different patterns of adherence and colonization of Actinomyces ssp. were found to be the result of cellular recognition mediated by different types of bacterial fimbriae: Type 1 fimbriae mediate adsorption to saliva-coated hydroxyapatite while type 2 fimbriae are responsible for a lectin-like interaction with galactose termini of surface glycoproteins and/or glycolipids.

Our initial interest was focused to understand the lectin-like properties of Actinomyces type 2 fimbriae as indicated by their ability to bind to carbohydrate termini of D-galactose or N-acetyl-D-galactosamine of mammalian cell surface components. For this purpose we examined the role of carbohydrates in cellular recognition with hemagglutination and found that agglutination between Actinomyces viscosus cells and erythrocytes occurs in two steps:

Lectins, Vol. V

The first step was shown to involve the action of bacterial sialidase resulting in the unmasking of galactose residues, the penultimate carbohydrate in many glycoproteins and glycolipids. The second step involves the binding of type 2 fimbriae of Actinomyces viscosus to the galactose termini leading to hemagglutination (4). This simple sequence of events was verified in several ways but for a more detailed study of the lectin specificity it became apparent that the use of erythrocytes as "model cells" was too complex. As a substitute for the complex erythrocyte surface we chose spherical latex beads of about the same dimensions as erythrocytes. This simplified aggregation system was developed whereby a single chemically well defined ligand can be adsorbed to latex beads (5). Using this assay lactose inhibitable, sialidase-dependent aggregation was demonstrated for ß-linked D-galactopyranose or α-linked N-acetyl-D-galactosamine. The use of radioactively labeled glycoproteins such as fetuin established the ligand density on the bead surface in the millimolar range. Structural features of model glycoproteins were examined in the latex bead aggregation assay and preferential binding to galactose ß(1-3)-N-acetylgalactosamine residues was noted (6).

These studies were extended to use hydroxyapatite instead of latex beads as the supporting matrix for glycoproteins (7). These studies revealed significant differences between the two matrix materials and their aggregating properties with bacteria. We decided to examine the molecular basis for these differences.

Materials and Methods

The various strains of Actinomyces viscosus T 14V J1, R 55-36, Q 59-S1 and 147 were obtained from Dr. J.O. Cisar, National Institute for Dental Research, Bethesda, Md. 20205. The proteins BSA and ASF were commercial preparations from Sigma Chemical Co., St. Louis, Ms. The aggregation assay conditions with latex beads as the matrix were carried out according to the procedure described by us earlier (5). In aggregation experiments with hydroxyapatite as the matrix 0.02 M sodium phosphate buffer, pH 6.8, was used and otherwise conditions similar to the one described for latex beads were employed (7).
The preparation of radiolabeled proteins was carried out by reductive alkylation (8). The radioactive labeled proteins were isolated and the specific activities were determined as follows:

(14C)-saliva: 1.9×10^5 DPM/mg protein, (14C)-asialofetuin (ASF): 2.2×10^5 DPM/mg protein, (3H)-bovine serumalbumin (BSA): 4.8×10^6 DPM/mg protein. For determination of the protein content of the matrices employed in aggregation assays the radioactive samples were counted in a Beckman liquid scintillation system LS 9000 that was programmed for double labeling measurements. The latex beads or hydroxyapatite matrices were individually incubated with the proteins at concentrations ranging from 3-5 mg protein/ml and washed 3 x according to the experimental protocols outlined below. A final wash with plain buffer was used to remove radioactive material not adsorbed to the matrix. Aliquots of the suspended matrix were used to determine the radioactivity. On the basis of the specific activity of the various proteins employed values for the quantities of protein adsorbed were calculated. To establish the order of magnitude for the molar quantities of the mixture of salivary glycoproteins a molecular weight of 100,000 was arbitrarily assigned.
Aliquots of the suspension of latex beads or hydroxyapatite particles were subjected to aggregation assays.

Results

A qualitative survey of the aggregation properties of our four test strains of A. viscosus are shown in Table 1. Latex beads or hydroxyapatite were coated with either salivary glycoproteins or asialofetuin (ASF) and examined with various bacterial strains bearing tpye 1 and type 2 fimbriae, type 1 only, or bearing neither fimbriae. Clearly, the aggregation pattern indicates that all interactions are fimbriae-dependent since strain 147 without fimbriae does not cause aggregation. The different functional properties

TABLE 1

Aggregation assays

Strain	Type Fimbriae	Latex Beads (LB)		Hydroxyapatite (HA)		Inhibition by Lactose	
		Saliva- coated	ASV- coated	Saliva- coated	ASF- coated	LB	HA
T 14V J1	1 and 2	strong	strong	strong	weak	weak	none
R 55-36	1 only	strong	none	strong	none	none	none
Q 59-51	2 only	strong	strong	strong	weak	strong	weak
147	none	none	none	none	none	none	none

between type 1 and type 2 fimbriae are also evident in Table 1: Type 1 fimbriae have apparently no lectin-like features, and do not respond to inhibition by carbohydrates. By contrast, type 2 fimbriae are clearly lectin-like and are lactose-inhibitable. A specific role for the matrix in the aggregation process is also apparent. For example, latex beads coated with ASF react strongly with type 2 fimbriae but only weakly with hydroxyapatite coated with ASF. One might suggest that these differences could be explained on the basis of differences in ligand density. For this purpose, the proteins were radioactively labeled with either (^{14}C) or (^{3}H) by reductive alkylation and the resulting ligand density was determined by measurement of radioactivity associated with each matrix. The experimental conditions were chosen to mimic the environment during the aggregation assay. In all these assay systems various concentrations of bovine serum albumin (BSA) are added to prevent unspecific interaction of bacteria with matrix areas that remained unoccupied by glycoprotein and that would otherwise cause unspecific attachment of bacteria. In all the experiments 0.1% BSA was added during the aggregation process, a concentration used by other investigators.

Sequential incubation of asialofetuin (ASF) and bovine serum albumin (BSA).

The sequential incubation of ASF and BSA was carried out to distinguish the properties of latex beads and hydroxyapatite. The results are summarized in Table 2.

TABLE 2

Sequential ASF and BSA incubations

	Incubation with ASF followed by three washes with BSA		Aggregation with T 14V J1	Incubation with BSA followed by three washes with ASF		Aggregation with T 14V J1
	ASF	BSA		ASF	BSA	
Latex beads	7.1	1.2	+++	10	9.7	++
Hydroxyapatite	1.2	0.3	+++	0.97	0.69	++

All results are expressed in nMoles x 10^{-1} per mg of matrix.

When the matrices are first exposed to ASF, followed by three washes with 0.1% BSA, apparently exposure to the first protein occupies the majority of sites and the subsequent washes with BSA cannot displace ASF adsorbed to the matrix. When the reversed sequence is carried out, first exposure to BSA followed by three washes with ASF it is apparent that latex beads are not as selective as hydroxyapatite: Eight times more BSA adsorbes to latex beads while first exposure of hydroxyapatite to BSA increases the amount of BSA only by a factor of 2. The adsorption of BSA seems to promote additional attachment of protein to latex beads. Nevertheless, the aggregation with A. viscosus is diminished by the higher density of BSA molecules on latex beads and hydroxyapatite and as a result shows less ability to aggregate.

Sequential incubation of saliva proteins and bovine serum albumin (BSA).

The same experimental design as just described was used to test salivary proteins and BSA. The data in Table 3 indicate that salivary proteins attach to the matrices under investigation to about the same extent as the model protein asialofetuin. When the order of exposure to proteins is reversed, hydroxyapatite is much more selective whereby only a limited number of sites can be occupied by BSA. By contrast, latex beads show little selectivity and fill available sites with BSA. The following exposure to salivary proteins increases significantly the amount of these proteins associated with latex beads.

TABLE 3

Sequential saliva and BSA incubations

	Incubation with saliva followed by three washes with BSA		Aggregation with T 14V J1	Incubation with BSA followed by three washes with saliva		Aggregation with T 14V J1
Matrix	saliva	BSA		saliva	BSA	
Latex beads	9.0	0.84	+++	28	6.8	+++
Hydroxyapatite	0.6	0.39	+++	1.2	0.53	+++

All results are expressed in nMoles x 10^{-1} per mg of matrix.

Simultaneous incubations.

Incubation of matrices with protein mixtures of various compositions were carried out whereby the percentage of BSA was varied from 0-40%. When latex beads were examined using mixtures of asialofetuin and BSA the amount of ASF remained almost constant while the amount of BSA was proportional to the incubation mixtures as can be seen from the data of Table 4. Hydroxyapatite showed specificity that incidates an increased affinity for BSA when compared to ASF. At all concentrations of BSA, a diminished ability to aggregate was noticed.

The same series of experiments was carried out with mixtures of salivary proteins and BSA. The data for these experiments are shown in Table 5. As can be seen latex beads show considerable specificity whereby salivary proteins show significant preferential adsorbed as compared to BSA. Hydroxyapatite adsorbs salivary proteins and BSA proportional to the concentrations found in the incubation mixtures. The ability to aggregate appear to be unaffected by the various BSA concentrations except for type 1 fimbriae at 40% BSA.

Discussion

To our knowledge, this represents the first quantitative study to examine the composition of proteins on the surface of particles to be examined in aggregation assays. The requirement to cover sites on the matrix that are not occupied by a glycoprotein with an "inert protein" such as albumin (BSA) is carried out routinely to avoid unspecific attachment of bacteria. For this reason 0.1% al-

TABLE 4

Simultaneous incubations with ASF-BSA mixtures

BSA as % of total protein	Latex beads ASF	Latex beads BSA	Aggregation with T 14V J1	Hydroxyapatite ASF	Hydroxyapatite BSA	Aggregation with T 14V J1
0	6.7	-	+++	1.5	-	+++
5	7.8	0.29	+++	1.4	0.1	++
10	7.5	0.8	+++	1.1	0.19	++
20	7.1	1.7	+++	0.91	0.32	++
40	5.8	3.2	+++	0.78	0.52	++

All results are expressed in nMoles x 10^{-1} *per mg of matrix.*

TABLE 5

Simultaneous incubations with saliva-BSA mixtures

BSA as % of total protein	Latex beads saliva	BSA	Aggregation with T 14V J1 R 55-36 Q 59-51	Hydroxy-apatite saliva	BSA	Aggregation with T 14V J1 Q 59-51	R 55-36
0	6.1	-	+++	0.67	-	+++	+++
5	6.9	0.07	+++	0.70	0.03	+++	+++
10	5.9	0.14	+++	0.71	0.05	+++	+++
20	5.4	0.35	+++	0.67	0.13	+++	+++
40	6.1	1.1	+++	0.46	0.21	+++	+

All results are expressed in nMoles x 10^{-1} per mg of matrix.

bumin is added to aggregation assays (2,5). We therefore decided to examine the influence of albumin upon the adsorption of proteins to latex beads and hydroxyapatite. For this purpose we studied the properties of asialofetuin (ASF) a defined model glycoprotein as well as a mixture of salivary glycoproteins in the presence of BSA. This study included the sequential as well as the simultaneous exposure to these proteins. All these experiments demonstrated that once a protein is adsorbed to a matrix it remains there and does not exchange with itself or with other proteins. In other words competitive displacement was not observed at least not during the time frame of the aggregation assay and when subjected to the shearing forces exerted by magnetic stirring during the assay procedure.

During sequential exposure of matrices to various proteins (Tables 2 and 3) it is apparent that the available sites are primarily occupied by the protein to which the matrix is first exposed. When the initial exposure of latex beads is carried out with BSA, a significantly higher quantity of ASF or salivary proteins adsorbs to the matrix. From this one has to conclude that BSA once attached to latex beads mediates further adsorption of other proteins such as ASF or salivary proteins. By contrast, the sequence is of less consequence for the adsorption of ASF to hydroxyapatite indicating a more specific adsorption process and suggesting specific sites for various proteins. The specificity of hydroxyapatite as compared to latex beads is clearly shown when the matrix is exposed to protein mixtures of ASV and BSA. In this instance latex beads show no specificity. By contrast, mixtures of salivary proteins and BSA exhibit strong preferential adsorption of salivary proteins to latex beads but apparently not to hydroxyapatite. From these studies one can conclude that the nature of the matrix and the various properties of different proteins have a significant effect on the ligand density on the surface of particles. This in turn will be a determining factor in lectin-mediated cellular recognition.

Acknowledgements

The authors wish to express their deep appreciation for excellent help in the preparation of this manuscript to Mrs. Elfriede Schauer, Christian-Albrechts-Universität, D-2300 Kiel, Federal Republic of Germany. This work was supported in part by Public Health Service grant DE 06744-02.

References

1. Jordan, H.V., Keyes, P.H., and Bellack, S.: J. Peridontal Res. 7, 21-28 (1972).

2. Clark, W.B., Wheeler,T.T., and Cisar, J.O.: Infect. Immun. 43, 497-501 (1984).

3. Cisar, J.O., Barsumian, E.L., Curl, S.H., Vatter, A.E., Sandberg, A.L., and Siraganian, R.P.: J. Immunol. 127, 1318-1322 (1981).

4. Costello, A.H., Cisar, J.O., Kolenbrander, P.E., and Gabriel, O.: Infect. Immun. 26, 563-572 (1979).

5. Heeb, M.J., Costello, A.H., and Gabriel, O.: Infect. Immun. 38, 993-1002 (1982).

6. Heeb, M.J., Marini, A.M., and Gabriel, O.: Infect. Immun. 47, 61-67 (1985).

7. Gabriel, O., Heeb, M.J., and Hinrichs, M.: Molecular Basis of Oral Microbial Adhesion (Mergenhagen, S. and Rosen, B., eds.) pp. 45-52, Amer. Soc. for Microbiology 1985.

8. Jentoff, N. and Dearborn, D.G.: J. Biol. Chem. 254:4359-4365 (1979).

ISOLATION, PURIFICATION AND SPECIFICITY OF THE LECTIN FROM PSEUDOMONAS AERUGINOSA BACTERIA (HABS STAIN H8)

Hafiz Ahmed, Rupali Pal, Arun K. Guha and Bishnu P. Chatterjee

Department of Macromolecules, Indian Association for the Cultivation of Science, Jadavpur, Calcutta, 700 032, India.

Lectins are a group of proteins having distinct affinity for carbohydrates. They have been found to occur in a wide variety of plants, invertebrates, bacteria, fungi, and even in vertebrates (1). Due to the carbohydrate binding specificity the lectins have been found to interact with numerous cells including erythrocytes, leukocytes, platelets, spermatozoa and to cause agglutination. The agglutination specificity allows them to be useful for blood group determination, purification of polysacharides and glycoproteins and for studying several biological events in cells. Some lectins agglutinate preferentially cancerous cells and thus they can be used as a marker of tumours and cancer cells.

In order to study such biological phenomenons, especially cellular growth and differentiation, we have made attempts to isolate the lectin from Pseudomonas aeruginosa - highly pathogenic bacteria known to cause infection after surgical operations and in burned patients. The presence of an agglutinin in P. aeruginosa was first reported by Neter (2). Gilboa-Garber purified this lectin by affinity chromatography using Sepharose 4B, determined its specificity for D-galactose (3, 4) and described its use in cancer research (5).

In the present communication we describe the purification of lectin from P. aeruginosa H8 (Habs strain) by affinity chromatography on melibiose-agarose and its broad-spectrum specificity.

Materials and Methods

Inhibitors: D-galactose was obtained from BDH, England. 2-acetamido-2 deoxy-D-galactose, O-nitrophenyl- α-D-galactopyranoside, p-nitrophenyl- α-D-galactopyranoside, methyl-β-D-galactopyranoside, fetuin IV were the products of Sigma, U.S.A. Methyl- α-D-galactopyranoside was obtained from Calbiochem, U.S.A. Ant egg glycoprotein was prepared according to the method of Chatterjee and Uhlenbruck (6). Bird nest glycoprotein and BSA-TF antigen were the gift of Prof. G. Uhlenbruck of Medical University Clinic, Cologne, FRG. Ovarian A and O substances were kindly supplied by Dr. T.K. Ghosh of The Johns Hopkins University, Baltimore, U.S.A. Glycophorin was prepared according to Uhlenbruck et al (7).

Chemicals: Melibiose-agarose and acrylamide were purchased from Sigma, Chemical Co., USA. Sephadex G-200 was the product of Pharmacia, Sweden. N-N'-methylene bis acrylamide and ammonium persulfate were obtained from Serva, Heidelberg, FRG. TEMED was purchased from BDH, England. All other chemicals used were of highest analytical grade from Merck and BDH.

Enzymes: Neuraminidase was obtained from Behringwerke AG, Marburg, FRG. Pronase P was the product of Serva, Heidelberg, FRG.

Isolation of P. aeruginosa, H8 Lectin: A pyocyanin producing strain of P. aeruginosa (H8) was grown in nutrient broth (Difco) with shaking at 28° for 3 days. The growth medium was provided daily with 0.2% acetyl choline. The cells were harvested, washed three times in isotonic saline and suspended to obtain a 20% suspension. The cell suspension was exposed to Raytheon 10 K.c. 250 W sonic oscillator for 5 minutes and centrifuged at 30,000 rpm for 15 min. The supernatant fluid, containing the bacterial haemagglutinin was kept at $-20^{o}C$ until used.

Affinity chromatography: To the crude hemagglutinin solid $(NH_4)_2SO_4$ was added to give 70% saturation. The precipitate obtained after centrifugation was dissolved in a minimum quantity of water and dialyzed against water till free from NH_4^+ ions, centrifuged and freeze dired. This hemagglutinin fraction (30 mg in 2 ml PBS) was applied to the melibiose-agarose column (1 x 3 cm) previously equilibrated with PBS and the material was slowly percolated thrice. The column was washed with the same buffer. The bound lectin was desorbed by eluting with the same buffer containing 0.2 M D-galactose. The fractions showing hemagglutinating activity were pooled, mixed, concentrated by Amicon membrane filtration and dialyzed against PBS untill free of galactose and stored at $-20^{o}C$.

Assay procedures: Hemagglutination assays were performed in Takatsy microtiter plate in PBS with two fold serial dilution of the lectin (8). The titer was expressed as the reciprocal of the highest dilution showing positive hemagglutination; specific activity was expressed as titer per mg of protein per ml. The hemagglutination inhibition test was carried out as described in (9). The protein content was measured by the method of Lowry using bovine serum albumin as standard (10).

Disc electrophoresis in 7.5% polycarylamide gel was performed according to Reisfeld et al (11) in β-alanine acetic acid buffer at pH 4.3 and at pH 8.9 by the method of Davis (12) in Tris-glycine buffer. Gels were fixed in 12% trichloroacetic acid, stained by 0.05% Coomassie brilliant blue in 12% trichloroacetic acid and destained and stored in 7% acetic acid.

Molecular weights were estimated by SDS-polyacrylamide gel electrophoresis and gel filtration. SDS-polyacrylamide gel electrophoresis was carried out on 10% polyacrylamide gel according to Laemmli (13). Dissociation and reduction of the protein was performed at 100°C for 5 min in 0.1% SDS with or without 0.1% 2-mercaptoethanol. Protein markers employed were bovine serum albumin (M_r = 68.000), ovalbumin (M_r = 43.000), Chymotrypsinogen A (M_r = 25.000), myoglobin (M_r = 17.200 and cytochrome C (M_r = 12.400). The molecular weight of the lectin was estimated by comparing its mobility with those of markers of known molecular weight. Staining for proteins was performed with 0.05% Coomassie brilliant blue and destained in 7.5% acetic acid containing 5% methanol. Gel filtration was performed on a Sephadex G-200 column (110 x 1.6 cm) (14). The molecular weight of the lectin was determined by comparison of its elution volume with those of molecular weight markers.

Ouchterlony double diffusion: Plates were prepared as described (9). The wells were filled with 10 µl samples to be tested (10 mg/ml) and the plates were developed overnight at 25°C.

Results

Purificatition to homogeneity was achieved by a combination of ammonium sulfate precipitation, affinity chromatography on melibiose-agarose and gel filtration on Sephadex G-100 (see Table 1 and Fig. 1) The hemagglutinating properties of the P. aeruginosa lectin are summarized in Table 2, where we show that the lectin agglutinated human erythrocutes of the A, B, and O blood groups to the same extent. The crude hemagglutinin tested with a panel of untreated and enzyme-treated erythrocytes showed a typical agglutination profile with human and rat erythrocytes while chicken and sheep cells were agglutinated only on exposure to the enzyme pronase and neuraminidase. This indicates a cryptic nature of the receptor de novo exposed by enzyme treatment. The erythrocytes of cow, pig, mouse and duck were neither agglutinated normally nor after enzyme treatment suggesting the absence of receptors for the agglutinin on these cells.

Molecular weight: The molecular weight of the P. aeruginosa H8 lectin was estimated to be about 61,000 dalton by gel filtration on sephadex G-200 (Fig. 2). However, SDS-polyacrylamide gel electrophoresis with or without 2-mercaptoethanol showed a single band cor-

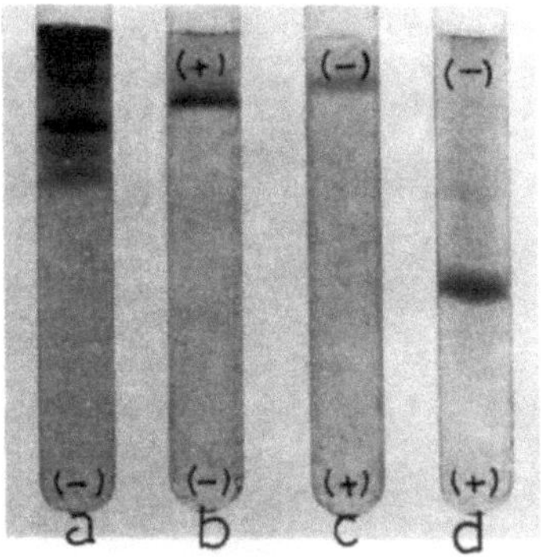

Figure 1: Disc gel electrophoresis of P. aeruginosa lectin (a) crude lectin at pH 4.3; (b) purified lectin at pH 4.3; (c) purified lectin at pH 8.9 - all run on 7.5% polyacrylamide gel; (d) SDS-gel electrophoresis on 10% polyacrylamide gel in presence of 0.1% SDS. The amount of protein in each gel was 70 μg.

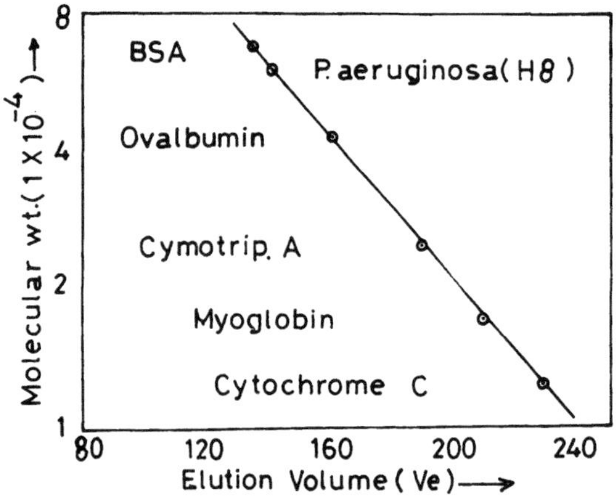

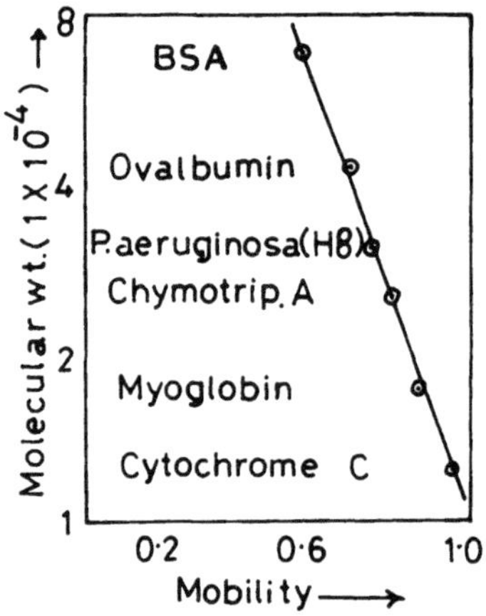

Figure 2. Determination of the molecular weight of P. aeruginosa by gel filtration. A column (1.6 x 110 cm) of Sephadex G-200 was used with PBS as eluent at 4°C. The protein solution (1 ml, 10 mg/ml in eluting buffer) was applied to the column and 5 ml fractions were collected at the rate of 10 ml/hr. Elution wolumes were determined from the position of the maxima of the elution profiles.

Figure 3. Determination of the subunit molecular weight of P. aeruginosa by 10% polyacrylamide gel electrophoresis in presence of 0.1% SDS. Standard protein samples (50 μg each) were applied together with P. aeruginosa (70 μg) and after staining with Coomassie brilliant blue R-250 their mobility was measured from the top of the gel to the leading edge of the protein band.

TABLE 1

Purification scheme of Pseudomonas aeruginosa H8 Lectin

Fraction	Protein mg/ml	Titer*	Specific activity	Purification fold
Crude extract	12.6	32	2.5	1
0-70% $(NH_4)_2SO_4$ fraction	2.55	256	100	40
Galactose eluate from affinity column	0.14	256	1829	731
Eluate from Sephadex column	0.10	256	2560	1000

*Titer was determined against pronase-treated human B cell. The specific activity is expressed as titer per mg protein per ml.

TABLE 2

Hemagglutination results of untreated and enzyme treated erythrocytes of human and animals by Pseudomonas aeruginosa crude lectin.

Erythrocytes	Untreated	Pronase treated	Neuraminidase treated
Human (O,A,B,AB)	2	32	8
Rat	8	64	16
Chicken	0	32	16
Sheep	0	16	8

responding to a molecular weight of 31,000 (Figs. 1 & 3). Thus the P. aeruginosa lectin is composed of two apparently identical subunits joined non-covalently in the same molecule.

Hemagglutination inhibition assays: The results of hemagglutination inhibition of the P. aeruginosa lectin with saccharides and glycoproteins are summerized in Table 3. Of the sugars tested in

TABLE 3

Hemagglutination-inhibition assay of Pseudomonas aeruginosa, H8 bacterial lectin by simple sugars and glucoproteins.

Inhibitor	Concentration
D-galactose	50
Methyl α-D-galactopyranoside	25
Methyl β-D-galactopyranoside	100
N-acetyl D-galactose amine	12.5
D-melibiose	3.12
Raffinose	6.25
O-nitrophenyl α-D-galactopyranoside	100
p-nitrophenyl α-D-galactopyranoside	100
p-nitrophenyl β-D-galactopyranoside	100
Ant egg glycoprotein	0.31
Bird nest glycoprotein	0.31
Ovarian A substance	0.31
BSA-TF antigen	0.62
Ovarian H-substance	0.62
Fetuin	1.25
Asialo Glypophorin	1.25

Sugars required for complete inhibition of two hemagglutinating doses of lectin are expressed in mM concentration for saccharides and in mg per ml for glycoproteins.

inhibition assay, melibiose was the most potent inhibitor and sixteen times more potent than D-galactose and twice more potent than raffinose which was eight times effective than D-galactose. Methyl Methyl β-D-galactopyranoside, o-nitrophenyl α-D-galactopyranoside, P-nitrophenyl α-D-galactopyranoside were poor inhibitors and four times less effective than α-D-galactoside which was twice more effective than its parent compound. N-acetyl-D-galactosamine was a moderate inhibitor. This suggest that the acetyl group in C-2 position is essential for binding since D-galactosamine did not inhibit the hemagglutination. D-glucose, L-arabinose, D-fucose, L-fucose, D-galactosamine and lactose had no inhibitory effect even at a concentration of 200 mM. The hemagglutination was also inhibited by glycoproteins. The glycoproteins from ant egg and edible bird nest, ovarian A substance were very strong inhibitor and twice more effective than BSA-TF antigen and ovarian H-substance. Fetuin and asialo glycoprotein were poor inhibitors among the glycoproteins tested.

In double diffusion the purified P. aeruginosa lectin precipitated galactomannans from fenugreek, Sesbania sesban and carob seeds as well as snail and bovine lung galactan. Moreover it produced a pre cipitin band with concanavalin A lectin thus showing the lectin be a glycoprotein.

Discussion

The extract of P. aeruginosa H8 bacteria contains a hemagglutinin which is inhibited strongly by melibiose and raffinose. This property was ultilized in purifying this bacterial lectin by affinity chromatography on melibiose-agarose. The lectin (MW = 61,000) splits into a single polypeptide chain of molecular weight 31,000 on treatment with detergent suggesting the lectin is a dimer. Like the D-galactose binding lectins of Ricinus communis, Abrus precatorius and Viscum album (15, 16) P. aeruginosa did not show any anomeric specificity as evidenced by the inhibition study, substantiated by precipition with polysaccharides. Plant galactomannans (17) which are composed of linear chains of $(1 \rightarrow 4)$-linked β-D-mannopyranosyl residues having single stubs of α-D-galactopyranosyl groups joined by $(1 \rightarrow 6)$ linkages along the chain, and the galactans (18) from invertebrate and vertebrate origin which are composed of β-$(1 \rightarrow 3)$ and β$(1 \rightarrow 6)$-linked galactose units precipitated well with P. aeruginosa lectin, showing the specificity for both α- and β-D-galactoside.

Our purified lectin differs from that purified by Gilboa-Garber on Sepharose 4B in being more strongly inhibited by α-D-galactoside in comparison to β-D-galactoside. Among the sugars tested melibiose (Gal α$(1 \rightarrow 6)$ Glc) has the strongest inhibitory effect. On the otherhand lactose (Gal β$(1 \rightarrow 4)$Glc) did not inhibit the hemagglutination even at 200 mM concentration. It was evidenced that our preparation neither was adsorbed on a lactose agarose nor was eluted by lactose from a Sepharose 4B column. This difference exhibited by our preparation may be due to different strain used for isolation. The increased affinity of the P. aeruginosa lectin for melibiose in comparison to D-galactose may indicate the presence of an extended combining site, the presence of which has also been proposed for concanavalin A (19), wheat germ agglutinin (20), peanut lectin (21) and potato lectin (22). A detailed quantitative precipitation study of the Pseudomonas lectin using sugars of various anomeric linkages to reveal the exact combining site and specificity is now in progress in this laboratory. Interesting applications of the lectin in mitogenic stimulation of lymphocytes, agglutination of transformed cells, and the effect on the tumor-producing activity of Lewis lung carcinoma cells in vivo has been shown by Gilboa-Garber at this meeting (5).

Acknowledgement: Ms Caroline Nielsen and Mr. Hjalmar Johansen are thanked for typing the manuscript.

References

1. Sharon, N., Lis, H. (1972), Science 4053, 949-959.
2. Neter, E. (1956), Bact. Rev. 20, 166-188.
3. Gilboa-Garber, N. (1972), FEBS Lett., 20, 242-244.
4. Gilboa-Garber, N., Mizzahi, L., Garber, N. (1972). FEBS Lett. 28, 93-95.
5. Gilboa-Garber, N., Avichezer, D., Leibovici, J. These proceedings.
6. Chatterjee, B.P., Uhlenbruck, G. (1982). Experientia 38, 1225-1226.
7. Hanisch, F.G., Farrar, G.H., Schmalisch, R., Uhlenbruck, G. (1983) Immunobiol. 165, 147-160
8. Chatterjee, B.P., Vaith, P., Chatterjee, S., Karduk, D., Uhlenbruck, G. (1979). Int. J. Biochem., 10, 321-327.
9. Chatterjee, B.P., Sarkar, N., Rao, A.S. (1982). Carbohydr. Res. 104, 348-353.
10. Lowry, O.H., Rosebrough, N.J., Farr, A.L., Randall, R.J. (1951). J. Biol. Chem. 193, 265-275.
11. Reisfeld, R.A., Lewis, U.J., Williams, D.E. (1962). Nature 195, 281-283.
12. Davis, B.J. (1964). Ann.N.Y. Acad, Sci. 121, 404-407.
13. Laemmli, U.K. (1970), Nature 227, 680-685.
14. Andress, P. (1965). Biochem, J. 96, 595-606.
15. Goldstein, I.J., Hayes, C.E. (1978). Adv. Carbohydr. Chem. Biochem. 35, 127-340.
16. Luther, P., Thiese, H., Chatterjee, B.P., Karduck, D., Uhlenbruck, G. (1980). Int. J. Biochem. 11, 429-435.
17. Dea, I.C.M., Morrison, A. (1975). Adv. Carbohydr. Chem. Biochem. 31, 241-312.
18. Uhlenbruck, G., Steinhausen, G., Baldo, B.A. (1975). Galactane und Antigalactane, Verlag Josef Stippak, Aachen, West Germany, p-36.
19. So, L.L., Goldstein, I.J. (1968). J. Biol. Chem. 243, 2003-2007.
20. Allen, A.K., Neuberger, A., Sharon, N. (1973). Biochem. J. 131, 155-162.
21. Lotan, R., Skutelsky, E., Danon, D., Sharon, N. (1975). J. Biol. Chem. 250, 8518-8523.
22. Allen, A.K., Neuberger, A. (1973). Biochem, J. 135, 307-314.

PART 5

BIOLOGICAL EFFECTS OF LECTINS

THE BIOLOGICAL EFFECTS OF LECTINS IN THE DIET OF ANIMALS AND MAN

A. Pusztai
The Rowett Research Institute, Bucksburn, Aberdeen AB2 9SB, Scotland, U.K.

Interactions of lectins with cells of the oral cavity

Most, if not all, foods of plant origin contain lectins in various amounts (1,2). Although lectins are reputed to be moderately sensitive to heat denaturation (3), heat treatment is expensive and therefore is kept to a minimum when plants are used for animal feeds. However human diets also contain lectins. Present surveys show that exposure of humans to dietary lectins is a frequent and widespread event (1,2). Some of these foods are eaten raw, such as tomatoes, peppers, fruits, etc., but some lectins survive heat treatment as is clearly illustrated by the finding of haemagglutinating activity in processed peanuts, wheatgerm and dry cereals (1). Furthermore over half of the hot or cold aqueous extracts prepared from commonly ingested fruits, vegetables and seeds react with normal rabbit sera or human saliva (4) and agglutinate human red cells or cells of *Streptococcus mutans* strains. Moreover these extracts also inhibit the attachment of *S. mutans* H12 or *S. sanquis* c1 to saliva-coated hydroxyapatite beads. This effect can be abolished by sufficiently high concentrations of specific sugars and thus the results strongly implicate the involvement of food lectins in host-parasite interactions in the mouth (4). As saliva and salivary mucins selectively inhibit the oral interactions between food lectins and teeth and buccal epithelial cells it was suggested that one of the main functions of these secretions may be to reduce the extent of this interaction (5). The general and high reactivity of lectins with oral mucosa and skin both *in vitro* and *in vivo* have now been fully substantiated (6-8). The attachment of increased numbers of streptococci to epithelial cells through polyvalent lectin-binding is thus a potentially important part of the overall effects of food on the bacterial ecology in the oral cavity.

Gastrointestinal effects

Several lectins are known to be resistant to proteolysis in the

Lectins, Vol. V

gastrointestinal tract (9-14) and are free to react with both the luminal content and the epithelial cells of the gut. Thus for example the dietary kidney bean lectin affects both the gross and microscopic morphology of the small intestine of the rat (15,16). There is an initially reversible increase of small intestinal tissue mass; its protein content increases by about 40-50% and the concentration of carbohydrate doubles in 10 days of feeding. These changes may indicate a stimulation of mucinous secretion. Increased DNA-content (30-40%) however also indicates mucosal hyperplasia (16). Lectins as histochemical tools have already been used extensively for the differential localization of cell surface and secretory components of intestinal epithelia (17). Similar reactions with lectins may also occur in the intestines *in vivo.* Thus both concanavalin A and wheat germ agglutinin when injected into intestinal loops cause increased shedding of brush border membranes, reduced absorptive area, accelerated cell-loss and shortened villi (18). Concanavalin A also affects leucine uptake and transmission of proteins in rat enterocytes (19-21). Orally introduced ricin or its galactose-specific B-subunit (the lectin-type subunit), but not its toxic A subunit, also induces severe changes, loss of villi and increased regeneration time of the absorptive epithelia of rat small intestine with a concomitant decrease of glucose absorption (22,23).

These observations of *in vivo* effects of lectins on the structure and the absorptive function of the small intestine support Jaffe's original suggestion that the toxic effects of ingested lectins were due to their ability to combine with specific receptor sites of cells lining the small intestine and to cause a non-specific interference with absorption and nutrient utilization (24-26). Thus, although in severe poisoning with the kidney bean lectin well over 80% of the dietary dry matter was still absorbed, there was a disproportionate effect on the absorption of protein which was utilized at an extent equivalent to less than 30% of that ingested (27). There was also extensive and severe damage caused by the kidney bean lectin to the structure of the brush borders of the proximal small intestine as demonstrated by both light and electron microscopy (28-30). Similar results were also obtained with a lectin from a Brazilian cultivar of kidney bean (31), a *Phaseolus acutifolius* (tepary bean) lectin (32) and also a lectin from the tropical plant *Psophocarpus tetragonolobus* (33). Binding of the kidney bean lectin to the disrupted brush border membranes was also demonstrated by immunofluorescent antibody technique (34). Similarly, tomato lectin was bound to the intestinal cells although it did not disrupt the normal structure of the brush border (13). An appreciable part of the bound kidney bean lectin penetrates the mucosa and can be traced to lysosome-like vacuoles of the small intestinal enterocytes (King, T.P., unpublished). The bound lectins interfere with the activities of brush border enzymes concerned with the final stages of nutrient hydrolysis and transport (35-38). Similar effects were also observed with the apparently non-toxic lectins from soya bean, peas or lentils (39-41).

According to more recent suggestions (42) the rather special digestive/absorptive disturbance, coeliac disease, which affects certain humans known to be sensitive to cereal foods containing gluten-type proteins, may also be lectin-related in its origin. Although the aetiology of the disease is complex and not fully understood, it is thought to involve the major histocompatibility complex and a cell-mediated (type-IV) immune hypersensitivity reaction to gluten proteins. In addition however the gluten-proteins are now also known to bind as polymannose-specific lectins to immature surface determinants of coeliac enterocytes (43-45). Moreover, gliadin peptides also bind to the small intestine of 17-18 day old rat foetuses and inhibit reversibly the development and morphogenesis of this tissue but they have no such effects on the differentiated jejunum from 21-day old foetuses or from newborn rats (46). Thus these gluten- or gliadin-lectins can bind to incomplete, high-mannose-type oligosaccharide chains of abnormal surface glycoconjugates of intestinal cells in which their conversion to the mature, complex carbohydrate chains is blocked. Some evidence for such an abnormal reactivity to concanavalin A-horse radish peroxidase complex of jejunal biopsies from coeliac patients has been reported previously (47). The net effect of this initial gluten-binding to immature enterocytes is that the structure of these cells is damaged and their absorptive potential is reduced. These may then be followed by more complex and injurious immune system-mediated lesions in coeliac individuals.

There are indications obtained from morphological studies that soya protein-induced damage to the small bowel mucosa in infants is consistent with a lectin-induced toxicity, similar to that postulated for coeliac disease (48). Moreover soya bean lectin was shown to disturb the apical membrane of rabbit jejunum by binding to reactive sites. This destabilization of the enterocyte membrane resulting in an incrased leakage of macromolecules, possibly amplified by a synergistic effect of soya saponins (49), may at least partly explain the gastrointestinal disturbances which occur when raw or incompletely processed soya bean products are fed to humans or animals (50-52).

The binding of a lectin to intestinal enterocytes however is not necessarily harmful *per se* as demonstrated by feeding tomato lectin to humans or rats (13). This lectin, in common with a number of others, resists digestion in the gut and may exert biological effects such as immunosuppression. Thus tomato lectin may have beneficial effects on humans suffering from autoimmune diseases or even in coeliacs.

The lectin, concanavalin A, was also shown to cause hypersecretion of mucus in both gut and nose (53). This mucotractive effect is apparently not related to the known stimulatory activity of lectins on mast cells but rather it is suggested to be the result of a direct interaction with goblet cells. Indeed, the beneficial effects of dietary fibre may be due in part to a lectin-related mechanism on gut goblet cells (53). The large increase in the small intestinal tissue weight of rats fed on kidney bean proteins (15,16) provides further support for such a lectin-induced increase of mucinous secretion. As these mucins are digested only by the bacteria of the large intestine this lectin-stimulated increase of endogenous nitrogen secretion makes a substantial contribution to the overall loss of nitrogen from the body.

Lectin absorption in the gut

Measurements of TCA-precipitable radioactivity in internal organs and circulating blood in rats which were given known amounts of ^{125}I-labelled lectins by stomach-tubing showed that variable but significant amounts of lectins can cross the intestinal barrier and become distributed in the tissues of the body. With the non-toxic tomato lectin these amounts are comparatively small; less than 0.01% of that introduced is found per ml of serum. Similarly, tissue-bound radioactivity is also low amounting to about 1% of that given (13). In contrast rat sera contain between 0.5 to 0.7% (per ml) of the kidney bean lectin introduced originally (Fig. 1) and the total TCA-precipitable radioactivity in internal tissues exceeds 5% of the total introduced (54). The initial rate of lectin-uptake is appreciably faster in rats pre-fed on diets containing kidney bean proteins, however 3 h after intubation the amounts of radioactivity are the same in all rats, including those control animals which were exposed to the lectin for the first time. It appears that the two types of bean isolectins are also absorbed at different rates initially; the L_4 lectin being taken up by about 2-3 times faster than the E_4 isolectin.

Local and systemic immune responses to dietary lectins

The immunosuppressive effects of parenterally introduced lectins for the development of antibodies to unrelated antigens are generally well-known. Thus for example intraperitoneally injected kidney bean lectin was a powerful suppressant of humoral antibody production in both rats and mice (55). However when kidney bean proteins were fed to various experimental animals they all developed a powerful and exclusive humoral antibody response of

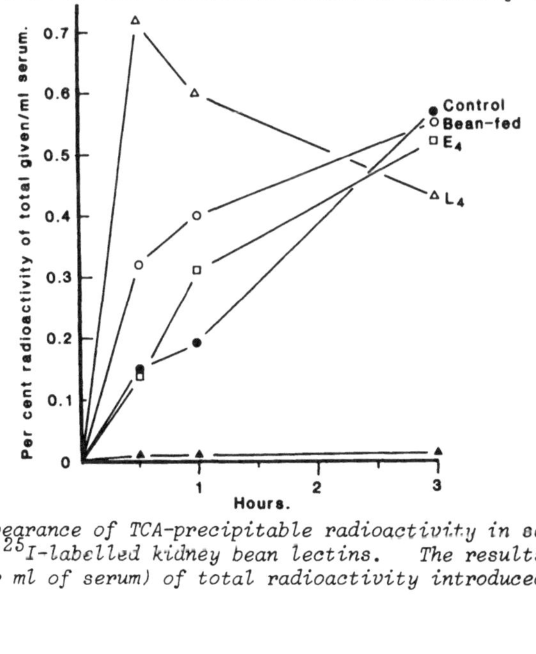

Fig. 1. The appearance of TCA-precipitable radioactivity in sera of rats incubated with ^{125}I-labelled kidney bean lectins. The results are expressed as per cent (per ml of serum) of total radioactivity introduced.

the IgG-type to the dietary lectin (56-58). The development of anti-lectin antibodies in the rat (Fig. 2) was dependent on the time-course of the oral-immunization. On lectin-free diets the anti-lectin titre declined slowly. However, re-introduction of the lectin in the diet produced an elevated secondary response (54). No other component of the diet produced an antibody response in these rats. This is probably an indication of the great specificity of the lectin uptake and/or the immuno-suppressive effects produced by the relatively large amounts of systemically absorbed kidney bean lectin. It also appears that the concentration of the local antibody, S-IgA, is not high enough to neutralize the dietary lectin and to prevent its crossing into the systemic circulation. Whether this insufficiency is confined to anti-lectin-sIgA levels in the gut or is a general suppression of local antibody production has not been established. However the well-known bacterial proliferation in the small

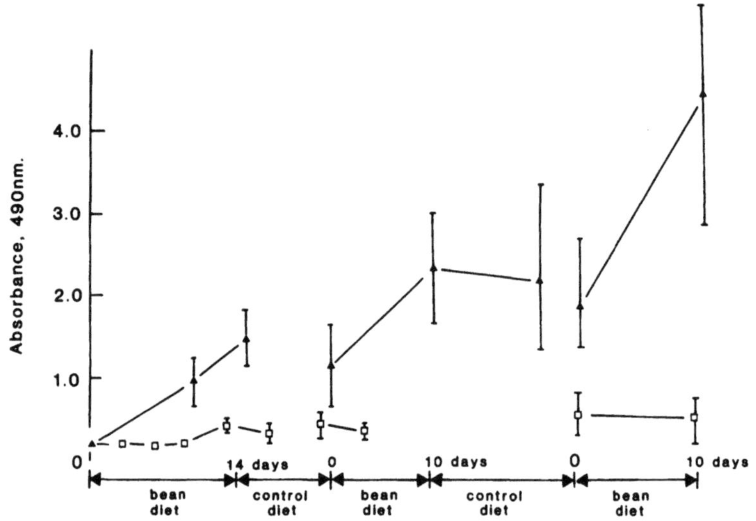

Fig. 2. The development of specific antibodies to the dietary lectin in rats fed alternately on bean proteins or on control diets.

intestine of bean-fed animals (59,60) suggests a more general suppression.

Rats fed on diets containing bean proteins also exhibit an increased mucosal, vascular permeability (15,61) with an increased leakage of circulating ^{125}I-labelled serum proteins into the gut. These results, when taken in conjunction with the known degranulation of mast cells by the kidney bean lectin, suggest that the increased permeability *in vivo* may be due to a direct or an IgE-mediated effect of the lectin on gut mast cells. The accompanying exudation of serum proteins into the gut may thus make a significant contribution to the overall loss of nitrogen and to bean toxicity.

Finally, it is likely that the rapid degeneration of the thymus is one of the systemic effects of the lectin absorbed from kidney bean diets (15,16). Although the thymus is usually regarded as an organ of major importance for the immune system, the effects of this lectin-induced thymus atrophy on immunity and on T-cell maturation have not been established.

General systemic effects of absorbed lectins

As an appreciable proportion of dietary lectins can pass through the intestinal wall and into systemic circulation, direct effects of these lectins on cells of various target tissues may be expected to occur. Thus dietary kidney bean lectin causes an accelerated rate of tissue protein catabolism (36). More recently it has been found that the breakdown of stored fat triglycerides and glycogen is also greatly increased (Grant, Oliveira and Pusztai, unpublished) in bean-fed rats (Table 1). These catabolic changes become very striking when compared to those obtained with pair-fed rats on identical total energy intake (Table 1). The results suggest a direct interference by dietary bean proteins/ lectins with the lipid and/or carbohydrate intermediary metabolism of rats. Although the reaction mechanism of this interference is unknown, lectins, including those from kidney bean (62), are known to possess insulin-like effects on fat cells *in vitro*. Moreover preliminary results show that, as a result of feeding rats on kidney bean proteins their blood insulin concentration is depressed. When these animals are challenged intragastrically with kidney bean proteins and lectins there is a further quick fall in blood insulin levels to very low values (Fig. 3). There is a smaller and transient decrease in blood insulin concentrations with control rats on a single exposure to kidney bean lectin for the first time (Grant, Oliveira and Pusztai, unpublished). The systemically absorbed kidney bean lectin may thus have a direct effect on the insulin-secreting cells of the pancreas leading to an impairment in the release of insulin to the blood stream with consequent reduction in lipogenesis.

Table 1. Body composition of rats (6 rats per group) fed on various diets for 10 days.

		Control diets				Bean protein (5%) + Egg albumin (5%)	
		Non-protein		10% Egg albumin			
			SE		SE		SE
Rat	Starting weight (g)	180.1	8.2	180.6	8.7	180.5	5.8
	End weight (g)	149.8	7.4	190.5	5.5	152.0	6.9
	Carcass dry weight (g)	45.7	2.8	55.8	1.7	39.8	2.4
Food intake/10 days (g)		112.9	3.1	104.5	3.2	103.9	3.2
Body composition	Protein (g)*	31.9	0.4	40.1	0.9	35.0	0.4
	% of total dry weight	65.8	0.3	69.6	1.5	78.0	1.7
	Fat (g)**	9.2	0.4	10.0	0.4	3.2	0.5
	% of total dry weight	19.0	1.0	17.4	0.5	7.2	1.3
	Glycogen (g)	0.63	0.04	0.74	0.02	0.35	0.02
	% of total dry weight	1.37	0.24	1.34	0.17	0.87	0.11

* Kjeldahl N x 6.25
**Chloroform-methanol (2:1) extracted material

Other disturbances in systemic metabolism have also been observed. One of the most striking effects caused by bean diets is the interference with the normal urinary mineral excretion (16). Thus there is an increased urinary loss of potassium in bean-fed rats probably due to an increased cellular catabolism. This is partly compensated for by increased retention of sodium. There is also a reduction in urinary phosphorus output, although no related trend is observed in the sera of these rats (16). These changes may thus again suggest a direct role for the systemically absorbed lectin, perhaps by mimicking hormonal effects.
In conclusion: dietary lectins of plant origin, if sufficiently resistant to digestion in the gut, will react with the appropriate carbohydrate surface components of various cells in the intestines

and this may result in interferences with their digestive, absorptive, protective or secretory functions. Due to the selective absorption of lectins systemic reactions may occur leading to the disruption of the normal protein-, fat- and carbohydrate intermediary metabolism possibly through hormone-mimicking effects of the lectins on endocrine cells. In addition,

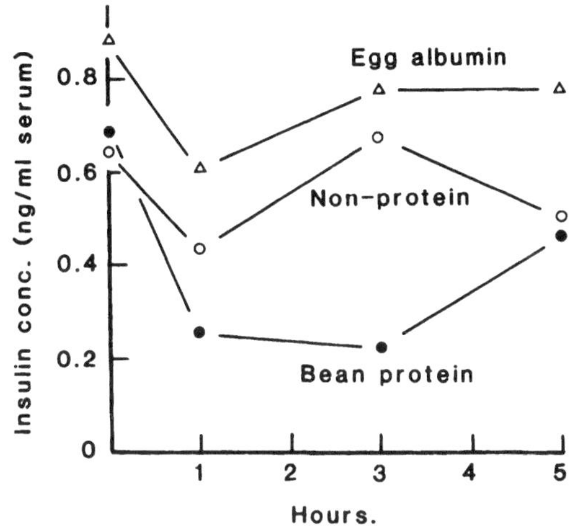

Fig. 3. Serum insulin concentration in rats fed on various diets (as indicated in the diagram) for 10 days. All rats were fasted overnight and then fed 3 g of an egg albumin-diet per rat in the morning. Each rat was given by stomach-tubing 50 mg of bean protein 2 h later (0-time) and blood samples were taken from the tail vein at times as indicated.

both local and systemic immune reactions to the lectins may also occur. A combination of all these effects may directly affect the nutritional utilization of diets and the growth and health of man and animals.

References

1. Nachbar, M.S., Oppenheim, J.D.: Am. J. Clin. Nutr. 33, 2338-2345 (1980).
2. Grant, G., More, L.J., McKenzie, N.H., Stewart, J.C., Pusztai, A.: Brit. J. Nutr. 50, 207-214 (1983).
3. Liener, I.E.: J. Agric. Food Chem. 22, 17-22 (1974).
4. Gibbons, R.J., Dankers, I.: Appl. Environ. Microbiol. 41, 880-888 (1981).
5. Gibbons, R.J., Dankers, I.: Am. J. Clin. Nutr. 36, 276-283 (1982).
6. Rittman, B.R., Mackenzie, I.C., Rittman, G.A.: Archs. Oral Biol. 27, 1013-1019 (1982).
7. Hietanen, J., Salo, O.P.: Scand. J. Dental Res. 92, 443-447 (1984).

8. Gibbons, R.J., Dankers, I.: Archs. Oral Biol. 28, 561-566 (1983).
9. Jaffe, W.G., Vega Lette, C.L.: J. Nutr. 94, 203-210 (1968).
10. Pusztai, A., Grant, G., Palmer, R.M.: J. Sci. Food Agric. 26, 149-156 (1975).
11. Nachbar, M.S., Oppenheim, J.D., Thomas, J.O. J. Biol. Chem. 2056-2061 (1980).
12. Banwell, J.G., Boldt, D.H., Meyers, J., Weber, Jr. F.L., Miller, B.: Gastroenterology 84, 506-515 (1983).
13. Kilpatrick, D.C., Pusztai, A., Grant, G., Graham, C., Ewen, S.W.B.: FEBS Lett. 185, 299-305 (1985).
14. Hata, T., Mukunaki, Y., Tsukamoto, I., Miyoshi, M. Hasegawa, K.: J. Nutr. Sci. Vitaminol. 30, 381-394 (1984).
15. Pusztai, A., Greer, F.: Protides Biol. Fluids 32, 347-350 (1984).
16. Greer, F., Brewer, A., Pusztai, A.: Brit. J. Nutr. 54, 95-103 (1985).
17. Etzler, M.E., Branstrator, M.L.: J. Cell Biol. 62, 329-343 (1974).
18. Lorenzsonn, V., Olsen, W.A.: Gastroenterology 82, 838-848 (1982).
19. Ichev, K., Ovtscharoff, W.: Acta Histochem. 69, 119-124 (1981).
20. Ichev, K., Chouchkov, C.: Acta Histochem. 72, 181-186 (1983).
21. Ovtscharoff, W., Ichev, K.: Acta Histochem. 74, 21-24 (1984).
22. Ishiguro, M., Mitarai, M., Harada, H., Sekine, I., Nishimori, I., Kikutani, M.: Chem. Pharm. Bull. 31, 3222-3227 (1983).
23. Ishiguro, M., Harada, H., Ichiki, O., Sekine, I., Nishimori, I., Kikutani, M.: Chem. Pharm. Bull. 32, 3141-3147 (1984).
24. Jaffe, W.G.: Arzneimittel. Forsch. 10, 1012-1016 (1960).
25. Jaffe, W.G.: Experientia 18, 76-77 (1962).
26. Jaffe, W.G., Camejo, G.: Acta Cient. Venez. 12, 59-61 (1961).
27. Pusztai, A., Clarke, E.M.W., King, T.P.: Proc. Nutr. Soc. 38, 115-121 (1979).
28. King, T.P., Pusztai, A., Clarke, E.M.W.: J. Comp. Path. 90, 585-595 (1980).
29. King, T.P., Pusztai, A., Clarke, E.M.W.: J. Comp. Path. 92, 357-373 (1982).
30. King, T.P., Begbie, R., Cadenhead, A.: J. Sci. Food Agric. 34, 1404-1412 (1983).
31. Rossi, M.A., Mancini Filho, J., Lajolo, F.M.: Brit. J. Exp. Path. 65, 117-123 (1984).
32. Sotelo, A., Licea, A.G., Gonzalez-Garza, M.T., Velasco, E., Feria-Velasco, A.: Nutr. Rep. Int. 27, 329-337 (1983).
33. Higuchi, M., Tsuchiya, I., Iwai, K.: Agric. Biol. Chem. 48, 695-701 (1984).
34. King, T.P., Pusztai, A., Clarke, E.M.W.: Histochem. J. 12, 201-208 (1980).
35. Rouanet, J.M., Besancon, P.: Ann. Nutr. Alim. 33, 405-416 (1979).
36. Pusztai, A., Clarke, E.M.W., Grant, G., King, T.P.: J. Sci. Food Agric. 32, 1037-1046 (1981).
37. Rouanet, J.M., Besancon, P., Lafont, J.: Experientia 39, 1356-1358 (1983).

38. Triadou, N., Audran, E.: Digestion 27, 1-7 (1983).
39. Jindal, S., Soni, G.L., Singh, R.: IRCS Med. Sci. 10, 214 (1982).
40. Jindal, S., Soni, G.L., Singh, R.: J. Plant Foods 4, 95-103 (1982).
41. Dhaunsi, G.S., Garg, U.C., Sidhu, G.S., Bhatnagar, R.: IRCS Med. Sci. 13, 469-470 (1985).
42. Weiser, M.M., Douglas, A.: Lancet 567-569 (1976).
43. Kottgen, E., Volk, B., Fluge, F., Gerok, W.: Biochem. Biophys. Res. Commun. 109, 168-173 (1982).
44. Concon, J.M., Newburg, D.S., Eades, S.N.: J. Agric. Food Chem. 31, 939-941 (1983).
45. Auricchio, S., De Ritis, G., De Vincenzi, M., Mancini, E., Minetti, M., Sapora, O., Silano, V.: Biochem. Biophys. Res. Commun. 121, 428-433 (1984).
46. De Ritis, G., Occorsio, P., Auricchio, S., Gramenzi, F., Morisi, G., Silano, V.: Pediatr. Res. 13, 1255-1261 (1979).
47. Vasmant, D., Feldman, G., Fontaine, J.L.: Pediatr. Res. 16, 441-445 (1982).
48. Poley, J.R., Klein, A.W.: J. Pediatr. Gastroenterol. Nutr. 2, 271-287 (1983).
49. Alvarez, J.R., Torres-Pinedo, R.: Pediatr. Res. 16, 728-731 (1982).
50. Barratt, M.E.J., Strachan, P.J., Porter, P.: Proc. Nutr. Soc. 38, 143-150 (1979).
51. Kilshaw, P.J., Sissons, J.W. Res. Vet. Sci. 27, 361-365 (1979).
52. Sissons, J.W., Smith, R.H., Hewitt, D., Nyrup, A.: Brit. J. Nutr. 47, 311-318 (1982).
53. Freed, D.L.J., Buckley, C.H.: Lancet 585-586 (1978).
54. Greer, F.: Ph.D. thesis, University of Aberdeen (1983).
55. Jennings, J.F., Oates, C.M.: Clin. Exp. Immunol. 2, 445-453 (1967).
56. Williams, P.E.V., Pusztai, A., Macdearmid, A., Innes, G.M.: Animal Feed Sci. Technol. 12, 1-10 (1984).
57. Grant, G., Greer, F., McKenzie, N.H., Pusztai, A.: J. Sci. Food Agric. 36, 409-414 (1985).
58. Begbie, R., King, T.P.: Lectins. Biology-Biochemistry, Clinical Biochemistry (Bog-Hansen, T.C., Breborowicz, J. eds.), vol 4, 15-27 W. de Gruyter, Berlin (1985).
59. Wilson, A.B., King, T.P., Clarke, E.M.W., Pusztai, A.: J. Comp. Path. 90, 597-602 (1980).
60. Banwell, J.G., Abramowsky, C.R., Weber, F., Howard, R., Boldt, D.H.: Dig. Dis. Sci. 29, 921-929 (1984).
61. Greer, F., Pusztai, A.: Digestion 32, 42-46 (1985).
62. Pusztai, A., Watt, W.B.: Biochim. Biophys. Acta 365, 57-71 (1974).

APPLICATION OF *PSEUDOMONAS AERUGINOSA* LECTIN (PA-I) FOR CANCER RESEARCH

Nechama Gilboa-Garber, Dodi Avichezer and Judith Leibovici*
Department of Life Sciences, Bar-Ilan University, Ramat-Gan 52100, and *Department of Pathology, Tel-Aviv University, Sackler Medical Faculty, Tel-Aviv 69978, Israel

The galactophilic lectin of *Pseudomonas aeruginosa* (PA-I) resembles plant lectins in its chemical and biophysical properties, biological activities and scientific and medical applications (1, 2). In addition to selective agglutination of various cells (3), it exhibits mitogenic stimulation in human peripheral lymphocytes (4) and stimulates various metabolic and functional systems in cells of higher, as well as lower organisms, including free living unicellulars. In the latter, e.g. *Tetrahymena pyriformis*, presence of PA-I increases growth rate and phagocytic activity (5). Treatment of bacteria with the lectin also increases their phagocytosis by human peripheral polymorphonuclear leukocytes *in vitro* (6). The medical potential of PA-I is multilateral. It was shown to be useful as a vaccine against *P. aeruginosa* infection in mice, fully protecting them from an otherwise lethal dose of the bacteria injected intraperitoneally (7). The lymphocyte stimulating activity of PA-I (4) may be useful in chromosomal examinations using human peripheral lymphocytes and in detection of disease states where lymphocyte blastogenic reaction is depressed (8,9). The present report describes the use of PA-I in three domains of cancer research: a) assay of mitogenic stimulation of human peripheral lymphocytes; b) agglutination of transformed cells, and c) effect on the tumor-producing activity of Lewis lung carcinoma cells (3LL) in C57Bl mice. It is well known that peripheral lymphocytes from cancer-bearing patients respond to mitogenic stimulation by *Phaseolus vulgaris* lectin (PHA) and several other plant lectins more weakly than those of healthy subjects (8,9). Reduced mitogenic reactivity is also observed in several other chronic diseases. This phenomenon, attributed to impaired immunological competence, aids in the diagnosis and prognosis of these diseases. Lectins are also widely used for detection of changes in the surface membrane of transformed cells (10-15). In studies on the effects of lectins on transformed cells *in vitro* and *in vivo* , diverse interesting results were obtained, which were dependent on the lectin and cell types, as well as on their levels and the mode of treatment. Robinson and Mekori (16) have shown that PHA-treated mice had lower ascites tumor volume than the control group, while Datta et al. (17) found that mice treated with PHA prior to inoculation with Ehrlich ascites cells demonstrated a shorter survival time than the control mice. Wagener (18) has also described an enhanced growth of fibrosarcoma in mice treated with PHA in large doses. Shoham et al. and others (19,20) have shown that Concanavalin A (ConA) reduced tumor cell viability *in vitro* and tumorigenicity *in vivo*. Other lectins, such as ricin, abrin and wheat germ agglutinin (WGA) were also shown to be effective in tumor depression (21-25) and tumor vaccination by lectins has been reported (23-25).

In the *in vivo* experiments, in addition to the direct effect of the lectins on the tumor cells, these lectins may also affect the animal's general immune state (18,26), i.e., activation of polymorphonuclear leukocytes, macrophages and lymphocytes, which lead to cytotoxic effects on the target tumor cells (25). Moreover, lectins were suggested as participating in macrophage-mediated tumor cell lysis (25,27-30). We have found that PA-I is similar to PHA and some other plant lectins in detecting impaired mitogenic response in lymphocytes of cancer patients and in agglutination of tumor cells, including 3LL cells. PA-I dramatically lowered the tumorigenic activity of 3LL cells inoculated subcutaneously (s.c.). The effect was exhibited in tumor incidence and size, as well as mortality of mice. It was specific to the sugar (inhibited by D-galactose). Intraperitoneal (i.p.) injection of PA-I concomitant with the s.c. inoculation of the 3LL cells did not prevent tumor development and mortality of mice. No inhibition was obtained by PA-I on the tumorigenicity of B16 melanoma cells with s.c. inoculation to the same host.

Materials and Methods

Lectins. PA-I was purified in our laboratory, as previously described (1). *Phaseolus vulgaris* lectin (PHA-p) and ConA were purchased from Difco and Sigma, respectively.

Mitogenic stimulation of human peripheral lymphocytes. Human peripheral lymphocytes were isolated from heparinized bloods by Lymphoprep and treated by sialidase, as previously described (4). Analysis of their mitogenic stimulation by the lectin was performed by suspending $1x10^6$ lymphocytes in 1ml of the culture medium (Eagle's Medium, EM, Gibco) containing 20% fetal calf serum (FCS) with or without the lectin (6-10 µg.ml^{-1}). Samples of 0.20 ml containing $2x10^5$ cells were grown in quadruplicate in a microtest plate (NUNC) at 37°C in an atmosphere of 95% air - 5% CO_2 for 72 h. At 20 h prior to termination of the incubation period, 1µCi [methyl ^3H]-thymidine in 10µl of phosphate buffered saline (PBS) was added to each well. After incubation, the cells were harvested with an automatic cell harvester and the radioactive content was assayed by liquid scintillation counting. The results are expressed as stimulation index, defined as SI = $(C_t-C_o)/C_o$ (C_t, cpm value in the presence of the lectin; C_o, the value in the absence of lectin.

Cancer-bearing patients. All the cancer-bearing patients were pre-operative and none received any treatment (chemotherapy or radiotherapy). Some patients were suffering from solid tumors of the lungs, stomach or colon.

Mice and tumors. C57Bl/6J mice (6-10-weeks old) were purchased from The Weizmann Institute of Science (Rehovot, Israel). The Lewis lung carcinoma and B16 melanoma cells were maintained by s.c. transfers and cell suspensions were prepared, as previously described (31).

Incubation of tumor cells with the lectins. Tumor-cell suspensions containing $1x10^6$ cells/ml were mixed with an equal volume of either saline or lectin preparation (ConA, PHA in 100 µg/ml concentration) and incubated at 37°C in a 5% CO_2 atmosphere for 1 to 2 h (for PA-I, time 0 was also examined). The effect of the presence of 0.3M D-galactose with the 100 µg/ml lectin concentration, as control for the specificity of the lectin effect, was examined. Cell viability was determined at the end of the incubation by the trypan blue exclusion test.

Agglutination test. Agglutination of the cells by the lectin was analyzed after 1 h of incubation. Intensity of agglutination was examined microscopically by phase contrast microscopy and photographed at magnification x400.

Assay of tumorigenicity. Tumor cells (2 - 2.5x10^5), non-treated or pretreated with lectins, were inoculated s.c. in 0.2 ml in the upper part of the back. Lectin concentration in the injected cell-lectin mixtures was 20 µg for PHA and ConA and 20, 10 and 1 µg, respectively, for PA-I. Each group of inoculated mice consisted of 5 animals. Incidence and size of the tumors were recorded two to three times per week; mortality was recorded daily up to 2 months. Following tumor cell inoculation, the hosts inoculated with non-treated cells generally survived no more than 40 days. Mortality correlated tumor incidence.

Treatment of 3LL-inoculated mice by PA-I. PA-I was administered i.p. at 10 or 200 µg doses in 0.2 ml just before s.c. inoculation of the tumor cells. Tumor development followed, as above.

Cytostatic activity of PA-I. Samples (0.2 ml) of the above tumor cells-lectin mixture in quadruplicate in microtest plates (containing 2.5x10^5 cells suspended in EM supplemented with 15% FCS and PA-I at the appropriate concentration) were incubated for 1 h in a CO_2 incubator. Then, [methyl-^{3}H]-thymidine was added at 5 µCi/ml to 200 µl of culture and incubated for 1 h. At the end of the labeling period, the cells were automatically harvested and their radioactivity counted in a Packard Tricarb scintillation counter.

Results

A. Mitogenic response of cancer patient lymphocytes to PA-I

The *in vitro* mitogenic response of the sialidase-treated peripheral lymphocytes of cancer patients to PA-I was found to be significantly lower (stimulation index 12.5 $\pm$ 7) than that of similarly treated healthy subject lymphocytes (stimulation index 22.5 $\pm$ 6) parallely examined (eight patients suffering from diverse types of solid cancer and eight healthy subjects examined in parallel pairs). The significance of the difference between the two groups was similar to that obtained when untreated lymphocytes of the same subjects were examined for mitogenic stimulation with PHA.

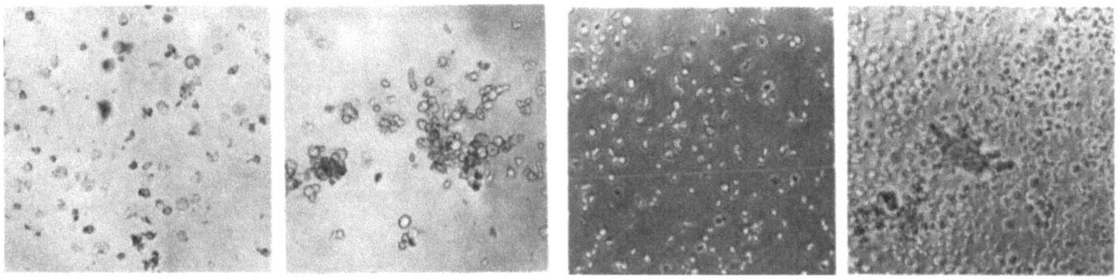

A B C D

Fig. 1. Agglutination of cells by PA-I. A - 3LL cells without PA-I, or with PA-I and 0.3M GAL; B - 3LL cells with 20 μg PA-I in 0.2 ml; C - B 16 melanoma cells without PA-I or with PA-I and 0.3M galactose; D - B 16 melanoma cells with 20 μg PA-I in 0.2 ml (phase contrast microscope, x400).

B. Agglutination of tumor cells by PA-I

3LL cells in freshly prepared suspensions and B16 melanoma cells similarly prepared were strongly agglutinated by the examined lectins in different concentrations. Agglutination of the 3LL cells and melanoma cells by 20 μg PA-I is presented in Fig. 1. In some preparations of the tumor cells without lectin, no clumping was observed, while others contained some clumping without lectin addition. The latter exhibited stronger agglutination after addition of the lectins. Agglutination induced by the lectin was D-galactose specific: presence of the sugar in the mixture abolished it completely.

C. Effect of PA-I on the tumorigenicity of 3LL cells

Incubation of 3LL cells with ConA, PHA and PA-I at 20 μg concentration for 1-2 h before s.c. injection, reduced their tumorigenicity (Fig. 2). PHA effect was the lowest, about 80% of the mice developed tumors at almost the same size as that in mice injected with 3LL alone. ConA effect was stronger and that of PA-I, the most pronounced. PA-I, at 20 μg level, fully inhibited tumor appearance in the mice. PA-I (10 μg) also inhibited tumor incidence more than did ConA.

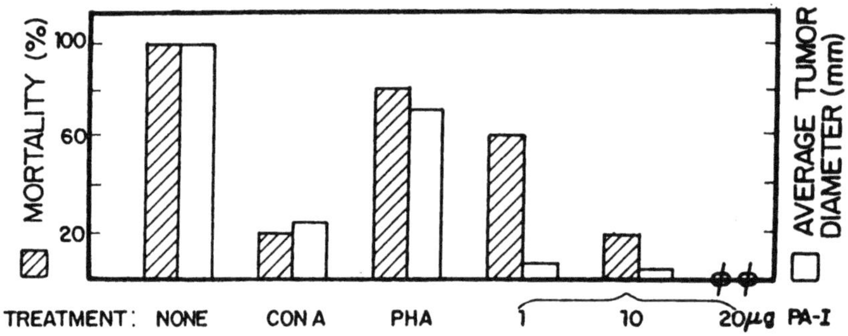

Fig. 2. Effect of ConA and PHA (20 μg), PA-I (1, 10 and 20 μg) on tumor size and mice mortality up to 40 days after 3LL inoculation.

Even 1 μg of PA-I was inhibitory (Fig. 2). Similar PA-I treatment with the melanoma s.c. injected cells to C57Bl mice was almost without any protective effect. Results presented in Fig. 3 show that 1 h incubation of the 3LL cells with PA-I was sufficient for its full effect and that even very short exposure of the cells to the lectin ('zero time') reduced tumorigenicity to 20%. However, the lectin had to be injected s.c. with the cells, since i.p. injection of 10 μg or even 200 μg PA-I concomitantly with the s.c.

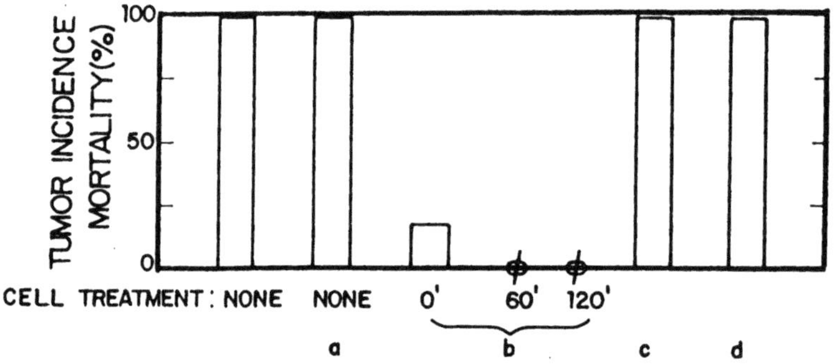

Fig. 3. Effects of diverse treatments by PA-I on the tumorigenicity (tumor incidence and mice mortality) of 3LL cells: a. PA-I injected i.p. (10-200 μg) immediately before s.c. injection of the tumor cells; b. PA-I (20 μg) incubated with the cells at 37°C for 0, 60 and 120 min before s.c. injection; c. PA-I (20 μg) and D-galactose (0.3 M for 60 min), and d. 0.3 M D-galactose incubated with the cells for 60 min before injection to the mice.

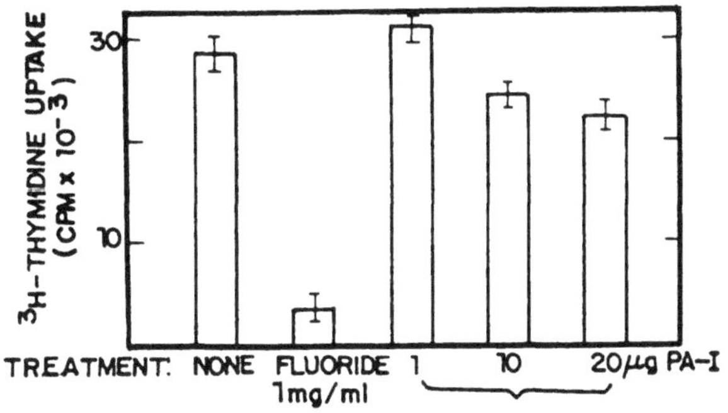

Fig. 4. Effect of PA-I (1, 10 and 20 μg in 0.2 ml) on viability of 3LL cells in vitro examined by ^{3}H-thymidine incorporation test (effect of fluoride is used as a control for the cytotoxic assay sensitivity).

injection of the tumor did not protect the mice (Fig. 3a). Addition of D-galactose to the lectin cell mixture totally abolished the PA-I effect (c). Galactose itself did not exert any effect on 3LL tumorigenicity (d).

Cytostatic activity of PA-I on 3LL cells

To determine the *in vitro* cytostatic activity of PA-I on the 3LL cells, we examined the DNA synthesis of the cells *in vitro* in the presence of PA-I. As shown in Fig. 4, PA-I weakly inhibited ^{3}H-thymidine incorporation into the tumor cells at 10 and 20 μg, but not at the 1 μg level.

Discussion

The present report demonstrates that PA-I may be very useful in diverse domains of cancer research. The results indicate that PA-I is useful as PHA and several other plant lectins for detecting impaired response of peripheral lymphocytes to mitogenic stimulation in cancer patients. Results presented in Fig. 1 show that PA-I, like the plant lectins, is reactive with transformed cells leading to their sugar-specific agglutination. The most remarkable finding of this study is that PA-I (10-20 μg) was considerably more effective than ConA and PHA (20 μg) in inhibiting 3LL s.c. tumorigenicity *in vivo* in C57Bl mice (Fig. 2). Even a very short exposure of the cells to the lectin was sufficient for inhibition (Fig. 3), although 1 h incubation leads to a greater effect. Reduction in tumorigenicity might be due to some cytotoxic effect of the lectin on the tumor cell, as seen in Fig. 4. Therefore, the tumor inhibition *in vivo* could be ascribed to selective toxicity of the lectin on the 3LL cells, as was suggested for the plant lectin anti-tumor effects (16,19,21, 24,28), which are based on the phenomenon of higher sensitivity of transformed cells to lectin (19,32). It is also possible that PA-I induces tumor inhibition in cooperation with macrophages by recognizing carbohydrate moieties on the cell membrane of both cells as reported in WGA with MM46 ascites tumors (27) and in *Griffonia simplicifolia* I lectin with Ehrlich ascites tumor (22,33).

The findings in the present study with 3LL cells may also be linked to several (previous) observations with these cells. Fauve et al. (34) have shown that Lewis lung carcinoma cells are able to repulse macrophages *in vitro*. Leibovici et al. (31,35) have shown that levan overcomes this property, leading to penetration of levan-treated macrophages to the tumor masses and concomitant inhibition of tumor growth (35). The lectin-dependent macrophage-mediated suppression of tumorigenicity *in vivo* (27,28,33) suggests that the lectin acts *in vivo* by bringing tumor cells and macrophages into close proximity by bridging the sugar groups in their respective glycoproteins. Such bridging and activation of the macrophage (36) could be a very plausible explanation for the suppression of s.c. development of 3LL cells by PA-I, in addition or even related to direct effect to the tumor cells. The inhibitory effect could also be attributed to binding of the lectins to glycoproteins or glycolipids present on the tumor cell membranes which are involved in their tumorigenicity. The added lectins could also function in competence with the endogenous membrane lectins of such cells. It was reported that tumor membranal lectins (some of them galactophilic) function to form cell aggregates in the circulation and mediate tumor cell homing and adhesion to target tissue (37-42). Recently, Kieda and Monsigny (43) reported possible involvement of membrane lectins in lung metastasis of Lewis lung carcinoma cells. They described two populations of 3LL cells isolated on the basis of surface glucose-binding lectins, differing in metastatic activity and adherence to lung endothelial cells in primary culture. However, the apparently contradictory lectin effects on cancer development reported by several groups (44) must be considered in planning animal treatment with lectins for cancer control purposes.

Acknowledgements. This report is part of a Ph.D. thesis of Miss Dodi Avichezer. The authors thank Mrs. Bluma Lederhendler for the skillful typing of the manuscript. This study was supported by The Mitzi-Dobrin Cancer Research Fund, Bar-Ilan University.

References

1. Gilboa-Garber, N.; in Colowick and Kaplan (Ginsburg, ed.). Methods in Enzymology 83, 378-385, Academic Press, New York (1982).
2. Gilboa-Garber, N.; in Bøg-Hansen and Spengler (eds.) Lectins 3, 495-502, Walter de Gruyter, Berlin (1983).
3. Gilboa-Garber, N.; in Mirelman (ed.) Microbial Lectins and Hemagglutinins, John Wiley & Sons, New York (1986) (in press)
4. Sharabi, Y., Gilboa-Garber, N.; FEMS Lett. 5, 273-276 (1979).
5. Gilboa-Garber, N., Sharabi, Y.; J. Protozool. 27, 209- 212 (1980).
6. Sudakevitz, D., Gilboa-Garber, N.; Microbios 34, 159- 166 (1982).
7. Gilboa-Garber, N., Sudakevitz, D.; in Levy (ed.) Advances in Pathology 1, 31-33, Pergamon Press, Oxford (1982).
8. Hersh, E.M. Oppenheim, J.J.; New Eng. J. Med. 273, 1006-1012 (1965).
9. Heilmann, E., Sasse, W., Berghaus, J.; Inn. Med. 6, 94 (1979)
10. Wallach, D.F.H.; Proc. Natl. Acad. Sci. USA 61, 868-874 (1968).
11. Burger, M.M.; Proc. Natl. Acad. Sci. USA 62, 994-1001 (1969).
12. Inbar, M., Sachs, L.; Proc. Natl. Acad. Sci. USA 63, 1418-1425 (1969).
13. Nicolson, G.L., Blaustein, J.; Biochim. Biophys. Acta 266, 543-547 (1972).
14. Burger, M.M.; Fed. Proc. 32, 91-101 (1973).
15. Pollack, R.E., Hough, P.V.C.; Ann. Rev. Med. 25, 431-446 (1976).
16. Robinson, E., Mekori, T,; Isr. J. Med. Sci. 7, 83-107 (1971).

17. Datta, S.P., Cerini, M., Ghose, T., Cerini, J.; Brit. J. Cancer 23, 616-621 (1969).
18. Wagener, D.J.T.; Cancer Res. 33, 1295-1300 (1973).
19. Shoham, J., Inbar, M., Sachs, L.; Nature (Lond.) 227, 1244-1246 (1970).
20. Inbar, M., Ben-Bassat, H., Sachs, L.; Int. J. Cancer 9, 143-149 (1972).
21. Lin, J.-Y., Tserng, K.-Y., Chen, C.-C., Lin, L.-T., Tung, T.-C.; Nature (Lond.) 227, 292-293 (1970).
22. Eckhardt, A.E., Malone, B.N., Goldstein, I.J.; Cancer Res. 42, 2977-2979 (1982).
23. Kataoka, T., Oh-hashi, F., Tsukagoshi, S., Sakurai, Y.; Cancer Res. 37, 964-968 (1977).
24. Shionoya, H., Arai, H., Koyanagi, N., Ohtake, S., Kobayashi, H., Kodama, T., Kato, H., Tung, T.-C. and Lin, J.-Y.; Cancer Res. 42, 2872-2876 (1982).
25. Esumi-Kurisu, M., Iwata-Dohi, N., Mizuno, D., Yamazaki, M.; Gann 74, 398-404 (1983).
26. Hoshi, H., Kamiya, K., Takemoto, R., Aijima, H.; Arch. Histol. Jap. 47, 505-518 (1984).
27. Kurisu, M., Yamazaki, M., Mizuno, D.; Cancer Res. 40, 3798-3803 (1980).
28. Nakajima, H., Komano, H., Esumi-Kurisu, M., Abe, S., Yamazaki, M., Natori, S., Mizuno, D.; Gann 73, 627-632 (1982).
29. Yamazaki, M., Okutomi, T., Sakakibora, F., Kawauchi, H., Mizuno, D.; J. Leuk. Biol. 35, 241-250 (1984).
30. Ohkuma, Y., Komano, H., Natori, S.; Cancer Res. 45, 288-292 (1985).
31. Leibovici, J., Borit, A., Sandbank, U., Wolman, M.; Br. J. Cancer 40, 597-607 (1979).
32. Tomita, M., Kurokawa, T., Onozaki, K., Osawa, T., Sakurai, Y., Ukita, T.; Int. J. Cancer 10, 602-606 (1972).
33. Maddox, D.E., Goldstein, I.J., Lobuglio, A.F.; Cell. Immunol. 71, 202-207 (1982).
34. Fauve, R.M., Hevin, B., Jacob, H., Gaillard, J.A., Jacob, F.; Proc. Natl. Acad. Sci. USA 71, 4052-4056 (1974).
35. Leibovici, J., Stark, Y., Eldar, T., Brudner, G., Wolman, H.; Recent Results Cancer Res. 75, 173-179 (1980).
36. Pick, E., Krejci, J., Turk, J.L.; Nature (Lond.) 225, 236-238 (1970).
37. Poste, G., Nicolson, G.L. (eds.): Cell Surface Reviews, Vol. 105, Elsevier, Amsterdam (1976).
38. Raz, A., Lotan, R.; Cancer Res. 41, 3642-3647 (1981).
39. Lotan, R., Raz, A.; Cancer Res. 43, 2088-2093 (1983).
40. Raz, A., Meromsky, L., Carmi, P., Karakash, R., Lotan, D., Lotan, R.; EMBO J. 3, 2979-2983 (1984).
41. Monsigny, M., Kieda, C., Roche, A.C.; Biol. Cell 47, 95-110 (1983).
42. Uhlenbruck, G., Schwonzen, M., Pissors, W., Diehl, V.; these proceedings.
43. Kieda, C., Monsigny, M.; Biol. Cell 51, 51a (1984).
44. Schell, R.F., Lawrence, D.A.; Cell. Immunol. 31, 142-154 (1977).

ADOPTIVE TRANSFER OF EXPERIMENTAL ALLERGIC ENCEPHALOMYELITIS WITH LECTIN-ACTIVATED SPLEEN CELLS

Ryoichi Mori, Akira Takenaka and Hiroko Minagawa

Department of Virology, School of Medicine, Kyushu University, Fukuoka 812, Japan

Experimental allergic encephalomyelitis (EAE) is a T cell-mediated autoimmune disease of the central nervous system which can be elicited by an injection of myelin basic protein (BP) in complete Freund's adjuvant (CFA) (1). The disease can also be adoptively or passively transferred from sensitized donors to naive recipients by means of lymph node cells but not by serum. Recently, EAE has been transferred by relatively small numbers of spleen cells following incubation in vitro with the encephalitogenic antigen BP (2) or concanavalin A (Con A) (3). The experimental system has proposed to study, a) T cell subsets responsible for the induction of EAE (4), b) the role of soluble factors in the differentiation of EAE-effector cells (5), and c) the direct actions of immunomodulating agents on EAE (6).

We previously reported the effects of various lectins on the adoptive transfer of EAE, and have shown that the ability of transferring EAE differs among lectins used in in vitro cultures (7). Here we reevaluated the lectin-induced EAE-effector cell activation, and examined a relationship between lectin-induced interleukin 2 (IL2) production and the disease transfer.

Materials and Methods

Female inbred Lewis rats, 7 - 10 weeks old, were used under specific pathogen-free conditions. Purified guinea pig BP was emulsified in CFA containing *Mycobacterium tuberculosis* H37Ra (M.tbc). Each rat was immunized with 0.1 ml of emulsion (BP/CFA) containing 50 ug BP and 200 ug M.tbc into the two hind footpads.

Spleen cells obtained from rats 12 days after immunization with BP/CFA were cultured at 2 x 10^6 cells/ml in RPMI 1640 supplemented with 5% fetal bovine serum, 25 mM HEPES, 50 uM 2-mercaptoethanol, and 50 ug/ml gentamicin plus lectins at 37°C in a humidified atmosphere containing 5% CO_2. All lectins, i.e., Con A, *Lens culinaris* (LCA),*Pisum sativum* (PSA), *Phaseolus vulgaris* (PHA), *Robinia pseudoacacia* (RPA), *Wisteria floribunda* (WFA), and *Triticum vulgaris* (wheat germ; WGA) agglutinins, were purchased from E.Y. Laboratories, San Mateo, CA. They are listed on Table 1. Lectin concentrations selected were those previously found to yield maximum lymphocyte proliferation. After 72 hr culture, the cells were washed and 5 x 10^7 viable cells were transferred to naive recipients by i.p. injection.

Clinical signs of EAE were graded on a scale of 0 to 6 according to the severities of tail and hind limbs paralysis (8). Histological sections of the brain and spinal cord stained with Luxol fast blue, PAS and hematoxylin were examined for perivascular infiltration of mononuclear cells, and were scored from 0 to 4 without knowledge of the regimen (9).

Lymphocyte proliferative responses to lectins were measured by ^{3}H-thymidine incorporation into DNA synthesis. The amount of IL2 activity in supernatant was assessed on the basis of IL2-dependent proliferation of a murine helper T cell line, HT-2 cells (10). The results were quantified by probit analysis (11), in which a sample obtained from a Con A-culture was assigned a value of 1 unit/ml.

Results

Transfer of EAE with lectin-activated spleen cells: Spleen cells obtained from rats immunized with BP/CFA were transferred after a) incubation with lectins, b) incubation with Con A or WGA in the presence of α-methyl-D-mannoside (αMM) or N-acetyl-D-glucosamine (GlcNAc), respectively, or c) incubation without added lectins (Table 2). BP-primed spleen cells cultured alone failed to induce EAE in the recipients. However, when stimulated with Con A, LCA,

TABLE 1

Lectins used

Source	Common name or abbreviation	Sugar specificity	Mäkäla's group
Phaseolus vulgaris	Phytohemagglutinin;PHA	GalNAc[b]	2[a]
Robinia pseudoacacia	RPA	GalNAc	2
Wisteria floribunda	WFA	GalNAc	2
Canavalia ensiformis	Concanavalin A; Con A	D-mannose	3
Lens culinaris	Lentil lectin; LCA	D-mannose	3
Pisum sativum	PSA	D-mannose	3
Triticum vulgaris	Wheat germ agglutinin; WGA	GlcNAc[c]	3

a. Mäkelä, O. (1957).
b. GalNAc: N-acetyl-D-galactosamine.
c. GlcNAc: N-acetyl-D-glucosamine.

PSA or WGA, spleen cells could transfer EAE. Recipient rats became sick 4 to 6 days after cell transfer and attained maximum scores at one week, showing weight loss, tail and/or hind limbs paralysis often accompanied by incontinence. In contrast, spleen cells stimulated with PHA, RPA or WFA did not transfer EAE. All lectins induced lymphocyte blast transformation, as evidenced by ^{3}H-thymidine incorporation. Con A-induced or WGA-induced lymphocyte proliferation and the EAE-transfer were inhibited by the hapten-specific inhibitor, αMM or GlcNAc, respectively. Normal non-immune rat spleen cells incubated with Con A or WGA produced no disease in the recipients, suggesting that any clinical signs of EAE were not the results of lectin toxicity in itself (data not shown). Histological assessments correlated significantly with clinical severities ($r=0.73$; $p<0.001$).

TABLE 2

Transfer of EAE with lectin-activated spleen cells[a]

Lectin in culture	^{3}H-thymidine incorporation (cpm)	EAE in recipients		
		Clinical[b]	Onset	Histologic[c]
None	4,626	0/8 (0.0)	-[d]	0.5 ± 0.2
Con A	104,067	21/27 (3.1)	4.9	1.2 ± 0.6
LCA	106,251	4/5 (2.8)	5.0	1.2 ± 0.5
PSA	60,100	6/6 (2.8)	5.3	1.5 ± 0.4
PHA	98,854	0/7 (0.0)	-	0.7 ± 0.3
RPA	56,377	0/4 (0.0)	-	0.6 ± 0.5
WFA	84,007	0/13 (0.0)	-	0.6 ± 0.4
WGA	21,465	5/5 (2.0)	6.0	1.0 ± 0.7
Con A + αMM[e]	10,927	0/4 (0.0)	-	0.5 ± 0.1
WGA + GlcNAc[f]	5,815	0/3 (0.0)	-	0.3 ± 0.1

[a] *Spleen cells obtained from rats 12 days after immunization with BP/CFA were cultured with lectins for 72 hr and 5x10^{7} cells were transferred to recipient rats by i.p. injection.*

[b] *Number positive/number in group; average score in parentheses. Clinical scale, 0-6.*

[c] *Average score ± SD. Histologic scale, 0-4.*

[d] *No clinical signs until 10 days after cell transfer.*

[e] *Cultured with Con A in the presence of 25 mM α-methyl-D-mannoside.*

[f] *Cultured with WGA in the presence of 50 mM N-acetyl-D-glucosamine.*

A part of the data was reported in ref. #7.

Lectin-induced IL2 production and its correlation with the ability to transfer EAE: IL2 activity present in the supernatants from spleen cell cultures incubated with a variety of lectins is shown in Fig. 1. Supernatants from cultures incubated with Con A, LCA

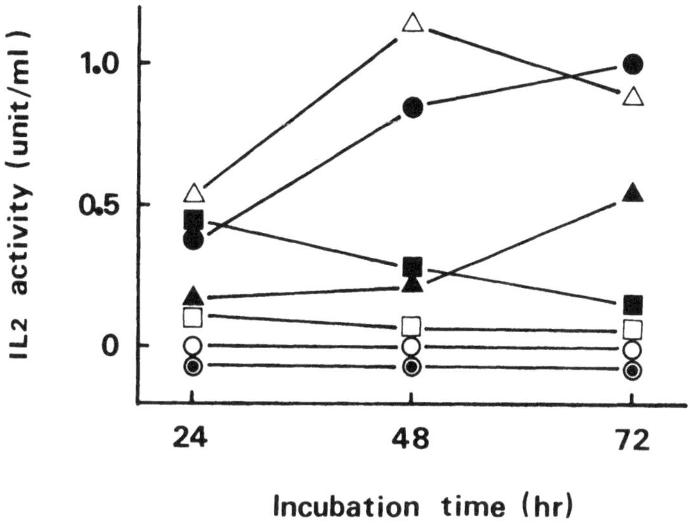

Fig. 1. Lectin-induced IL2 production. IL2 activity was determined in a standard microassay (11) based upon the IL2-dependent proliferation of HT-2 cell line (10). ○; None, ●; Con A, △; LCA, ▲; PSA, □; PHA, ■; WFA, and ◉; WGA.

and PSA contained effective levels of IL2, but those from cultures with PHA, RPA and WFA had less IL2 activity than the above cultures. Supernatant from WGA-stimulated culture did not contain any IL2 activity during any incubation periods from 24 to 72 hr.

Discussion

In our previous observations (7) and the present study, we demonstrated that spleen cells stimulated in vitro with lectins belonging to Mäkelä's group 3 (12) whose bindings are specifically inhibited by D-mannose (e.g. Con A, LCA and PSA) or GlcNAc (WGA) were capable of transferring EAE, while lectins of Mäkelä's group 2 whose hapten specificities are D-galactose or N-acetyl-D-galactosamine (PHA, RPA and WFA) were incapable of transferring EAE.

The reasons for the differential ability of lectins to transfer EAE are still unclear. In this study, we attempted to give an explanation for this difference by measuring IL2 activity produced by various lectins, since it has been reported that the transfer of EAE was dependent on the generation of IL2 in the culture supernatant (5, 13). The results indicated that the ability of spleen cells to transfer EAE after in vitro stimulation with D-mannose-binding lectins appeared to correlate with the induction of IL2. However, WGA-stimulated spleen cells did not produce IL2 but could transfer EAE to the recipients, suggesting that some factors distinct from IL2 (14, 15) might be involved in the differentiation of EAE-effector precursors. The exact role of IL2 in the pathogenesis of EAE should be documented by use of recombinant IL2, and the study is in progress.

Acknowledgements

The authors wish to thank Dr. George A. Hashim, St. Luke's Hospital Center, New York, N.Y. for his kind gift of purified BP, and Dr. Yasuto Itoyama, Department of Neurology, Neurological Institute, School of Medicine, Kyushu University, for his histological assessments.

References

1. Paterson, P.Y. (1977). In: Autoimmunity (Talal, N., Ed.), Academic Press, New York, pp. 643-692.

2. Richert, J.R., Dricoll, B.F., Kies, M.W., Alvord, E.C., Jr. (1979). J. Immunol. 122, 494-496.

3. Panitch, H.S., McFarlin, D.E. (1977). J. Immunol. 119, 1134-1137.

4. Swanborg, R.H. (1983). J. Immunol. 130, 1503-1505.

5. Ortiz-Ortiz, L., Weigle, W.O. (1982). J. Immunol. 128, 1545-1550.

6. Abreu, S.L. (1985). Int. Archs Allergy appl. Immun. 76, 302-307.

7. Takenaka, A., Kaneko, K., Mori, R., Itoyama, Y. (1984). J. neurol. Sci. 66, 255-261.

8. McFarlin, D.E., Blank, S.E., Kibler, R.F. (1974). J. Immunol. 113, 712-715.

9. Alvord, E.C., Jr., Kies, M.W. (1959). J. Neuropath. exp. Neurol. 18, 447-457.

10. Watson, J. (1979). J. exp. Med. 150, 1510-1519.

11. Gillis, S., Ferm, M.M., Ou, W., Smith, K.A. (1978). J. Immunol. 120, 2027-2032.

12. Mäkelä, O. (1957). Ann. Med. exp. Biol. Fenn. 35 (suppl.11).

13. Hayosh, N.S., Simon, L.L., Swanborg, R.H. (1984). J. Immunol. 133, 1943-1945.

14. Raulet, D.H., Bevan, M.J. (1982). Nature 296, 754-757.

15. Wagner, H., Hardt, C., Rouse, B.T., Röllinghoff, M., Scheurich, P., Pfizenmaier, K. (1982). J. exp. Med. 155, 1876-1881.

FACTOR DEPENDENT CONCANAVALIN A ACTIVATION OF B LYMPHOCYTES

L. Danielsson, S.A. Möller and C.A.K. Borrebaeck
Department of Biotechnology, University of Lund, P.O.Box 124, S-221 00 Lund, Sweden.

The plant lectin Concanavalin A (Con A) induces mouse T cells but not B cells to proliferate, even though it binds equally to both cell types (1). Insolubilized Con A has been reported to directly induce B cell proliferation (2) although this seems difficult to reproduce (3). Recently it was, however, demonstrated that Con A had the ability to induce purified murine B cells to depolarize, enlarge and display increased levels of I-A antigens, but not to induce DNA synthesis (3). The activation of B cells in vitro can be obtained through different pathways (for review see ref. 4). Cells or competent ligands stimulate small, resting B cells to enter G_1 phase of the cell cycle. These induced cells have now become responsive to soluble B cell-specific growth and differentiation factors and are committed to clonal proliferation and maturation to Ig-secreting cells.

The results presented here demonstrate that Con A together with allogeneic helper factors, produced by a mixed lymphocyte culture, stimulated small, resting murine B lymphocytes to DNA synthesis and proliferation. This dose-dependent induction of cellular proliferation could be completely inhibited by methyl α-D-mannopyranoside. The stimulation did not seem to induce a maturation of the B cells to Ig-secreting cells. An Ig-secretion was however, induced when cells were stimulated with Con A in combination with helper factors derived from both a mixed lymphocyte culture and a PMA stimulated EL-4 thymoma cell line.

Lectins, Vol. V

Materials and Methods

Experimental animals. Female BALB/c and C57Bl/6 mice (Gl. Bomholtgård Ltd., Ry, Denmark) 5-7 weeks of age, were used.

Culture media. Dulbecco's modification of Eagles medium or RPMI 1640 were supplemented with streptomycin (50 µg/ml), penicillin (50 IU/ml), 4 mM L-glutamine, 1 % (v/v) 100x non-essential amino acids (Flow Laboratories Inc., Ayrshire, U.K.), and 10 % (D-10) or 5 % (R-5) fetal calf serum, respectively.

Spleen cell preparation. A spleen cell suspension was prepared as recently described (5). The spleen cells were depleted of adherent cells by incubation in culture flasks as described (6). Normally 80-90 % of the spleen cells were recovered in the non-adherent fraction.

Density separation of splenocytes. Small, dense, resting splenocytes were prepared on a discontinous Percoll gradient (Pharmacia Fine Chemicals AB., Uppsala, Sweden), consisting of 2 ml layers with densities 1.09, 1.08, 1.075, 1.065, and 1.055 g/ml (7). Cells on the 1.09 g/ml layer were harvested and this fraction con tained 0.1 % of the number of antibody secreting cells as determined using the filter immuno plaque assay (FIPA)(8).

Separation of Ig-positive cells. The preparation of Ig-positive cells was performed by the indirect panning technique (9) with some modifications (6). The surface Ig-positive cell population represented 15-20 % of the total spleen cells. When analyzed by a fluorescence-activated cell sorter, Epics C cytofluorometer (Coulter Electronics Ltd., Hertz, U.K.) 98-99 % of the cells were Ig-positive and <0.5 % Thyl.2 positive.

Preparation of allogeneic helper factors. Thymocytes from BALB/c and C57Bl/6 were co-cultivated at a final cell concentration of 5×10^6 cells/ml for 48 h, as recently described (5, 10). The supernatant of the mixed lymphocyte culture (sMLC) was collected by centrifugation at 800 g for 10 minutes, sterile filtered (0.2 µm) and frozen at -20°C.

Cell proliferation assay. Ig-positive splenocytes (2.5×10^5) were seeded in 96-wells microtiter plates at 200 µl of R-5 medium per well together with the following stimulators or various combinations of them (final concentrations): 2.5 µg Con A/ml (Miles-Yeda Ltd., Rehovot, Israel), 25 µg lipopolysaccharide (LPS)/ml (E.coli, serotype 0127:B8, Sigma Chem. Co., St. Louis, MO, USA), 25 % supernatant from a mixed lymphocyte culture (sMLC), 25 % supernatant from phorbol myristate acetate (10 ng/ml) stimulated EL-4 mouse thymoma cells prepared according to Howard et al. (11), 5 µg PHA-E_4 or PHA-L_4 /ml (12), human IL-2 (10 U/ml) (Collaborative Research, Lexington, MA, USA), and human IL-1 (1:50,000) prepared according to Schmidt (13) and kindly supplied by Dr. Lennart Lögdberg. Methyl α-D-mannopyranoside (α-mm) (100 mM) (Sigma Chem. Co.) was used to specifically inhibit the binding of Con A. The cell suspensions were cultured for 3-5 days in a 37°C humidified incubator with an 8 % CO_2 and 92 % air gase phase, and pulsed with 1 µCi of (methyl-^{3}H) thymidine (5 Ci/mmol, Amersham Int. Ltd., Amersham, UK) 8 h before the cell cultures were harvested. Samples were run in triplicate.

Ig-ELISA. A competitive immunoassay was used to determine the total Ig secretion in the culture supernatants (6). The sensitivity was approximately 70 ng Ig/ml.

Fractionation of sMLC. Serum free sMLC was concentrated 50 times using an ultrafiltration unit (membrane YM5, Amicon Corporation, Danvers, MA.) where the buffer also was changed to 50 mM phosphate buffer, pH 7.2 containing 0.15 M sodium chloride (PBS) and 0.05 % PEG 4000. A 300 µl sample was applied to a FPLC Superose 12 column (Pharmacia Fine Chemicals AB) with a flow rate of 0.3 ml/min and fractions of 0.6 ml were collected.

RESULTS

The DNA synthesis of cultured Ig-positive cells was determined on day 1 to 5 (Table 1). The cells were stimulated by a T-cell mitogen (Con A), a B cell mitogen (LPS), allogeneic helper factors (AHF) derived from a sMLC, PMA stimulated EL-4 supernatant (sEL-4) and different combinations of these. Con A alone had no ability to stimulate B cell proliferation but together with AHF a strong proliferation was recorded with the maximum on day 4. Cells stimulated with sEL-4 alone or in combination with Con A gave less than 1 % of the proliferation obtained with Con A supported by sEL-4 in combination with AHF, which induced the strongest recorded proliferation in our system. The kinetics of B cell proliferation differed somewhat depending on the triggering stimulant.

TABLE 1

Mitogen stimulation of Ig-positive lymphocytes during 1-5 days

	^{3}H-thymidine uptake (dpm) on day				
Mitogen*	1	2	3	4	5
-	570	260	100	100	350
Con A	860	780	310	270	350
AHF	1,840	4,450	3,250	3,360	2,230
Con A+AHF	4,820	38,570	181,340	334,040	50,870
LPS	13,360	36,080	1,810	150	170
LPS+AHF	43,300	262,430	308,660	133,150	135,890
sEL-4	1,770	1,630	320	180	not done
Con A+sEL-4	1,860	2,150	480	210	- " -
AHF+sEL-4	6,990	25,780	27,030	26,580	- " -
Con A+AHF+sEL-4	11,050	71,230	256,350	611,590	151,580

**For concentrations see Materials and Methods. Abbreviations used; AHF= allogeneic helper factor, LPS= lipopolysaccharide, EL-4= supernatant from stimulated EL-4 thymoma cells.*

Small, dense, resting cells were then obtained by Percoll density separation before the enrichment of Ig-positive cells, using the indirect panning technique, was performed. Con A induced proliferation of these small, dense, resting Ig-positive B cells was also investigated on day 3 (Table 2). The cells were unresponsive to LPS, indicating extremely low levels of accessory cells (14), whereas they exhibited a responsivness to AHF supported Con A activation which was comparable to previous experiments. Addition of IL-1 or IL-2 did not change the proliferative response. The morphology of the activated cells on day 3 of the proliferation

TABLE 2

Mitogen stimulation of small, dense, resting Ig-positive lymphocytes on day 3

Mitogen*	^{3}H-thymidine uptake (dpm)
-	80
Con A	320
Con A+IL-1	210
AHF	600
Con A+AHF	280,020
LPS	100
LPS+AHF	183,730
IL-2	110
Con A+IL-2	210
Con A+AHF+IL-2	278,800

**For concentrations see Materials and Methods. Abbreviations used; IL-1= interleukin 1, IL-2= interleukin 2, LPS= lipopolysaccaride, AHF= allogeneic helper factors.*

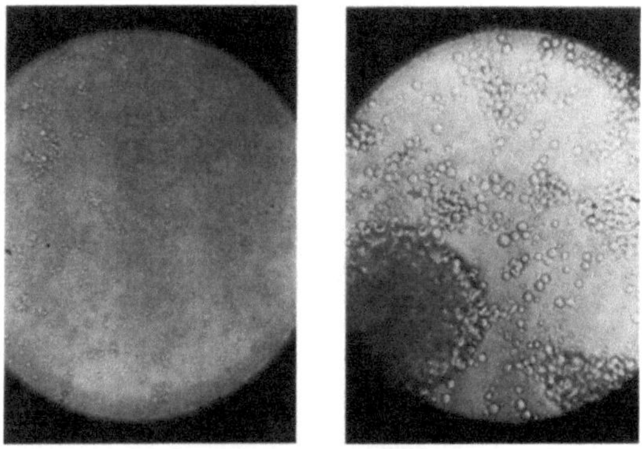

2.5 µg Con A/ml 2.5 µg Con A/ml+AHF

Fig. 1. Ig-positive cells stimulated with Con A and Con A+AHF (25 % supernatant from a mixed lymphocyte culture) on day 3 of culture (64 x magnification).

assay is shown in Fig 1. Stimulation with Con A+AHF gave a significant enlargement and polyclonal expansion of the cells as compared to stimulation with Con A alone. The isolectins, L_4 and E_4, of the T cell mitogen PHA, also induced a cell proliferation of Ig-positive cells when supported by AHF (data not shown).

The triggering of surface Ig-positive cells to enter S phase of the cell cycle was dependent on the presence of AHF at the initiation of the cell culture. Con A could, however, be added as late as 6 h after the initation without diminishing the proliferative response (Fig.2). Addition of α-mm completely inhibited the cell proliferation up to 6 h after the initiation as measured on day 4 (Fig. 2). The induction time needed for Con A+AHF to induce cells to some DNA-synthesis was 14-18 h. The dose-response of Con A induced B

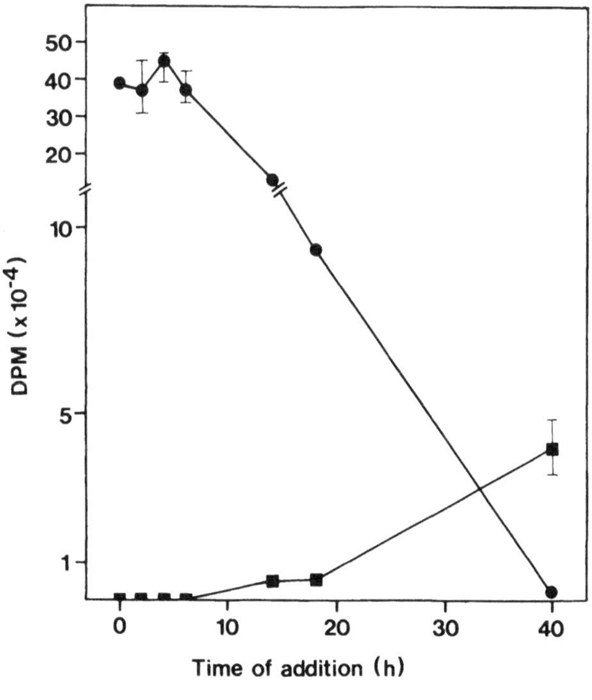

Fig. 2. Addition of 2.5 μg Con A/ml (●) to cells cultivated with AHF (25 % sMLC) after 0, 2, 4, 6, 14, 18, and 40 h. The cells were cultivated for 4 days Activation of B cells cultivated with 2.5 μg Con A/ml and AHF (25 % sMLC) was inhibited by addition of α-mm (100 mM) (■) at the time intervals mentioned above and with a culture period of 4 days. Abbreviations used; AHF= allogeneic helper factors, sMLC= supernatant from a mixed lymphocyte culture, α-mm= methyl α-D-mannopyranoside.

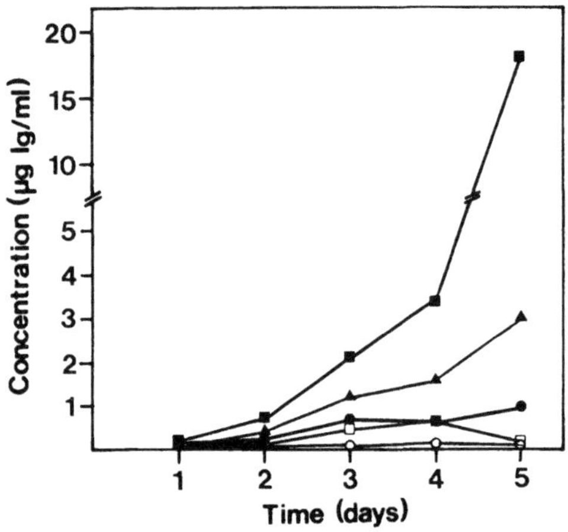

Fig. 3. Ig-secretion of Ig-positive cells stimulated with Con A (○), Con A + AHF (●), LPS (□), LPS+AHF (■) or Con A+AHF+sEL-4 (▲) determined on day 1-5 of the stimulation period. Con A was used at a concentration of 2.5 µg/ml and AHF at 25 % of supernatant. Abbreviations used; AHF= allogeneic helper factors, LPS= lipopolysaccharide, sEL-4= supernatant from stimulated EL-4 thymoma cells.

cell proliferation was measured using 0-5 µg Con A/ml and 0-50 % sMLC. At a concentration of 2.5 µg Con A/ml optimal cell proliferation was obtained using 25 % sMLC (data not shown).

The total amount of secreted immunoglobulins in stimulated culture supernatants was determined on day 1-5 using the competitive Ig-ELISA (Fig.3). Cells stimulated with Con A+AHF did not secrete significant amounts of Ig, while stimulation with Con A+AHF+sEL-4 or LPS+AHF induced a significant secretion of Ig.

Preliminary attempts have been made to fractionate serum-free sMLC by gel filtration on a Superose 12 column attached to a FPLC. The ability of the collected fractions to support a Con A induced B cell proliferation was measured and the results indicate that allogeneic helper activity eluted in the low molecular weight range (≤5000 Da). The possibility of interactions with the gel matrix can, however, at this stage of the investigation not be excluded.

Discussion

Con A has for many years been used as an inducer of T cell proliferation without any apparent effects on B cells. It was, however, recently shown that Con A induced purified B cells to depolarize, to enlarge and to display increased levels of Ia antigens (3). It was therefore concluded that Con A can activate resting B lymphocytes i.e. to stimulate them to enter the cell cycle but not to induce cell proliferation. We have taken this further and have been able to induce B cell proliferation, using a lymphokine supported Con A stimulation. The lymphokines were derived from a bidirectional mixed lymphocyte reaction and is a heterogeneous mixture of helper factors, e.g. B cell growth factor (BCGF) and B cell differentiation factor(s) (15, 16, 17).

This Con A-AHF induced B cell proliferation did not seem to progress through a maturation step yielding a secretion of immunoglobulins. However, if the proliferation was further supported by a supernatant from a PMA-stimulated EL-4 thymoma cell line we obtained a significant secretion of immunoglobulins. EL-4 derived supernatant contains B cell stimulatory factors such as BSF-pl (11, 18), EL-BCGF-swa (19), EL-BCDF-nak (20), IL-2 (21) and BCDFu (16). These factors were alone unable to support a Con A induced cell proliferation. The complexity of the system is thus illustrated by the synergistic effect of two or more factors yielding a strong proliferative response followed by a late differentiation step, resulting in production of immunoglobulins.

The murine Ig-positive cells used in this study were isolated by the indirect panning procedure which resulted in a cell population where 98-99% expressed surface immunoglobulins. This positive selection gives in our hands a higher enrichment of cells, than negative selection procedures such as adherence. We are aware that even positively selected small resting B cells might contain minute amounts of T cells and accessory cells. The highly purified B cells were, however, unresponsive to both Con A and LPS indicating very low amounts of both T cells and macrophages. The addition of exogenous IL-1 and IL-2 did not change the unresponsivness of the same cells to Con A , further suggesting that the lectin was acting directly on B cells and not via residual T cells. The proliferating B cells were still >95% Ig-positive after 5 days in culture.

In conclusion, we have demonstrated that Con A dependent proliferation of small resting B lymphocytes can be mediated by allogeneically derived helper factors. Further maturation of proliferating B cells is also shown to be dependent on additional factors found in supernatant from PMA stimulated EL-4 thymoma cells. The lectin induced proliferation is dependent on the carbohydrate binding specificity of the lectin and might be due to a crosslinking of the surface IgM molecules eventually making the cell responsive to T cell derived helper factors.

Acknowledgements

We would like to thank Drs. Roland Carlsson and Håkan Eriksson for performing the flow cytometric analysis. This investigation was supported by the Swedish Medical Research Council (B85-13X-06831-02B), the medical faculty (University of Lund) and in part by the Swedish Cancer Society (1901-B85-01X).

References

1. Stobo, J.D., A.S. Rosenthal, W.E. Paul. 1972. J. Immunol. 108, 1.

2. Andersson, J., G.M. Edelman, G. Möller, O. Sjöberg. 1972. Eur. J. Immunol. 2, 233.

3. Hawrylowicz, C.M., G.G.B. Klaus. 1984. Immunology 53, 703.

4. Möller, G. (Ed). 1984. Immunol. Rev. 78, 1.

5. Borrebaeck, C.A.K. 1983. Scand. J. Immunol. 18, 9.

6. Möller, S.A., L. Danielsson, C.A.K. Borrebaeck. Immunology (in press).

7. Ratcliffe, M.J.H., M.H. Julius. 1982. Eur.J.Immunol. 12, 634.

8. Möller, S.A., C.A.K. Borrebaeck. 1985. J.Immunol.Methods 79, 195.

9. Wysocki, L.J., V.L. Sato. 1978. Proc.Natl.Acad.Sci., USA 75, 2844.

10. Borrebaeck, C.A.K. 1984. Mol.Immunol. 21, 841.

11. Howard, M., J. Farrar, M. Hilfiker, B. Johnson, K. Takatsu, T. Hamaoka, W.E. Paul. 1982. J.Exp.Med. 155, 914.

12. Glad, C., C.A.K. Borrebaeck. 1984. J.Immunol. 133, 2126.

13. Schmidt, J.A. 1984. J.Exp.Med. 160, 772.

14. Corbel, C., F. Melchers. 1983. Eur.J.Immunol. 13, 528.

15. Muraguchi, A., A.S. Fauci. 1982. J.Immunol. 129, 1104.

16. Puré, E., P. Isaksson, J. Kappler, P. Marrack, P. Kramer, E. Vitteta. 1983. J.Exp.Med. 157, 583.

17. Kishimoto, K., K. Yoshizaki, M. Kimoto, M. Okada, T. Kuritani, H. Hikutani, K Shimizu, T. Nakagawa, N. Nakagawa, Y. Miki, H. Kishi, K. Fukanaga, T. Yoshikubo, T Taga. 1984. Immunol.Rev. 78, 97.

18. Paul, W.E. 1983. Cell. Immunol. 84, 461.

19. Swain, S., M. Howard, J. Kappler, P. Marrack, J. Watson, R. Booth, R. Dutton. 1983. J.Exp.Med. 158, 822.

20. Nakanishi, K., M. Howard, A. Muraguchi, J. Farrar, K. Takatsu, T. Hamaoka, W.E. Paul. 1983. J. Immunol. 130, 2219.

21. Farrar, J.J., J. Fuller-Farrar, P.L. Simon, M.L. Hilfiker, B.M.Stadler, W.L. Farrar. 1980. J. Immunol. 125, 2555.

BLOCKING AND STIMULATION OF POKEWEED MITOGEN-INDUCED BLASTOGENESIS BY ANTI-POKEWEED ANTISERUM

E. de Vries, J.P. van der Weij, G. Doekes and A. Cats. Dept. of Rheumatology, University Hospital, Leiden, The Netherlands.

The use of lectins as non-specific stimulators of human immune competent cells has been the object of many studies. In particular pokeweed mitogen (PWM) isolated from the roots of Phytolacca americana, induces lymphocyte mitogenesis as well as the synthesis of immunoglobulins by B lymphocytes (1). It was demonstrated that the PWM-driven stimulation of human peripheral blood mononuclear cells requires the presence of both monocytes acting as accessory cells (2-4) and T cells regulating B cell activation (5-7). However, little is known about the mechanism by which PWM interacts with cells. Stevenson et al. (8) reported that after monocytes had bound PWM on their membranes, the mitogen remained detectable on the surface for as long as 12 hr of culture. They concluded that endocytosis of PWM by monocytes was not a prerequisite of stimulation. Thiele and Lipsky (9) found that both B and T cell activation require phagocytic monocytes as accessory cells, but proliferating T cells became rapidly independent of monocytes, whereas B cell response could be inhibited by blocking the monocytes with silica up to 4 days after the start of the stimulation. Linthicum et al. (10) showed that lymphocytes bear a receptor on their surface to which mitogens can bind, but they did not show specific blocking of these receptors resulting in prevention of blastogenesis. The present study was undertaken to investigate the way in which PWM interacts with mononuclear cells during stimulation. In this paper, we report on the use of an anti-PWM antiserum, raised in rabbits, in the mitogen-driven stimulation of human mononuclear cells. This antiserum reacts specifically with PWM.

Lectins, Vol. V

MATERIAL AND METHODS

Cells
Human mononuclear cells were obtained by Ficoll/Isopaque density gradient centrifugation (11) and consisted of about 20-25% monocytes and 75-80% lymphocytes as judged by non-specific esterase activity (12). Lymphocyte-enriched suspensions were prepared by suspending 10^7 mononuclear cells in 1 ml Percoll (Pharmacia, Sweden, specific density 1.064 gr/cm) at 4°C, followed by centrifugation at 200xg for 10 min. A second run on Percoll resulted in a monocyte-enriched fraction containing on average 85-90% monocytes, whereas the lymphocyte-enriched fraction included less than 0.1% monocytes. In reconstitution experiments, monocytes and lymphocytes were mixed in a ratio of 1:4, with a final cell concentration of 1.5×10^6 cells/ml culture medium.

Anti-PWM antiserum.
Three rabbits were injected subcutaneously with PWM in Freunds' complete adjuvant (0.3 ml PWM, stock solution and 0.3 ml FCA) and boostered with the same amount of PWM in Freund incomplete adjuvant bimonthly. Blood was drawn and serum tested for the presence of anti-PWM antibodies by double immunodiffusion. PWM antibodies were obtained by salt precipitation of rabbit serum, followed by DEAE-chromatography and gelfiltration on Sephadex G-200. $F(ab')_2$ parts of IgG were prepared by pepsin digestion of anti-PWM antibodies (13) and purified by gelfiltration. Normal rabbit serum was obtained from either non-immunized or antigen-immunized rabbits.

Culture and stimulation.
Mononuclear cells, lymphocyte and monocyte-enriched suspensions were cultured as described before (14). In brief, 1.5×10^6 cells per ml culture medium (RPMI 1640, supplemented with fetal calf serum, gluthamin and antibiotics) were cultured for 3 days. PWM (Gibco, USA) was added at a final concentration of 1/500 of stock solution, per ml culture medium for time intervals as indicated in the experiments. In all experiments, cell suspensions were incubated at 37°C in a humidified CO_2 incubator, centrifuged at day 3, resuspended in mitogen-free medium and cultured for another 3 days.

Scoring of the stimulation.
At day 6, cells were harvested by centrifugation, the total cell number counted and viability tested by trypan blue exclusion. Smears were made and processed for the demonstration of non-specific cytoplasmic esterase activity (12). As a measure of stimulation the percentage of blast cells present at day 6 were counted and expressed in absolute numbers per ml culture medium. Blast cells were identified as large (9-15 μm in diameter), rounded cells, either negative for or containing one or a few spots of esterase activity, in contrast to monocytes which appeared to be strongly positive for the enzyme throughout the cytoplasm.

RESULTS

To investigate the efficacy of rabbit-anti-PWM antiserum, serial dilutions of the antiserum were added to mononuclear cells, concomitantly with different concentrations of PWM (Table 1). For control, each of the different PWM concentrations was tested also in the presence of 1/10 diluted normal rabbit serum. Anti-PWM antiserum was shown to block the PWM-driven stimulation in a concentration-dependent way. Maximal stimulation was obtained at a 1/100 dilution of PWM and this could be blocked completely by 1/10 diluted anti-PWM. In the following experiments, anti-PWM

Table 1. Effect of addition of different concentrations of anti-PWM antiserum to human peripheral blood mononuclear cells stimulated by three different dilutions of PWM.

Added at day 0		Day 6	
PWM (dilution)	anti-PWM (concentration)	blast cells ($x10^6$/ml)	total cells ($x10^6$/ml)
-	-	0.09	0.80
-	1/100	0.12	1.12
-	1/30	0.09	1.08
-	1/10	0.16	1.21
-	NRS*	0.08	0.99
1/2500	-	0.71	1.83
1/2500	1/100	0.02	0.73
1/2500	1/30	0.09	0.97
1/2500	1/10	0.10	1.03
1/2500	NRS*	0.53	1.32
1/500	-	1.01	1.80
1/500	1/100	0.20	1.17
1,500	1/30	0.09	0.78
1/500	1/10	0.08	0.91
1/500	NRS*	0.63	1.62
1/100	-	1.20	2.07
1/100	1/100	1.13	2.31
1/100	1/30	1.27	2.54
1/100	1/10	0.05	0.94
1/100	NRS*	1.03	2.06

*NRS: normal rabbit serum

Table 2. Enhancement of blast cell formation induced by addition of anti-PWM antiserum (1/10 diluted) to PWM-stimulated human mononuclear cells. Time indicated reflects hrs after initiation of the stimulation.

Added to mononuclear cells at day 0	Presence of anti-PWM (hrs)	Day 6	
		blast cells ($x10^6$/ml)	total cells ($x10^6$/ml)
PWM	48-72	1.43	2.31
PWM	24-72	1.08	2.30
PWM	2-72	0.06	0.84
PWM	0-72	0.07	0.91
Medium	0-72	0.11	0.98
PWM	absent	0.52	1.96
Medium	absent	0.04	0.78

antiserum was used in a 1/30 dilution and PWM was 1/500 diluted. Inhibition of blast cell formation occurred when anti-PWM was added up to 8-10 hr after the start of the stimulation. Thereafter, the number of blast cells present at day 6 increased gradually with increasing time of addition of anti-PWM antiserum to the stimulation, up to a number of blast cells comparable to PWM stimulation in the absence of anti-PWM.

If anti-PWM antiserum was added to PWM-stimulated mononuclear cells, 24 or 48 hr after the initiation of the stimulation, the number of blast cells present at day 6, exceeded the number found after stimulation with PWM alone (Table 2). This effect was most pronounced when anti-PWM was added at 48 hr. As shown in Table 2, the number of blast cells at day 6 after addition of anti-PWM antiserum at 48 hr, had increased almost three-fold over PWM stimulation alone (0.52 to $1.43x10^6$ cells/ml). Normal rabbit serum added 24 or 48 hr after the start of the stimulation, had no effect on the number of blast cells.

To investigate whether the effect of anti-PWM antiserum on mitogen-driven stimulation was due to an immunospecific interaction of the antibodies with the mitogen, IgG was isolated from the rabbit serum and $F(ab')_2$ parts were prepared by pepsin digestion. Addition of this $F(ab')_2$-anti-PWM, either concomitantly with or 48 hr after PWM

Table 3. Effect of anti-PWM antiserum (1/30 diluted) on monocyte and lymphocyte-enriched suspensions. Cell populations were reconstituted at day 1 with or without anti-PWM antiserum.

Preculture for 24 hrs		Mixed	Day 6	
Monocytes	Lymphocytes	at day 1	blast cells ($x10^6$/ml)	total cells ($x10^6$/ml)
PWM	Medium	-	0.83	1.87
PWM+anti-PWM	Medium	-	0.04	0.79
PWM	Medium	anti-PWM	0.16	0.98
PWM	PWM	-	0.72	1.27
PWM+anti-PWM	PWM	-	0.78	1.41
PWM	PWM	anti-PWM	0.82	1.50
PWM	PWM+anti-PWM	-	0.16	0.87
Medium	PWM	-	0.80	1.40
Medium	PWM+anti-PWM	-	0.06	0.86
Medium	PWM	anti-PWM	0.91	1.69
CONTROLS				
Mononuclear cells in medium			0.04	0.82
Mononuclear cells + PWM at day 0			1.07	1.58
Mononuclear cells + PWM + anti-PWM at day 0			0.12	0.80

stimulation of mononuclear cells, resulted in a blastogenic response fully comparable with the effects of either anti-PWM antiserum or anti-PWM antibodies (results not shown). Again a blockade of the stimulation was found after addition at day 0 and enhancement of the stimulation if added 48 hr after the initiation of the stimulation.

Blastogenesis of mononuclear cells upon PWM stimulation has been reported to be monocyte-dependent. Therefore we investigated whether anti-PWM antiserum affected monocyte or lymphocyte function or both. Monocyte-enriched suspensions were precultured in the presence of PWM with or without anti-PWM and after 24 hr, mixed with autologous lymphocytes either cultured or stimulated with PWM with or without anti-PWM antiserum. In some experiments, reconstitution was done in the presence of anti-PWM antiserum (Table 3). It was found that whenever lymphocytes were prestimulated with PWM, blastogenesis was comparable with PWM stimulation of mononuclear cells, with the exception of the presence of anti-PWM antiserum during prestimulation (Tablc 3, line 7 and 9). Even reconstitution of these latter lymphocyte suspension with PWM-prestimulated monocytes did not restore blastogenesis (Table 3, line 7). By contrast, the blocking effect of anti-PWM antiserum on monocyte prestimulation (Table 3, line 2) could be overruled by reconstitution with PWM-primed lymphocytes (Table 3, line 5). Additional evidence that anti-PWM antiserum interfered with lymphocyte activation was derived from PWM-prestimulated monocytes reconstituted with non-stimulated lymphocytes. Since the cell suspensions had not been washed before reconstitution, blastogenesis was found to occur. However, if anti-PWM antiserum was added at the time of reconstitution, blast cell formation was blocked.

DISCUSSION

In the present study, two divergent effects of anti-PWM antiserum are shown on the PWM-driven stimulation of mononuclear cells. Firstly, anti-PWM antiserum prevented blast cell formation, if added concomitantly with the mitogen. Secondly, considerable enhancement of blast cell formation was observed upon addition of anti-PWM antiserum to mitogen-stimulated mononuclear cells at 48 hr after the start of the stimulation. Both these effects were due to specific interaction of anti-PWM antibodies, since anti-PWM antibodies as well as $F(ab')_2$-anti-PWM fragments exerted the same effects.

Blocking experiments of mitogen-induced cell stimulation have been described before for Concanavalin A stimulation, using antibodies against ConA (15-17) or using polysaccharides (8,18). The latter compounds are reported to interchange with mitogens present on the monocyte surface-membrane (10), whereas antibodies against the mitogen might prevent interaction with lymphocytes (18,19). That lymphocytes have to be triggered directly by PWM, comes from the experiments using enriched cell suspensions. In all experiments where interaction between PWM and lymphocytes was prevented by anti-PWM antiserum, blastogenesis was greatly reduced. By contrast, mitogen-presenting function of monocytes, being inhibited initially by anti-PWM antibodies, could be restored after reconstitution with PWM-stimulated lymphocytes. Earlier experiments (14) have demonstrated that thoroughly washing of monocytes, prestimulated with PWM prior to reconstitution with non-primed lymphocytes, give rise to blast cell formation, whereas the supernatant of such a monocyte suspension does not. Therefore, it may be argued that the inhibition of the blastogenesis observed upon addition of anti-PWM antiserum is due to binding of the antibodies with either monocyte-derived or monocyte-associated mitogen. As shown in our experiments, a blockade appears to be independent of the time of addition of anti-PWM antiserum, provided that lymphocytes have not seen PWM before reconstitution. This suggests that an interaction of PWM with lymphocytes is a prerequisite of lymphocyte activation. Once activated, lymphocyte blastogenesis can no longer be prevented by the addition of anti-PWM antibodies.

More difficult to explain is the enhanced blast cell formation, found after addition of anti-PWM antiserum 48 hr after the initiation of the stimulation. It seems likely that there is a conversion from a suppressive effect of anti-PWM antibodies during the initial 24 hr of stimulation to an enhancement of the stimulation if the antiserum is given after 24 hr. Enhanced blastogenesis has been reported of the rabbit system (20), probably due to cross-linking of already bound anti-allotypic antibodies. Such a mechanism of cross-linking may take place also in our system, in that PWM or PWM fragments already bound to the lymphocyte surface membrane, are cross-linked by anti-PWM antibodies. This may induce an additional signal leading to enhanced blast cell formation and/or proliferation.

In conclusion, our results demonstrate that anti-PWM antibodies exert divergent effects on the mitogen-driven stimulation of human mononuclear cells. In addition, it is demonstrated that, besides the necessity of monocytes as presenting cells, blastogenesis of mononuclear cells depends also on a direct interaction of the mitogen with lymphocytes. Activated in this way, lymphocytes are able to respond on further stimuli, resulting in cell proliferation.

REFERENCES

1. Fauci, A.S. (1979). Immunol. Rev. 45, 93-98.

2. De Vries, J.E., Caviles, A.P., Bont, W.S., Mendelsohn, J. (1979). J. Immunol. 122, 1099-1107.

3. Gerrard, T.L., Fauci, A.S. (1982). J. Immunol. 128, 2367-2372.

4. Gerrard, T.L., Bonnard, G.D., Fauci, A.S. (1983). Cell. Immunol. 78, 64-72.

5. Janossy, G., Greaves, M.F. (1971). Clin. exp. Immunol. 9, 483-498.

6. Shin, H.-S., Wang, C.-Y., Choi, Y.S. (1981). J. Immunol. 126, 2485-2489.

7. Puck, J.M., Rich, R.R. (1984). J. Immunol. 132, 1106-1112.

8. Stevenson, H.C., Miller, P.J., Waxdal, M.J., Haynes, B.F., Thomas, C.A., Fauci, A.S. (1983). Immunol. 49, 633-640.

9. Thiele, D.L., Lipsky, P.E. (1982). J. Immunol. 129, 1033-1040.

10. Linthicum, D.A., Elson, E., Mendelsohn, J., Sell, S. (1977). Exp. Cell Res. 110, 237-250.

11. Boyum, A. (1968). Scand. J. Clin. Lab. Invest. 21, 77-82.

12. Yam, L.T., Li, C.Y., Crosby, W.H. (1971). Am. J. Clin. Pathol. 55, 283-290.

13. Nisonoff, A., Markers, G., Wissler, F.C. (1961). Nature 189, 293-295.

14. De Vries, E., Lafeber, G.J.M., Van der Weij, J.P., Van Buysen, A.C., Leijh, P.C.J., Cats, A. (1980). Immunol. 40, 177-182.

15. Byrd, W.J., Finley, W.H., Finley, S.C., McCluse, S. (1964). Lancet ii, 420-421.

16. Kay, J.E., Oppenheim, J.J. (1971). Proc. Soc. exp. Biol. Med. 138, 112-117.

17. Sell, S., Shepperd, W.H. Jr. (1974). Exp. Cell Res. 84, 153-158.

18. Greaves, M.F., Bauminger, S., Janossy, G. (1972). Clin. exp. Immunol. 10, 537-554.

19. Raffel, C., Sell, S. (1982). J. Cell Sci. 56, 141-156.

20. Sell, S., Skaletsky, E., Holdbrook, R., Linthicum, D.S., Raffel, C. (1980). Immunol. Rev. 52, 141-179.

A HIGH MOLECULAR WEIGHT MITOGENIC FACTOR IN CULTURE SUPERNATANT OF POKEWEED MITOGEN-STIMULATED HUMAN PERIPHERAL BLOOD MONONUCLEAR CELLS

H.G. Uiterdijk, H.J. Kornman-van den Bosch*, F. Klein*, A.M. de Bruijn* and E. de Vries. Dept. of Rheumatology, University Hospital, Leiden, and Institute of Epidemiology*, Erasmus University, Rotterdam, The Netherlands

Lectins can induce mitogenesis in peripheral blood mononuclear cells and are used to study the activation of the humoral and cellular immune system (1). Phytohaemagglutinin (PHA), Concanavalin A (ConA), and Pokeweed mitogen (PWM) are widely used mitogens. PHA and ConA are stimulators of mainly the T lymphocytes, whereas PWM obtained from the roots of the *Phytolacca americana* is a monocyte mediated, T-cell dependent, polyclonal B cell activator (2).

Besides investigations on mitogen stimulation of cells, several studies on the production of biologically active substances generated after antigen or mitogen stimulation of mononuclear cells have been published (3,4,5,6,7,8). Among these substances are the interleukins and a number of specific T- or B-lymphocyte stimulating factors. Most of these factors have been demonstrated not to be mitogenic but are described as helper factors in antigen- or mitogen-driven stimulation (5,6,8), and some of them act directly on cells, being independent stimulators of B and T lymphocytes (3, 4,7).

The present study concerned the generation of an independent mitogenic factor in cultures of mononuclear cells stimulated with PWM. This mitogenic factor proved to be capable of stimulating freshly isolated mononuclear cells inducing blast cell formation and immunoglobulin production in the absence of antigens or mitogens. Experiments were performed to study the generation of this mitogenic factor as well as to investigate some biological and biochemical characteristics.

Lectins, Vol. V

MATERIALS AND METHODS

Cell isolation and culture

Peripheral blood mononuclear cells were obtained from heparinized venous blood of healthy blood donors and stimulated as described by De Vries et al. (9).

T lymphocyte enriched cell populations were prepared by rosetting PBMC with sheep red blood cells (10) and contained on average over 95% lymphocytes.

Cell cultures were grown in Teflon bags (11) for the mass production of mitogenic factor and in 24-well culture plates (Costar, Cambridge, Mass.) for study of the effect of mitogenic factor.

Preparation of culture supernatant containing mitogenic factor.

PWM (Gibco; 1/100 final dilution of stock solution) was added to mononuclear cells on day 0. After 3 days of culture the cells were thoroughly washed and resuspended in fresh mitogen-free culture medium. This medium was harvested on day 6. Cell-free supernatant was tested for mitogenic activity on freshly isolated mononuclear cells, at a final concentration of 30% (v/v) in culture. To assess the mitogenic capacity, the results of stimulation with mitogenic factor were compared with the effect of PWM (1/500 final dilution) stimulation of the same mononuclear cells.

Cell monitoring

To assess the mitogenic response of cell cultures to stimulation, samples were taken on day 6, the total cell number was determined, and viability was tested by trypan-blue exclusion. Smears were made and stained for non-specific cytoplasmic esterase activity (12), as counterstain haemalum was used. Stimulation of cells was assessed from counts of monocytes, blast cells, and lymphocytes, according to the following criteria: monocytes: large villous cells (12-20 µm in diameter), strongly esterase-positive, throughout the cytoplasm; lymphocytes: small cells with a narrow rim of cytoplasm that are either negative for or show few positive spots of enzyme activity; blast cells: large rounded cells (10-20 µm in diameter) with prominent nuclei and an esterase pattern similar to that of lymphocytes. On day 6, non-stimulated cultures of mononuclear cells were composed of 10-20% monocytes, 0-5% blast cells, and 70-80% lymphocytes.

The synthesis of immunoglobulins is a parameter for the activation of B lymphocytes (13). In fluids of all cultures the IgM, IgA, and IgG content was measured by an ELISA method.

ELISA

Immunoglobulins were measured by a sandwich-ELISA technique (14). The rabbit anti-human IgM, IgA, and IgG used and their horseradish peroxidase conjugates were obtained from Dakopatts (Denmark).

The Dutch reference preparation HOO-02 (Central Laboratory Bloodtransfusion, Amsterdam) was used as a standard, using the WHO values for IgG and IgA, and the values found earlier by de Bruijn et al. (15) for IgM. The detection limit for all three assays was approximately 0.01 µg/ml. The values given for the amounts of IgA were not corrected for the presence of polymeric IgA (16).

Fractionation of culture supernatant containing mitogenic factor by gelfiltration
Supernatants were fractionated according to molecular weight on an Ultrogel ACA 34 column (LKB, Sweden) equilibrated with 0.1 M phosphate (pH 7.4). The column was 90 cm long and 2.5 cm wide and had a void volume of about 150 ml. The collected fractions (50 ml each, fraction 1 started after 75 ml) were concentrated to one-third of the original volume and tested for mitogenic capacity on cell cultures as described above, at a final concentration of 10% (v/v) fraction in culture medium. Human IgM (polymeric), IgG (monomer and dimer) and serum albumin were used as markers for molecular weight estimations.

For the assessment of non-covalent binding of mitogenic factor fractionation of a supernatant was performed under dissociating conditions with 1 M potassium thiocyanate buffered to pH 7.4 with phosphate.

Immunoglobulin depletion
Since supernatants containing mitogenic factor were found to contain immunoglobulins a possible mitogenic activity of IgM, IgA, and IgG was investigated. Supernatant or ACA34 fractions containing mitogenic factor were depleted of IgM, IgA, and IgG, using rabbit anti-human IgM, IgA, or IgG (Dakopatts, Denmark). Anti-human antibodies were coupled to CNB-activated Sepharose 4B (Pharmacia, Sweden) according to standard procedures advised by the manufacturer. The immunoglobulin content before and after absorption was measured by ELISA. After each absorption the concentration of the depleted immunoglobulins was reduced to below detection level.

Incubation with erythrocytes
Although the supernatants are supposed to be mitogen-free, a possible contamination with residual PWM cannot be excluded. To deplete (17) supernatants of PWM (or PWM-subunits) we added about $5x10^8$ erythrocytes/ml to supernatant before incubation at 37^oC for one hour. The erythrocytes were then spun down and the supernatants were tested for mitogenic activity as described above. For control, an identical procedure was applied to PWM (1/100 dilution) in culture medium.

RESULTS

Effect of the mitogenic factor on mononuclear cells
Several supernatants were tested for mitogenic activity on freshly isolated mononuclear cells. The results of three representative experiments are shown in Table 1. All supernatants collected after PWM stimulation of various suspensions of mononuclear cells proved to be capable of inducing blast cell formation, immunoglobulin production, and, often, proliferation of lymphoid cells in cultures of freshly isolated mononuclear cells. The magnitude of the response, as measured by these parameters of lymphocyte activation, was comparable to the results obtained with PWM stimulation of the same cell populations, although the amount of immunoglobulins

Table 1. Effect of three different supernatants containing mitogenic factor on freshly isolated mononuclear cells in relation to the stimulation induced by PWM (1/500 dilution). Control = culture medium only, MF = mitogenic factor, PWM = pokeweed mitogen.

	Cells $\times 10^6$/ml	Blast cells $\times 10^6$/ml	IgM µg/ml	IgA µg/ml	IgG µg/ml
Control	1.21	0.08	<0.1	<0.1	0.1
MF	1.67	0.57	2.6	0.4	4.0
PWM	1.62	0.71	2.6	1.1	9.1
Control	1.12	0.10	0.3	0.1	0.3
MF	0.88	0.32	0.4	0.3	0.6
PWM	1.18	0.33	0.3	0.5	0.8
Control	0.98	0.02	<0.1	<0.1	<0.1
MF	1.54	0.52	0.3	<0.1	0.5
PWM	1.15	0.47	0.5	0.5	2.2

produced between the third and the sixth day of culture was smaller than that seen after PWM stimulation (culture supernatants from cultures of non-stimulated cells did not induce activation).

For further characterization of the nature of this mitogenic activity, culture supernatants were applied to gelfiltration columns.

Gelfiltration of culture supernatants

Filtration of supernatant over an ACA34 column led to the recovery of the mitogenic factor in fractions with a molecular weight between 400,000 and 900,000 (Fig. 1). These fractions included the peak amounts of IgM and IgA as well as traces of IgG, as measured by ELISA. As can be seen from Fig. 1a, IgA was present in monomeric as well as polymeric configurations. However, corrections for the presence of polymeric IgA as suggested by Delacroix et al. (16) could not be applied since there are no data available on such correction factors when using an ELISA technique.

As can be derived from Fig. 1, blast cell formation after stimulation of mononuclear cells with fractions 3 and 4 roughly equaled that by non-fractionated supernatant, and declined rapidly with ascending fraction number. A similar pattern was found for the immunoglobulin production induced by these fractions.

Involvement of immunoglobulins in mitogenic activity

For the investigation of a possible relationship between the mitogenic factor and the immunoglobulins, supernatants containing mitogenic factor were depleted successively of IgA, IgM, and IgG before

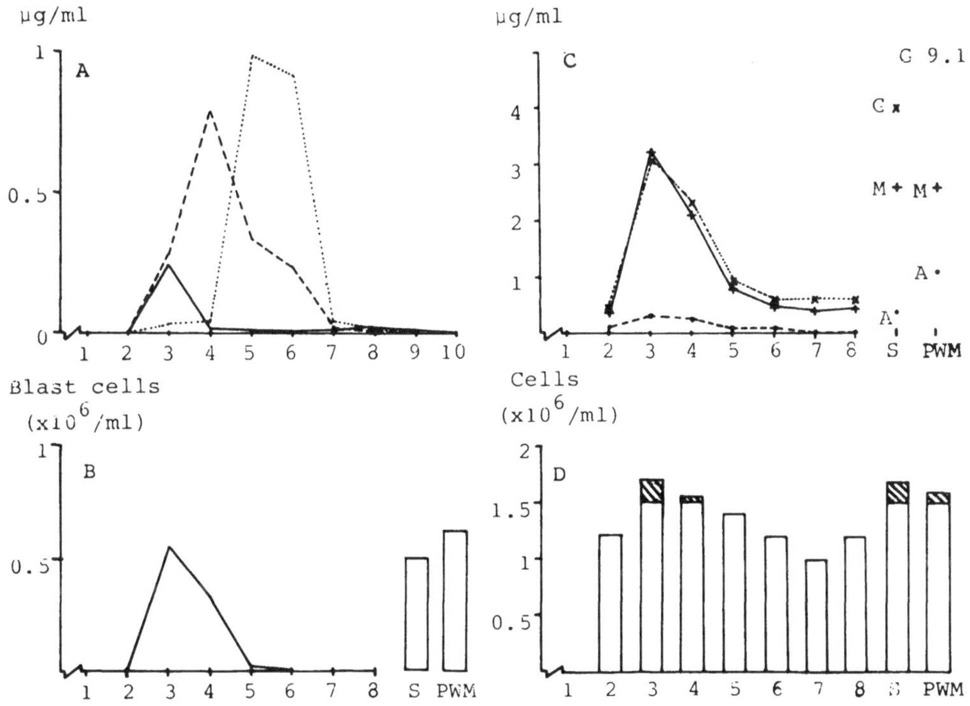

Fig. 1 a-d. Results of gelfiltration of culture supernatant containing mitogenic factor (fraction 1 started after 75 ml, each fraction contained 50 ml, see Materials and Methods)

a. Immunoglobulin content of the fractions.

b. Absolute number of blast cells induced by each fraction and by the original supernatant (S).

c. Immunoglobulin synthesis induced by each fraction and S.

d. Total cell number. Fractions 3 and 4 and the original supernatant (S) all induced proliferation.

gelfiltration. After each consecutive step and after depletion of all three immunoglobulins, supernatants showed mitogenic activity of the same potency as was found for untreated supernatants (Fig. 2).

Fractionation of immunoglobulin-depleted supernatants on an ACA 34 column resulted in the recovery of mitogenic factor present in exactly the same fractions as found for non-depleted supernatants (Fig. 2).

Furthermore, when the ACA34 fractions were depleted of immunoglobulins, no effect on the mitogenic activity of the respective fractions was found either (Fig. 3).

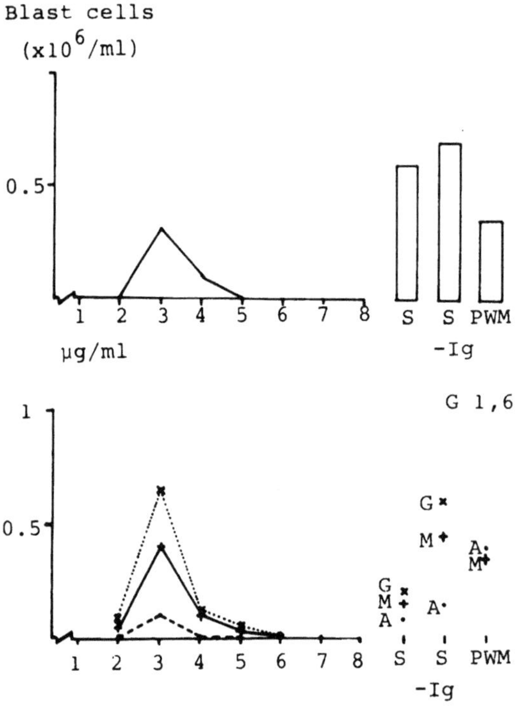

Fig. 2. Gelfiltration of immunoglobulin-depleted culture supernatant containing mitogenic factor (fractions collected as described for Fig. 1).
a. Absolute number of blast cells induced by each fraction and by the original supernatant (S) and the immunoglobulin-depleted supernatant (S-Ig).
b. Immunoglobulin synthesis induced by each fraction, S and S-Ig.

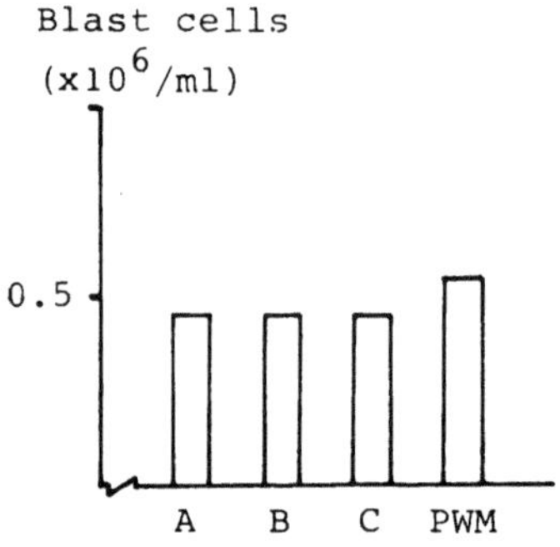

Fig. 3. The result of immunoglobulin depletion of an ACA 34 fraction containing mitogenic factor.
A = original supernatant,
B = fraction,
C = immunoglobulin-depleted fraction.

Involvement of PWM in mitogenic activity

For the investigation of the possibility that the mitogenic factor consisted of PWM or PWM subunits noncovalently linked to carriers with a high molecular weight, supernatant was fractionated under dissociating conditions. The mitogenic factor was found in the same high molecular weight fractions.

Another way to investigate the involvement of PWM in the formation of mitogenic factor following the stimulation of mononuclear cells was also used, as follows. Several supernatants were incubated with a high concentration of erythrocytes, which bind lectins present in conditioned media or solutions (17). As control, a PWM solution was incubated with erythrocytes. The mitogenic activity of the PWM solution disappeared completely after the incubation, whereas all supernatants were found to be still mitogenic after the incubation with erythrocytes.

DISCUSSION

In this paper we have described a mitogenic activity present in culture supernatants of PWM-stimulated peripheral blood mononuclear cells. Unlike several other biologically active factors generated upon mitogenic stimulation (5,6,8), this mitogenic factor induces blastogenesis of freshly isolated, non-sensitized mononuclear cells, in the absence of any mitogen or antigen.

When measured according to parameters of activation of lymphoid cells, such as generation of blast cells and the synthesis of immunoglobulins, this mitogenic factor was found to be comparable in activity to fluid-phase PWM. The results concerning the response of mononuclear cells to mitogenic factor stimulation point strongly to a polyclonal activation, as indicated by the synthesis of different classes of immunoglobulins.

The observation that the mitogenic factor was found in the same fractions containing immunoglobulins, especially IgM and IgA (Fig. 1), prompted us to investigate a possible role of these immunoglobulins in the mitogenicity of those fractions. Mitogenic activity of immunoglobulins or aggregated immunoglobulins has been described (3). However, immunoglobulin-depletion of supernatants or of high molecular weight fractions of supernatants revealed that the mitogenic capacity is not related to immunoglobulins produced after PWM stimulation of mononuclear cells.

An important question is whether the mitogenic factor consists entirely or partially of PWM or PWM fragments. In our culture system PWM can form high molecular weight complexes with fetal calf serum components. Several arguements contradict merely carry-over of PWM. In the first place, the culture system used to generate mitogenic factor includes a thorough washing procedure at the end of the initial three-day stimulation with PWM, and the washed cells are resuspended in fresh mitogen-free culture medium. The mitogenic factor threfore has to be the result of a passive or active interaction between PWM and cells. Gelfiltration of supernatants revealed that the mitogenic factor was present in fractions containing high molecular weight components. However, after fractionation by gelfiltration with a high salt concentration solution, expected to induce dissociation of non-covalently bound components, the mitogenic factor was recovered in the same high molecular weight fractions. Finally, the incubation with erythrocytes in a concentration known to adsorb PWM (17) did not alter the mitogenic capacity of the mitogenic factor while the mitogenic activity of a PWM solution in culture medium was adsorbed. From these results it seems likely that the mitogenic factor differs from a PWM-fetal calf serum component complex.

Preliminary results of experiments with T cell-enriched cell populations indicate that these T cells are stimulated by the mitogenic factor. This is consistent with our observation that the blast-cell formation induced by mitogenic factor is comparable to that obtained after PWM stimulation, whereas immunoglobulin production is only moderate. This distinguishes our mitogenic factor from the B cell-stimulating capacity found in supernatants of PWM-stimualted peripheral blood lymphocytes described by North and Brenner (4).

Further studies are needed to establish the precise nature of this mitogenic factor.

ACKNOWLEDGEMENTS

This study was supported by the Dutch League against Rheumatism. The authors wish to thank Professor Dr. A. Cats for his critical reading of the manuscript, Mr. J.P. van der Weij for excellent technical assistance, and Mrs. C.M.M. Zock for secretarial assistance and preparation of the Figures.

REFERENCES

1. Scott, M.G., Nahm, M.H. (1984). J. Immunol. 135, 2454-2460.

2. Greaves, M.F., Bauminger, S., Janossy, G. (1972). Clin. exp. Immunol. 10, 537-554.

3. Morgan, E.L., Weigle, D.W. (1981). J. exp. Med. 154, 778-790.

4. North, M.E., Brenner, M.K. (1983). Immunology 48, 157-163.

5. Kurnick, J.T., Gronvik, K., Kimura, A., Lindblom, J., Skoog, V.T., Sjoberg, O., Wigzell, H. (1979). J. Immunol. 122, 1255-1260.

6. Schmidt, J.A. (1984). J. exp. Med. 160, 772-787.

7. Kasakura, S., Taguchi, M., Watanabe, Y., Okubo, T., Murachi, T. Uchino, H., Hanaoka, M. (1984). J. Immunol. 133, 3084-3090.

8. Ralph, P., Jeong, G., Welte, K., Mertelsman, R., Rabin, H., Henderson, L.E., Souza, L.M., Boone, T.C., Robb, R.J. (1984). J. Immunol. 133, 2442-2445.

9. Vries, E. de, Weij, J.P. van der, Doekes, G., Cats, A. This volume.

10. Putte, L.B.A. van de, Meijer, C.J.L.M., Lafeber, G.J.M., Kleinjan, R., Cats, A. (1976). Ann. Rheum. Dis. 35, 451-455.

11. Meer, J.W.M. van der, Gevel, J.S. van de, Elzenga-Claassen, I., Furth, R. van (1979). Cell. Immunol. 42, 208-212.

12. Yam, L.T., Li, C.Y., Crosby, W.H. (1971). Am. J. Clin. Pathol. 55, 283-290.

13. Waldman, T.A., Durm, M., Broder, S., Blackman, M., Blase, R.M. Strober, W. (1974). Lancet ii, 609-613.

14. Voller, A., Bidwell, D.E., Bartlett, A. (1976). Bull. World Health Organ. 53, 55-65.

15. Bruijn, A.M. de, Klein, F., Neumann, H., Sandkuyl, L.A., Vermeeren, R., Le Blansch, G. (1982). J. Immunol. Meth. 48, 339-348.

16. Delacroix, D.L., Dehennin, J.P., Vaerman, J.P. (1982). J. Immunol. Meth. 48, 327-337.

17. Ohno, R., Kodera, Y., Ezaki, K., Tanimoto, M., Iwamoto, K., Yamada, K. (1979). J. Immunol. Metho. 31, 65-69.

MORPHOLOGICAL STUDIES OF THE INTERACTION OF HUMAN WHITE BLOOD CELLS WITH MISTLETOE LECTIN I BY MEANS OF SCANNING ELECTRON MICROSCOPY

Hartmut Franz

Staatliches Institut für Immunpräparate und Nährmedien
DDR-1120 Berlin, German Democratic Republic

Kurt Augsten

Zentralinstitut für Mikrobiologie und Experimentelle Therapie der Akademie der Wissenschaften
DDR-6900 Jena, German Democratic Republic

Gerhard Metzner

Institut für Klinische Immunologie der Universität
DDR-6900 Jena, German Democratic Republic

The interactions of mistletoe lectin I (ML I) and its A and B chains with different kinds of cells result in various effects such as increased cytotoxicity (2, 3), mitogenicity by A chain (4), lymphokine release by both A and B chains (5) and increased phagocytic activity by B chain (6).

It is the aim of this paper to study the morphological alterations by means of the scanning electron microscopy (SEM) which will be correlated to functional and biochemical reactions (2, 4, 5, 6). The cell types used are human mononuclear cells (MNC) and polymorphonuclear leukocytes (PMN) after ficoll density gradient centrifugation (7).

Materials and methods

ML I and its A and B chains, respectively, are prepared as described earlier (2). The preparation of MNC and PMN follows the prescription of May et al. (8) and is demonstrated in fig. 1.

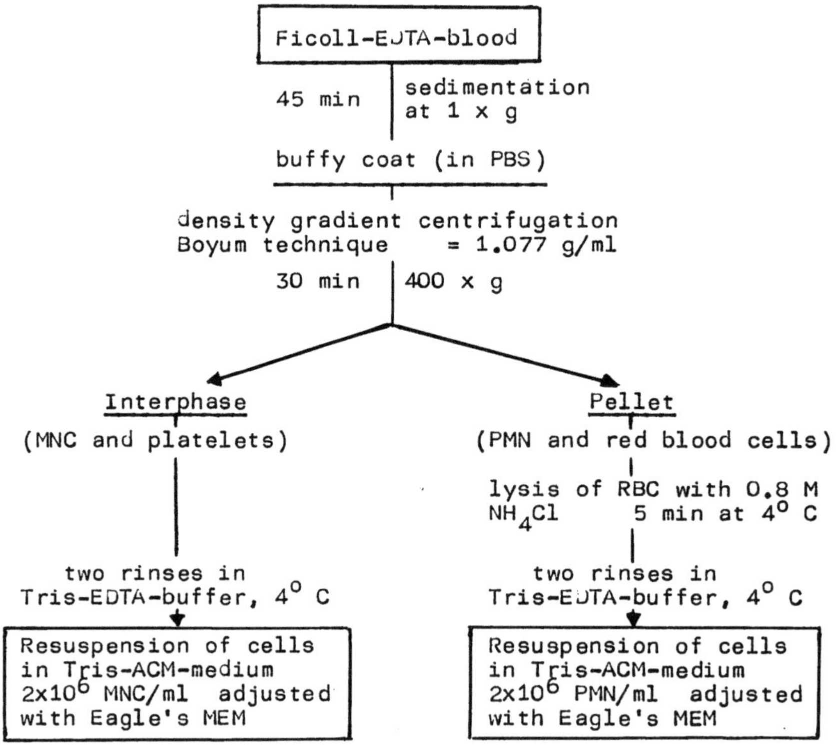

Fig. 1: Flow diagram for preparation of MNC and PMN from human peripheral blood.

ACM = isotonic buffer medium, containing human serum albumine (300 µg/ml), $CaCl_2$ (10^{-3} M), $MgCl_2$ ($5 - 10^{-4}$ M).

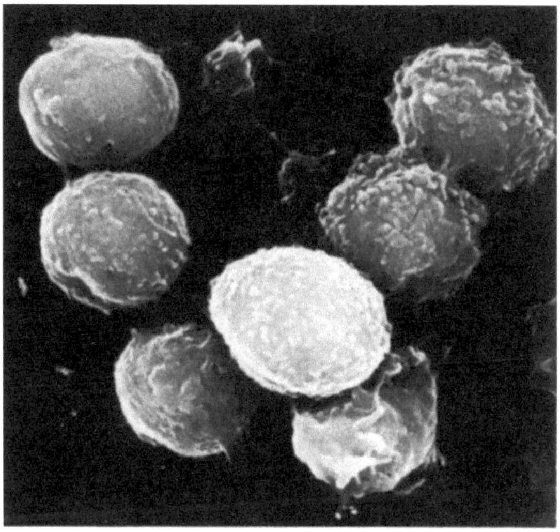

Fig. 2: Scanning electron micrograph of cell surfaces of untreated mononuclear cells characterized by some microvilli and smooth areas. Control. Magnification 45000x.

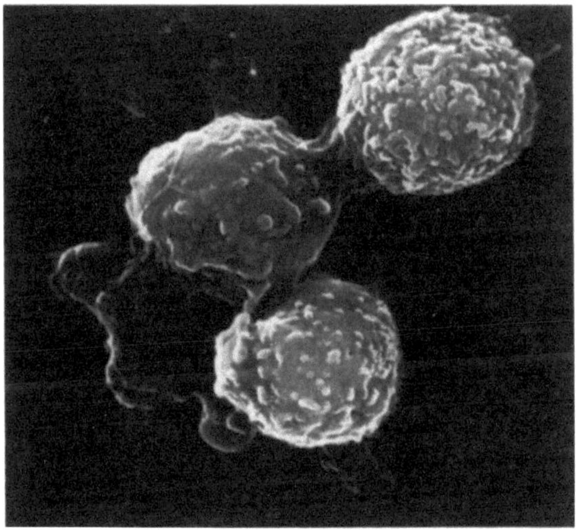

Fig. 3: Incubation of human mononuclear cells with ML I, 10^{-7} M, for 15 h at room temperature induced an agglutination and a reduction of the fine structure. Magnification 5400x

The constituents of the MNC population are lymphocytes and monocytes (ratio 3:1 after peroxidase staining) as well as platelets (MNC : platelets = 1 : 8). The platelets are regulary found in the ficoll interphase and are without special attention in immunological trials. The PMN consist of 98 percent polymorphonuclear leukocytes and of 2 percent lymphocytes. The erythrocyte contamination is very low (one third of PMN).

Scanning electron microscopy (SEM)
200 µl of MNC and PMN were incubated in Eagle's MEM without and with solutions of as well ML I as the isolated A and B chains in PBS. The final concentrations were 10^{-7} M and 10^{-8} M which correspond 10 µg/ml and 1 µg/ml (ML I) or 3 µg/ml and 0.3 µg/ml (A and B chains), respectively. The preparation was performed according to Augsten et al. (9). Briefly, during keeping for 15 hours at room temperature in a humidified chamber cells were allowed to settle on glass surface. After repeatedly rinsing with PBS the cells were fixed using 0.25% (v/v) glutaraldehyde in PBS, dehydrated in graded series of ethanol for 10 min in each step and finally air-dried. Thereafter, the cells were sputtered with platinum and observed in the scanning electron microscope (JEOL JSM 35) at 25 kV without tilting.

Results and discussion

The morphological findings without and with the mentioned mistletoe lectin I constituents are demonstrated in figures 2 to 7.
At a concentration of 10^{-7} M ML I induces mainly agglutination of MNC and PMN. But it is not possible to exclude a cell fusion which is shown in fig. 3. At the concentration of 10^{-8} M ML I induces no more aggregation but only a reduction of the surface structures on the MNC. Morphological alterations like blebs formation or spherical cells without ruffles were not found.
After incubation of MNC with A chain solutions (10^{-8} M) some of the lymphocytes show an orientation of the fine structural elements to one or two poles of the cell (fig. 4). The concentration of A chain used was in the same range as such which causes mitosis (4). MNC express under

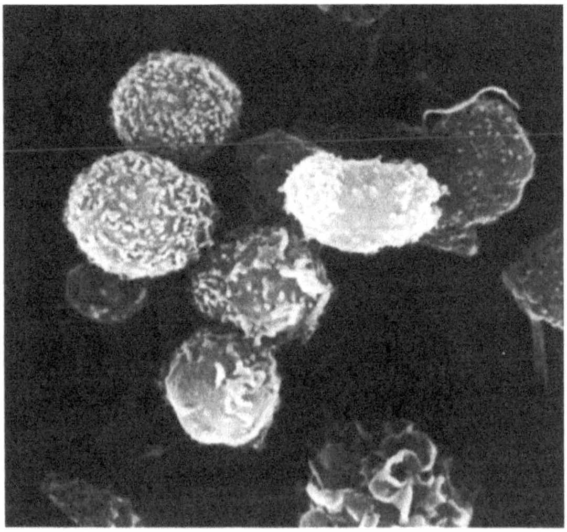

Fig. 4: Incubation of human mononuclear cells with the A chain of ML I, 10^{-8} M, for 15 h at room temperature was in most cases without any effect on the surface; in one case a polar orientation of the microvilli could be demonstrated. Magnification 3600x.

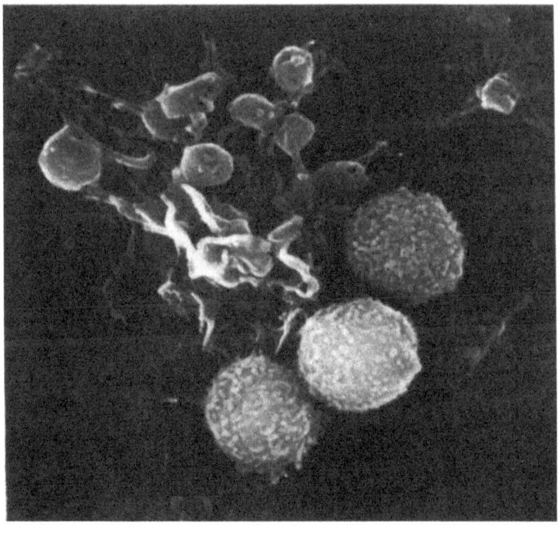

Fig. 5: Incubation of human mononuclear cells with the A chain of ML I, 10^{-7} M, for 15 h at room temperature induced attachment of lymphocytes and platelets to monocytes. The platelets are arranged in tight contact with monocytes. Magnification 3600x.

these conditions many ruffles as well as smoothed areas. After incubation of MNC with the B chain (fig. 5) the lymphocytes showed no significant alterations at their surface. Monocytes are flat and surrounded mainly with platelets and lymphocytes, respectively. Monocytes and platelets are very tightly connected.
The interaction between the A chain and PMN cells lead to partial damaging of the cells (not shown here). The mechanism of this cytotoxicity is not completely understood. Perhaps, the A chain can bind to the cell membrane and penetrates it. After incubation with the solutions of B chain PMN form rosettes (fig. 7). Conceivably, the formation of morulalike protuberances (fig. 7) indicates an activation of the cells. This could be correlated with an increase of phagocytic activity as demonstrated previously (6). In conclusion, the morphological findings reported here demonstrate as well as earlier functional and biochemical investigations multiple effects of ML I and its chains on MNC and PMN. In some cases (e.g. morula-like structures on lymphocyte poles and mitogenic effect) seems to exist an immediate correspondence of the observed morphological changes and functional findings.

Summary

The morphological effects of the interactions between mistletoe lectin I and its A and B chains with mononuclear cell and polymorphonuclear cells were investigated. The observed cell alterations are mostly in agreement with earlier reported functional (activation of as well lymphocytes as phagocytes by ML I and both chains) and biochemical (inhibition of protein synthesis at ribosomal level by A chain) effects.

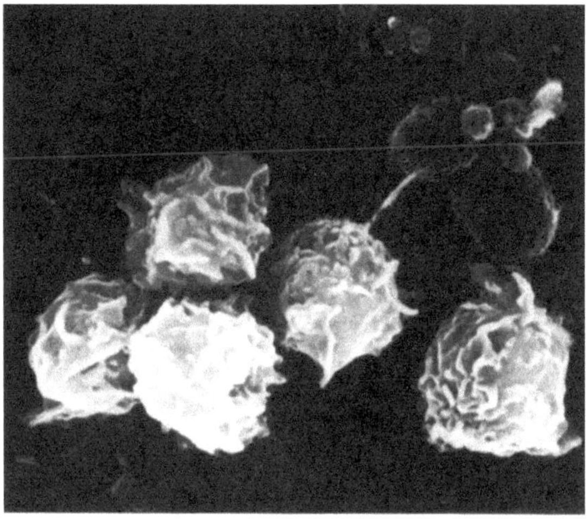

Fig. 6: The cell surface of untreated human polymorphonuclear cells are characterized by many ruffles and folds. Control. Magnification 3000x.

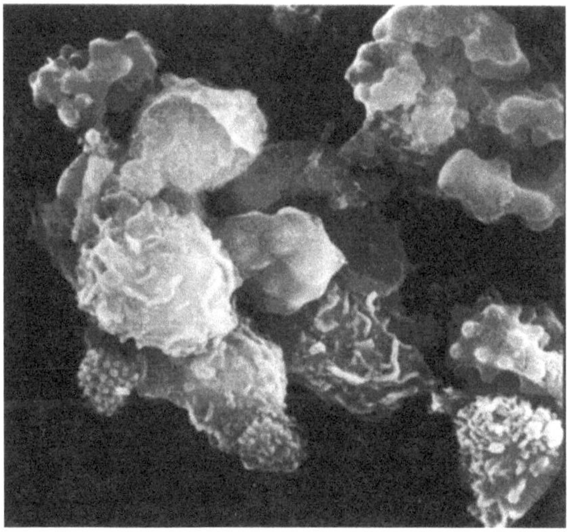

Fig. 7: Incubation of human polymorphonuclear cells with the B chain of ML I, 10^{-7} M, induced a morula-like alteration of the surface. The other cells were unaffected. Many erythrocytes are seen in this field. Magnification 3000x.

References

1. Franz, H., Ziska, P., Kindt, A.: Biochem. J. 195, 481 - 484 (1981)
2. Franz, H., Kindt, A., Ziska, P., Benndorf, H., Venker, L.: Acta biol. med. germ. 41, K9 - K14 (1982)
3. Stirpe, F., Sandvig, K., Olsnes, S., Pihl, A.: J. Biol. Chem. 257, 13271 - 13277 (1982)
4. Metzner, G., Franz, H., Kindt, A.: in this book
5. Metzner, G., Franz, H., Kindt, A., Schumann, I.: in preparation
6. Metzner, G., Franz, H., Kindt, A., Fahlbusch, B., Süss, J.: Immunbiol. 169, 461 - 471 (1985)
7. Boyum, A.: Scand. J. Clin. Lab. Invest. 21, Suppl. 97, 1 - 8 (1968)
8. May, C.D., Lyman, M., Alberto, R., Cheng, J.: J. Allergy 46, 12 - 19 (1970)
9. Augsten, K., Strauss, D.G.: Studia biophys. 104, 219 - 226 (1984)

MITOGENICITY OF A CHAIN OF MISTLETOE LECTIN I (ML I)

Gerhard Metzner

Institut für Klinische Immunologie der Universität
DDR-6900 Jena, German Democratic Republic

Hartmut Franz and Annemarie Kindt

Staatliches Institut für Immunpräparate und Nährmedien
DDR-1120 Berlin, German Democratic Republic

Lectins are characterized by their ability to bind specifically to sugars and as a consequence to compounds on the cell membranes (1, 2). With respect to their effects on lymphoid cells, the lectins are divided into mitogenic, indifferent and inhibitory ones (3, 4). Well known mitogenic lectins are Con A and PHA. However, the mechanism of action of mitogenic lectins is not fully understood. Several lectins have been isolated from mistletoe (Viscum album). Of these the mistletoe lectin I is the best characterized (5). It consists, like ricin, of covalently bound A and B chains and occurs in a monomeric (AB) and a dimeric $(AB)_2$ form (Mol.weight: ML I 60 000 or 120 000 D, A chain 29 000 D, B chain 34 000 D). The B chain fulfills the lectin function and is D-galactose-specific whereas the A chain acts in a manner similar to the A chain of ricin (and abrin). In cell-free systems it inhibits protein synthesis at the ribosomal level (6). Recently we reported that surprisingly the A chain of ML I possesses mitogenic activity (7) which raised the following questions: 1. how is the A chain fixed to the cell membrane and 2. how does the A chain act as a mitogen?

Lectins, Vol. V

To this end, we tested the inhibition of mitogenicity of such sugars which are known to inhibit mitogenic lectins.

Materials and methods

1) Preparation of ML I, isolation of A and B chains
ML I was prepared by a previously described procedure (5). Briefly, ML I was bound to the partially hydrolyzed sepharose and eluted with 0.2 M D-galactose. Cleavage of ML I into A and B chains was performed according to Franz et al. (6) by treatment of sepharose-bound ML I with ß-mercaptoethanol. Solutions of ML I and A and B chains in PBS are stable at 4°C for at least several weeks.

2) Evaluation of mitogenic activity
Mononuclear cells (MNS) taken from peripheral blood of 25 healthy donors were isolated using the Boyum technique (7). 2×10^5 cells in culture medium (100 µl) were added to serial dilutions (1:10 steps) of ML I and its A and B chains to a final concentration between 10^{-7}M and 10^{-13} M. Tests were performed on round bottom microtitre plates (Titertek Linbro) in triplicate for each concentration. MNC were incubated for 72 h at 37°C in Eagle's MEM supplemented with 10% (v/v) newborn calf serum. The cell culture was kept in humidified atmosphere containing 5% CO_2. 16 hrs before harvesting the culture was pulsed with 37 kBq ^{3}H-thymidine. The ^{3}H-thymidine incorporation was measured by means of a liquid scintillation counter (LKB Bromma). From the number of desintegration per minute (dpm) the stimulation index was calculated as follows:

$$\text{Stimulation index (SI)} = \frac{dpm_{test}}{dpm_{control}}$$

$SI \geq 1.5$ indicates a significant mitogenic effect whereas $SI \leq 0.8$ suggests a significant inhibition.

3) Influence of monosaccharides on the mitogenic effect of A chain and Con A, respectively
MNC of 10 healthy blood donors were incubated (2×10^5 cells per well) in 100 µl of the culture medium described above containing one of the following sugars: α-methyl-D-mannoside (αMM), α-L-fucose (L-Fuc), D-galactosamine (D-Gal-NH_2), N-acetyl-galactosamine (D-GalNAc)

and N-acetyl-glucosamine (D-Glc-NAc). The concentrations of the sugars were chosen between 10^{-4} M and 10^{-1} M depending on their effect on MNC in the absence of mitogens (fig. 3). In the inhibition test the concentration of the A chain and Con A were 10^{-8} M (0.3 µg per 10^{-6} cells/ml) and 5 µg per 10^{-6} cells/ml, respectively. The inhibition of the proliferation was calculated using the SI as defined above. The effects of the corresponding sugars on nonstimulated MNC served as controls.

$$\text{Inhibition of proliferation (\%)} = \frac{SI_{without} - SI_{with}}{SI_{without\ sugar} - 1} \times 100$$

Values of more than 20% considered to indicate a significant inhibition of proliferation.

Results and discussion

As seen from fig. 1 the A chain acts as a mitogenic. It effects in a limited concentration range (about 10^{-8} M) a significant stimulation (stimulation index 3.5 ± 1.0). Morphologically we also found typical immunoblasts in the A chain-treated MNC culture. Treatment by Con A resulted in a more evident stimulation (stimulation index 20.5 ± 5.9). In contrast, the intact ML I and the B chain showed a strong inhibitory effect which disappeared with decreasing concentrations of ML I and the B chain. The inhibitory effect of ML I was about 30 times higher than that of the B chain. Using the trypan blue exclusion test cytotoxic effects have not been observed. According to Licastro et al. (3) ML I and the B chain belong to the group of inhibitory lectins. Considering the well known cancerotoxic properties of ML I we expected such a behaviour. However, it was very surprising that the A chain acts in a narrow concentration range as mitogen.
In order to answer the two questions formulated in the introduction we tested whether the mitogenic effect mediated by the A chain can be inhibited by different sugars.

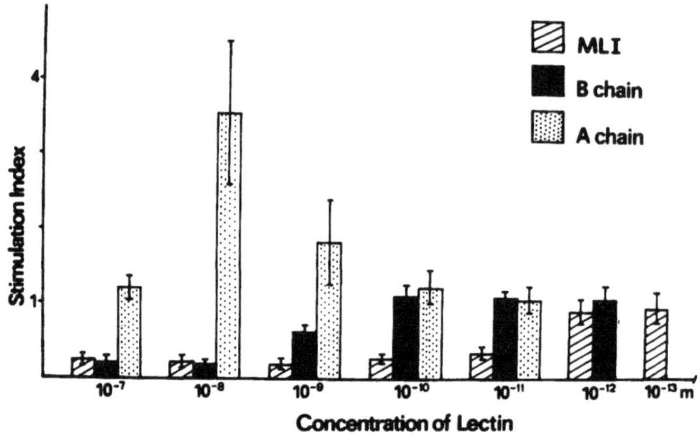

Fig. 1: Influence of ML I and its isolated A and B chains on the proliferation of lymphocytes from 25 healthy persons.

Because the A chain is a glycoprotein (10% carbohydrate content) it would be conceivable that it is bound to the MNC surface by membrane lectins. Some years ago Barzilay et al. (9) demonstrated mannose-specific membrane lectins on human lymphocytes. If fixation of the A chain was based exclusively on binding to membrane lectins we would expect complete inhibition by analogy with the Con A/αMM-system. On the other hand, the results of the sugar inhibition tests argue against the exclusive role of membrane lectins for binding the A chain (fig. 2).

Therefore, we cannot eliminate the possibility that hydrophobic interactions are involved in this phenomenon. The formation of aggregates in aqueous solutions of the A chain indicates the existence of hydrophobic areas within this molecule. The mechanism of the mitogenic activity of the A chain has been poorly understood until now. It would be necessary to test A chains from other sources (gelonin, A chains from ricin and abrin) for mitogenic activity. Partial inhibition was demonstrated with the sugars D-Gal-NH_2, D-Glc-NAc and D-Gal-NAc (fig. 2) not only for Con A but also for the A chain. These effects may be caused by the fact that cell proliferation is a multistep process.

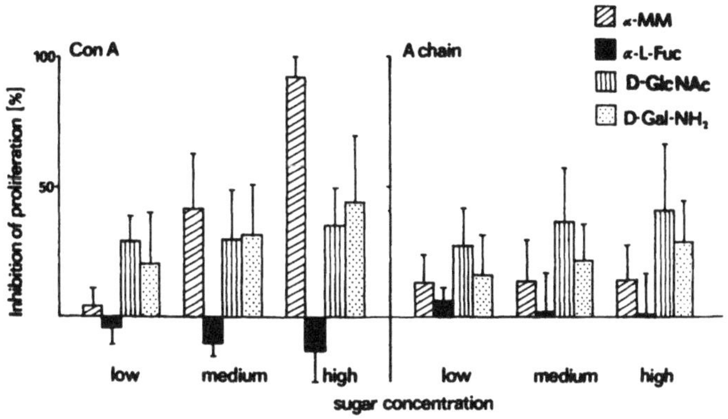

Fig. 2: Inhibitory effects on the mitogenic activity of Con A and the A chain of ML I by four sugars.
Con A: 5 µg per 10^6 MNC/ml
A chain of ML I: 10^{-8} M, i.e. 0.3 µg per 10^6 MNC/ml.

During this process at least two signals are necessary (10). Perhaps the sugars tested influence the second (or third) signal which is necessary for mitogenesis (11).

Fig. 3 demonstrates that sugars alone show concentration-dependent effects on nonstimulated ^{3}H-thymidine uptake. The mitogenic effect of the A chain might be in accordance with immunostimulating activities of mistletoe preparations like Iscador®, Helixor or Plenosol used in cancer therapy. Probably, our findings are in accordance to those of Coeugniet (12) who described a mitogenic activity of Plenosol on T-lymphocytes.

Summary

Mistletoe lectin I (ML I) and its B chain act inhibiting the thymidine uptake into mononuclear cells. Surprisingly, the A chain of ML I which inhibits the protein synthesis at the ribosomal level show significant mitogenicity. Investigating the nature of the interaction of A chain with

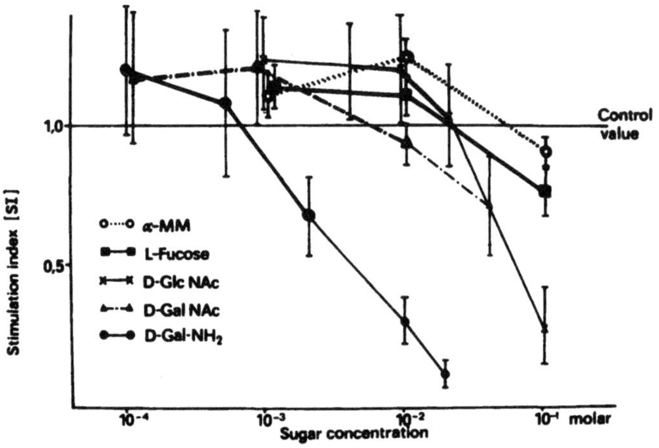

Fig. 3: Monosaccharides posses bidirectional effects on the nonstimulated MNC cultures. Low concentrations are stimulating the cell growth whereas high concentrations result in a marked inhibition.
The concentrations of sugars tested for their ability to inhibit the Con A and A chain-induced proliferation are indicated by stressing the lines.

the cell membrane the characteristic inhibition by monosaccharides was not found. Therefore, an exclusive involvement of membrane lectins for the binding of the A chain to MNC is unlikely. Rather hydrophobic interactions are possibly involved. The mitogenic activity of the A chain could be one reason for immunostimulating effects of mistletoe preparations in cancer therapy.

References

1. Goldstein, I.J., Hughes, R.C., Monsigny, M., Osawa, T., Sharon, N.: Nature (London) 285, 66 (1980)

2. Kocourek, J., Hořejši, V.: Nature (London) 390, 188 (1981)

3. Licastro, F., Chiricolo, M., Barbieri, L., Stirpe, F., Falasca, A., Rossi, C.A., Franceschi, C.: In "Lectins. Biology, biochemistry, clinical biochemistry. Vol. 3". Ed. by T.C. Bøg-Hansen, G.A. Spengler. Walter de Gruyter Berlin.New York 1983, pp. 293-302

4. Jillner-Centerlind, M.L., Axelsson, B., Hammarström, S., Hellström, U., Perlmann, P.: Eur. J. Immunol. 10, 434-442 (1980)

5. Franz, H., Ziska, P., Kindt, A.: Biochem. J. 195, 481-484 (1981)

6. Franz, H., Kindt, A., Ziska, P., Benndorf, H., Venker, L.: Acta biol. med. germ. 41, K9 - K14 (1982)

7. Metzner, G., Franz, H., Kindt, A., Schumann, I.: in preparation

8. Boyum, A.: Scand. J. Clin. Lab. Invest. 21, Suppl. 97, 1 - 8 (1968)

9. Barzilay, M., Monsigny, M., Sharon, N.: In "Lectins. Biology, biochemistry, clinical biochemistry. Vol. 2" Ed. by T.C. Bøg-Hansen, G.A. Spengler. Walter de Gruyter, Berlin New York 1982, pp. 67 - 81

10. Klaus, G.G.B., Hawrylowicz, C.M.: Immunology Today 5, 15 - 19 (1984)

11. Muchmore, A.V., Decker, J.M., Blaese, R.M.: J. Immunol. 125, 1306 - 1311 (1980)

12. Coeugniet, E.: Erfahrungs-heilk. 34, 104-109 (1985)

DOLICHOS LABLAB AGGLUTININ: A POTENT T LYMPHOCYTE MITOGEN AND A HIGH INTERLEUKIN-2 PROMOTER

Jean Favero, Françoise Miquel*, Jacques Dornand, Marguerite Janicot* and Jean-Claude Mani.
Laboratoire de Biochimie des Membranes, ER CNRS 228, 8 rue de l'Ecole Normale, F-34075 Montpellier, and
*Laboratoire d'Anatomie et Physiologie Humaine, Faculté de Pharmacie, 15 Avenue Charles Flahaut, F-34060 Montpellier.

Mitogenic plant lectins have been used for a long time to study lymphocyte stimulation (1). On one hand their sugar binding specificity have made these compounds useful probes in the determination of cell surface glycoconjugates involved in the mechanism of lectin-induced lymphocyte proliferation (2-4); on the other hand, by mimicking the *in vivo* lymphocyte activation by antigens, mitogenic lectins have been proved suitable tools to study the early and latter events occuring in the process of stimulation. It is now established that lymphocyte stimulations induced by different lectins display differences with regard to animal species, lymphocyte subpopulation and lymphokine production. Moreover all lectins do not act through the same cell surface receptors. It is therefore important to study many mitogenic lectins in order to correlate the the differences in their abilities to induce blastogenesis with their specific cell surface receptors (5). These reasons led us to evaluate the mitogenic properties of a lectin isolated from *Dolichos lablab* L. with respect to concanavalin A (Con A) a well-known T lymphocyte mitogen

Lectins, Vol. V

Materials and Methods

Materials: RPMI 1640 was purchased from GIBCO, fetal calf serum and interleukin-2 from Boehringer, α-methyl-D-mannopyranoside from Fluka, concanavalin A and Agarose-p-aminophenyl-D-mannopyranoside from PL Biochemicals, Sephadex G-100 from Pharmacia, [^{3}H]-thymidine from CEA.

Methods:

Isolation and purification of Dolichos lablab agglutinin (DLA): 1g cotyledons from the seeds of *Dolichos lablab* L. were ground in 9 ml phosphate buffered saline (PBS), pH 7.4; after one night at room temperature the suspension was centrifuged at 1500*g* for 10 min and the supernatant recovered. 8 ml supernatant was poured on a Sephadex G-100 column (90x3 cm). Two fractions (F-I, F-II) were eluted with phosphate buffered saline alone and a third one (F-III) with 0.2 M glucose in PBS. After concentration with Diaflow PM-10 membranes and dialysis, F-III was added to a D-mannopyranoside-Agarose column. After washing with PBS, bound material was eluted with an α-methyl-D-mannopyranoside gradient (0-1 M) (F-IV).

Sodium dodecyl sulfate-polycarylamide gel electrophoresis (SDS-PAGE): the different fractions were concentrated on PM-10 membranes and their protein content determined by the Lowry method. Sample electrophoresis buffer containing sodium dodecyl sulfate and 2-mercaptoethanol was added and the fractions were boiled for 5 min. SDS-PAGE was performed according to Laemmli (6) in a 15% polyacrylamide gel.

Cell culture: spleen and thymus lymphocytes from 3 week old male Swiss mice, and human peripheral blood lymphocytes were prepared as described previously (7). In some experiments splenocytes from congenitally athymic *nu/nu* mice were used. Cells were cultured in RPMI 1640 medium supplemented with 10% fetal calf serum at the concentrations of 3.10^6 splenocytes, $2.5.10^6$ thymocytes and 1.10^6 human lymphocytes per ml during 48 hr (mouse cells) or 72 hr (human cells) at 37°C in 5% CO_2 wet atmosphere. In some experiments the stimulation was blocked by adding 20 mg/ml α-methyl-D-mannopyranoside, except in the first blocking experiment in which the minimal sugar inhibiting concentration was determined.

Preparation of conditioned media and determination of interleukin-2 amount: supernatants from 24 hr lectin stimulated mouse or human cultures were collected and used as conditioned media to support the growth of an interleukin-2 dependent cytotoxic T cell line. The amount of interleukin-2 (expressed as U/ml) was evaluated by comparison with commercial interleukin-2 which was arbitrarily assigned a value of 1 U/ml as in (8).

Splenocyte membrane preparation and determination of ecto 5'nucleotidase activity: preparation of mouse splenocyte membranes and measurement of ecto-5'nucleotidase activity were performed according to Dornand *et al* (9).

Results

DLA isolation: the crude extract poured on a Sephadex G-100 column led by elution with phosphate buffered saline (pH 7.4) to two major fractions (F-I and F-II) (Figure 1); F-I contained proteins which passed unretarded through the column and F-II, which was completely separated from F-I, showed a broad asymetrical elution profile. This second fraction showed, like DLA (10,11), the capacity to strongly agglutinate human erythrocytes in a non specific way with respect to blood groups.
Further elution of the column with 0.2 M glucose in PBS resulted in a single symetrical peak (F-III) displaying also high erythroagglutinating activity. SDS-PAGE analysis of F-II and F-III showed the same profile, F-II presenting one higher molecular weight extra band which disappeared in F-III (Fig. 2).
Different sugars were tested for the inhibition of DLA erythroagglutinating capacity; α-methyl-D-mannopyranoside was far the most effective (α-methyl-D-mannopyranoside>α-methyl-D-glucopyranoside>N-acetyl-glucosamine>D-glucose>D-glucosamine>galactose= N-acetyl-galactose). This result led us to further purify the F-III fraction by affinity chromatography on immobilized D-mannopyranoside which present a rather good affinity for the lectin. The whole material (F-III) was retained on the D-mannopyranoside-agarose column. Elution with a 0-1 M α-methyl-Dmannopyranoside gradient yielded a single peak (F-IV) accounting only for 10% of the starting protein. F-III and F-IV presented the same electrophoretic pattern: four bands (Fig. 2), the apparent molecular weights of which (14400, 24000, 28000, 32000 daltons) were consistent with those found by Rao *et al* (11) who reported four major bands and two faint ones ranging from 10200 to 35000 daltons.DLA is a glycoprotein as evidenced by the sulfuric acid hydrolysis and orcinol coloration, and all the four bands of SDS-PAGE were stained by Schiff reagent

Induction by DLA of mouse lymphocyte stimulation: in these experiments mouse spleen lymphocytes were stimulated with increasing concentrations of DLA (dialysed F-III). The leucoagglutinating activity of the lectin, observed by optical microscopy, was accompanied by lymphocyte proliferation as assessed by [^{3}H] -thymidine incorporation. These phenomena were inhibited by 25 mM

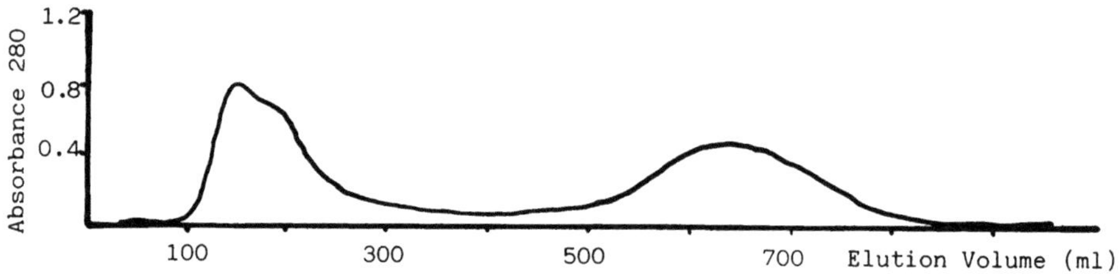

Figure 1: chromatogram of Dolichos lablab crude extract on Sephadex G-100: elution with 20 mM phosphate buffered saline, pH 7.4.

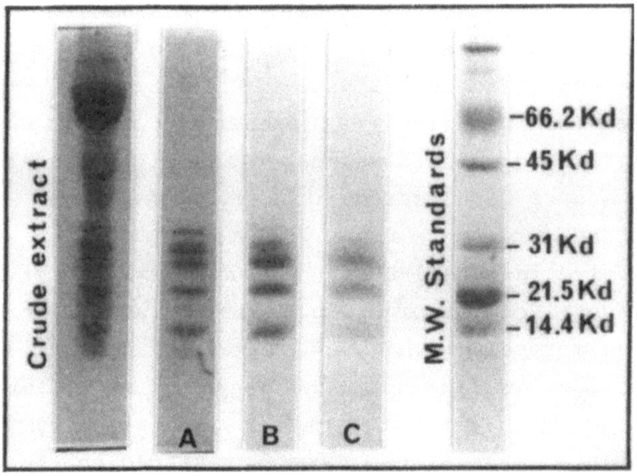

Figure 2: SDS-PAGE of Dolichos lablab crude extract and fractions F-II (A), F-III (B) and F-IV (C).

α-methyl-D-mannopyranoside, provided the sugar was added during the first 8 hours. Parallel experiments with Con A as stimulating agent showed (Fig. 3) that the DLA stimulation level was in the same range as that obtained with Con A, but the maximal activity was reached at lower lectin doses (2 μg/ml for DLA, 10 μg/ml for Con A). At supraoptimal concentrations, the response decreased faster for Con A than for DLA.

DLA was found to highly stimulate mouse thymocytes (T-cells), which confers to this lectin a T cell mitogenic character. This T-cell specific mitogenic capacity was confirmed by negative stimulation assays with splenocytes from congenitally athymic *nu/nu* mice (B-cells).

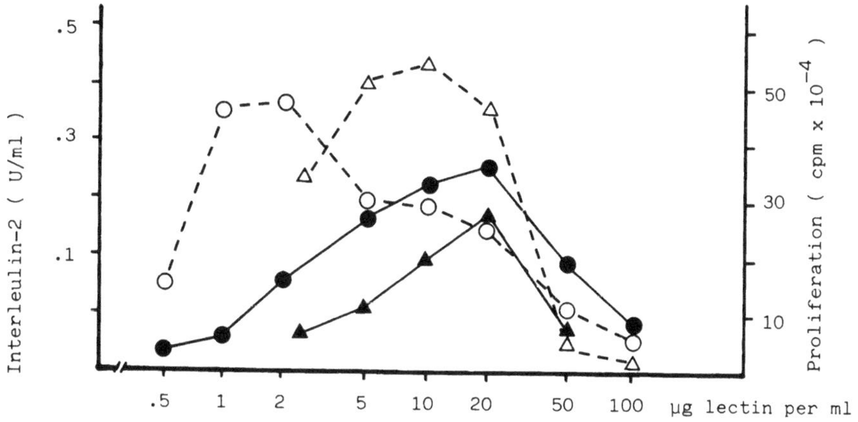

Figure 3: mitogenic response (cpm) displayed by 48 hr stimulated mouse splenocytes in the presence of various DLA (○-○) or Con A (△-△) concentrations and corresponding amounts of interleukin-2 released in the culture medium (U/ml) by these cells upon 24 hr stimulation with the same DLA (●-●) or Con A (▲-▲) concentrations.

Interleukin-2 synthesis in mouse splenocytes upon DLA stimulation: interleukin-2 discovered as a T-cell growth factor by Morgan *et al* (12) plays a central role in the phenomenon of lymphocyte proliferation. We compared the interleukin-2 amount present in supernatants of 24 hr splenocyte cultures stimulated by either DLA or Con A. These supernatants were used as conditioned media to support the growth of an interleukin-2 dependent cytotoxic T cell line.

Figure 3 shows that the interleukin-2 level in DLA stimulated cultures reached higher values when compared to Con A. Moreover the proliferation curves, also plotted in Figure 3, revealed that in DLA cultures the maximal interleukin-2 synthesis occured at doses found supraoptimal for proliferation, while in Con A cultures maximal interleukin-2 synthesis and maximal proliferation were obtained at the same lectin concentration.

Effect of DLA on the ecto-5'nucleotidase activity of mouse lymphocyte membranes: we have previously shown (13) that the glycoprotein ecto-5'nucleotidase, a lymphocyte cell surface enzyme converting 5'AMP into adenosine, is a Con A receptor; the interaction of the lectin with this enzyme through its mannose residues leads to the inhibition of its enzymatic activity.

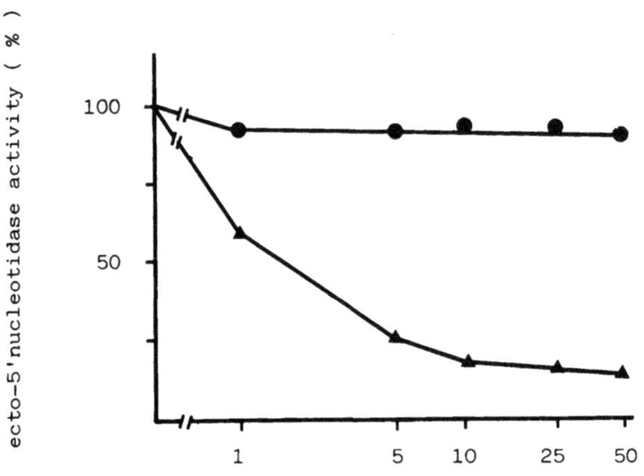

Figure 4: effect of increasing concentrations of DLA (●-●) and Con A (▲-▲) on the ecto-5'nucleotidase activity of mouse splenocyte membranes.

Further more we have established a rather good correlation between the capacity for a compound to induce lymphocyte proliferation and its ability to inhibit 5'nucleotidase activity. It was therefore hypothesized that 5'nucleotidase inhibition could be an early event in lymphocyte stimulation. Because DLA displays, like Con A, high lymphocyte mitogenic activity inhibited by αmethyl-D-mannopyranoside, we checked the effect of DLA on 5'nucleotidase activity. Figure 4 confirms that Con A inhibits 5'nucleotidase activity while DLA, even at high concentrations (50 μg/ml) has no effect. These results rule out the hypothesis of a direct relationship between lymphocyte stimulation and 5'nucleotidase inhibition.

DLA-induced stimulation of human peripheral blood lymphocytes: DLA was found highly stimulatory for human peripheral blood lymphocytes with a stimulation level as high as that obtained with Con A, for an optimal concentration similar for the two lectins (10 μg/ml), but, unlike Con A (14), DLA did not inhibit lymphocyte proliferation at higher doses, as evidenced by the high mitogenic response observed at 50 μg/ml. DLA- or Con A-induced interleukin-2 production, unlike that obtained with mouse lymphocytes, was found rather low in human peripheral blood lymphocyte cultures.

Discussion

Contrarily to Aurich *et al* (15) who reported a rather low activity of DLA and to Rao *et al* (16) who claimed that it was devoided of mitogenic and leucoagglutinating activity, we have demonstrated that the lectin isolated from the seeds of *Dolichos lablab* L. (DLA) is a potent mitogenic agent specific for T lymphocytes as shown by positive results on mouse splenocytes and thymocytes and negative ones on splenocytes from congenitally athymic nude mice. The mitogenic response is similar to that obtained with Con A and can be inhibited by 25 mM α-methyl-D-mannopyranoside. The DLA-induced mouse lymphocyte stimulation peaks at an optimal concentration lower than that obtained with Con A and decreases slower. DLA maximally stimulates human peripheral blood lymphocytes at 10 μg/ml as in the case for Con A, but we found no decrease between 10 to 50 μg/ml, unlike in Con A activation.

Interleukin-2 is released in the culture medium of DLA stimulated mouse lymphocytes in higher amounts than in Con A cultures. Moreover the maximal interleukin-2 production accurs at a DLA dose found supraoptimal for proliferation, while there is a good correlation between the optimal Con A doses for interleukin-2 release and blastogenesis. This phenomenon is similar to that reported by Ravid *et al* (17) for human peripheral blood lymphocytes which produce more interleukin-2 at supraoptimal Con A doses than at optimal mitogenic concentrations, possibly through an extensive cross-linking of cell surface sites, which may be important for interleukin-2 production (18).

We have shown that ecto-5'nucleotidase is inhibited by Con A binding and not by DLA; this result evidences the possibility for two similarly potent lymphocyte activators to act through different cell surface receptors. The characterization of DLA receptors related to its mitogenic properties, and their comparison with those of other mitogenic or non-mitogenic lectins (5) will be of great interest, making DLA an useful tool in the study of early events occuring in lymphocyte mitogenic stimulation.

References

1. Sharon, N.: Mitogens in Immunobiology (Oppenheim, J.J., Rosentreich, D.L., Eds.) Academic Press, N. Y. 1976, pp 31-41.

2. Gordon, I.L., Coluzzi, P.H., Favero, J.J., Parker, J.W., O'Brien, R.L.: Cell. Immunol. **68**, 25-37 (1982).

3. Favero, J.J., Gordon, I.L., Parker, J.W., O'Brien, R.L.: Cell. Immunol. **68**, 38-52 (1982).

4. Favero, J., Bonnafous, J.C., Dornand, J., Mani, J.C.: Cell. Immunol. **86**, 439-447 (1984).

5. Dillner-Centerlind, M.L., Axelsson, B., Hammarström, S., Hellström, U., Perlmann, P.: Eur. J.,Immunol. **10**, 434-442 (1980).

6. Laemmli, U.K.: Nature (London) **227**, 680-685 (1970).

7. Dornand, J., Favero, J., Bonnafous, J.C., Mani, J.C.: Cell. Immunol. **83**, 351-359 (1984).

8. Gillis, S., Ferm, M.M., Ou, W., Smith, K.A.: J. Immunol. **120**, 2027-2032 (1978).

9. Dornand, J., Bonnafous, J.C., Mani, J.C.: Eur.J.Biochem. **87** 459-465 (1978).

10. Güran, A., Ticha, M., Filka, K., Kocourek, J.: Biochem. J. **209**, 653-657 (1983).

11. Rao, D.N., Hariharan, K., Rao, D.R.: Lebens.-Wiss. Technol. **9**, 246-250 (1976).

12. Morgan, D.A., Ruscetti, F.W., Gallo, R.C.: Science **193**, 1007-1008 (1976).

13. Dornand, J., Bonnafous, J.C., Mani, J.C.: Biochem. Biophys. Res. Commun. **82**, 685-692 (1978)

14. Cunningham, B.A., Sela, B.A., Yahara, I., Edelman, G.M.: Mitogens in Immunobiology (Oppenheim, J.J., Rosentreich, D.L., Eds.) Academic Press, New York 1976, pp 13-30.

15. Aurich, G. von, Cerpinsky, G., Plenert, W.: Z. Med. Labotechn. **12**, 32-40 (1971).

16. Hariharan, K., Rao, D.R.: Indian J. Biochem. Biophys. **18**, 201 (1981).

17. Ravid, A., Rubin, A.L., Novogrodky, A., Stenzel, K.H.: Exp. Cell Res. **145**, 239-244 (1983).

18. Favero, J., Marti, J., Dornand, J., Bonnafous, J.C., Mani, J.C.: Lectins, Biology, Biochemistry, Clinical Biochemistry (Bøg-Hansen, T.C., Breborowicz, J., Eds.) Walter de Gruyter, Berlin 1985, pp.45-52.

ACTIVATED PNA^+ PERIPHERAL IMMUNOCYTES IN INFLAMMATORY BOWEL DISEASES

A. Raedler, R. Kelm, H.G. Thiele, H. Greten

Department of Medicine, University of Hamburg, FRG

The functional properties of PNA binding sites on lymphocytes are a matter of controversy. It has been proposed that PNA receptors may represent markers for immaturity (1,2) and indeed it is well established that immature T cells (1,3,4) as well as pre-B cells (5-7) express binding sites for the PNA lectin. The use of PNA receptors as markers of immaturity was put into question by the observation that certain mature cells can display also PNA binding cell surface determinants (8).

Studying activated $T9^+$, Tac^+ T cells in inflammatory bowel disease as well as other auto-immune disorders (9,10), we observed that PNA binding molecules are coexpressed on activated T cells. In the following, we present data in regard to the phenotypic characteristics of PNA^+ peripheral T and B cells activated in vivo and in vitro.

Lectins, Vol. V

Material and Methods

Preparation of lymphocytes. Patients with Crohn's disease (CDAI 150) (n=15), active ulcerative colitis (n=10), Septemic lupus erythematodes (n=5), immune-vasculitis (n=5), sarcoidosis (n=5) and normal controls (n=15) were taken samples of lo ml of blood and added to 10 ml of 0.9% saline supplemented with 2000 U of heparin. Two and one-half milliliters of 'Lymphoprep' (Nyegaard) was overlayered with 5.0 ml of the diluted blood and centrifugated for 30 min (20°C, 560 g). The lymphocytes were recovered from the interphase and washed twice with phosphate-buffered saline (Gibco Laboratories, St. Lawrence, Mass.). The pellet was suspended in 15ml of cell culture medium (RPMI 1640, Gibco) supplemented with 10% fetal calf serum (Medac), and cultured for 2 h at 37°C for removal of monocytes. Nonadherent cells were then collected and their number was determined.

Enumeration of $T9^+$, Tac^+ and PNA^+ cells. For double labelling and cocapping experiments 1x10 cells were incubated in 5 µl OKT9/TRITC (Ortho-mune, AB against the transferrin receptor = activated cells), or anti-Tac/TRITC (Becton, Dickinson, AB against the Il-2 receptor = activated cells) and simultanously in PNA/FITC (Medac, 1 mg/ml, diluted 1:10). Cells were incubated for 30 min at 4°C and washed three times in culture medium. Cells were then resuspended, examined, and counted (2-3x 100 cells) under a fluorescence microscope (Carl Zeiss). For inhibition experiments cells were incubated in OKT9 or anti-Tac and then labelled with PNA/FITC or vice versa as described above.

Enumeration of sIg^+ B cells and PNA^+ cells. $1x10^6$ cells were incubated in PNA/FITC as described and simultanously reacted with goat-anti-human Ig $(Fab)_2$ (1:20, Tago) or anti-Leu12 (1:10, Becton, Dickinson). TRITC-coupled goat $(Fab)_2$ anti-mouse IgG (Tago, diluted 1:60) were used as second antibody.

Phenotypic characterization of PNA^+, sIg- T cells. After removal of monocytes cells were layered on petri dishes (Ø=5.5 cm, $5x10^6$ cells per plate) coated with goat antihuman Ig (4 ml, 25 µl/ml, Tago). After incubation for 3o min at room temperature, non-adherent cells were recovered and $1x10^6$ cells incubated in PNA/FITC and simultanously reacted with TRITC-coupled OKT3,4 and 8 (Orthomune, 5 µl, diluted 1:10, AB against pan-T-cell marker, T-helper and to T-suppressor determinants), anti-Leu 7 (Becton, diluted 1:10, AB recognizing natural killer cells) goat-anti-hu-IgA (Medac, FRG, diluted 1:50), after incubating the pellet in heat-aggregated hu-IgA (Behring, FRG).

Lectin mediated agglutination experiments. $5x10^6$ non-adherent, sIg cells were layered on petri dishes (Ø= 5.5 cm) coated with PNA (4ml, 0.001%, 12h, 4°C). After incubation for 30 min, at room temperature non-adherent cells were recovered. Adherent cells were harvested after supplementing D-galactose (0.2mmol, 10 min, 4°C).
PNA^+ and PNA^- cells were then analysed for lectin-mediated panning using taxonolectin assays (Medac). The following lectins were used: CONA (CONCANAVALIN A), TKA (TRICHOSANTHES KINLOWII), PNA (ARACHIS HYPOGAEA), DBA (DOLICHOS BIFLORUS), STA (SOLANUM

TUBEROSUM), LFA (LIMAX FLAVUS), UEA I (ULEX EUROPAEUS I), LCL (LENS CULINARIS), GSA I (GRIFFONIA SIMPLICIFOLIA I), SBA (GLYCINE MAX), GSA II (GRIFFONIA SIMPLICIFOLIA II), WGA (TRITICUM VULGARIS), PVL (PHASEOLUS VULGARIS), LOTUS (LOTUS TETRAGONOLOBUS).

Pokeweed mitogen (PWM) driven T cell culture. $1x10^6$ non-adherent sIg^- cells were cultured for 8d (RPMI 1640, Gibco, plus 10% fetal calf serum and 50 µg/ml PWM, Gibco and antibiotics). Cells were then harvested and analysed for PNA binding sites and and T9 antigens as described above.

Controls of labelling as well as panning procedures were performed as described elsewhere (9).

Fluorescence Cell-Sorter Analysis. After removal of glassadherent and (sIg^+) B-cells, the remaining lymphocytes were labelled with PNA/FITC. Cells were measured for fluorescence and small-angle scatter (cell size) with a FACS IV cell sorter (Becton, Dickinson). The blue laser line (488 nm, 0.2 W) was used for excitation. The green fluorescence was measured using the filter combination KP 650 and LP 530 and stored correlated with the scatter signal in a 4x64 channel histogram.

R e s u l t s

Comparison of $T9^+$, Tac^+ and PNA^+ peripheral lymphocytes.

Patients with active Crohn's disease (Crohn's disease activity index 150) exhibit $T9^8$, Tac^+ T cells in about 19% (S.D. 5,1). The vast majority of these cells coexpress T9 and PNA binding sites (Fig.1). **A minority of these cells express PNA binding** sites or the T9 antigen only. A minor subfraction of PNA^+ peripheral immunocytes express surface Ig and is thus thought to immunocytes belong to the B cell lineage. The majority of $T9^+$ PNA^+ express the T3 antigen and thus belong to the T cell lineage. Only a small subfraction of $T9^+$ PNA^+ cells display surface Ig (or Leu12). Activated peripheral lymphocytes from patients with ulcerative colitis or other auto-immune disorders like systemic lupus erythermatodes, immune-vasculitis, sarcoidosis show a similar pattern concerning the distribution of activation associated antigens and PNA binding sites.

Inhibition and cocapping experiments using OKT9, anti-Tac and PNA. Labelling of activated lymphocytes with OKT9 or anti-Tac cannot be prevented by PNA. Vice versa preincubation of activated lymphocytes with OKT9 or anti-Tac does not inhibit binding of PNA. Cocapping experiments reveal that OKT 9 and PNA cap separately as do anti-Tac and PNA.

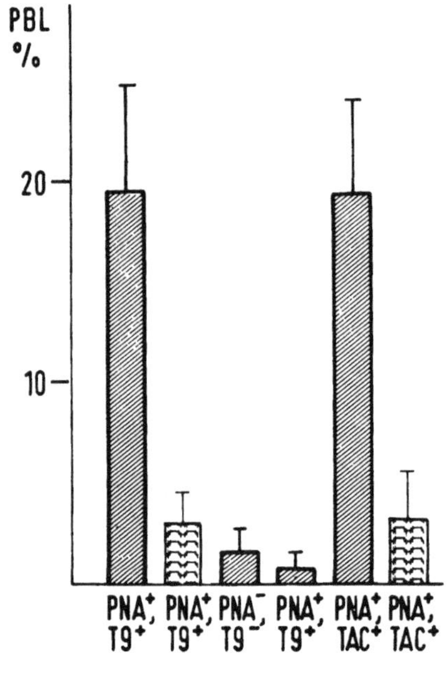

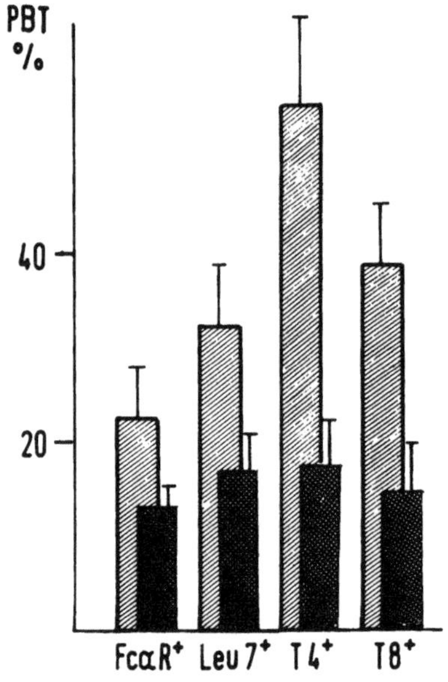

Fig.1: (left) Expression of PNA binding sites TAC and T9 on peripheral blood lymphocytes from patients with Crohn's disease and normal controls

Fig.2: (right) Subsets of peripheral T cells (PBT) coexpressing PNA binding sites from patients with Crohn's disease. FcαR = receptors for IgA = T helper and suppressor cells for IgA - B cells, Leu7 = natural killer cells, T4 = helper cells, T8 = suppressor T cells. cells coexpressing PNA receptors.

Phenotypic characterization of PNA^+ peripheral T cells. After removal of monocytes and B cells the remaining T cells were analysed in double labelling experiments using preincubation of the cells in heat-aggregated human IgA. It was found that in patients with inflammatory bowel disease 6% of the $Fc\alpha R^+$ T cells were also positive for PNA (Fig.2). The total amount of $Fc\alpha R^+$ cells was 22% (S.D. 5.7). $Leu7^+$ T cells in patients with inflammatory bowel disease amounted 33% (S.D. 6.5), 56% of these cells additionally react with PNA. In patients with other autoimmune disorders the fraction of T cells coexpressing PNA and $Fc\alpha R$ was significantly lower (11%, S.D. 4.4). The subset of $T4^+$ cells in these experiments was about 58% (S.D. 8.2) whereas 4o% (S.D. 6.5) express the T8 determinant. 3o% of the $T4^+$ and 37% of the $T8^+$ cells express PNA binding molecules.

FACS-analysis of PNA positive T cells. These experiments show, that T cells exhibiting binding sites for PNA are of medium size to large size. There is no exact limit between medium or large sized T cells (Fig.3).

Lectin-mediated agglutination of PNA^+ T cells. PNA^+ T cells were prepared by PNA-mediated panning after removal of monocytes and B cells. It was shown that PNA^+ T cells from patients with inflammatory bowel disease can be agglutinated by TKA, STA, LFA and PHA but not by ConA, LcH, GSA-I and II, WGA, UEA-I and HPA. PNA^- cells from patients with inflammatory bowel disease and PNA^+ T cells from patients with other diseases show an identical behaviour towards lectin mediated agglutination.

In vitro generation of activated T cells. Peripheral T cells from normal controls were cultured in the presence of pokeweed mitogen or interleukin 2 and the cells analysed for the expression of T9 and PNA binding sites (Fig.4). It could be demonstrated in double labelling experiments that PNA and T9 positive cells appeared after 2d. After 8d almost 35% (S.D. 4.6) of the cultured cells react with OKT 9 and PNA.

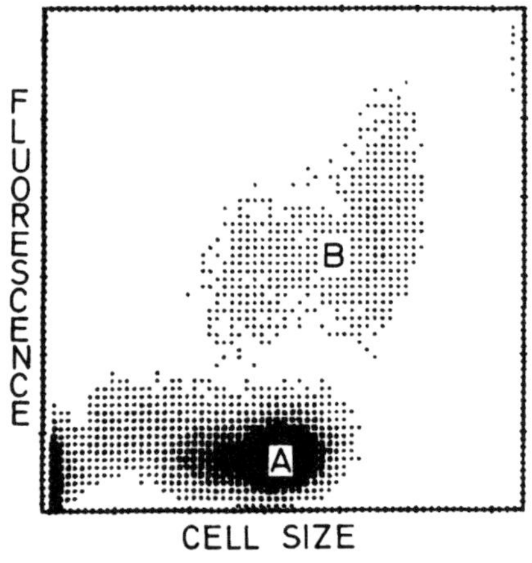

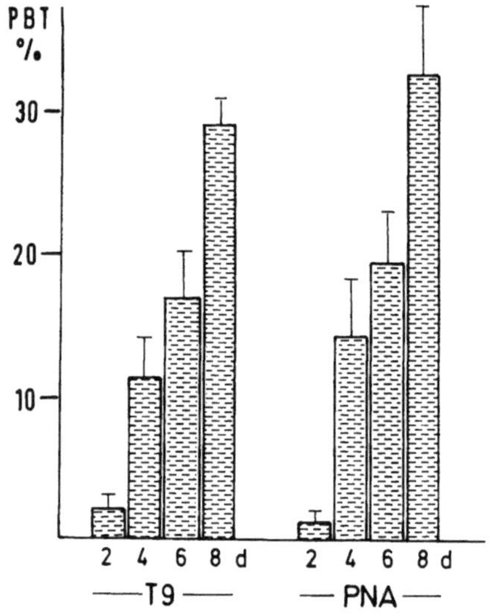

Fig.3: (left) Non-adherent, sIg-(T) cells labelled with FITC-PNA simultaneously measured for fluorescence intensity and light scatter (cell size) by means of flow cytometry. A = population of FITC-PNA positive medium size to large T cells, B = autofluorescent T cells.

Fig.4: (right) Percentage of T9⁺ and PNA⁺ peripheral T cells (PBT) in PWM-stimulated - T cell cultures of normal controls.

Discussion

The first observation that stressed the importance of PNA binding molecules on lymphocytes was achieved by agglutination experiments (1,2). Histochemical analysis revealed that the thymus can be separated into two functional different compartements i.e. the thymic cortex populated by PNA^+ thymocytes and the thymic medulla where mainly PNA^- thymocytes reside (1,3,5). Cortical PNA^+ thymocytes are found to be functionally immature, not to express MHC gene products and to be steroid-sensitive, whereas medullary PNA^- thymocytes proved to possess the opposite properties. Moreover it was corroborated that PNA can also be found to react with a subfraction of murine and human B cells (5-7). Germinal centers within stimulated lymph nodes display large quantities of PNA binding sites. Ultrastructurally these determinants could be localized on the outer cell surface of centrocytes and centroblasts (6). Additionally PNA binding molecules could be detected on certain B or T cell lymphomas (9-13). Recently pre-B cells with cytoplasmatic μ chains have been shown to bind PNA whereas more mature B cells do not (7).

From these experiments was deduced that the expression of PNA receptors is correlated with certain steps during the course of maturation. However this was jeopardized by the observation that PNA binding structures are expressed on mature cells i.e. on certain smooth muscle cells and epithelial cells of the gut (14). The experiments presented here revealed that activated T cells and less frequently B cells coexpress PNA receptors and the activation associated antigens T9 and Tac. It is further shown by inhibition as well as cocapping experiments that PNA binding molecules are not displayed on the same cell surface structures as the T9 or the Tac antigen.

Generation of activated T cells in vitro using pokeweed mitogen or interleukin 2 additionally demonstrates that expression of T9 antigens coincides temporally with the expression of PNA binding sites (Fig.4). On the other hand the population of $T9^+$ cells and of PNA^+ cells in patients with inflammatory bowel disease or other auto-immune diseases is almost identical in regard to phenotypical characteristics, i.e. both PNA^+ and $T9^+$ subsets contain Leu7, $Fc\alpha R^+$, T4 and T8 positive cells in nearly the same ratio. However there is a significant difference between the number of PNA^+, $Fc\alpha R^+$ T cells between inflammatory bowel disease and other auto-immune disorders.

Thus we have shown that PNA^+ T cells are not immature but equipped with functionally important cell surface determinants that point to functional maturity. Furthermore, expression of PNA binding sites is not correlated with immaturity even in the immune system. The data presented here suggest that at least in the human lymphatic system the expression of PNA binding sites seem to be correlated to functional activity rather than to immaturity. This is further stressed by the fact, that PNA binding sites are coexpressed with the T9 antigen supposed to be identical with the transferrin receptor. This determinant is exhibited particularly on proliferative cells (17,18). Furthermore functional and/or mitotic activation can explain the presence of PNA receptors on mature, activation-associated antigens expressing peripheral lymphocytes as well as the display of PNA binding sites on germinal center cells and cortical thymocytes. Whether this concept is true also for other cell systems remains to be investigated.

Summary

Starting from the observation that PNA binds to immature cortical thymocytes and germinal center cells, pointing to a correlation between PNA binding and immaturity at least in the immune system, peripheral human lymphocytes were analysed for the expression of PNA binding sites.

It was found that normal controls only exhibit a low number of PNA^+ lymphocytes. In patients with auto-immune diseases, in particular Crohn's disease and ulcerative colitis, exhibiting an enlarged subset of activated peripheral immunocytes, the number of PNA^+ cells is high. PNA binding sites and the activation-associated antigens T9 and Tac are shown to be co-expressed on activated T cells and on a minor fraction of B cells. However the PNA binding molecule is not identical with the cell surface structures displaying Tac or T9 as shown in inhibition and cocapping experiments. PNA^+ cells exhibit epitopes that are correlated to functional maturity i.e. T4, T8, T3 and Leu7 in the same relation as $T9^+$ T cells.

It is concluded that the data available are in accordance with the hypothesis that PNA receptor expression is a marker of functional and/or mitotic activation.

References

1. London J, Barrih S, Bach JF (1978), J Immunol 121: 438-443

2. Reisner Y, Biniaminov M, Rosenthal E, Sharon N, Ramot B (1979), Proc Natl Acad Sci USA 76: 447-451

3. Reisner Y, Linker-Israeli M, Sharon N (1976), Cell Immunol 25: 129-134

4. Irlé C, Piquet PF, Vassalli P (1978), J Exp Med 148: 32-45

5. Rose ML, Birbeck MSC, Wallis VJ, Forrester JA, Davies AJS (198O), Nature 284: 364-366

6. Raedler A, Raedler E, Arndt R, Thiele HG (1981), Immunology Letters 2: 335-338

7. Osmond DG, Melchers F, Paige CJ (1984), J Immunol 133, 1: 86-91

8. Rose ML (1982), Immunol today 3: 6-7

9. Raedler A, Fraenkel S, Klose G, Seyfarth K, Thiele HG (1985), Gastroenterol 88: 978-983

lo. Raedler A, Fraenkel S, Klose G, Thiele HG (1985), Clin exp Immunol 61, in press

11. Reisner Y, Biniaminov M, Rosenthal E, Sharon N, Ramot B (1979), Proc Natl Acad Sci USA 76: 447

12. Levin S, Russel EC, Blanchard D, McWilliams NB, Maurer HM, Mohanakumar T (198o), Blood 55: 37

13. Rose ML, Habeshaw JA, Kennedy R, Sloane J, Wiltshaw E, Davies AJS (1981), Brit J Cancer 44: 68

14. London J, Perrot JY, Berrih S, Laroche L, Niaudet P (1979), Scand J Immunol 9: 451

15. Galili U, Galili R, Polliack A (1981), Clin Exp Immunol 43:311

16. Watanabe K, Muramatsu T, Shirane H, Ugai K (1981), J Histochem Cytochem 29: 779

17. Sutherland R, Delia D, Schneider C, Newmann R, Kemshed J, Graeves J (1981), Immunology 78: 4515

THE CON A CONJUGATE OF BOWMAN-BIRK SOYBEAN TRYPSIN INHIBITOR IS AN ANTICARCINOGEN.

Jung-Yaw Lin and & Mon-Hwei Hu

Institute of Biochemistry, College of Medicine, National Taiwan University, Taipei Taiwan, R.O.C.

It has been shown by many investigators that protease inhibitors suppress the growth of tumor (1-6). The fundamental principle of the antitumor properties of protease inhibitors could be due to the inhibition of protease activities of tumor cells which play an important role in the tumorigenicity of transformed cells (8-10).

Recently, the target delivery technique has been employed to enhance the antitumor activity of drugs, and one sucessful example is the application of an immunotoxin for specific inhibition of the growth of tumor cells (7). A chain fo phytotoxin, abrin or ricin, or diphtheria toxin was covalently linked to an antibody against the tumor-associated antigen to form an immunotoxin which brought the toxin to the target-tumor cells but not normal cells.

The purpose of present investigation was to enhance the antitumor activity of the Bowman-Birk trypsin inhibitor (BBTI) through conjugation to Con A. The physical and chemical character, and the antitumor activity of the conjugate was studied, and the medhanism of the antitumor activity was demonstrated at the molecular level.

Lectins, Vol. V

MATERIALS AND METHODS:

N-succinimidyl-3-(2-puridyl dithio) propionate, SPDP, Sepharose 6B-CL, and Sephadex G-100, were purchased from Pharmacia Fine Chemicals. ^{125}I (17 Ci/mg) was obtained from New England Nuclear. The Bowman-Birk trypsin inhibitor (BBTI), trypsin, bovine serum albumin (BSA), and lactoperoxidase were purchased from Sigma. BAPNA was obtained from E. Merck, Germany. Dulbecco's minimum essential medium, DMEN, and fetal bovine serum were the products of GIBCO. *Canavalia ensiformis* was purchased from a private source in Taiwan.

Isolation of Con A. Affinity column chromatography was employed for isolation of Con A according to (16).

Preparation of the BBTI-Con A Conjugate. The conjugate was synthesized as described previously with several modifications (2). The 2-pyridyl disulfide derivative of BBTI, PDP-BBTI, was prepared as follows: BBTI (4mg) was dissolved in 0.1M NaCl and 0.1M phosphate buffer, pH 8.0, and a 25 fold molar excess of SPDP was added to the reaction mixture (13). After the reaction was kept at 25°C for 30 min, the reaction was terminated by applying the reaction mixture on a Biogel P.2 column to remove the excess reagent and N-hydroxysuccininmide. PDP-Con A was synthesized as described above except that 0.1 M glucose was included in the reaction mixture. PDP-Con A was converted into thiolated Con A by reduction with 50 mM DTT in PBS at 4°C for 36 hours. BBTI-Con A was prepared by addition of 11 mg PDP-BBTI to 13 mg thiolated Con A, in the presence 0.1 M glucose. The reaction was carried out at 4°C for 24 hours. The product, BBTI-Con A, was purified by gel filtration with a Sephadex 6B-CL column.

Hemagglutination. 0.1 ml of 2% washed red blood cells of NCI mice in 0.1 M phosphate buffer, pH 7.4 containing 0.14 M NaCl was used with 0.1 ml of conjugates at 28°C for 2 h (19).

Evaluation of tumor growth. Male NCI mice of 20$\pm$ 1 g body weight were used for this purpose. 0.4 ml of ascites fluid containing 4x10 cells was injected i.p. for tumor inoculation. A single dose of conjugates, BBTI-Con A, BSA-Con A or BSA-BBTI was injected i.p. on the day of tumor inoculation. The growth of sarcoma-180 tumor cells was evaluated by measuring the increase of life span. The dose of the mixture of BBTI and Con A was equivalent to the individual amount of BBTI and Con A present in BBTI-Con A conjugate.

Assay of trypsin inhibitor. The trypsin inhibitory activity was determined by using 2-N-benzoyl-D-L-arginine-D-nitroanilide, BAPNA, or casein as substrate (17).

Iodination of protein. 250 µg of protein in 0.05 M phosphate buffer pH 7.4 containing 0.15 N NaCl was added to 1.6 mg lactoperoxidase, 16 µg NaI and 250 µci of Na^{125}I. 2 µl og 8.5 mM hydrogen peroxide was added to the reaction mixture, and the reaction mixture was kept at 25°C for 30 min (18). The iodinated protein was isolated by gel filtration with a Biogel P-2 column prequlibrated and wluted with 0.01 M PBS (ph 7.4) containing 15 µg BSA/ml buffer.

SDS-PAGE. Slab gels (14x17.5 cm) were used at 40 mA according to (14). After electrophoresis, the gel was stained with 5.5% Coomassie brilliant blue.

Stability of ^{125}I-BBTI-Con A in the Cells. To $1x10^7$ sarcoma cells (in BSA-coated Eppendorff tubes) $3x10^5$ cpm of ^{125}I-BBTI-Con A was added, and the reaction mixtures were kept at 37° for various periodes of time or at 0°C for 2 hours. The reaction was terminated by centrifugation at 8000 rpm for 2 min. The cells were washed with HEPES buffer containing 0.2 M glucose, BSA (1 mg/ml) 2mM $CaCl_2$ and 0.5 µg *Streptomyces griseus* protease type XIV thrice to remove BBTI-Con A on the cell membrane. The cells were finally washed with HEPES buffer containing BSA once. The washed cells were lysed by freezing and thawing. 75 µl of combined extracts was subjected to SDS-PAGE analyses, and autoradiograms were performed with intensifying screens at 70°C for 7 days.

Effects of Conjugates on the Growth of Cell Culture. LTK⁻, Sarcoma 180 and NIH 3T3 cells were used this purpose. The cell cultures were treated with 1 ml of 0.25% trypsin at 37°C for 3 min. After the addition of 10 ml of Dulbecco's modified Eagle's medium, they were separated by centrifugation. 0.025 ml of Dulbecco's modified Eagle's medium containing 500 S-180 cells, 100 LTK⁻ cell and 100 3T3 cells was added to 24 well plates, and the cells were incubated in a CO_2 incubator at 37°C for 24 hr. 0.1 ml of various amounts of conjugates was placed in each well and the treated cells were incubated at 37°C in a CO_2 incubator for one week. After removing the incubation medium, the surival cells were stained with 0.125% Coomassie blue in 45% methanol and 9 % acetic acid.

RESULTS:

Physicochemical and Biological activities of the BBTI-Con A Conjugate.

The molecular weight of our Con A-conjugate with BBTI was measured by SDS-PAGE and Gel filtration with a Sepharose 6B CL-column, to be 200.000 ± 2000, probably consisting of four subunits, each of BBTI and Con A as shown by scanning of the polyacrylamide gels containing reduced conjugates and a mixture of Con A and BBTI at a molar ratio of 1:1. The BBTI-Con A had the same hemagglutinating activity as the native Con A and 42.7% of its inhibitory activity towards trypsin.

Stability of the BBTI-Con A Conjugate to Tryptic Digestion:

By using ^{125}I-labelled proteins we could demonstrate that BBTI-Con A was more resistant to tryptic digestion than a mixture of the BBTI and Con A. 80% of the conjugate was bound to tumor cells after treatment at 37°C for one hour with trypsin while only 25% of the free form of BBTI and Con A was bound on the tumor cells, Fig. 1.

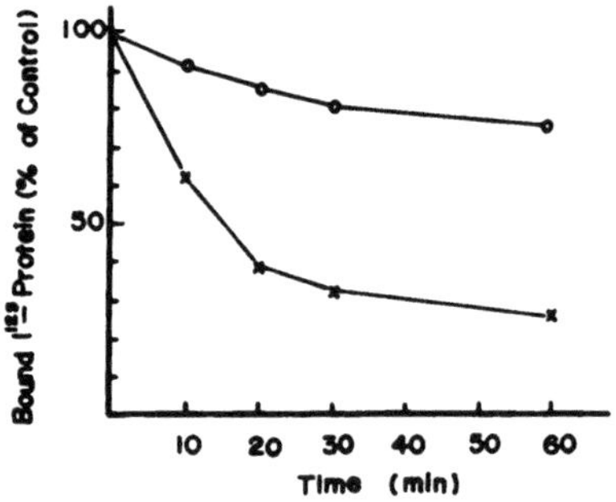

Fig. 1. Effect of tryptic digestion on the binding of Bowman-Birk trypsin inhibitor, BBTI-Con A conjugate and their components. 5 µg of ^{125}I-BBTI-Con A (5x10^4 cpm) or 2.8 µg of ^{125}I-Con A (3x10^4 cpm) and 2.2 µg BBTI in 0.1 ml of 0.1 M phosphate buffer saline, pH 8.0, were treated with 12.5 µg of trypsin in 25 µl at 37°C for various periods of time. The reaction was terminated by addition of 50 µg BBTI-Con A conjugate (0) and a mixture of BBTI and Con A, molar ratio 1:1 (x). Results are averages of two separate experiments.

Specific Binding of the BBTI-Con A Conjugate to Tumor Cells:

^{131}I-BBTI-BSA had no specific binding to sarcoma-180 cells, but ^{131}I-BBTI-Con A conjugate had, Fig. 2. The free form of ^{131}I-BBTI did not bind to the cells, and the addition of Con A did not increase the binding significantly. When the ^{131}I-BBTI-Con A conjugate was reduced with DDT, the product lost its specific binding activity.

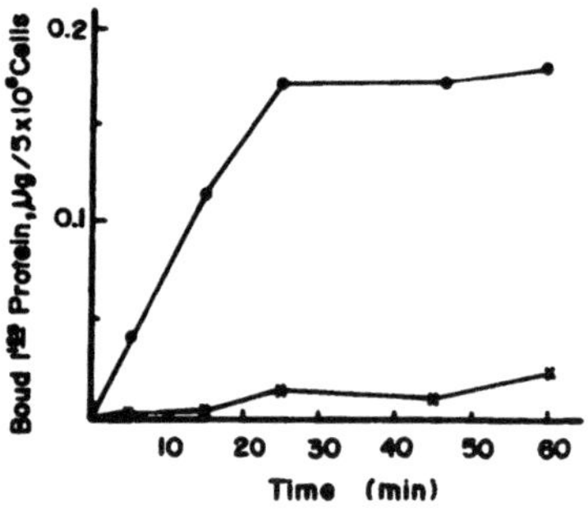

Fig. 2. Binding of BBTI-Con A conjugate and its components on Sarcoma-180 cells. 1 µg ^{125}I-BBTI-Con A conjugate (●) or a mixture of ^{125}I-BBTI and Con A, molar ratio, 1:1 (x) was added to 5x10^6 sarcoma 180 cells. Experiments were carried out as noted in "Materials & Methods.

TABLE 1

The effect of Con A-BBTI conjugate on the growth of Sarcoma-180 cells in ascites.

Proteins	Dose (ug/mouse)	Number of mice	Survival Time (days)	%increase in life span
Con A-BBTI	300	15	16.5±1.4	12
	500	15	21.3±2.6	45
	300+500[a]	5	25.6±1.1	75
Con A+BBTI	165+135[a]	15	14.3±1.7	*
	275+225[a]	15	15.3±1.2	*
BSA-BBTI	500	5	14.0±1.4	*
	910	5	14.0±0.7	*
BSA-Con A	500	5	14.6±1.1	*
	980	5	15.2±0.8	*
Control	0	15	14.7±1.2	

10^7 Sarcroma-180 cells were transplanted per mouse. After one hour a dose of protein was injected i.p. a=a second dose was injected on the third day. * = The results have no significant difference compared with control by t-test ($p>0.05$). Survival time was expressed as mean±S.D.

The BBTI-Con A conjugate not only bound to the tumor cells but also entered into the cells as measured by the intracellular trypsin inhibitory activity. It took about 20 min for the BBTI-Con A conjugate to get into cells, whereas the free form of BBTI and Con A did not increase the intracellular trypsin inhibitory activity. The uptake of the BBTI-Con A conjugate was temperature dependent, Fig. 3. By using an autoradiogram technique, we demonstrated that the ^{125}I-BBTI-Con A conjugate entered into the tumor cells at 37°C but not at 4°C, and that the conjugate was not reduced or degraded by incubation at 37°C for 60 min.

Effects of the BBTI-Con A Conjugate on Culture Cells:

By using S-180, LKT^- and NIH3T3 cells, we could show that the BBTI-Con A conjugate (4 µg/ml) inhibited 95% the growth of the cell cultures whereas a mixture of the free form of BBTI and Con A (at the same concentration) inhibited about 35% Fig. 4.

Antitumor Activity of the BBTI-Con A Conjugate:

The control group of mice bearing Sarcoma-180 tumor cells died after 14.7 ± 1.2 days. The mice treated with 500 µg of BBTI-Con A conjugate showed 45% increaese of life span while the mice receiving 275 of µg Con A plus 225 µg of BBTI did not show any significant difference from the control group (Table 1). Likewise did neither BBTI-BSA nor the BSA-Con A conjugate exhibit any significant difference.

DISCUSSION:

Using immunotoxins for cancer chemotherapy is a newly developed approach which has been demonstrated *in vitro* by many investigators (14).

The antitumor activity of protease inhibitors is well documented. Inhibition of tumorigenesis in mouse skin (11,12), lowering of breast tumor in irradiated rats (13), and suppression of chemical transformation in $C_3H/10T1/2$ cells (6). The molecular mechanism of antitumor activity of a protease inhibitor may be that it inhibits post-translational processing which is one of the processes for cell transformation, and that it suppresses modification of cell surface structures by proteases, which is involved in contact inhibition of normal cells.

Our present investigation is designed and based on these observations. BBTI and Con A was covalently linked to form a conjugate which was demonstrated to be an effective antitumor agent. We found the conjugate to be very stable probably due to the protection by Con A. We suggest that if immunotoxins contain an antibody against a tumor-associated antigen, toxins and proteases inhibitors will be new and ideal antitumor agents. Furthermore we suggest that a conjugate may be more effective against tumor growth than an immunotoxin without

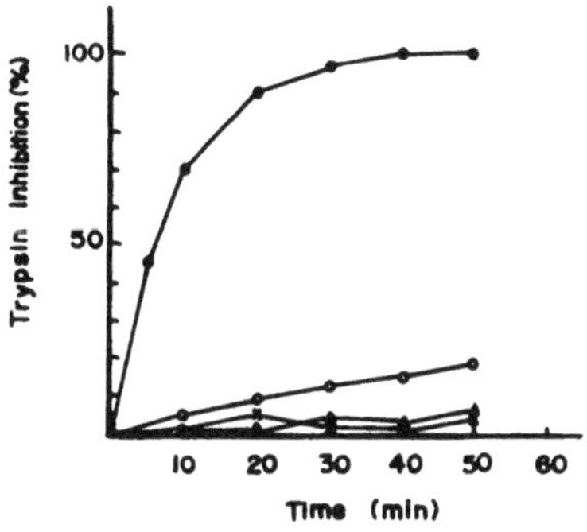

Fig. 3. Uptake of BBTI-Con A conjugate by Sarcoma-180 cells determined by trypsin inhibitor activity. 300 μg of BBTI-Con A conjugate (●), the mixture of BBTI and Con A, (molar ratio 1:1 (▲), or 546 μg BBTI-BSA (O) each in 0.4 ml phosphate buffer saline, 0.1 M, pH 8.0, was mixed with 0.4 ml of 1x10^7 Sarcoma-180 cells at 37°C. BBTI-Con A at 4°C (x).

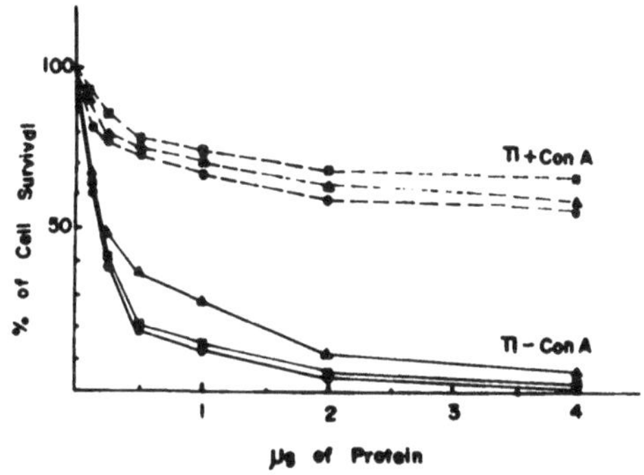

Fig. 4. Effect of BBTI-Con A conjugate and its components on the growth of cell cultures. Various amounts of BBTI-Con A conjugate or the mixture of BBTI and Con A (molar ratio 1:1) was added to the cell cultures, and the experiments were performed as in Materials Methods: ■ = Sarcoma 180 cells; ▲ =LTK cells; ● = NIH 3T3 cells.

a protease inhibitor because the protease inhibitor not only makes the conjugate stable against the proteases *in vitro*, but also acts synergestically with the toxin to inhibit the growth of a tumor.

REFERENCES:

1. Borek, C., Miller, R., Pain, C. and Troll, W.: Proc. Natl. Sci. USA *76*, 1800-1803 (1979).

2. Carlesson, J., Drevin, H. and Axen, R.: Biochem. J. *173*, 723-737 (1978).

3. Hozumi, M., Ogawa, M., Sugimura, T., Takuchi, T. and Umezawa, H.: Cancer res. *32*, 1725-1728 (1972).

4. Kennedy, A.R. and Little, J.B.: Nature *276*, 825-828 (1978).

5. Kennedy, A.R. and Little, J.B.: Cancer Res *41*, 2103-2108 (1981).

6. Kuroki, J. and Drevon, C.: Cancer Res. *39*, 2755-2761 (1979).

7. Olsnes, S. and Pihhl, A.: Pharmacol ther. *15*, 355-381 (1982).

8. Ossewski, L., Quigley, J.P., and Reich, E. In Reich, E., Rifkin, D. and Shaw, E. eds. Protease and biological control. Vol. 2. Cold Spring Harbor, NY: Cold Spring Harbor Laboratory 901-903 (1975)

9. Stranli, P., Barrett, A.J. and Baici, A., eds. Protease and tumor invasion. In: Monograph series of the European organization for research and treatment of cancer. Vol. 6 New York: Raven Press (1980).

10. Troll, W., Belman, S., Wiesner, R. and Shellabarger, C. In: Jolzer, H. and Tschoche, J. eds. Biological function of proteases. Berlin Springer-Verlag, 165-170 (1979).

11. Troll, W., Klassen, A. and Janoff, A. Science *169*, 1211-1213 (1972).

12. Troll, W., Wiesner, R., Belman, S. and Schellabarger, C.J. Proc. Am. Assoc. Cancer Res *20*, 265 (1979).

13. Troll, M., Wiesner, R., Schellabarger, C.L., Holtzman, S. and Stones, J.P. Carcinogeneses (Lond) *1*, 469-472 (1980).

14. Weber, K. and Osborne, M. J. Biol. Chem. *244*, 4406 (1969).

15. Yavelow, J., Finlay, T.H., Kennedy, A.R. and Troll, W. Cancer Res. *43*, 2454s-2459s (1983).

16. Agrawal, B., Goldstein, I.J. Bioc.Biophys.Acta 147, 262-268 (1976).

17. Laskoski, M., Kato, I. Ann.Rev.Biochem. 49, 593-626 (1980).

18. Marchalonis,J.J. Biochem.J. 113, 299-301 (1954).

19. Stavisky, A.B. J.Immunol. 72, 36o-370 (1954).

PART 6

HISTOCHEMISTRY WITH LECTINS

LIGHT AND ELECTRON MICROSCOPIC LOCALIZATION OF CELLULAR GLYCOCONJUGATES WITH LECTIN-GOLD COMPLEXES: DEMONSTRATION OF GLYCOCALYX SPECIALIZATION AND INTRACELLULAR GLYCOSYLATION COMPARTMENTS

Jürgen Roth

Interdepartmental Electron Microscopy, Biocenter, University of Basle, CH-4056 Basle, Switzerland

Lectins are multivalent carbohydrate binding proteins which have been extensively used in the various fields of cell biology research (24, 27). Because of their ability to recognize sugar residues they have been widely applied in the purification of polysaccharides and glycoproteins and examination of certain aspects of cell surface structure and function. At the same time lectins have found particular interest in carbohydrate cytochemistry and their validity as probes for in situ visualization of sugar sequences in oligosaccharides of glycoconjugates in cells and tissues has been unequivocally demonstrated (24, 27). The binding of lectins to sugar residues can be detected microscopically supposed appropriate markers are used. These are fluorescent dyes and peroxidase for light microscopy and ferritin, peroxidase, hemocyanin, latex beads and polysaccharide-iron complexes for electron microscopy (24, 27). Particles of colloidal gold have proven to be useful markers in cytochemistry (29) and can be used for labelling of lectins. Earlier studies with lectin-gold complexes have focussed on the detection of cell surface binding sites and their dynamic state (18, 19, 33-35, 46, 47). More recent applications were aimed at the visualization of sugar residues in intact tissues and inside cells. This goal can be easily achieved in light microscopy (21, 22, 30) but is a much more difficult task in electron microscopy (3, 12, 15, 26, 42, 48). However, recent progress in the tissue processing by the

Lectins, Vol. V

low temperature embedding technique with Lowicryl K4M (6, 38) has helped to overcome the main difficulties such as limited penetrability and accessibility (31).

In the following a brief review is given dealing with the use of lectin-gold complexes to detect cellular heterogeneity in plasma-membrane binding, regional specializations of the plasma membrane in a given cell type, and sites of intracellular glycosylation steps. Limitations in space are my excure to report mainly about our own work in which the investigations from other groups are properly mentioned and discussed.

Methodical Aspects

The preparation of particles of colloidal gold and the conditions for complex formation with lectins are described in great detail elsewhere (22, 29, 30, 31). It is important to note that lectin-gold complexes are very stable. Equally important is the fact that lectins when bound at the surface of particles of colloidal gold fully retain their specific sugar binding properties for long periods of time. Lectin-gold complexes are easily applied in cytochemical one-step techniques. But certain lectins pose problems for complex formation with gold particles and, as an alternative, cytochemical affinity (two-step) techniques (as inaugurated by Bernhard and Avrameas (2) with the use of glycoprotein-gold complexes have been introduced. The latter include horseradish peroxidase-gold for Concanavalin A, ovomuccoid-gold for wheat germ agglutinin and fetuin-gold for Limax flavus lectin (40).

The application of these reagents in light microscopy is possible on routinely fixed and processed tissues. Fixation with (para-) formaldehyde, glutaraldehyde or mixtures of both aldehydes is recommended since nonspecific binding reactions have been observed with other fixatives such ad Bouin's or Carnoy's fluid (30). Such fixation protocols have the advantage that the tissue can be processed for electron microscopy as well. Incubation can be performed on frozen sections or sections from paraffin-embedded tissues (Fig. 1) although the section thickness is a limiting factor for

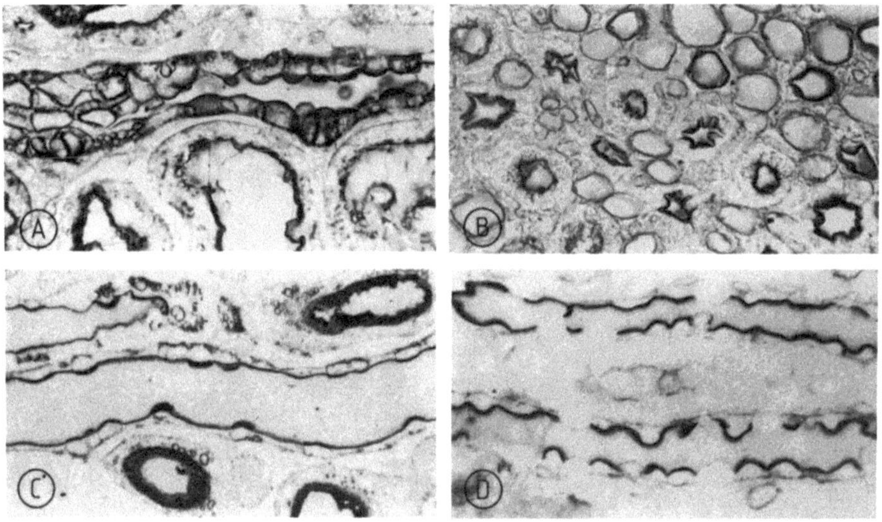

Figure 1. Rat kidney, paraffin sections (A, B) and semithin resin sections (C, D) incubated with Helix pomatia lectin-gold complex. In paraffin sections all the cells in a cortical collecting duct (A) and an outer medulla collecting duct (B) are positive. In semithin sections all cortical collecting duct epithelia are positive as well (C) but in the outer medulla collecting duct positive and negative cells can now be recognized (D) due to the higher resolution of this approach. x 300 (A-D).

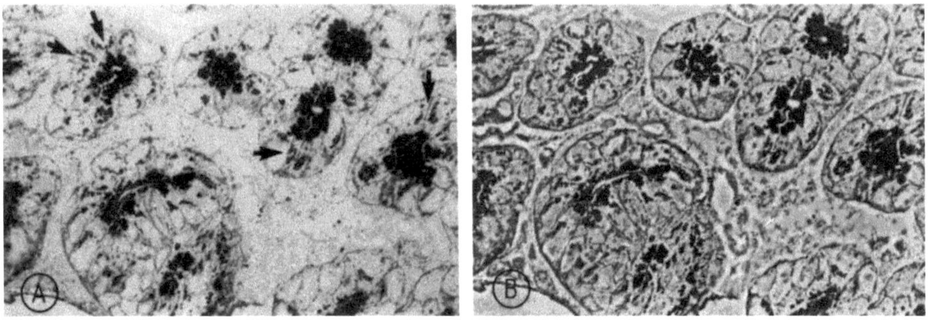

Figure 2. Rat stomach, semithin resin sections incubated with soyabean agglutinin-gold complex. Neck mucous cells are intensely stained over the mucous droplets, and supranuclear structures indicative of Golgi apparatus (arrows). Cell surface staining is also clearly visible. A - bright field transmitted light, B - phase contrast. x 600.

structural resolution and image information which is much improved when semithin (0.5-1.5 μm) sections from plastic embedded tissues (Figs. 1 and 2) are used (21, 39). The cytochemical reaction on such sections can be seen by a pink to red coloration of the posi-

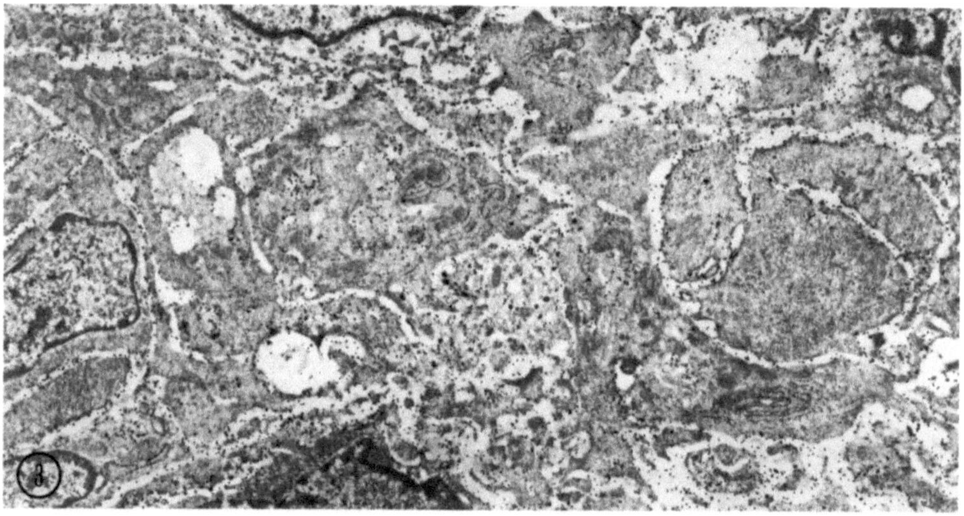

Figure 3. Chick duodenum, Lowicryl K4M thin section incubated with Ricinus communis lectin I-gold complex. Part of the villus stroma with smooth muscle cells and fibroblasts is seen. Gold particle label appears along the plasma membrane of the various cells, the extracellular matrix and lysosomes inside the cells demonstrating the good accessibility of lectin binding sites. x 11,000.

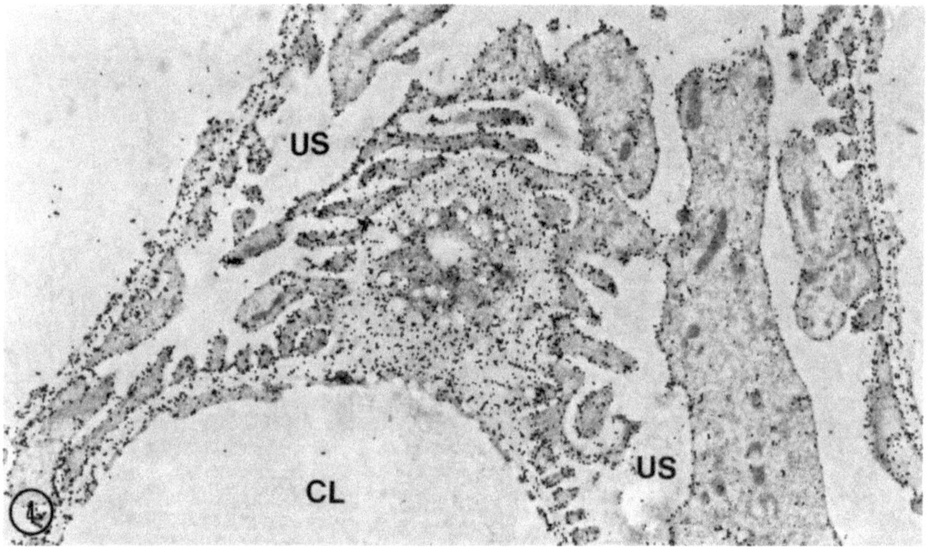

Figure 4. Lowicryl K4M thin section showing part of a glomerular loop from rat kidney glomerulus labeled with wheat germ agglutinin Intense labeling by gold particles is visible over all tissue structures. The urinary space (US) and the capillary lumen (CL) are free of gold particles showing the virtually absent background. x 11,000.

tive structures. The color is due to the reflection of transmitted visible light by colloidal gold and not produced as result of a revealing procedure. Sections can be examined not only by bright field transmitted light illumination but also by phase contrast and Nomarski differential interference contrast.

Until recently localization of lectin binding sites in intact tissues was severely limited mainly due to problems of penetration of the reagents. A major advantage was made possible by the use of thin sections from low temperature Lowicryl K4M embedded tissues (31). On such thin sections sugar residues present in different intracellular organelles as well as at the cell surface and in the intercellular matrix are accessible for interaction with lectin-gold complexes (Figs. 3 and 4). In the electron microscope colloidal gold appears as highly electron dense particles which can be easily recognized over the cellular structures. Since gold particles can be prepared in different sizes, double labeling can be performed (35). But care must be taken to exclude reactions of the lectins with each other because of their glycoprotein nature. Another important advantage provided by the particulate nature of colloidal gold is that quantification of the label becomes possible (31). In the following some results are presented to demonstrate the potential these techniques have for the localization of lectin binding sites in cellular membranes and various intracellular compartments.

Plasma Membrane Regionalization and Cellular Heterogeneity of Lectin Binding

Two different lectins, the *Helix pomatia* lectin (HPL), specific for terminal non-reducing α-N-acetyl-D-galactosamine (GalNAc) residues (1, 16), and the *Limax flavus* lectin (LFL), specific for N-acetyl and N-glycolylneuraminic acid (23) were used to study the distribution of their binding sites in rat kidney. In the glomerulus, binding of HPL-gold complexes was extraordinarily restricted (39). Only the plasma membrane at the base of the foot process of the podocytes was labeled (Fig. 5). The gold particle label extended from here for various depths into the glomerular basement membrane. A quantitative evaluation of the distribution of the label showed that about 94% of the total gold particles were overlying the lamina rara externa and the lamina densa. The complete absence of labeling in preparations of isolated glomerular basement membranes suggested strongly that a component of the glycocalyx was detected rather than a glycocomponent of the basement membrane. No other structures of the glomerulus showed binding of HPL-gold complexes. Therefore, these data show that the glycocalyx of the foot process base is a highly specialized cell surface domain with respect to its carbohydrate composition. When the LFL was applied to thin sections from rat glomerulus to detectec sialic acid residues the pattern of labeling was in striking contrast to the above described (7). All the components of the glomerulus were labeled. Nevertheless, the distribution of sialic acid residues in the prodocyte plasma membrane again was remarkably regionalized (Fig. 6). The plasma membrane of the podocyte main body exhibited the highest labeling intensity when compared to the foot process region above the slit diaphragm and the foot process base below the slit diaphragm, which were 2.5 times and 4 times less intensely labeled respectively. Based on these data three plasma membrane domains can be defined in podocytes which differ for cytochemically detectable sialic acid: the foot process base, the foot process region above the slit diaphragm and the body of podocytes. Taken together these data add new parameters to the description of the podocyte plasma membrane which appears an exeptionally differentiated structure in keeping with the complex functions taking place at this level. Under experimental nephrosis in which the characteristic shape of the podocytes is lost these domains are perturbed: HPL-gold binding is significantly reduced on the foot process bases and no more binding of LFL can be detected (7). These data suggest a role for these doamins in the maintainance of the normal podocyte shape.

Various studies have demonstrated heterogeneity of lectin binding to tubular cells in kidney (4, 17, 20, 43). Two such examples from our recent work are described in the following. In the collecting duct of rat kidney two cell types can be principally distinguished: intercalated cells and clear cells. The plasma membrane of both cell types is heavily labelled with HPL-gold complexes in collecting ducts in the kidney cortex and outer stripe of

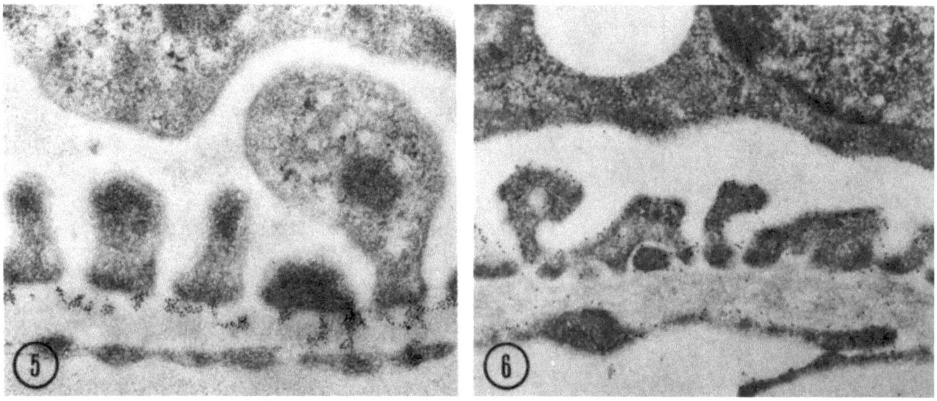

Figures 5 and 6. Rat kidney glomerulus, Lowicryl K4M thin section incubated with Helix pomatia lectin-gold complex (Fig. 5) or processed with the Limax flavus lectin - fetuin-gold technique (Fig. 6). Binding sites for Helix pomatia lectin are detectable only at the plasma membrane of the foot process base of the podocyte and in the lamina rara externa and lamina densa of the basement membrane. Binding sites for Limax flavus lectin can be seen along the entire plasma membrane of podocyte but note the low degree of labeling at the foot process base. In addition labeling is present in the basement membrane and endothelial cell. x 40,000 (Fig. 5); x 30,000 (Fig. 6).

outer medulla (Fig. 1C) (5). The intercalated cells present in collecting ducts of the inner stripe of outer medulla and outer medulla, however, were negative or very weakly labeled whereas the clear cells were strongly positive (Fig. 1D). This indicates that intercalated cells, in terms of their glycocalyx composition, represent a heterogeneous cell population along the collecting duct.

With the LFL-gold complex another impressive example for glycocalyx heterogeneity along the rat kidney urinary tubule was found (37). Sialic acid residues, as detected with LFL could be found almost along the entire length of the nephron with the remarkable exception of the thick ascending limbs of Henlé (Fig. 7 and 8). In its medullary portion the tubular epitelium was not labeled. In the cortical portion of the thick ascending limbs, however, about 35% of the tubular cells were positive. These data demonstrate the existence of a gross difference in glycocalyx composition between thick ascending limbs of Henlé and the rest of the nephron segments. In addition, a fine heterogeneity in glycocalyx composition between

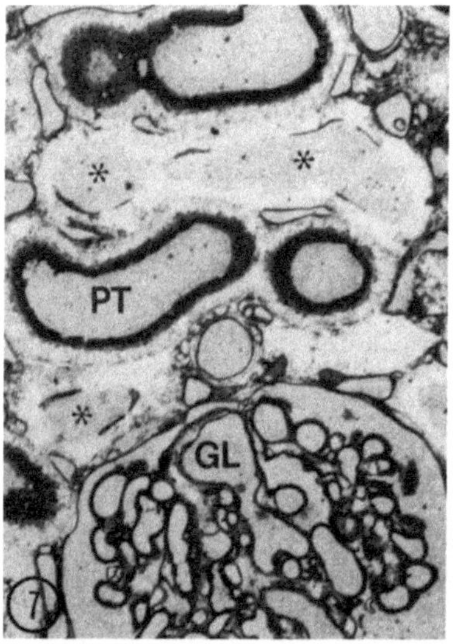

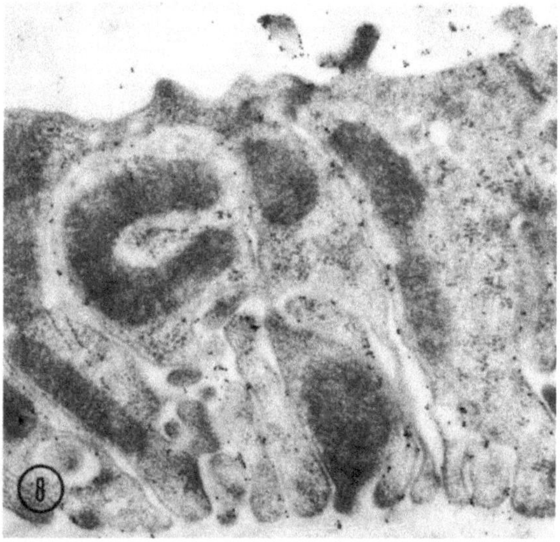

Figures 7 and 8. Rat kidney, Limax flavus lectin - fetuin-gold technique. In a semithin section from the kidney cortex (Fig. 7) intense staining is present over a glomerulus (GL) and proximal tubules (PT). Some tubular profiles (asterisks) show positive and negative cells and were identified as cortical portion of thick ascending limb of Henlé by electron microscopy (Fig. 8). The electron micrograph shows part of a negative and a positive cell in this tubular region. x 300 (Fig. 7); x 40,000 (Fig. 8).

medullary and cortical portion of thick ascending limbs was detected. It is unknown at present if and how the observed heterogeneous sialic acid distribution is related to the functional diversity exhibited by these tubular regions.

Studies on Compartmentation of Intracellular Glycosylation

There is now strong evidence from cellular fractionation and immunoelectron microscopic studies that the various steps in glycosylation of proteins take place in distinct subcellular structures in certain cells (8, 9, 13, 14, 36, 44, 45). The synthesis of N-linked glycoproteins which are characterized by an asparagine-to-N-acetyl-D-glucosamine score linkage is considered to involve an initial *en block* transfer of an oligosaccharide from a lipid carrier to the nascent protein. This is followed by the processing of the oligosaccharide as the protein passes through the endoplasmic reticulum and Golgi apparatus. Certain reactions, the addition of the

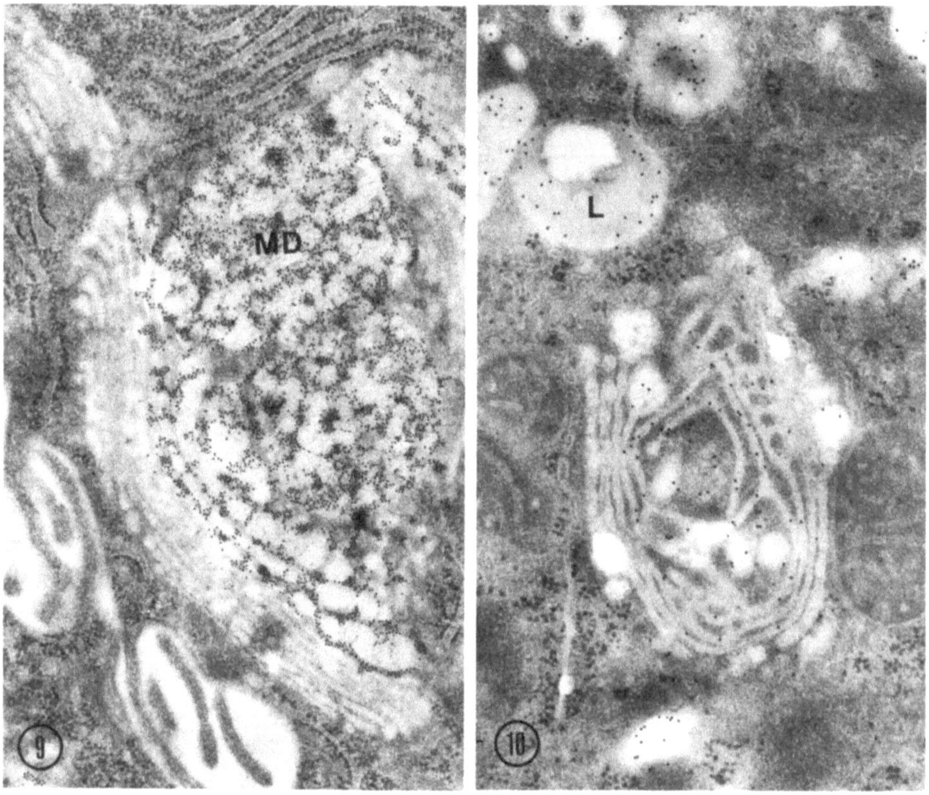

Figures 9 and 10. Limax flavus lectin - fetuin-gold technique for detection of sialic acid residues in the Golgi apparatus. In a rat colonic goblet cell only about two trans cisternae are positive. The rest of the Golgi cisternal stack and the rough ER are not labeled (Fig. 9) Similar labeling pattern is found in rat hepatocytes (Fig. 10) with labeling of two trans cisternae and the trans-tubular network. MD-mucin droplets, L-lysosomes. x 28.000 (Fig. 9); x 38,000 (Fig. 10).

terminal sugars galactose, fucose and sialic acid, take place in the Golgi apparatus. Direct evidence for subcompartmentation of the Golgi apparatus came from the immunolocalization of galactosyltransferase which codistributed with thiamine pyrophosphatase positive trans cisternae (36) and from cytochemically localization of galactose residues with Ricinus communis lectin I to trans Golgi apparatus cisternae (14). In intestinal goblet cells, sialic acid residues became first detectable in one or two trans Golgi apparatus cisternae (Fig. 9) (39). Very recently, the site of sialylation within the Golgi apparatus of rat repatocytes was revealed (37). In this study sialic residues were localized with the LFL and sialytransferase was detected with a specific antibody. It could be shown that sialylation occurs not only as expected in trans cisternae of the Golgi apparatus but apparently also in a complex structure, the trans-tubular network (Fig. 10). This trans-tubular network was found to be partially positive for acid phosphatase, a cytochemical marker for GERL and - most important to note - was contiguous with the cisternal stack of the Golgi apparatus. In contrast to the GERL concept (25) these data demonstrate that in rat hepatocytes acid phosphatase positive elements form an integral part of the trans region of the Golgi apparatus and appear to belong to the terminal glycosylation machinery of the Golgi apparatus.

The subcellular localization of terminal GalNAc residues with HPL-gold complexes has provided some more information about the sites of synthesis of O-linked glycoproteins (32). The initial glycosy-

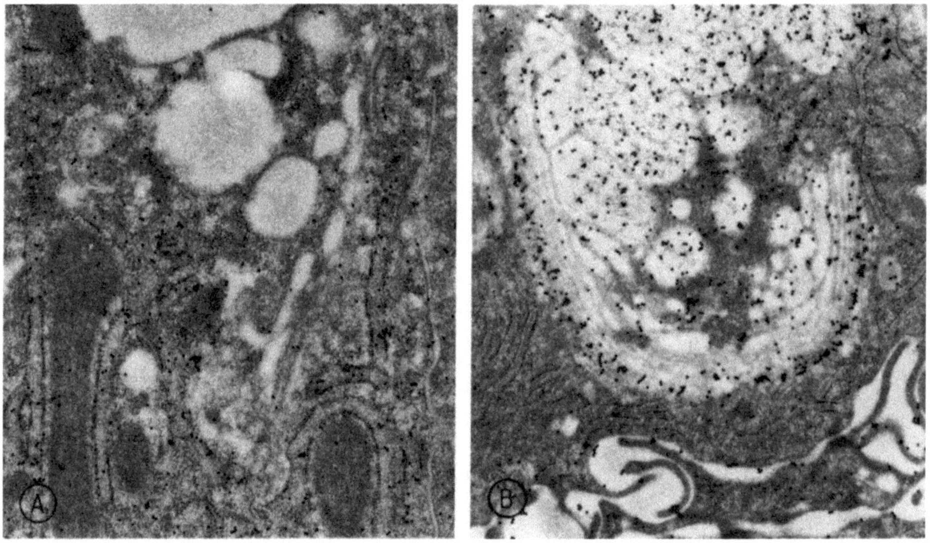

Figure 11. Rat colonic goblet cells. Mannose and glucose residues as visualized with Concanavalin A are present over the rough endoplasmic reticlumu (A). Label greatly diminshes over the Golgi apparatus and is practically absent over mucin droplets (MD). Terminal GalNAc residues are not detectable in the rough endoplasmic reticulum (B) but intense labeling starts at the cis side of the Golgi apparatus indicating onset of O-glycosylation. x 24,000 (A); x 11,000 (B).

lation reaction involves the transfer of GalNAc to serine and/or threonine. But the subcellular location of this reaction was not firmly established. No labeling for GalNAc residues could be detected over nuclear envelope and smooth as well as rough endoplasmic reticulum but intense labeling occured over the first cis cisterna and the several subsequent Golgi apparatus cisternae. (Fig. 11). These data together with results from a cellular fractionation study (10) are consistent with the idea that O-linked glycosylation is a posttranslational event that begins in the cis region of the Golgi apparatus and proceeds further as the newly synthesized glycoproteins pass through the Golgi apparatus.

Acknowledgement The authors work summarized in this review was supported by the Swiss National Science Foundation (grant no. 3.443.-0.83), the Sandoz Foundation, Basle, and the Kanton Basel-Stadt. I thank Ms Caroline Nielsen and Ms Erica Oesch for preparing the manuscript and Daniel Wey for photographic assistance.

REFERENCES

1. Baker, D.A., Sugii, S., Kabat, E.A., Ratcliffe, R.M., Hermentin, P., Lemieux, R.U.: Biochemistry *22*, 2741, 1983.
2. Bernhard, W., Avrameas, S.: Exptl. Cell Res. *64*, 232, 1971.
3. Bretton, R., Bariety, J.: J. Ultrastruct. Res. *48*, 396, 1974.
4. Brown, D., Kunz, A., Wohlwend, A., Vassalli, J.-D-. Orci. L.: C.R. Acad. Sci. (Paris) *297*, 501, 1983.
5. Brown, D., Roth, J., Orci, L.: Am. J. Physiol. *248*, C348, 1985.
6. Carlemalm, E., Garavito, M., Villiger, W.: J. Micr. (Oxford) *126*, 123, 1982.
7. Charest, P.M., Roth, J.: Proc. Natl. Acad. Sci. USA, in press, 1985.
8. Dunphy, W.G., Fries, E., Urbani, L.J., Rothman, J.E.: Proc. Natl. Acad. Sci. USA *78*, 7453, 1981.
9. Dunphy, W.G., Rothman, J.E.: J. Cell Biol. *97*, 270, 1983.
10. Elhammer, A., Kornfeld, S.: J. Cell Biol. *98*,327, 1984.
11. Geoghegan, W.D., Ackerman, G.A.: J. Histochem. Cytochem. *25*, 1187, 1977.
12. Guillouzo, A., Fledman, G.: J. Histochem. Cytochem. *25*, 1303, 1977.
13. Goldberg, D.E., Korfeld, S.: J. Biol. Chem. *258*, 3159, 1983.
14. Griffiths, G., Brands, R., Burke, B., Louvard, D., Warren, G.: J. Cell Biol. *95*, 781, 1982.
15. Gros, D., Obrenovitch, A., Challice, C.E., Monsigny, M., Schrevel, J.: J. Histochem. Cytochem. *25*, 104, 1977.
16. Hammarström, S., Kabat, E.A.: Biochemistry *8*, 2696, 1969.
17. Holthöfer, H., Virtanen, J., Pettersson, E., Törnroth, T., Alfthan, O., Linder, E., Miettinen, A.: Lab. Invest. *45*, 391, 1981.
18 Horisberger, M., Rosset, J., Bauer, H.: Experinetia *31*, 1147, 1975.
19. Horisberger, M., Vonlanthen, M.: J. Histochem. Cytochem. *26*, 960, 1978.
20. Le Hir, M., Kaissling, B., Koeppen, B.M., Wade, J.B.: Am. J. Physiol. *242*, C117, 1982.
21. Lucocq, J.M., Roth, J.: J. Histochem. Cytochem. *32*, 1075, 1984.
22. Lucocq, J.M., Roth, J.: In Thecniques in Immunocytochemistry (eds. Bullock, G.R., Petrusz, P.), Academic Press, London, vol. 3, 203, 1985.
23. Müller, R.L., Collawan, J.F., Fish, W.W.: J. BIol. Chem. *257*, 7574, 1982.

24. Nicolson, G.L.: In Advanced Techniques in Biological Electron Microscopy II (ed. Koehler, J.K.), Springer, Berlin - Heidelberg - New York, p. 1, 1978.

25. Novikoff, A.B., Novikoff, P.M.: Histochem. J. *9*, 525, 1977.

26. Pinto da Silva, P., Parkinson, C., Dwyer, N.: J. Histochem. Cytochem. *29*, 917, 1981.

27. Roth, J.: Exp. Pathol. Suppl. *3*, 1, 1978.

28. Goldstein, I.J., Hayes, C.E.: Adv. Carbohydr. Chem. Biochem. *35*, 127, 1978.

29. Roth, J.: In Techniques in Immunocytochemistry (eds. Bullock, G.R., Petrusz, P.), Academic Press, London, vol 2, 217, 1983a.

30. Roth, J.: J. Histochem. Cytochem. *31*, 547, 1983b.

31. Roth, J.: J. Histochem. Cytochem. *31*, 987, 1983c.

32. Roth, J.: J. Cell Biol. *98*, 399, 1984.

33. Roth, J., Wagner, M.: Exp. Pathol. *14*, 311, 1977a.

34. Roth, J., Wagner, M.: J. Histochem. Cytochem. *25*, 1181, 1977b.

35. Roth, J., Binder, M.: J. Histochem. Cytochem. *26*, 163, 1978.

36. Roth, J., Berger, E.G.: J. Cell Biol. *93*, 223, 1982.

37. Roth, J., Taatjes, D.J.: Eur. J. Cell Biol. in press, 1985.

38. Roth, J., Bendayan, M., Carlemalm, E., Villiger, W., Garavito, M.: J. Histochem. Cytochem. *29*, 663, 1981.

39. Roth, J., Brown, D., Orci, L.: J. Cell Biol. *96*, 1189, 1983.

40. Roth, J., Lucocq, J.M., Charest, P.M.: J. Histochem. Cytochem. *32*, 1167, 1984.

41. Roth, J., Taatjes, D.J., Lucocq, J.M., Weinstein, J., Paulson, J.C.: Cell in press, 1985.

42. Sato, A., Spicer, S.S.: Histochemistry *73*, 607, 1982.

43. Schulte, B.A., Spicer, S.S.: Am. J. Anat. *168*, 345, 1983.

44. Tartakoff, B.A., Vassalli, P.: J. Cell Biol. *97*, 1243, 1983.

45. Torrisi, M.R., Pinto da Silva, P.: J. Cell Biol. *98*, 29, 1984.

46. Wagner, M., Wagner, B.: Z. Immun. Forsch. *151*, 117, 1976.

47. Wagner, M., Roth, J., Wagner, B.: Exp. Pathol. *12*, 277, 1976.

48. Wood, J.G., McLaughlin, B.J., Barber, R.P.: J. Cell Biol. *63*, 541, 1974.

LECTINS AND CANCER: CHARACTERIZATION OF HODGKIN-DERIVED CELL LINES BY VARIOUS LECTINS

G. Uhlenbruck, M. Schwonzen[1], W. Pissors and V. Diehl

Medical University Clinic I of Cologne, Joseph-Stelzmann-Straße 9, D-5000 Cologne 41, Federal Republic of Germany

[1]Stipendiate of the Deutsche Forschungsgemeinschaft

The changes in glycosylation of glycoproteins and membrane glycolipids occur as quickly and dramatically as the process of phosphorylation at all stages of differentiation, development and oncogenesis, as Hakomori has pointed out (1). Accordingly, stage specific, tissue-specific or tumor-associated carbohydrate antigens are expressed.

Lectins are (glyco)proteins which bind specifically to simple or complex carbohydrates and lectin binding sites may provide a new pattern of cell markers which could be applicable to the identification of cells.

The uncertain nature of the Hodgkin (H) and Sternberg-Reed (SR) neoplastic cells has stimulated decades of research, but this question still remains unanswered. However, many different theories have been proposed: The more recent ones have related H and SR cells to B-cells (2), macrophages (3), interdigitating reticulum cells (4), and also for instance various granulocytic cell lines (5) or myelomonocytic precursor cells (6).

In the last years Diehl and co-workers have established long term cultures from tumor material of Hodgkin's disease (7). There is no doubt that the in vitro cells represent the in vivo Hodgkin and Sternberg-Reed cells as shown by the identity of multiple properties, which are characteristic, but could not identify relationship of H and SR cells to a known cell type (8).

We tried to classify the three Hodgkin's disease-derived cell lines L428, L540 and L591 with a panel of lectins. Eleven other cell lines were tested in comparison to the H and SR cells: promyelocytic (HL60), histiocytic (U937), lymphoblastoid, B-, T- and Burkitt-lymphoma cell lines.

Lectins, Vol. V

MATERIALS AND METHODS

Cells

Establishment of the in vitro Hodgkin's disease-derived cell lines and culture conditions are described elsewhere (9).

L428KSA is a bovine serum adapted variant of L428 which grows as an adherent monolayer after exposure of phorbol ester (TPA) for a period of 3 weeks (6).

Cell lines were maintained in RPMI-1640 medium with 5 - 10% fetal calf serum, glutamine and penicillin/streptomycin in a 5% CO_2 atmosphere, passaged twice weekly.

Analysis of cell surface carbohydrates

Cells were harvested on day 3, washed 2 times in 10 mM Tris-buffered saline (TBS: pH 7.2, 2 mM $MgCl_2$, 5 mM $CaCl_2$, 0.01% NaN_3, 154 mM NaCl).

1×10^6 cells/0.1 ml were incubated with 100 µl FITC-conjugated lectin (50 µg/ml) for 30 min at 4° C with intermittent shaking. Geodia cydonium lectin was purified and FITC-labeled as previously described (10). Geodia and other FITC-conjugated lectins (Medac, Hamburg, FRG) were stored at -20° C in aliquots. Aliquots were diluted in TBS supplemented with 1% bovine serum albumine (BSA) before use. After incubation excess lectin was removed by washing in TBS two times. Sedimented cells were examined with a Leitz Fluorescence Microscope equipped with appropriate filters. Two hundred cells were examined in each assay and the percentage of cells demonstrating membrane fluorescence was determined.

Similar procedure was done for monoclonal anti-Lewis^x antibody staining (Leu M1, Becton&Dickinson, Heidelberg, FRG; diluted 1:40 in 5% BSA). Instead of TBS phosphate-buffered saline (PBS) was used. Secondary fluorescinated antibody (anti-mouse IgG, Medac, Hamburg, FRG) was diluted 1:400 in PBS containing 5% BSA.

Influence of neuraminidase or pronase on lectin binding

Before incubation with lectin or antibody 0.1 ml of neuraminidase (0.1 U/ml) from Vibrio cholerae (Behringwerke AG, Marburg, FRG) or 0.1 ml of pronase (0.1 mg/ml) (Calbiochem, Frankfurt, FRG) was added to cell sediments (10^6/0.1 ml) for 30 min at 37° C and washed twice.

RESULTS AND DISCUSSION

Despite characteristic and striking morphological features the origin of Hodgkin and Sternberg-Reed cells is still controversial. In order to clarify this question the distribution of lectin receptors of H and SR cells was studied in comparison to other human cell lines. As useful tools in preliminary and easy study of carbohydrate structures we chose to use fluorescinated lectins. Native Con A (Concanavalin A), Succinyl-Con A, LCA (Lens culinaris), PHA (Phaseolus vulgaris), RCA I (Ricinus communis I), RCA II (Ricinus communis II), WGA (wheat germ, Triticum vulgaris) and PEA (Pisum sativum) strongly labeled all tested cells up to a concentration of 5 µg/ml. Differences in binding with other lectins are shown in Table I, Table II and Table III. The specificity of the membrane fluorescence was demonstrated by inhibition with various saccharides except VVA (Vicia villosa; no inhibition could be achieved with 0.1 M D-GalNAc).

If proteolytic digestion of intact cells with pronase decreases lectin binding drastically, one must assume that the carbohydrate receptor is mainly expressed by glycoproteins as shown with Lotus (ASP)-, Maclura pomifera (MPA)-, Wistaria floribunda (WFA)- and Peanut (PNA)-receptors. In case of similar or constant binding the lectin receptor is supposed to be predominantly located on glycolipids (Con A, RCA, WGA, PWM (Pokeweed Mitogen), ECA (Erythrina cristagalli), PHA). But in most cases the carbohydrate structures are important parts of both classes, and neoplasia-associated changes in glycoprotein carbohydrates seem to parallel those of glycolipids (1).

The histological binding of PNA to H and SR cells is very controversial and is discussed in the literature in these days. In a prior study (11) PNA and Lotus lectin were reported to bind only weakly or not at all to H and SR cells. Möller (12) found a prominent, constant and characteristic pattern in cytoplasmatic staining of H and SR cells. These differences cannot be due to the PNA concentration used. Our results do agree with those of Hsu et al. (4) who stated that PNA does not reveal a constant reaction with H and SR cells (Table II).

SBA only stained L428 and L540 although all cell lines were positive after neuraminidase treatment. This behaviour could be due to low metastatic potential of these cell lines because it was pro-

Table I. Lectin binding pattern of different cell lines before and after neuraminidase treatment

CELL LINE	D S A		V A A		P W M		E C A		W F A	
	b.	a.	b.	a.	b.	a.	b.	a.	b.	a.
Hodgkin										
L428	++(100)	+++(100)	-	+(50)	++(100)	+++(100)	+++(100)	+++(100)	+(100)	+(100)
L428KSA	+(90)	++(100)	-	+(100)	+(60)	+(60)	+++(100)	+++(100)	+(50)	+(50)
L540	+(90)	+++(100)	-	++(90)	++(100)	+(40)	++(90)	+++(100)	(+)(5)	+(70)
L591	++(100)	++(100)	-	-	-	+(100)	++(100)	+++(100)	-	-
Promyelocytic										
HL60	++(80)	++(100)	-	+(90)	-	+(90)	+(60)	+++(100)	-	+(90)
Histiocytic										
U937	+(80)	++(100)	-	-	(+)(5)	+(90)	-	+(100)	-	+(90)
Lymphoblastoid										
LCL725/538	++(100)	++(100)	++(100)	++(100)	-	-	+(80)	++(100)	+(40)	+(80)
LCL662	-	+(100)	+(80)	+(70)	+(100)	++(90)	+(100)	++(100)	-	+(80)
T-cell										
CEM	+(100)	+(100)	+(100)	++(90)	-	-	++(100)	++(100)	-	-
B-ALL										
660Vi	+(80)	+(80)	-	-	+(90)	+(100)	++(100)	+++(100)	-	-
Myeloma										
363	+(80)	(+)(10)	-	-	-	-	(+)(10)	++(100)	-	-
U266H	+(100)	+(100)	+(90)	+(80)	-	+(100)	+(90)	+++(100)	-	-
Burkitt										
Raji	+(90)	+(90)	-	-	+(100)	+(80)	++(100)	+++(100)	-	-

b. = before neuraminidase treatment; a. = after neuraminidase treatment; (+) = very weak membrane fluorescence; + - +++ = intensity of positive membrane fluorescence; numbers in brackets () designate percent of cells counted with positive membrane fluorescence

Table II. Lectin binding pattern of different cell lines before and after neuraminidase treatment

CELL LINE	S B A		T K A		P N A		P H A		H P A	
	b.	a.	b.	a.	b.	a.	b.	a.	b.	a.
Hodgkin										
L428	+(30)	+++(100)	+(80)	+(90)	++(10)	++(100)	++(90)	+++(100)	+(20)	++(90)
L428KSA	+(30)	++(100)	+(80)	+(80)	+(5)	++(100)	++(90)	+(90)	-	-
L540	++(70)	+++(100)	+(70)	+(10)	++(10)	+++(100)	(+)(5)	-	+(60)	++(90)
L591	-	++(100)	+(80)	+(100)	+(100)	++(100)	++(90)	+++(100)	-	-
Prcmyelocytic										
HL60	-	+++(100)	-	+(50)	-	+++(100)	+(90)	++(100)	-	+(80)
Histiocytic										
U937	-	+(100)	-	+(80)	-	+++(100)	+(90)	+(100)	-	-
Lymphoblastoid										
LCL725/538	-	-	+(70)	+(80)	+(30)	++(100)	++(90)	++(90)	-	-
LCL662	-	+(100)	-	-	-	++(100)	n.t.	n.t.	+(20)	+(80)
T-cell										
CEM	-	+(100)	+(50)	+(80)	+(40)	+++(100)	+(90)	++(90)	-	+(100)
B-ALL										
660Vi	-	++(100)	-	n.t.	-	+(100)	++(80)	++(90)	-	-
Myeloma										
363	-	+(100)	-	-	-	+(100)	+(90)	++(90)	-	-
U266H	-	++(100)	+(100)	+(90)	(+)(5)	++(100)	+(10)	+(80)	-	+(100)
Burkitt										
Raji	-	++(90)	+(80)	(+)(5)	+(70)	+++(100)	+(90)	+(100)	-	+(100)

b. = before neuraminidase treatment; a. = after neuraminidase treatment; (+) = very weak membrane fluorescence; n.t. = not tested ; numbers in brackets () designate percent of cells counted with positive membrane fluorescence; + – +++ = intensity of positive membrane fluorescence

Table III. Lectin binding pattern of different cell lines before and after neuraminidase treatment

CELL LINE	A S P		L e u M1 comparat. ab	M P A		V V A		G e o d i a	
	b.	a.	b.	b.	a.	b.	a.	b.	a.
Hodgkin L428	+(90)	+(100)	++(70)	+(90)	+(100)	-	-	++(80)	++(100)
L428KSA	-	+(70)	++(80)	-	-	-	+(80)	+(50)	++(80)
L540	+(30)	++(80)	+(50)	+(30)	++(70)	+(50)	-	+(80)	++(100)
L591	-	-	-	+(100)	+(90)	+(100)	++(100)	-	+(70)
Promyelocytic HL60	(+)(5)	+(90)	+(70)	+(90)	+(100)	(+)(5)	+(80)	+(40)	+(100)
Histiocytic U937	-	+(80)	+(80)	-	+(90)	-	++(100)	n.t.	n.t.
Lymphoblastoid LCL725/538	-	-	-	-	+++(90)	-	-	-	+(70)
LCL662	-	-	n.t.	-	+(100)	-	+(100)	-	+(80)
T-cell CEM	-	-	-	+(100)	+(100)	-	++(100)	+(50)	+(80)
B-ALL 660Vi	-	-	n.t.	+(100)	+++(100)	-	-	+(100)	++(100)
Myeloma 363	-	-	-	+(50)	+(90)	-	++(100)	+(100)	++(100)
U266H	-	-	n.t.	+(100)	+(100)	+(10)	++(100)	-	++(100)
Burkitt Raji	-	-	-	+(100)	+(80)	-	++(80)	+(90)	+(100)

b. = before neuraminidase treatment; a. = after neuraminidase treatment; (+) = very weak membrane fluorescence; n.t. = not tested; numbers in brackets () designate percent of cells counted with positive membrane fluorescence; + - +++ = intensity of positive membrane fluorescence posed that the difference in positioning of sialic acid on the cell surface leading to masking or unmasking of terminal sugars could influence the metastatic potential of tumor cells (13). Normal peripheral blood lymphocytes were also shown to bind SBA only after neuraminidase treatment (14,15). Why the bovine serum adapted variant of L428 which grows as an adherent monolayer after exposure of phorbolester (TPA) for 3 weeks (L428 KSA) has lost MPA- and HPA-receptors cannot be explained today. This finding might be due to different quantitative glycosylation.

Although L-fucose is an important determinant for PEA and LCA binding (all lines positive) (16), differences were seen with the L-fucose binding lectin from Lotus tetragonolobus (ASP). The Lotus lectin and a monoclonal anti-Le^x antibody (Leu M1) showed similar binding patterns before and after neuraminidase treatment (L428, L540, HL60, U937). In a previous study a correlation was noted between the reactivity of Lotus lectin, DR and Fc receptor expression (17). An increasing intensity of Lotus lectin binding has also been reported to parallel the normal myeloid maturation sequence (18). Enhancing of binding after neuraminidase treatment in our results might be due to steric effects.

Anti-Leu M1 reacted with nearly all cases of Hodgkin's disease tested histologically. However, in contrast to H-RS cells a positive anti-Leu M1 reaction in histiocytes and interdigitating reticulum cells requires pre-digestion of sections with neuraminidase (4).

Recent work has shown an identical population of branching cells in four histiocytic lymphomata which possess a very similar lectin staining pattern to the one of L428 and L540 cells (Con A, LCA, PSA, PHA, WGA, SBA, PNA, ASP).

In summary the origin of H and SR cells is still unknown, but there is obviously some relationship in the glycoconjugate profile to the myelo-histiocytic lineage.

ACKNOWLEDGEMENT

The work has been supported by the Minister für Wissenschaft und Forschung des Landes Nordrhein-Westfalen. We are greatly indebted to MTA Cornelia Böttinger for her skilful technical assistance, and for her excellent help in the preparation of the manuscript.

REFERENCES

1. Hakomori, S.-i. (1985) Cancer Res. 45, 2405-2414.

2. Poppema, S., De Jong, B., Atmosoerodjo,J., Idenburg, V., Visser, L., and De Ley, L. (1985) Cancer 55, 683-690.

3. Kaplan, H.S., and Gartner, S. (1977) Int. J. Cancer 19, 511-525.

4. Hsu, S.-M., Yang, K., and Jaffe, E.S. (1985) Amer. J. Pathol. 118, 209-217.

5. Stein, H., Uchanska-Ziegler, B., Gerdes, J., Ziegler, A., and Wernet, P. (1982) Int. J. Cancer 29, 283-290.

6. Diehl, V., Kirchner, H.H., Burrichter, H., Stein, H., Fonatsch, C., Gerdes, J., Schaadt, M., Heit, W., Uchanska-Ziegler, B., Ziegler, A., Heinz, F., and Sueno, K. (1982) Cancer Treat. Rep. 66, 615-632.

7. Diehl, V., Burrichter, H., Schaadt, M., Kirchner, H.H., Fonatsch, C., Stein, H., Gerdes, J., Heit, W., and Ziegler, A. (1983) Hematol. Oncol. 1, 139-147.

8. Schwab, U., Stein, H., Gerdes, J., Lemke, H., Kirchner, H., Schaadt, M., and Diehl, V. (1982) Nature 299, 65-67.

9. Diehl, V., Kirchner, H.H., Schaadt, M., Fonatsch, C., Stein, H., Gerdes, J., and Boie, C. (1981) J. Cancer Res. Clin. Oncol. 101, 111-124.

10. Hanisch, F.-G., Saur. A., Müller, W.E.G., Conrad, J., and Uhlenbruck, G. (1984) Biochim. Biophys. Acta 801, 388-395.

11. Strauchen, J.A. (1984) Amer. J. Pathol. 116, 370-376.

12. Möller, P. (1982) Virchows Arch.(Pathol. Anat.) 396, 313-317.

13. Altevogt, P., Fogel, M., Cheingsong-Popov, R., Dennis, J., Robinson, P., and Schirrmacher, V. (1983) Cancer Res. 43, 5138-5144.

14. Newman, R.A., Uhlenbruck, G., Schumacher, K., van Mil, A., and Karduck, D. (1978) Z. Immun.-Forsch. 154, 451-462.

15. Farrar, G.H., Holz, G., and Uhlenbruck, G. (1981) Eur. J. Biochem. 121, 237-241.

16. Kornfeld, K., Reitman, M.L., and Kornfeld, R. (1981) J. Biol. Chem. 256, 6633-6640.

17. Elias, L., and Van Epps, D.E. (1984) Blood 63, 1285-1290.

18. Nicola, N.A., Morstyn, G., and Metcalf, D. (1980) Blood Cells 6, 563-579.

19. Freemont, A.J., Matthews, S., Stoddart, R.W., and Jones, C.J. P. (1985) J. Pathol. 146, 139-150.

HISTOCHEMICAL INVESTIGATIONS ON LECTIN BINDING IN NORMAL AND NEOPLASTIC ENDOMETRIAL TISSUE AND IN CELL CULTURES OF ENDOMETRIAL CARCINOMAS *

H.H. Zippel, R. Hackenberg, F. Benzenberg, H. Hölzel*, K.-D. Schulz
P.J. Klein**

Zentrum für Frauenheilkunde und Geburtshilfe der Philipps-Universität Marburg, Pilgrimstein 3, D-3550 Marburg
Universitäts-Frauenklinik Hamburg*
Pathologisches Institut Freiburg**

Earlier we have shown that the expression in the mamma lactans of binding sites for peanut lectin (PNA) is correlated to the secretion, so that the detection of binding sites may have a hormone-dependent functional significance (3). Vierbuchen et al. (6) have demonstrated in animal experiments that in breast cancer the PNA-associated secretory phenomena are the end product of a functional process which is controlled by steroid hormones. This also appears to be confirmed in human breast cancer by the relatively good correlation between PNA binding and the cytoplasmactic steroid hormone receptor as well as to the efficiency of endocrine treatment (1, 4, 5).

In the present study, we investigated the extent to which the histochemical results obtained with lectins in breast tissue can be related to other organs which are likewise steroid hormone-dependent. The endometrium appeared to be especially suitable, since extensive knowledge has been obtained especially on the endocrine regulation of manifest endometrial carcinoma (review in 8).

Lectins, Vol. V

Material and Methods
The histochemical investigations were performed on formalin-fixed curettage material embedded in paraffin as well as on hysterectomy material. PNA binding to the endometrium was investigated in the proliferation phase (n = 10) as well as in the secretion phase (n = 10) of the normal cycle. Furthermore, endometrium with glandular-cystic hyperplasia (n = 10) and adenomatous hyperplasia (n = 10) was available. Besides these benign and premalignant alterations, the PNA behavior was examined in 59 adenocarcinomas of the endometrium. The carcinomas were subdivided in accordance with their degree of histological differentiation into purely glandular (= grade I), partly glandular, partly solid (= grade II) and mainly solid forms (= grade III).

The staining with the peanut agglutinin (PNA) was performed with the peroxidase technique using the method described previously (3). In all cases, a pretreatment with neuraminidase was performed. In the evaluation the number of PNA-binding cells, binding type and staining intensity were considered. The criteria for appraisal of PNA binding are as follows: I. percentage of PNA binding cells: more than 60% - between 30% and 60% - under 30%. II. predominant binding type: secretion binding - apical binding - vacuolar binding - specific cytoplasmatic binding - diffuse cytoplasmatic binding. III. staining intensity: intense - moderately strong - weak.

Furthermore, the PNA binding behavior was investigated in three endometrial carcinoma cell lines. The cell lines were established as already described in a previous publication (2). In the cell lines, besides the peroxidase technique, the fluorescence technique with FITC-labelled PNA was carried out to determine the lectin capacity per cell as well. For this investigation we used a microscope MPV Compact (Leitz).

Results

In the normal cycle, a different PNA binding behavior was found in the proliferative as well as in the secretory phase. In the proliferation phase (n = 10), PNA binding could always be demonstrated, whereas in the secretory phase (n = 10) it was detected only in one case. In this case, signs of early secretory transformation were found in the endometrium. In the glandular-cystic hyperplasia (n = 10) and in the adenomatous hyperplasia (n = 10), PNA binding could be demonstrated in nine of ten cases in each group (Table 1). In normal, benign and premalignant endometrium, an apical binding along the cell membrane predominated in most cases. Occasionally, secretion staining was also found intraluminal. However, a cytoplasmatic reaction hardly could be observed; merely isolated or sporadic cells gave rise to suspicion of a homogeneous cytoplasmatic reaction.

state		n	PNA pos.	PNA neg.
cycle	proliferation	10	10	-
	secretion	10	1	9
glandular-cystic hyperplasia		10	9	1
adenomatous hyperplasia		10	9	1

Table 1
The histochemistry of PNA binding in benign and premalignant alterations of the endometrium

Of the 59 endometrial carcinomas investigated, 45 (76%) showed PNA binding. Different binding patterns were found:

1. Staining of the apical cell membranes
2. Staining of varying proportions of the cytoplasm
3. Staining of the intraluminal secretion
4. Combination of 1-3

The staining pattern varied from case to case and even within the carcinomas concerning both the pattern as well as the intensity of staining. Frequently, only a minimal stainability could be detected in circumscribed tumor areas. Since the staining appeared to be relatively uncharacteristic, it was not considered in the further study. The quantitative analysis revealed an unequivocal correlation between the degree of histological differentiation and the staining intensity (Table 2). In tumors with grade I, distinct PNA binding (tumor cells >30% PNA positive) could be detected in 67%, whereas in tumors of grade II only in 42%. Three tumors of grade III did not show any PNA binding.

		PNA binding tumor cells		
grading	n	< 30%	30% - 60%	> 60%
I	39	13 (33%)	21 (54%)	5 (13%)
II	17	10 (58%)	7 (42%)	-
III	3	3	-	-
	59	26 (44%)	28 (47%)	5 (9%)

Table 2
Correlation between PNA binding and histological grading in endometrial carcinoma

PNA binding could be investigated in three endometrial carcinoma cell lines. In all cases, the primary tumor was PNA-positive. In the corresponding cell cultures, PNA binding could likewise be demonstrated with the peroxidase technique in all cases. More than half of the tumor cells showed PNA staining: a diffuse cytoplasmatic staining predominated, but a more intensive membraneous binding could be detected occasionally as well.

In addition, the mean PNA content per tumor cell was determined with an instrument for measuring the intensity of fluorescense. The measurements are listed in table 3. As compared with the PNA staining with the peroxidase technique, much higher values (Mu 17) could be detected in one case with quantitative analysis.

cell line	primary tumor	cell culture PNA		passage
	peroxidase	peroxidase	FITC m s	
Mu 13	pos.	pos.	59.7 ± 23.5	35
Mu 17	pos.	pos.	97.8 ± 36.2	20
Mu 50	pos.	pos.	52.9 ± 24.7	14

Table 3

Detection of PNA binding in primary endometrial carcinomas and in cell cultures.

The PNA content in cell culture has been determined by peroxydase- and fluorescence technique (right part of the table). In the primary tumor PNA binding was only demonstrated by peroxydase technique (left part of the table).

Quantitative analysis also made it possible to detect differences in the PNA staining intensity before and after treatment of the cell lines with 17-ß-estradiol. Whereas only a slight increase of staining intensity was shown after endocrine treatment in two cell lines (Mu 13, Mu 65), in the cell line Mu 17 a marked rise in PNA could be observed after treatment with 17-ß-estradiol (10^{-6} mol/l) (Fig. 1).

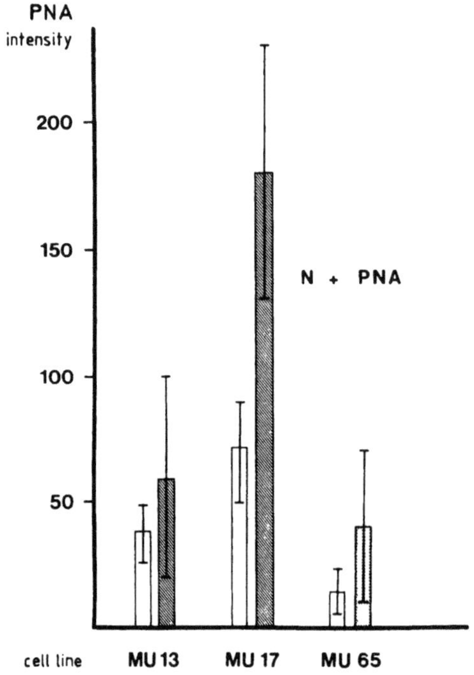

Fig. 1
Intensity of PNA binding in cell lines derived from endometrial cancer ☐ without treatment, ▨ = with estradiol treatment

Diskussion

In the endometrium, the detection of PNA binding appears to depend on proliferation, i.e. on an estrogen stimulus. In our investigations on the normal, benign and premalignant alterations, binding sites were mainly found in the apical gland areas and in intraluminal secretion. Thus, roughly the same binding behavior was shwon as in normal mammary gland epithelium (1, 3). In the endometrial carcinoma, the PNA binding rate ist 76%. Experience in breast cancer has shown that only a good response to hormonal treatment can be expected when 30% of the tumor cells are PNA positive (4).

The correlation for the endometrium cancer has not yet been done. But previous studies have shown, that the histological grading is a possible prediction test for response to hormonal therapy of progressive endometrium cancer. In our investigation a correlation could be seen to the histological grading when at least 30% of tumor cells were PNA positive.

When the limit is set under 30% of required tumor cells, we have not seen any correlations as already reported (7). Further investigations are required to establish whether the different PNA binding in tumors with the same histological grading has prognostic significance. Lectin histochemistry reveals that tumor cells in culture also have PNA binding sites. So far, the determination of the lectin binding capacity per cell with FITC-labelled PNA has proved to be more sensitive respectively more objective than in the peroxydase technique.

To summarize, the investigations allow the following conclusions: In the endometrium, the detection of PNA binding appears to be associated with proliferative, estrogen-dependent processes and can also be demonstrated in estrogen-stimulated endometrial carcinoma cells in culture. Thus cell culture appears to provide an interesting experimental model for comparative histochemical investigations with lectins.

Supported by the Kempkes Stiftung and SFB 215 (Tumor and Endokrinium)

References

1. W. Böcker, A. Klaubert, J. Bahnsen, G. Schweikhart, K. Pollow, M. Mitze, R. Kreienberg, Th. Beck, H.E. Stegner: Virchows Arch (Pathol. Anat.) 403: 149-161 (1984)

2. F. Hölzel, M. Albrecht, W.E. Simon, M. Hänsel, R. Metz, J. Schweizer, M. Dietel: J. Cancer Res Clin Oncol. 109: 217-226 (1985)

3. P.J. Klein, R.A. Newman, P. Müller, G. Uhlenbruck, P. Citoler, H.E. Schaefer, K.J. Lennartz, R. Fischer: J. Cancer Res. Clin Oncol. 93: 205-214 (1979)

4. P.J. Klein, M. Vierbuchen, J. Fischer, K.-D. Schulz, G. Farrar, G. Uhlenbruck:
J. Steroid-Biochem. Vol. 19, No 1, 839-844 (1983)

5. J. Martin (a.G.), P.J. Klein, H. Kalbfleisch, H.J. Künzig (a.G.)
Verh. Dtsch. Krebs Ges. 5: 525 (1984)

6. M. Vierbuchen, P.J. Klein, G. Uhlenbruck, R. Fischer:
Tumordiagnostik 2, 235 (1981)

7. H.H. Zippel, I. Haiden, H. Würz, P.J. Klein:
Klinische Forschung in der Gynäkologie und Geburtshilfe,
X. Akad. Tagung, Köln im Mai 1984, Sonderdruck, S. 52 (1985)

8. H.H. Zippel, K.-D. Schulz:
Medroxyprogesteronacetat (MPA) in der Onkologie, Schmidt, Schmidt-Matthiesen, Schattauer Verlag Stuttgart, New York (1984)

LECTIN HISTOCHEMISTRY OF COLORECTAL CARCINOMA COMBINED WITH LOCALIZATION OF CARCINOEMBRYONIC ANTIGEN AND MUCOPOLYSACCHARIDES

Jan Breborowicz, Alicja Michalska and Danuta Breborowicz.

Department of Pathomorphology, Karol Marcinkowski Academy of Medicine, 60 355 Poznań, Poland.

During the last decade many histochemical techniques applicable for histopathological sections have been developed. Demonstration of lectin binding sites with labeled lectins is one of these techniques, currently attracting the attention of many histopathologists. The technique is simple and inexpesive as compared to immunohistochemistry. However, so far lectin histochemistry is empirical as the chemical identity of the binding sites usually remains obscure. Although the specificity of lectins with respect to oligosaccharides is well characterized (8), the reaction of lectins with complex glycoconjugates cannot be predicted easily. The reaction of two lectins: concanavalin A (Con A) and Lens culinaris agglutinin (LCA) with glycoproteins can be cited as an example. These lectins have a similar sugar specificity but their reaction with individual microheterogeneity forms of a glycoprotein like α-fetoprotein is quite different (1). In spite of the difficulties in predicting the reaction of lectins with complex glycoconjugates, lectins remain an indispesable immunochemical tool. For example the microheterogeneity forms of α-fetoprotein cannot be distiguished with either polyclonal or with monoclonal antibodies while these forms are seen as different molecules by lectins.

Binding of lectins to normal and to cancerous mucosa of the large intestine has been studied by various workers (3-6). One of the lectins used by these authors was peanut agglutinin (PNA). It is interesting that localization of PNA on sections of colorectal carcinomas corresponds to the localization of carcinoembryonic antigen,

Lectins, Vol. V

CEA (2). CEA is a glycoprotein containing about 60% of carbohydrates is known to bind with Con A and other lectins. In the present work we decided to compare the localization of lectin binding sites with those of CEA and of mucopolysaccharides in normal and in cancerous colorectal mucosa.

MATERIAL AND METHODS

Primary colorectal tumours with adjacent normal mucosa were obtained at surgery from 11 patients. Patients were 45 to 72 years old and included 7 men and 4 women. Samples of tissues were fixed in 10% formalin, embedded in paraffin and processed according to the routine histological procedure. Tumours were diagnosed as villous adenoma (2 patients), well differentiated adenocarcinomas (6 patients) and poorly differentiated adenocarcinomas (3 patients). Paraffin sections were used for lectin histochemistry, for demonstration of CEA and for demonstration of mucopolysaccharides. The activity of endogenous peroxidase was blocked when necessary with methanol and H_2O_2.

Peanut agglutinin, PNA, Griffonia simplicifolia agglutinin (GS 1), Soyabean agglutinin (SBA) and Pisum sativum agglutinin (PSA) all labeled with horseradish peroxidase (HRP) were a gift from Kem-En-Tec, Hellerup, Denmark; Con A was obtained from Pharmacia (Sweden); peroxidase-labeled rabbit anti CEA antiserum (HRP-anti CEA), rabbit anti Con A antiserum (anti Con A), swine antiserum against rabbit immunoglobulins (swine anti Ig) and PAP complex made in rabbit were obtained from Dako Immunoglobulins (Denmark).

Binding sites for PNA, GS 1, SBA and PSA were detected by a direct technique using lectin-HRP conjugates. Lectin-HRP conjugates as supplied by the producer were diluted 50 times by PBS and applied on top of sections. After 60 min the slides were washed and peroxidase was demonstrated with DAB as a substrate. Localization of the Con A binding sites was done according to the PAP technique: Slides were incubated in Con A solution (100 µg per ml) for 60 min and after 30 min washing, incubated in anti Con A antiserum diluted 2000 times. Further processing was the same as for CEA. We used Ca^{++} and Mg^{++} supplemented buffers during all steps of the procedure. Localization of CEA was also done by the PAP technique. Sections were first incubated in 2000 times diluted rabbit anti CEA and then treated with swine anti Ig (1:20). Final incubation was in PAP complex (1:100). Alternatively CEA was detected by direct technique using HRP-anti CEA.

Neutral mucopolysaccharides, acid mucopolisaccharides and mucins were detected by periodic acid Schif reaction (PAS), by Alcian Blue staining (AB) and by mucicarmin staining (MUC) respectively, according to the routine histological methods used in the Department of Pathomorphology, Academy of Medicine in Poznań.

RESULTS

We studied the distribution of lectin binding sites, see Table 1. A positive histological staining was observed in one of the following locations: the glandular content, the apical parts of the epithelial cells,the supranuclear regions of the epithelial cells, goblet cells vacuoles, the epithelial cell cytoplasm (diffuse staining), the basal parts of the glandular crypts, and in the connective tissue. The results obtained with all staining procedures are listed in Table 1.

CEA was present in the glandular content and in apical regions in all tumors but only focally and in the same location as in the normal mucosa. The staining in the apical regions of adjacent cells fused thus forming a constant lining of the glandular lumina (Fig. 1b and f). In poorly differentiated tumours the reaction product was also visible in cell cytoplasm (Fig. 2d).

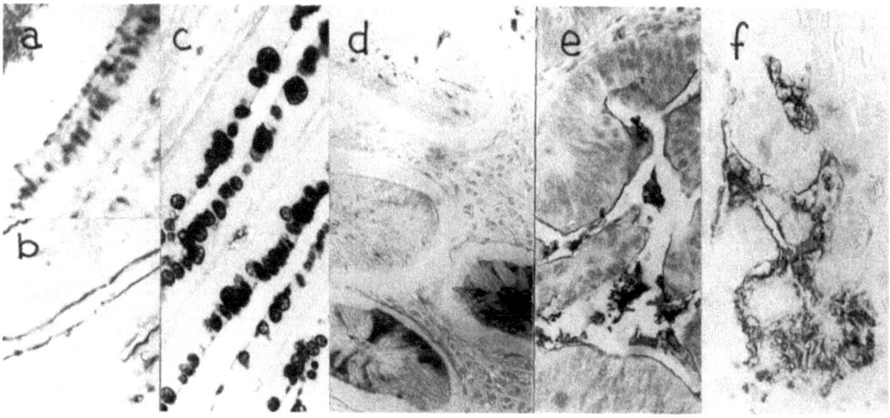

Fig. 1. Histochemical staining of glycoconjugates with HRP-lectins, PAS, and anti CEA. Results obtained with villous adenomas (a-d) and with well differentiated papillary adenocarcinomas: a = discreate apical and supranuclear staining with PNA-HRP; b = localization of CEA as a continuous apical lining; c = goblet cell vacuoles stained with PAS; d = positive staining of the basal crypts with GS 1-HRP; e and f = positive staining of the glandular content and luminal lining with PNA-HRP and with CEA-PAP respectively; e = counterstained with hematoxylin. Magnification 100 times.

TABLE 1.
Localization and semiquantitative estimation of the results obtained with various staining procedures

Method	Localization	Normal mucosa	Villous adenoma	Adenocarcinoma well diff.	Adenocarcinoma poorly diff.
CEA	Gl content	+f	+	+++	+++
	Apical	+f	++	+++	+++
	Diff cytopl				++
PNA	Gl content			+++	+++
	Apical	+	+++	+++	+++
	Supranuclear	++	+++	++f	
GS 1	Gl content			++	++
	Apical			++	++
	Supranuclear	++	+		
	Crypt bases		+++		
SBA	Gl content			++	+
	Apical	++f	++	++	+
	Supranuclear	++f	++		
PSA	Gl content	++	++	++	++
	Diff cytopl	++	++	++	++
	Con tissue	+	+	+	+
Con A	Gl content	+	+	+	+
	Goblet vac	+++	+++	+++	
	Diff cytopl	++	++	++	++
	Con tissue	+	+	+	+
PAS	Gl content	+	+	+	+
	Apical	+	+	+	+
	Goblet vac	+++	+++	+++	
	Diff cytopl	++	++	++	++
	Con tissue	+	+	+	+
AB	Apical	+	+	+	+
	Goblet vac	+++	+++	++	
MUC	Gl content	+	+	+	+
	Goblet vac	+++	++	++	
	Diff cytopl	+	+	+	+
	Con tissue	+	+	+	+

Abbreviations: AB = Alcian Blue staining; Apical = apical parts of epithelial cells; con = connective; Diff = diffuse staining; f = focal; Gl = glandular; MUC = mucicarmine; vac = vacuoles.

The results obtained with PNA-HRP in all adenocarcinomas were very similar to that obtained with anti CEA. However, in villous adenomas and in normal mucosa, PNA binding sites were demonstrated in supranuclear and in apical regions (Fig. 1a). The localization of the reaction product in the apical regions differed from the localization of CEA. Apical PNA binding sites in both normal mucosa and in villous adenoma epithelial cells was discrete, and a continuous luminar lining was not found (Fig. 1a).

Patterns observed in experiments with GS 1-HRP and with SBA-HRP were similar to those observed in experiments with PNA-HRP (Fig.2a and e). GS 1 binding sites were detected additionally in the basal parts of glandular crypts in one case of villous adenoma (Fig. 1d). A similar staining was neither observed in normal nor in cancerous mucosa from other patients.

The location of PSA and of Con A binding sites differed from those of PNA, GS 1 and SBA. The PSA-HRP and Con A-PAP experiments showed mainly a location in the epithelial cell cytoplasm and in the glandular content, but we also observed binding in the connective tissue

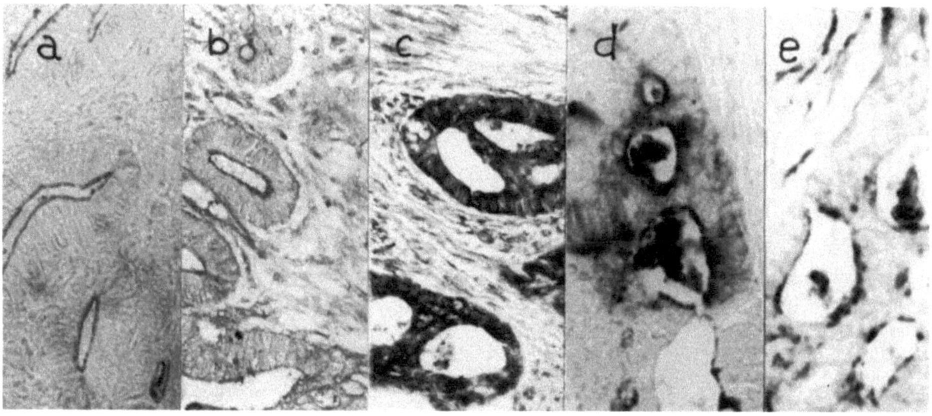

Fig. 2. Histochemical stainings for glycoconjugates with HRP-lectins, PAS, and anti CEA. Results obtained with well differentiated tubular adenocarcinoma (a and b) and with poorly differentiated adenocarcinoma (c-e); a and b = positive staining of the glandualr lining with SBA-HRP and with PAS respectively; c = diffuse cytoplasmic staining with PSA-HRP; d and f = the positive glandular content and apical staining with anti CEA-HRP and GS 1-HRP respectively; d = CEA is also focally present in the cytoplasm of epithelial cells. Magnification 150 times.

(Fig. 2c). In addition, Con A reacted with the content of goblet cells vacuoles. Results obtained with the periodic acid Schiff reaction, PAS, the Alcian Blue staining AB, and the mucicarmine staining MUC, were the same as those observed in the routine histological studies of colorectal mucosa and tumors. The glandular content was positive in PAS and MUC techniques while apical staining was observed with AB and PAS (Fig. 2b). Goblet cell vacuoles were positive with all the three techniques (Fig. 1c). The staining intensity of normal mucosa and of tumors differed depending on the technique used and on the differentiation of the lesion. Supranuclear staining of the epithelial cells similar to that obtained with PNA-HRP was not observed with PAS, AB or MUC.

DISCUSSION

Most of the demonstrated lectin binding sites were also positive by PAS, AB and MUC staining. Therefore it seems that most of the lectin binding sites in normal and in cancerous intestinal epithelium are glycoproteins and mucopolysaccharides. Both substances are first produced in the endoplastic reticulum and then processed in the Golgi apparatus. Our understanding of this process is far from being complete, but new data are emerging especially with respect to processing of glycoprotein through the Golgi apparatus (7). It is important to note that the location of PNA binding sites in the supranuclear region of the intestineal epithelial cells corresponds to the Golgi apparatus (4). Binding of PNA to the supranuclear region of normal and adenomatous cells of colonic mucosa has also been demonstrated by other workers (2, 5, 6). It was suggested that binding of PNA to the Golgi area might be due to presence of nonsialylated MN blood group substance, TAg (4). Nervertheless, it is impossible to exclude other glycoproteins processed in the Golgi apparatus as responsible also for the PNA binding. Precursors of CEA and of CEA-related substances are possible candidates as these glycoproteins are produced in large amounts by both normal and cancerous mucosa (2).

Binding sites of PNA, SBA and GS 1 in the lining of the cancerous glandular lumina overlaps with the location of CEA. CEA is a constituent of the glycocalyx, and therefore PNA binding sites in colonic adenocarcinomas may also correspond to the glycocalyx. Whether or not PNA, GS 1 and SB 1 binds to the CEA itself or to other glycoproteins present in the glycocalyx is not yet known. The sugar branches of the CEA molecule are only partially characterized. Partial desialylation of CEA isolated from various sources is well documented, and desialylation is to some extent responsible for the CEA heterogeneity. Desialylation might also result in binding of PNA to CEA. This possibility is supported by the observation that treatment of tissue sections with neuraminidase increases PNA binding (3). It is alos woth to note that the PNA binding sites in villous adenoma differ from those of CEA. This observation indicates that even if PNA binds to CEA it does also bind to other glycocompunds.

It is intriguing why supranuclear regions in cells of adenocarcinomas do not bind PNA. As suggested by Cooper (4), processing of glycoproteins in cancerous cells may differ from that of normal cells. An increase in the sensitivity of the methods for detection of PNA binding would be helpful for demonstrating wether PNA "receptors" in supranuclear region of the cancer cells are completly abolished of only reduced in number.

The location of PSA binding sites was similar to that of Con A binding sites. This observation is not supprising as both lectins have similar sugar specificity (8). However, contrary to Con A, PSA binds also the content of goblet cells, as does PNA after treatment of the histological slides with neuraminidase. A similar staining pattern was also observed with PAS:

GS 1 and SBA were similar binding to the same locations as PNA. Differences in the staining intensity with different lectins may be due simply to the technical reasons. It is interesting to mention the unusual binding of GS 1 to bases of the glandular crypts in one case of villous adenoma. The molecular basis of this binding is unknown.

The results obtained in this work confirm the applicability of lectins for histochemical examination of pathological specimens. Although the molecular basis of lectin binding to tissue sections is not yet understood, the results so far obtained might be used already for differentiation of colorectal tumours.

ACKNOWLEDGMENT: We are indebted to Kem-En-Tec, Pharmacia and Dako Immunoglobulins for a supply of lectins and antisera. Mr. Hjalmar Johansen is thanked for typing the manuscript.

REFERENCES

1. Breborowicz, J. and Bøg-Hansen, T.C.: Lectins, Biology, Biochemistry, Clinical Biochemistry (Bøg-Hansen, T.C., ed) vol 2, pp 445-456, Walter de Gruyter, Berlin, New York 1982.

2. Breborowicz, J., Easty, G., Birbeck, M., Robertson, D., Nery, R. and Neville, A.M.: Br. J. Cancer *31*, 559-569 (1975).

3. Coggi, G., Dell'Orto, P., Bonoldi, E., Doi, P. and Viale, G.: Lectins, Biology, Biochemistry, Clinical Biochemistry (Bøg-Hansen, T.C. and Spengler, G.A., ed) vol 3, pp 87-103, Walter de Gruyter, Berlin, New York 1983.

4. Cooper, H.S.: Human Pathology *15*, 904-906 (1984).

5. Fischer, J., Klein, P.J., Vierbuchen, M., Fischer, R. and Uhlenbruck, G.; Lectins, Biology, Biochemistry, Clinical Biochemistry (Bøg-Hansen, T.C. and Spengler, G.A., ed) vol 3, pp 167-178, Walter de Gruyter, Berlin, New York 1985.

6. Ørntoft, T.F. and Petersen, S.E.: Lectins, Biology, Biochemistry, Clinical Biochemistry (Bøg-Hansen, T.C. and Breborowicz, J., ed) vol 4, pp 90-93, Walter de Gruyter, Berling, New York 1985.

7. Rothman, J.E.: Scientific American *253*, 84-95 (1985).

8. Wu, A.M. and Herp, A.: Lectins, Biology, Biochemistry, Clinical Biochemistry (Bøg-Hansen, T.C. and Breborowicz, J., ed) vol 4, pp 629-636, Walter de Gruyter, Berlin, New York 1985.

HISTOCHEMICAL FEATURES OF RENAL ONCOCYTOMAS AS STUDIED WITH LABELED LECTINS

Harry Holthöfer

Department of Bacteriology and Immunology, University of Helsinki, Helsinki, Finland

Oncocytes are thought to be altered, aging epithelial cells found usually among the normal cells of salivary glands, thyroid, parathyroid and kidneys (1). The cellular appearance of oncocytes is changed, and they typically present a granular eosinophilic cytoplasm, increase of cellular volume, and numerous tightly packed mitochondria in the cytoplasm.
In kidneys, tumors composed entirely of concocytes (oncocytomas) are found increasingly (2). The increase is only partially due to better knowledge of this tumor, and there are reports of the oncocytogenic effects of cigarette smoke and some lead compounds (3,4). Clinically oncocytomas are benign, and microscopically the cells are identified by the typical appearance. Atypical mitoses are usually not found (1).

Although morphologically this tumor is well documented (2,5), there is lack of studies describing its immunohistochemical features. By enzymatic similarities renal oncocytomas are thought to originate from the senescent cells of the proximal tubules (6). In search of diagnostic aids and for a better characterization of the biogenesis of renal oncocytomas, this study was undertaken, describing the glycoconjugate histochemical features of this tumor, using fluorochrome labeled lectins as probes.

Lectins, Vol. V

Materials and methods

Four samples of renal oncocytomas were obtained from the files of Department of Pathology, University of Helsinki. The samples were routinely fixed in formalin and embedded in paraffin. The light microscopic appearance of all the samples was that of a typical oncocytoma.

Commercially available fluorochrome labeled lectins used were Helix pomatia (HPA), Maclura pomifera (MPA), Glycine maximum (SBA), Arachis hypogaea (PNA), Ricinus communis (RCA I), Ulex europeaus (UEA I), Triticum vulgaris (WGA) and Concanavalin A (ConA) (EY Laboratories, San Mateo, Ca). Prior to lectin stainings, a deparaffination with xylene and graded alcohols was performed. Spcificity testing of lectin binding was performed as described earlier (7). The 3μm deparaffinized sections were washed and flooded with the appropriate lectin dilutions, washed and mounted in polyvinylalcohol (7).

A Zeiss Standard microscope equipped with a filter system for FITC-fluorescence was used for microscopy.

Results

The results show that the cells of renal oncocytomas express only a few characteristic accessible glycoconjugates as studied by lectin binding. The normal renal tissue in the same cut section surrounding the tumor area served as a control, and presented the typical compartmentation of lectin binding sites along the nephron

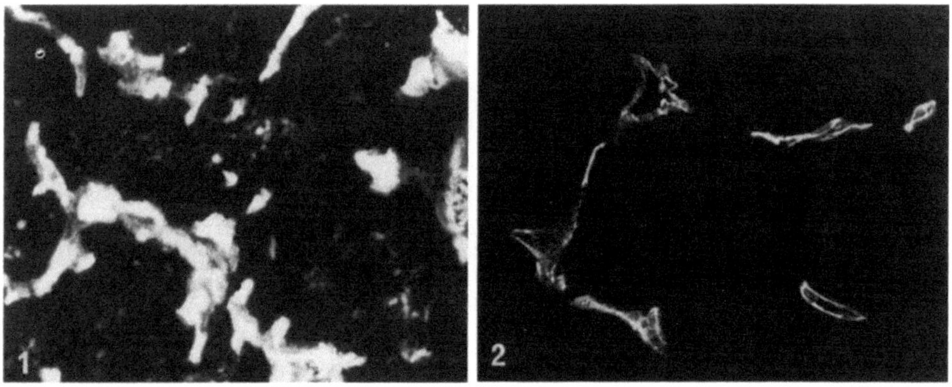

Fig. 1. PNA-staining of a sample of renal oncocytoma. Only the connective tissue elements surrounding tumor cell islets are visualized.
Fig. 2. The capillary network of oncocytoma tissue revealed by FITC-UEA I conjugate.

in all samples (7). HPA, MPA, SBA and PNA lectins failed to stain any of the tumor cells (Fig. 1) but gave a bright fluorescence of the connective tissue elements surrounding the tumor cell islets. The rich vascularization of the tumor tissue could be revealed with UEA I lectin (Fig. 2). WGA and Con A lectins gave a specific granular staining pattern of tumor cells (Fig. 3) in all samples studied, whereas RCA I specifically reacted with the plasmalemma of the tumor cells (Fig. 4). In addition to tumor cells, WGA, ConA and RCA I also stained the surrounding connective tissue.

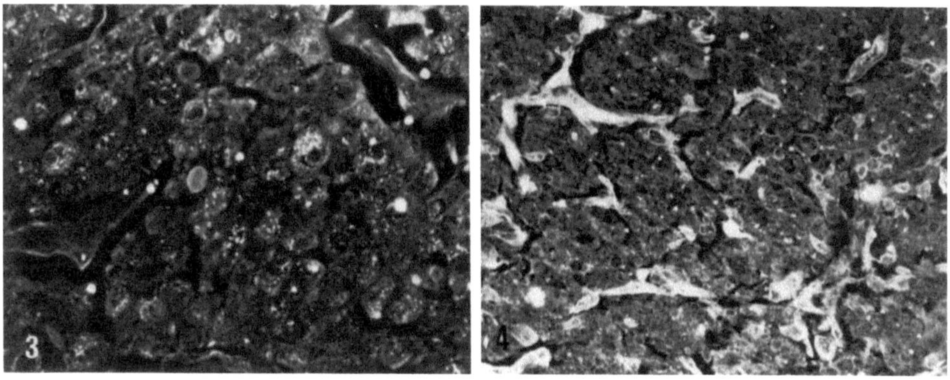

Fig. 3. ConA conjugate gives a bright granular staining at the tumor cells but also at the connective tissue elements.
Fig. 4. A preferentially plasmalemmal fluorescence of the tumor cells is seen after staining with FITC-RCA I.

Discussion

The results of this study show that renal oncocytomas express only a few accessible glycoconjugate structures as revealed with labeled lectins. However, these tumors are able to synthesize also specific glycoconjugates which can be detected with lectins. As the clinically benign renal oncocytomas may be difficult to distinguish from renal carcinomas, lectin histochemistry may thus offer an additional aid in the differential diagnosis of these tumors.

The morphologic features of oncocytomas are currently reasonably well characterized (2,5), but the histochemical features for further characterization of this tumor type are poorly known. Arnold et al (8) have shown that ATPase and other mitochondrial enzymes can be found in oncocytomas.

We have earlier shown the strict distribution of lectin binding sites along the nephron, and lectins are thus suitable as markers for the specific nephron segments (7). Any of the segment specific lectins did not react with the oncocytoma tissue, showing the lack of segment specific saccharides. Thus the lectin binding studies could not verify the proximal tubular origin of this tumor. This suggests that the cells of some other nephron segment may give rise to oncocytomas. Alternatively, the tumor cells may not show enough differentiation to express the segment specific saccharides. Technical failure of the lectin staining can be excluded, as all the lectins used gave their typical binding pattern on the normal part of the samples. The rich vasculature of the oncocytomas could be revealed by using UEA I, a marker for vascular endothelial cells (9).

WGA, ConA and RCA I, found to react with the tumor cells, bind to glucosyl, mannosyl and galactosyl residues of glycoconjugates (11). Of these, WGA and ConA share nominal sugar specificity of glucosyl residues (11). These lectins can also identify mid-chain saccharides (11), whereas most of the used lectins react preferentially with the ferm chain sugar residues. It is thus possible that ConA and WGA detect the same core-region saccharide structures, typically exposed in oncocytomas, but as the sugar chain elongation requires the precisely functioning glycosyltransferases, the termical structures can not be expressed.

This study revealed typical features of carbohydrate histochemistry of renal oncocytomas by showing the specific reaction of the tumor cells with RCA I, WGA and ConA lectins. These lectins may thus be used as additional probes for further characterization of oncocytomas and as an aid in the diagnosis of this tumor type.

REFERENCES

1. Hamperl, H. (1962). Virchows Arch. 335, 452.

2. Klein, J.M., Valensi, R.J. (1976). Cancer 38, 906-914.

3. Yu, G.S.M., Rendler, S., Herskowitz, A., Molnar, J.J. (1980). Cancer 45, 1010-1018.

4. Bannash, P., Krech, R., Zerban, H. (1978). Z. Krebsforsch. 92, 87-104.

5. Barnes, C.A., Beckman, E.N. (1983). Am. J. Clin. Pathol. 79, 312-318.

6. MacIves, A.G. (1982). Diagn. Histopathology 5, 231-281.

7. Holthöfer, H., Virtanen, I., Pettersson, G., Törnroth, T., Alfthan, O., Linder, E., Miettinen, A. (1981). Lab. Invest. 45, 391-399.

8. Arnold, B.M., Kovacs, K., Horvath, E., Murray, T., Higgins, H.P. (1977). J. Clin. Endocrin. Methods 38

9. Holthöfer, H., Virtanen, I., Kariniemi, A.-L., Hormia, M., Linder, E., Miettinen, A. (1982). Lab Invest. 47, 60-66.

LECTIN BINDING TO NEOPLASTIC HUMAN UROTHELIUM.

M.J.S. Nielsen, T.F. Ørntoft, L. Bendtsen and H. Wolf.

Dept. of Surgery, Aarhus County Hospital, and The Institute of Cancer Research, Radiumstationen, DK-8000 Aarhus C., Denmark.

Neoplasia of human urothelium is associated with profound changes of glycoconjugate structures. These structures belong primarily to the cell membrane and are present as glycolipids and glycolipids and glycoproteins. The most often published alterations are the deletion of isoantigens A, B, H[1] and recently the expression of PNA-receptors - considered to be Thomsen-Friedenreichs antigens (T-antigens)[2]. The mechanism responsible for this alteration is thought to be an incomplete synthesis of the proper carbohydrate structures with accumulation of precursors[3]. The importance of these findings is that apart from tumor biological information, they give the clinician information about the biological potential of the tumor. The deletion of A, B, H antigens or the accumulation of PNA-receptors is related to an invasive tumor behaviour[2]. To further characterize the alterations of carbohydrate structures associated with malignant transformation of the human urothelium, we examined the binding of a panel of lectins to various degrees of malignant transformed urothelium.

Lectins, Vol. V

MATERIALS AND METHODS.
Invasive bladder tumors from 12 patients were selected from the pathology files. These had been routinely processed in 10% formalin, pH 7,2 and embedded in paraplast. They were less than 5 years old. The blood groups of the patients were assessed in a routine blood bank. They were: 7 group A, 3 group B, and 2 group O. Biotinylated lectins were purchased from Kem-En-Tec (Hellerup, Denmark). Direct peroxidase technique was used to visualize the lectin binding on 4my tissue sections. Following incubation with biotinylated lectin, the sections were washed and incubated with avidin-biotin-peroxidase complex (Vector, Ulingame, Ca., USA). The color was developed by the use of 3-amino-9-ethylcarbazole. Nuclei were counterstained with hematoxylin. Controls included omitting the lectin and absorption of lectin with simple sugars as follows: 0,3M D-GluNAc (WGA, wheat germ agglutinin), 0,3M D-Nannose (LCA, Lens culinaris agglutinin), 0,2 L-fucose (UEA_1, Ulex europeaus agglutinin), and 0,4M D-GalNAc (HPA, Helix pomatia agglutinin).
The staining was abolished by these procedures, or drastically reduced.
Individual groups of cells in each tissue section were graded according to Bergkvist (4). The different gradings of the sections and the corresponding type of staining (cytoplasmic or membranous) was noted with approximate percentage by two independent observers.

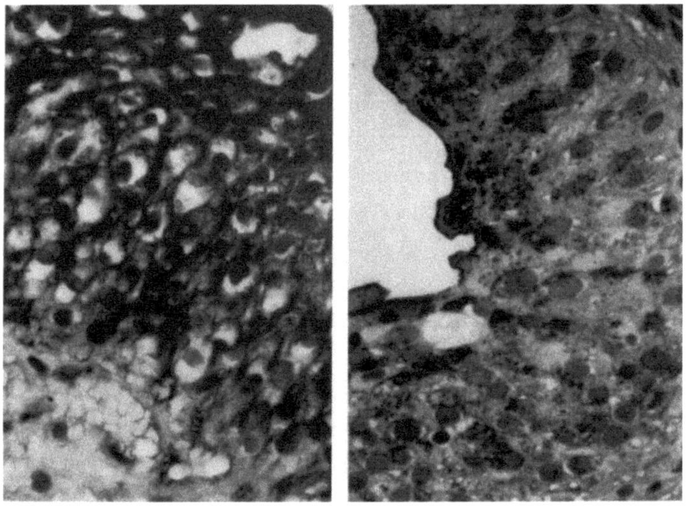

FIG.1: LEFT: NORMAL (GRADE O) UROTHELIUM SHOWING HPA BINDING TO CELL MEMBRANES IN ALL LAYERS AND TO CYTOPLASM IN UPPER LAYERS.X400. RIGHT: GRADE 2 ATYPIA SHOWING HPA BINDING AS GRANULES IN THE CYTOPLASM.X400.

RESULTS.

Lectins WGA and LCA.

These lectins, which identify D-GlcNAC and D-Mannose, showed the same staining pattern (Fig. 1). In parts of specimens containing normal-looking tissue areas the exhibited cytoplasmic as well as membranous staining. With increasing dedifferentiation the staining of the cell membranes disappeared, first from the basal layers, then from the intermediate layers, and with grade II tumors from the superficial layers. Invasive grade II or III tumors showed no membrane staining at all. The cytoplasmic staining was reduced in a similar way, but a minority of invasive cells still retained cytoplasmic staining.

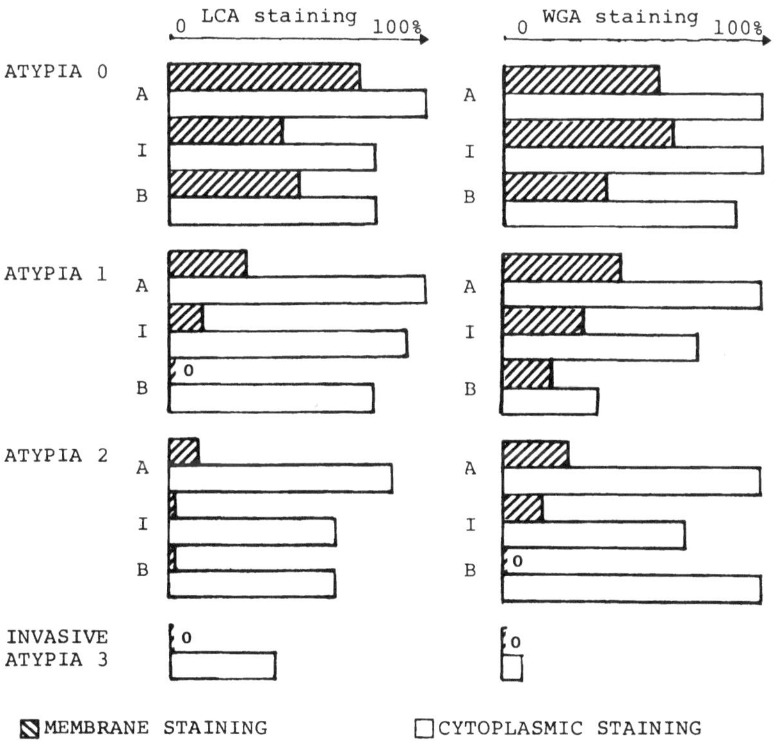

A: APICAL LAYER I: INTERMEDIATE LAYER B: BASAL LAYER

FIG. 2: CYTOSTRUCTURAL LOCALISATION OF LECTIN BINDING IN HUMAN BLADDER CANCER IN DIFFERENT DEGREES OF ATYPIA.

Lectins UEA_1 and HPA.

UEA_1 identifies L-Fucose which is a terminal structure in blood group H. Its staining was blood group dependent, as it showed cytoplasmic staining in all layers of urothelium from blood group O persons. This staining remained in all degrees of atypia. In normal urothelium of blood group A persons it was only present in the cytoplasm of luminal cells. In situ dedifferentiation was associated with staining of cytoplasm in all layers and occurrence of a membranous staining in the superficial cell layer. All invasive cells were completely negative. HPA has a high affinity for D-GalNAC structures, which are present in blood groups A and T_N. In blood group A persons it was present on membranes and in cytoplasm

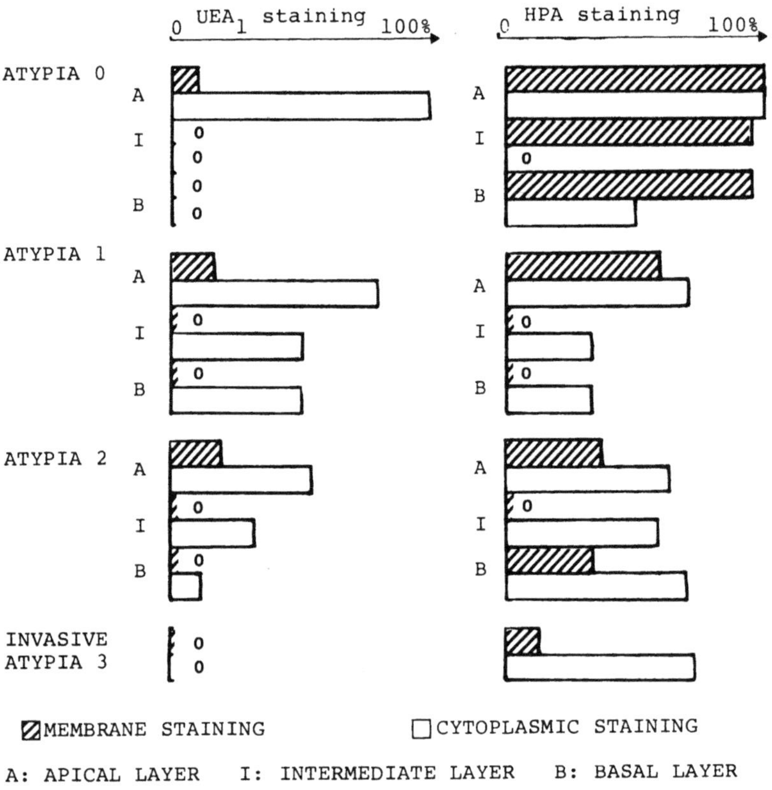

FIG. 3: CYTOSTRUCTURAL LOCALISATION OF LECTIN BINDING IN HUMAN BLADDER CANCER IN DIFFERENT DEGREES OF ATYPIA FROM PATIENTS WITH BLOODGROUP A.

of normal urothelium. Dedifferentiation resulted in the loss of membranous staining in all intermediate and basal layers and accentuation of cytoplasmic staining in these. In invasive cells, 60% had cytoplasmic staining and 40% membranous staining.

DISCUSSION.

Malignant transformation seems to be associated with dramatic changes of cell membrane glycoconjugates. All examined lectins, identifying different carbohydrates, showed a reduction of staining of cell membranes in dedifferentiated cells. The structures identified by WGA and LCA are probably not terminal saccharides, but located in the backbone and core of glycoconjugates [5,6].
The blood group H specific lectin UEA_1 was not found on membranes at all in the one tumor containing high degrees of differentiation. The reason for this is unknown. In blood group A persons, UEA_1 increased its staining in cytoplasm and appeared on membranes in grade II in situ atypia. This could be explained by a block in synthesis of blood group A [3] with accumulation of its immediate precensor - blood group H - which binds UEA_1. Invasive cells did not bind UEA_1 and thus might have lost the ability to synthesize the H-structure. HPA identifies both A and T_N structures. Its diminished reaction with cell membranes fits the above hypothesis of a block in blood group A synthesis. The staining wich remains on membranes, might be explained bt T_N antigens. This is in accordance of T_N antigens on tumor cell membranes [7]. Our studies are in accord with other studies of the accumulation of T_N antigens on tumor cell membranes.
The characteristic alterations of lectin binding associated with malignant transformation of human urethelium might prove to be a tool in pathological practice, when evaluating bladder biopsies and urinary cytology specimens.

REFERENCES.

1. Decenzo, J.M., Howard, P., Irish, E.: J.Urol. 114, 874-878 (1975).

2. Summers, J.L., Coon, J.S., Wand, R.M., Falov, W.H., Miller III, A.W., Weinstein, R.S.: C.Res. 43, 934-939 (1983).

3. Stellner, K., Hakomori, S., Warner, G.A.: Biochem. Res. Commun. 55, 439-445 (1973).

4. Bergkvist, A., Ljungkvist, A., Mobergh, G.: Acta. Chir. Scand. 130, 371-378 (1965).

5. Feizi, T.: Conti. Oncol. 19, 51-63 (1984).

6. Steane, S.M.: In: A seminar on blood cells and body fluids. Bell, C.A. (ed.). American Association of blood banks, Washington D.C., 1-24 (1980).

7. Springer, G.F. 1984 Science 224, 1198-1206.

LECTIN HISTOCHEMICAL FEATURES OF MIDDLE EAR EPITHELIUM IN CHRONIC SECRETORY OTITIS MEDIA

Seppo Meri[1], Harry Holthöfer[1] and Tauno Palva[2]

[1]Department of Bacteriology and Immunology, University of Helsinki, Finland and [2]Department of Otolaryngology, University Central Hospital of Helsinki, Finland

Infection of the middle ear with all its sequelae presents one of the major problems in everyday pediatric otolaryngologic practice. Chronic secretory otitis media is characterized by secretion of highly viscous, mucoid effusion into the middle ear cavity thus allowing the designation "glue-ear disease" for this condition. The established precipitating factors of secretory otitis media are dysfunction of the Eustachian tube, inflammation and irritation within the middle ear (1,2). As a result, a disturbance in the function of the middle ear epithelium develops. Morphologically, the epithelium undergoes a transformation from one-layered cuboidal epithelium into a multilayered, often pseudostratified, form (3,4). The goblet cell density increases and mucous glands secreting mucosubstance-rich material appear.

In this work we studied the distribution and pathologic changes of cellular glycoconjugates within and on the surface of the middle ear epithelium in secretory otitis media using fluorochrome-conjugated lectins, which allow a more detailed characterization of tissue mucosubstances than the ordinary histopathological staining procedures.

Lectins, Vol. V

Materials and methods

14 biopsies of the middle ear mucosa were taken from the ears of 10 secretory otitis media patients during myringotomy and the installation of ventilation tubes. All patients had a history of a prolonged disease. Bacterial cultures were negative from all middle ear effusion samples.

FITC- or TRITC-conjugated lectins (WGA, UEA I, RCA, GSA I and WFA) were obtained from Miles laboratories, Inc., (Elkhart, Indiana) and from Vector Laboratories, Inc., (Burlingame, California). Working dilution of the lectins was 1:10 (in Dulbecco's balanced salt solution, pH 7.2; Orion, Finland) corresponding to 100 g lectin/ml. In some experiments the sections were treated with Vibrio cholerae neuraminidase (Behringwerke, Marburg, FRG) at a concentration of 0.1 U/ml (in Dulbecco) for 30 minutes at +37°C before staining with lectins.

Fluorescence microscopy. The biopsy specimens were frozen in isopentane cooled with liquid nitrogen. Cryostat sections were fixed with 3.5 % paraformaldehyde (in phosphate buffered saline, PBS, pH 7.2) for 10 minutes and the staining was performed as earlier described (5). For immunofluorescence microscopy (IFL) the slides were mounted with polyvinyl alcohol and studied uisng a Zeiss Standard Microscope equipped with a II FL epiilluminator and filters for FITC and TRITC fluorescence. Positive staining was assessed separately for the thick, mucous secretion (glue) above the epithelium, superficial and basal epithelial layers and for the epithelial basement membrane (BM, Table I).

Results

A specific typical binding pattern was observed for each of the lectins used (Table I). The mucus blanket on the surface of the middle ear was readily stained with WGA, UEA I and RCA lectins revealing differences in mucus formation between samples. The mucus usually appeared as a thick, irregular layer above the epithelium (Figs. 1 a and b). Air-dried mucoid secretions contained a network of mucosubstance material with occasional clusters of epithelial and inflammatory cells within it. In some cases the mucus blanket was present only as a thin layer above the superficial epithelium, the apical part of which was also stained with WGA, UEA I and RCA. Without glue this apical staining was less intense or absent.

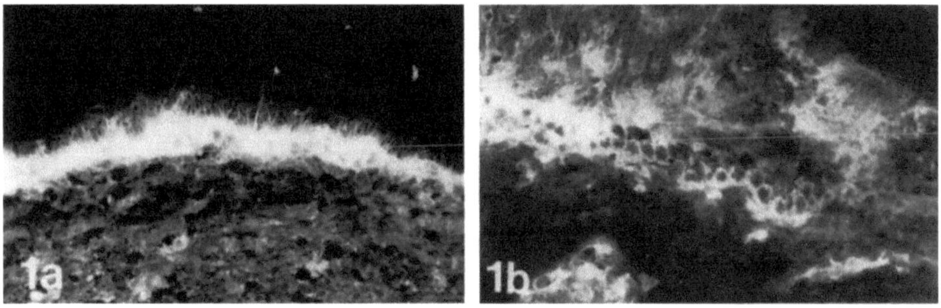

Fig. 1. Binding of FITC-WGA (a) and FITC-UEA I (b) to middle ear glue and epithelium in a frozen section of a middle ear biopsy from a patient with the glue-ear disease.

In the lamina propria (Fig. 1b) UEA I lectin stained constantly the vascular endothelia. In cases with marked hyperplasia, the binding pattern of UEA I and WGA lectins allowed a subdivision of the epithelium into 3 layers, an apical layer with intense staining, an intermediate layer with weak or absent staining and a basal layer of goblet cells, where the staining most clearly appeared at cell surfaces with a preponderance towards the apical

TABLE 1

Binding of selected fluorochrome-conjugated lectins to frozen sections of middle-ear biopsies from SOM patients

Lectin	Glue	Epithelial layer		BM	other
		Superficial	Basal		
WGA	+++	+++	+	-	-
RCA	++	++	+	+	diffuse staining of lamina propria
UEA	++	++	++	+	vascular endothelia
GSA I	-	+	-	-	mononuclear cells in the lamina propria
WFA	-	-	++	+	-

Glue refers to mucous secretion above the epithelium.
BM = basement membrane

parts of the cells (Figs. 2 a and b). In cases with less severe hyperplasia an equal staining of all the layers in the epithelium was seen with WGA and UEA I. GSA I lectin stained weakly the apical section of the epithelium and some single cells in the lamina propria (Fig. 3b). A selective staining of the basal goblet cell layer of the middle ear epithelium, sometimes accompanied by staining of the basement membrane, was seen with WFA (Fig. 3c). In inhibition studies using specific sugars, added in 0.2M to the working dilutions of the respective lectins, a dis-

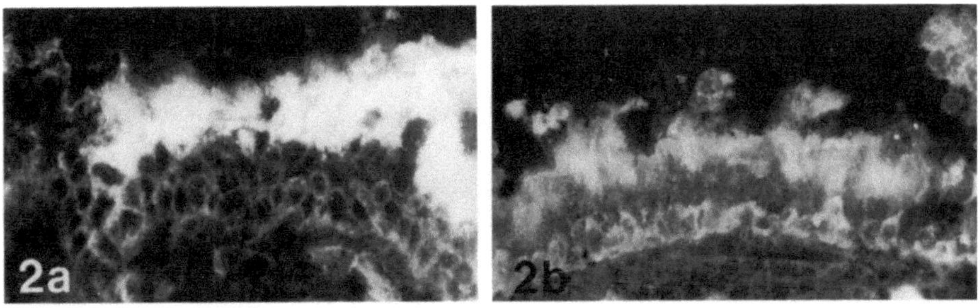

Fig. 2. Three-layered structure of the hyperplastic middle ear epithelium in secretory otitis media demonstrated with FITC-WGA (a) and FITC-UEA I (b). The glue substance and the basal goblet cell layer show bright fluorescence.

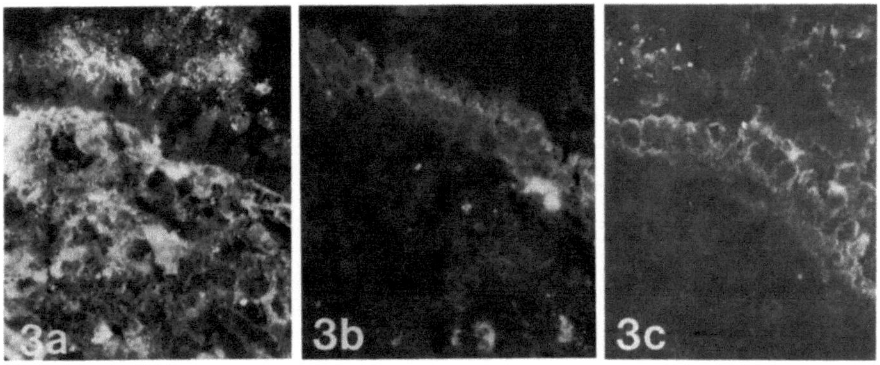

Fig. 3. FITC-RCA (a) binds to the glue substance and to lamina propria in a frozen section of a middle ear biopsy from a secretory otitis media patient. FITC-GSA I (b) binds to the superficial epithelium and FITC-WFA (c) to the basal epithelial layer or goblet cells.

appearance of the lectin binding was usually seen. Binding of RCA, however, was not totally inhibited with D-galactose. Prior neuraminidase treatment of the samples inhibited the specific binding of WGA.

Discussion

The middle ear epithelium undergoes prominent pathological changes during chronic secretory otitis media. As the secretion of muco-substances into middle ear is the principal pathophysiologic event in this disease, we choose to investigate changes in the epithelium with lectins of known carbohydrate specificity.

The appearance of the sticky glue on the surface of the middle ear epithelium is a characteristic feature of secretory otitis media (Fig. 1). The origin of the secretion is the hyperplastic epithelium in response to local irritation and a preceding bacterial infection, which is caused by pneumococci in about 60 % of cases. On the other hand, the clearance of the secretion is dependent on the absorptive capacity of the epithelium, especially when the Eustachian tube does not function. To perform these processes the epithelium must operate in a directional fashion. The polarity in the distribution of saccharide moieties in the epithelium could thus be regarded as an anatomic counterpart of this directional function, similarly as in other types of epithelia (6).

The hyperplastic basal cell layer of goblet cells showed distinct properties suggesting the expression of GalNAc/GlcNAc, NeuNAc and fucose residues on the cell surfaces (Figs. 2 a, b and 3c). When treated with neuraminidase, the staining of the basal cell layer with WGA disappeared. The differential staining properties of the goblet cells obviously reflect their specialized function in mucus secretion. During secretory otitis media the number of these cells increases considerably.

The staining of the superficial cell layer with several of the used lectins seems to be due to adsorption and coating of the cells with the glue substance, known to be rich in glycocompounds

(7). This is supported by the homogenous appearance of the cellular staining and concomitant staining of the glue. In cases without glue there was much less, if any, observable heterogeneity in the epithelium, Thus, during the glue-ear disease a spectrum of changes occurs in the mucosubstances suggesting of their important pathogenetic role in this disease.

Acknowledgements

We wish to thank Mrs. Tuija Kutvonen ane Mrs. Riitta Väisänen for technical assistance and Ms Eeva Kuusalo for expert secretarial work. This study was supported by the Duodecim Society.

REFERENCES

1. Sade, J. (1966). Arch. Otolaryngol. 86, 79-87.

2. Tos, M., Pulsen, G., Borch, J. (1979). Arch. Otolaryngol. 105, 582-588.

3. Tos, M. (1980). Arch. Otolaryngol. 106, 593-597.

4. Palva, T., Mäkinen, J., Rinne, J. (1981). ORL J. Otorhinolaryngol. Relat. Species 43, 241-247.

5. Holthöfer, H., Virtanen, I., Pettersson, E., Törnroth, T., Alfthan, O., Linder, E., Miettinen, A. (1981). Lab. Invest. 45, 391-399.

6. Spicer, S.S., Baron, D.A., Sato, A., Schulte, B.A. (1981). J. Histochem. Cytochem. 29, 994-1002.

7. Vered, J., Eliezer, N., Sade, J. (1972). Ann. Otol. 81, 394-500.

The Effect of Prostaglandin E_2 (PGE_2) on the Lectin Binding Pattern in Rat Gastric Mucosa Injured by Ethanol or Sodium Taurocholate.

B. Fringes, D. Lorenz, W. Oehlert and P.J. Klein
Institute of Pathology, University of Freiburg, Albertstrasse 19, D-7800 Freiburg i.Br., FRG

Prostaglandins, especially those of the A, E and I types which are endogenously synthesized within the gastro-intestinal wall, have the ability to protect the gastric mucosa against different noxious agents (12). This maintenance of the cell's integrity may be enhanced additionally by exogenous administration of natural prostaglandins or by their methylated analogues. By definition this cytoprotective effect of prostaglandins is unrelated to their ability to inhibit gastric acid secretion (12, 13). The dosage for maximal cytoprotection of the gastric mucosa by prostaglandins is some orders of magnitude lower than the dose that inhibits gastric acid secretion (13).

Cytoprotection proved to be effective in animals and humans, and the methylated prostaglandin analogues are in clinical use to prevent ulcers and erosions in human gastric mucosa (14). The mechanism of action underlying cytoprotection induced by prostaglandins is unknown and is still a matter of debate. The following hypotheses have been proposed: 1. Modification of the gastric mucosal circulation (8). 2. Secretion of mucus and bicarbonate (5, 16). 3. Stabilizing of cell membranes by stimulation of the sodium pump and accumulation of cyclic AMP (18), for review consult (9). The cytoprotective effect of prostaglandins was first discovered to work on mucus-producing and excreting epithelial cells (12). More recent papers also mention cytoprotective effects of prostaglandins on liver parenchyma and pancreas (10, 15). Based on these results, one would have to assume a very unique cellular process underlying the mechanism of action of the cell's protection by prostaglandins.

In a recent paper we have described longer (i.e. deeper) zones of cells within the foveolae of the gastric glands storing neutral and acid mucins in the prostaglandin-protected animals, compared with untreated animals and with animals treated with noxious agents (3). This report describes the application of a battery of biotinylated lectin conjugates to localize glycoconjugates in the

Lectins, Vol. V

gastric mucosa of rats injured by ethanol or sodiumtaurocholate compared with animals pretreated and protected with Prostaglandin E_2 (PGE_2) at cytoprotective dosages.

Materials and Methods

Animals: Male Wistar rats (Ivanovas, Kisslegg) with a mean body weight of 210 g (± 15 g) were first fed on standard diet. Animals were fasted for 24 hours, but were allowed water ad libitum. They were placed in special stainless steel cages constructed to prevent coprophagy. All substances were administered topically by esophageal gavage in order to avoid the systemic side effects of prostaglandins. Intubations were applied to conscious rats.

Substances: Prostaglandin E_2 (PGE_2) was from Sigma Chemicals, Munich, and was dissolved in 5% ethanol and phosphate buffer (pH 7.4). Sodiumtaurocholate, pKa 1.8 (Fluca, Buchs) was solved in 0.2n hydrochloric acid (pH 1.0). The purity of sodiumtaurocholate was analysed by DEAP-Sephadex-column chromatography and proved to be close to 90%. Biotinylated lectins were from Vector Laboratories Inc. (Burlingame). The conjugated lectins used in this study were: Arachis hypogea agglutinin (PNA), Ricinus communis agglutinin (RCA I), Ulex europaeus agglutinin (UEA I), Canavalia ensiformis (ConA), Dolichos biflorus agglutinin (DBA), Glycine max agglutinin (SBA), and Triticum vulgaris agglutinin (WGA). Avidin-Biotin, Standard Kit (Vector Laboratories Inc.). Inhibitory monosaccharides: D-Galactose (Gal) was from Merck, Darmstadt; α-Methyl-D-Manoside from Fluca; N-Acetyl-D-Galactosamine (GalNAc); N-Acetyl-D-Glucosamine (GlcNAc) and α -L-Fucose (Fuc) from Sigma Chemicals, Munich. Neuraminidase (Vibrio cholerae), Behring-Werke, Marburg (FGR).

Experimental set-up: Two groups of rats (15 animals each)received a single topical application of PGE_2 (200 µ/kg body weight) by esophageal gavage, 30 min prior to the topical application of either 1 ml concentrated ethanol (group A), or acidified sodiumtaurocholate, 80mM, pH 1.0 (group B). Two further groups received equal amounts of either absolute ethanol (group C), or sodiumtaurocholate alone (group D). Two control groups were given the respective vehicles, i.e. 1 ml 5% ethanol or 1 ml 0.2n HCl. Exact survival time for all animals was 4 hours. One group of rats received topical PGE_2 exclusively (200 µg/kg). Animals were killed in ether anaesthesia by exsanguination. The stomachs were immediately excised, opened along the greater curvature, fixed on cork plates and immersed in phosphate-buffered 10% formalin. After fixation for 24 hours, 2-4 mm thick tissue slices were taken from the fundus and the antral wall and were embedded in paraffin according to routine procedures. Tissue sections were stained with hematoxilin-eosin, PAS and Alcian blue (pH 2.5).

Quantitative assessment of gastric lesions: The gastric mucosal erosions were counted in a binocular magnifier (Photomicroscope 400 M, Fa. Wild). Their severity (number and size of the hemorrhagic erosions) was transformed into a point scale, the lesion score, modified according to Carmichael et al., 1978 (2). The significance of the differences of the lesion score in the different experimental groups was evaluated by the Student t-test and the chi-square test.

Lectin immunohistochemistry: For light microscopic histochemistry, 4-5 um thin sections were cut from the formalin-fixed, paraffin-embedded tissues. The sections were deparaffinized with xylen, rehydrated in graded ethanols and carried through the following steps at room temperature.

1. Exposure to 3% hydrogen peroxide in concentrated methanol for 20 min to destroy endogenous peroxidase activity. 2. A brief rinse in Tris buffer (pH 7.4). 3. Preincubation with a solution of normal swine serum (10%), bovine serum albumin (20%), gelatine (5%) in the above buffer. 4. Incubation for 30 min in the respective biotinylated lectins at the following dilutions in Tris buffer: 1:200 for PNA, SBA, UEA; 1:500 for RCA, WGA, DBA; 1:300 for ConA. 5. Two rinses in buffer (5 min each). 6. Incubation for 30 min with the avidin-biotin-complex. 7. Two rinses in buffer. 8. Incubation in diaminobenzidine (DAB) hydrogen peroxide medium for 10 min (5 mg DAB and 10 µl H_2O_2 in 10 ml Tris buffer). 9. Two rinses in Tris buffer and one water rinse. 10. Counterstain with hematoxilin-eosin followed by dehydration through graded ethanols, clearing in xylene and mounting with permount.

Digestions of sections: one set of tissue section from the same animal groups was incubated (between steps 3 and 4) with neuraminidase (70 µl per 1 ml buffer) for 30 to 45 min at 37°C in a moist chamber. The effect of the reactions was proved by the acquired PNA expression of erythrocyte membranes.

Controls included:
1. Omission of the biotinylated lectin conjugates for the detection of endogenous peroxidase. 2. Omission of the avidin-biotin-complex. 3. Blocking of lectins with their inhibitory sugars (lectins were incubated with 0.1 M of their respective inhibitory monosaccharides for 30 min at room temperature before incubation to the tissue sections).

For the quantification of the lectin-positive cells within the rat gastric corpus mucosa the following conditions were met:
1. Tissue slices of the gastric wall were cut perpendicularly to the surface; 2. Measurements were performed exclusively at sites of the non-eroded areas of the corpus mucosa (in addition peri-erosive areas were excluded); 3. A "fold-biased" sampling was introduced (3), i.e. measurements were restricted to the intermediate areas between two gastric mucosal folds; 4. Estimations of lectin-positive epithelial cells were restricted to gland profiles visibly cut at the total of their long axes; 5. Numbers of lectin-positive epithelial cells were recorded per gastric gland; 6. A total of 15-30 single measurements per rat was performed (i.e. 90-120 single data per group); 7. Grading intensity of lectin-positive cells is based on a subjective scale from (+) faint to (++) moderate and (+++) intense.

Gastric ulcers and lectin histochemistry were scored by a treatment-blinded observer.

Results

No macroscopic evidence of mucosal injury was noted in any of the control stomachs and only in one of the stomachs exposed to PGE_2 alone. In contrast, all animals exposed to either ethanol or taurocholate alone (group C + D) showed multiple hemorrhagic erosions that were preferentially located in the corpus fundus region, the mean lesion score being 90.6 ± 58.5 and 24.6 ± 11.4 points respectively. This score was significantly reduced by 86% and 93% within the PGE_2-pretreated, protected rats ($p < 0.005$ and < 0.0005, with the chi-square test for the taurocholate and PGE_2 + taurocholate group significant at the 99.5 percentile).

Light microscopically, the ethanol-induced erosions were broad-based with a tendency to exfoliate great sheets of the mucosa. In contrast, the taurocholate-induced erosions were smaller and preferred the area between the gastric folds. Further histologic data on erosions and mucus stains have been detailed elsewhere (3). Each of the seven lectins exhibited characteristic histologic staining patterns (data not shown). Two of them were investigated semi-quantitatively (PNA and SBA). In the controls and in the animals treated with PGE_2 alone, the biotinylated PNA conjugate disclosed a striking and selective affinity for a sharply delined band of parietal cells in the upper 1/3 (isthmus) of the corpus glands (Fig. 1a). These epithelial cells exhibited a defined grading from faint to intense, continuously increasing the higher the cell's position within the gland gets. A second zone of PNA-positive cells corresponded to mucus neck cells. Semi-quantitative evaluations of the PNA-positive parietal cells in the taurocholate- or ethanol-injured rats (see also Fig. 1b) revealed a significant ($p < 0.005$) numerical decrease of the mean number of intensely (+++) PNA-reactive cells from about 4.9 ± 0.9 cells per gland profile in the controls to 1.8 ± 1.6 taurocholate- and 1.9 ± 0.6 cells in the ethanol-treated rats (Table 1). If the total number of PNA-positive parietal cells per gland was taken into account, this corresponded to a relative decrease of 26% and 27% respectively. In contrast, within the PGE_2-pretreated, protected animals (group A + B) the number of intensely stained parietal cells per gland profile was close to control values, in spite of the application of equal amounts of injurious substances (see Table 1). The absolute and the relative numbers of parietal cells with moderate and faint affinity for PNA conjugates were higher in both the injured (group C + D) and the PGE_2-pretreated ethanol animals (group A), when compared with controls.

Fig. 1 a

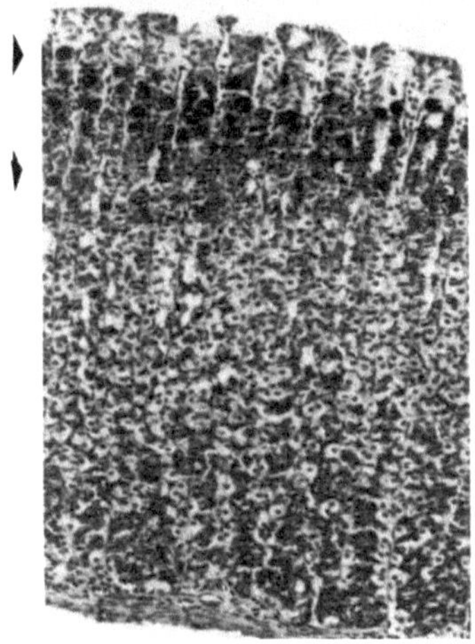

Fig.1b

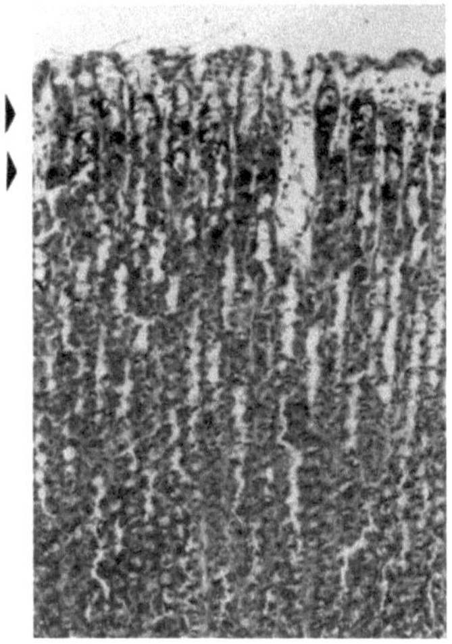

Fig. 1a: Lectin histochemistry of the corpus mucosa of a control animal showing PNA-positive parietal cells in the isthmus zone (arrow heads). Terminal D-galactose is expressed by parietal cells in the upper fifth of the corpus glands giving a clear-cut band of cells. Staining intensity increases the closer the cell's position gets to the opening of the gastric pits.

Fig. 1b: Parietal cell profiles that express terminal D-galactose, as shown by the positive PNA reaction, are reduced in number per gastric corpus gland profile in a taurocholate-treated rat.

The total number of all PNA-reacting parietal cells per gland profile in all experimental groups was obviously unchanged (Table 1). The same was true for the PNA-positive mucus neck cells. *Based on a subjective survey, the amount of PNA-positive parietal cells after sialidase digestion did not differ from the one on the undigested slides. *(see Table 2).

Semi-quantitative estimations of the mucus-producing SBA-positive surface epithelial cells, located within the gastric pits, revealed significantly ($p < 0.005$) higher amounts of (++)-stained cells per gland profile in the PGE_2-pretreated, protected taurocholate and ethanol stomachs (36% and 66% respectively), when compared with the corresponding injured animal groups (28% and 34%).

In contrast, absolute and relative amounts of the (+) faintly SBA-positive surface epithelial cells per gland profile were 8% and 30% higher in the taurocholate- and ethanol-injured stomachs respectively when compared with the PGE_2-protected ones (dato not shown). The mean total number of SBA-expressing cells, measured per gland profile, was reduced by 50% in the injured but only by one third in the protected stomachs, when compared to controls.

Omission of the lectin conjugates totally abolished the staining; the same was true after application of the corresponding inhibitory sugars, except for GlcNAc that markedly decreased the staining with WGA.

Discussion

The property of prostaglandins to maintain cellular integrity, in spite of the administration of noxious agents, has been termed cytoprotection, but the exact mechanism has not been defined (9). We used both sodium taurocholate and ethanol to reproducibly in-

Table 1

Number of PNA-positive cells in rat gastric corpus mucosa injured either by ethanol or sodium taurocholate compared with controls and PGE_2-pretreated, protected stomachs.

staining intensity parietal cells	controls (n=4)	PGE_2 (n=4)	tauro-cholate (n=3)	PGE_2 + tauro-cholate (n=4)	ethanol (n=3)	PGE_2 + ethanol (n=5)	
+++	4.89	4.07	1.77	*3.80	1.95	*3.98	m.
	0.90	0.20	1.65	0.69	0.68	1.02	s.d.
++	4.33	4.30	5.17	*4.72	5.46	*4.53	m.
	0.28	0.96	1.18	1.42	1.47	0.49	s.d.
+	1.65	1.71	2.44	1.72	3.23	3.15	m.
	0.72	0.62	0.33	1.31	1.6.	1.10	s.d.
total number	10.87	10.08	9.34	10.24	10.63	11.66	

*Table 1 gives the number of PNA-positive parietal cell profiles estimated per gastric gland profile in rat corpus in the different treatment groups. The grading in the staining intensity is based on a subjectively estimated scale from (+) weakly via (++) moderate to (+++) most reactive. The numbers give the mean ± s.d. of 30 estimations per animal, i.e. 90 to 120 estimations per group. (*p values of $<$ 0.005, in the pretreated, protected groups compared with their corresponding non-protected group*

Table 2

Number of PNA-positive mucus neck cells in rat gastric corpus mucosa injured either by ethanol or sodium taurocholate compared with controls and PGE_2-pretreated rats.

number of mucus neck cells	controls	PGE_2	tauro-cholate	PGE_2 + tauro-cholate	ethanol	PGE_2 + ethanol	
+++	3.17	2.17	2.50	2.55	2.74	3.14	m.
	1.03	0.58	0.57	0.49	0.66	0.45	s.d.

The numbers of intensely PNA-stained mucus neck cell profiles were estimated per gastric corpus gland profile. Data represent mean and standard deviations of 30 estimations per animal and 90-120 single estimations per group. The number of PNA-positive mucus neck cells is higher in the protected stomachs compared with the injured rats.

duce gastric erosions in the rat since both substances frequently cause gastric lesions in humans. A role of mucus glycoproteins and gastric intracellular glycoconjugates in the defense mechanism against ulcerations has long been suggested (1, 6, 20). Our investigation was directed to the mucosal secretions and glycoconjugates still being in situ. The results of our present study using two different noxious agents that reproducibly induced multiple gastric erosions closely agree on significantly reduced numbers of intensely stained PNA-positive epithelial (parietal) cells in histologically undamaged areas of the otherwise eroded gastric corpus mucosa of rats. In addition, our results show a significantly less reduced number of the instensely PNA-expressing cells in the PGE_2-pretreated, protected animal groups. The additional observation of a higher amount of moderately and faintly PNA-stained epithelial cells in the injured as well as in the protected animals, as compared to controls, together with the obviously unchanged thus constant number of PNA-positive epithelial cells per gland profile allows the following tentative suggestions:

(i) Glycoprotein synthesis in the gastric mucosa may be quantitatively reduced or slowed down in the injured stomachs, but seems not totally blocked.

(ii) This slowing down seems less intense in the PGE_2-protected rats, thus prostaglandin cytoprotection may be associated with a tendency of cells to maintain biosynthesis of glycoconjugates at a level closer to controls.

(iii) The reduction in number of the intensely PNA-positive cells

could emerge as a loss of cells out of the gastric pits.
(iiii) This loss of cells is significantly less in the PGE_2-protected animals.

The same holds true for the SBA-expressing surface epithelial cells although the total loss of these cells was higher. The type of glycoprotein in the PAS-negative, PNA-positive parietal cells demonstrating terminal ß-galactosyl residues is not fully understood (19). The strongly PAS-positive but PNA-negative, sialidase PNA-negative surface epithelial cells with affinity for SBA seem to have mucosubstances with internal galactose and terminal n-acetyl-galactosamine (19).

Recent in vitro incorporation studies using ^3H-glucosamine mention a significant prevention of the ethanol-induced decreases in incorporation activity for gastric macromolecular glycoproteins by pre-treatment with PGE_2 (4, 11). Slomiany suggests that PGE_2 may provide protection for the enzymes involved in the biosynthesis of mucus glycoprotein (17).

It this context it may also be of interest to note that the time for maximal cytoprotective action of prostaglandins subsequent to their administration averages the time intracellular glycoconjugates post-translationally need to appear at the cell membranes (7).

In conclusion, our histochemical investigation supports the hypothesis that the mechanism of action of cytoprotective prostaglandins at the cellular level may be mediated by the maintenance of normal amounts of intracellular glycoconjugates, thus increasing the defensive, anti-ulcerogenic factors.

Acknowledgments:
The authors would like to thank Ms. B. Wild, Ms. G. Roth and Ms. E. Nietzsche for their excellent technical assistance in the performance of the experiments and the histochemistry. The patience of Mrs. U. Müller and Mrs. H. Ehret in typing this manuscript is gratefully acknowledged.

References

1. Allan, A., A. Garner: Gut 21, 249-262 (1980)
2. Carmichael, H.A., L.M. Nelson, R.J. Russell: Gastroenterol. 74, 1229-1232 (1978)
3. Fringes, B., D. Lorenz, W. Oehlert: Path.Res.Pract. 179, 478-486 (1985)
4. Fujimoto, N., S. Murakami, T. Suwa, Y. Mori: J.Pharm.Dyn. 7, 864-871 (1984)
5. Garner, A., J.R. Heylings: Gastroenterol. 76, 497-503 (1979)
6. Glass, G.B.J., B.L. Slomiany: In: Mucus in Health and Disease, ed. by M. Elstein and D.V. Parke; Plenum, New York, 1977, pp. 311-347
7. Hannover, J.A., W.J. Lennarz: Arch.Biochem.Biophys. 211, 1-19 (1981)
8. Konturek, S.J., A. Robert, A.J. Hanchar, J.E. Nezamis: Digest. Dis.Sci. 25, 673-679 (1980)
9. Konturek, S.J.: In: Magen und Magenkrankheiten: Pathophysiologische Aspekte. W. Domschke, K.G. Wormsley (eds.), G. Thieme Verlag, Stuttgart-New York, 1981, pp. 175-184
10. Martin, D.M., A.O. Someren, S.M. Nasrallah: Gastroenterol. 81, 736-741 (1981)
11. Quadros, E., E. Ramsamooj, D.E. Wilson: Digestive Disease Week, May 12-15, 1985, New York, Abstr. 674
12. Robert, A., A.J. Hanchar, C. Lancaster, J.E. Nezamis: Fed. Proc. 37, 460 (1978)
13. Robert, A.: Gastroenterology 77, 761-767 (1979)
14. Ruppin, H., B. Person, A. Robert, W. Domschke: Scand.J.Gastroenterol. 16, 647-652 (1981)
15. Ruwart, M.J., B.D. Rush, N.M. Friedle, R.C. Piper, G.J. Kolaja: Prostaglandins 21, 97-102 (1981)
16. Selling, J.A., D.L. Hogan, M.A. Koss, J.I. Isenberg: Digestive Disease Week, New York, May 12-15, 1985, A. 57, 225
17. Slomiany A., A. Takagi, M. Szymanska-Kosmala, B.L. Slomiany: Digestive Disease Week, New York, May 12-15, 1985, A. 169, 677
18. Soll, A.H., B.J.R. Whittle: Prostaglandins 21, 39-45 (1981)
19. Stoward, P.J., S.S. Spicer, R.L. Miller: J.Histochem.Cytochem. 28, 979-990 (1980)
20. Szabo, S., J.S. Trier, P.W. Frankel: Science 214, 200-202 (1981)

LECTIN - BINDING PATTERNS OF THE GOLGI APPARATUS

Margit Pavelka and Adolf Ellinger

Institute of Micromorphology and Electronmicroscopy,
University of Vienna, Schwarzspanierstrasse 17
A - 1090 Vienna, Austria

The Golgi apparatus, representing a complex system of stacked cisternae (saccules), tubules and vesicles, is a main crossroad in the intracellular traffic (for recent reviews 1,2,3); Golgi elements are main sites of posttranslational processing and targeting of newly synthesized molecules and function as recipients for molecules retrieved from the cell surface and other cell compartments. Main questions as to the organization of this complex organelle are directed to possible functional subcompartments and their location in the Golgi area. Among morphological techniques, demonstration of binding sites for lectins of different sugar specificities (4) represents one of the promising tools to get further knowledge of Golgi organization (e.g. 5-10). In the present work, we report on the reaction patterns of the Golgi complex of diversely specialized cells, covering rat duodenal goblet and absorptive cells and acinar cells of the embryonic pancreas, obtained with 4 different lectins: Concanavalin A (ConA), binding to α -D-mannose and α -D-glucose; Ricinus communis I agglutinin (RCA I), preferentially interacting with terminal ß-D-galactosyl residues; Helix pomatia lectin (HPA), binding to N-acetyl-D-galactosamine; and Pisum sativum lectin (PSA), recognizing α-D-mannose and α -D-glucose, for high affinity-binding to glycopeptides presence of a fucose residue attached to the asparagine-linked N-acetyl-glucosamine being essential (11,12).

Lectins, Vol. V

Materials and Methods

Female albino rats, including mated animals, were obtained from the Research Institute for Test Animal Breeding Himberg of the University of Vienna.
The tissues, - segments of the duodenum and pancreatic rudiments of rat embryos of 13 to 20 days of gestation, - were fixed in a mixture of 4% paraformaldehyde (freshly prepared) and 0.5% glutaraldehyde (electronmicroscopy grade, Merck, Darmstadt, FRG) for 30min at 4°C and collected in 0.1M sodium cacodylate buffer. In order to block free aldehyde groups, the tissues were treated with 0.05M ammonium chloride for 2hr at 20°C.

For pre-embedding lectin-labeling, 10μm thick cryocut-sections, thawed in phosphate-buffered saline (PBS), were incubated in solutions in PBS of the respective lectins or lectin-horseradish peroxidase (HRP) conjugates for 4hr at 20°C: ConA - 200μg/ml, followed by incubation in HRP (200μg/ml, 4hr, 20°C); RCA I-HRP, HPA-HRP, and PSA-HRP (all E-Y Laboratories, San Mateo, Ca.) each 100μg/ml. After several rinses in PBS, the peroxidase activity was visualized by means of the diamine benzidine reaction; the sections were postfixed in 1% osmium ferrocyanide, dehydrated in ethanol and embedded in Epon.

For post-embedment lectin-staining, the tissues were embedded in "London Resin White"; thin sections, mounted on gold or nickel grids, were incubated in ferritin or colloidal gold conjugates of the respective lectins (E.Y. Laboratories, San Mateo, Ca.) for 60min at 20°C (ConA-ferritin - 1mg/ml; RCA I-gold - 80μg/ml; HPA-gold - 35μg/ml); after several rinses in bidestilled water, the sections were counterstained in 1% aqueous phosphotungstic acid.

For cytochemical controls, the lectin-incubation step was omitted prior to incubation in HRP; the specificity of the binding reactions was tested by addition to the incubation media of the respective competitive sugars: 0.2M methyl-α-mannopyranoside (Con A PSA), 0.4M D-galactose (RCAI), and N-acetyl-D-galactosamine (HPA).

Results

The lectin-binding reactions were studied by means of two methods: pre-embedment incubation of cryosections of prefixed tissues by using HRP for electronmicroscopic visualization, and post-embedment labeling of thin sections by making usage of lectin-ferritin conjugates and the colloidal gold system. The results obtained with the two different techniques corresponded to each other.

ConA (Fig. 1)
In the goblet cells and the majority of the absorptive enterocytes ConA binding sites were confined to cisternae of the cis side of the Golgi stacks. By contrast, in the embryonic

acinar cells the reaction patterns were variable: in any stage of differentiation, the ConA reaction either was restricted to the cis Golgi side (Fig. 1a) or distributed across the entire set of stacked cisternae (Fig. 1b). In addition, ConA intracellularly reacted with the endoplasmic reticulum and stained several vesicles and lysosomes.

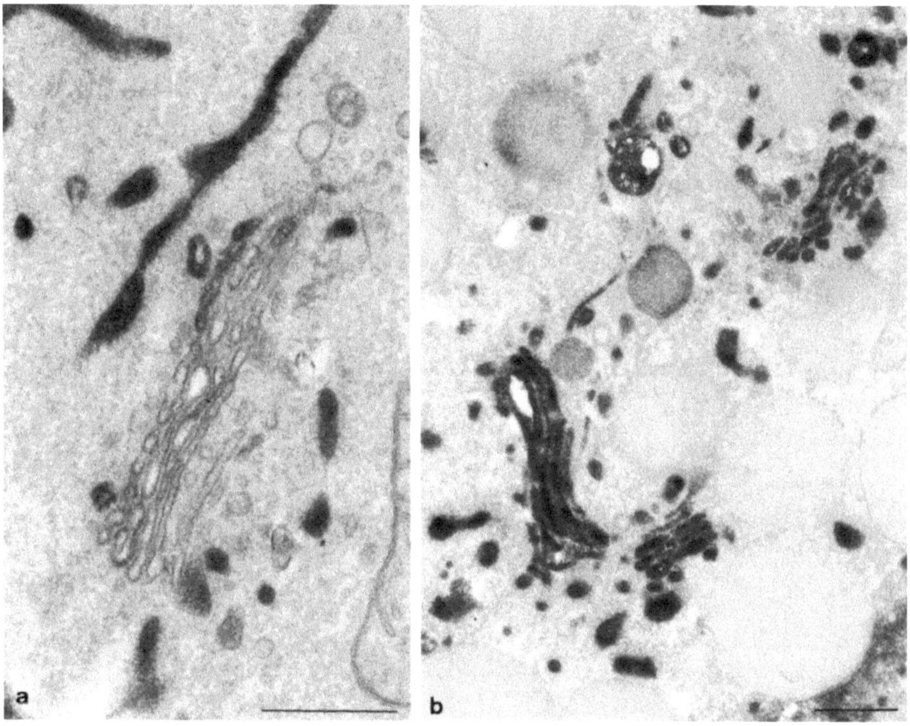

Fig. 1: ConA-HRP Embryonic acinar cells - days 16 (a) and 18 (b) of gestation. ConA label either is confined to the cis Golgi side (a) or distributed across the entire set of cisternae (b). bars 0.5µm

RCA I (Fig. 2)

The goblet cells exhibited intense RCA I reaction of cisternae of the trans Golgi side; in the absorptive cells, in addition to staining of trans saccules, RCA I binding sites were concentrated in medial cisterns of the stacks (Fig. 2a). The RCA I reaction was variable in the embryonic acinar cells, the label either being confined to trans and / or medial saccules (Figs. 2b,c) or distributed across elements of all compartments of the stacks (Figs. 2d,e). The RCA I reaction included several Golgi-associated vesicles and lysosomes.

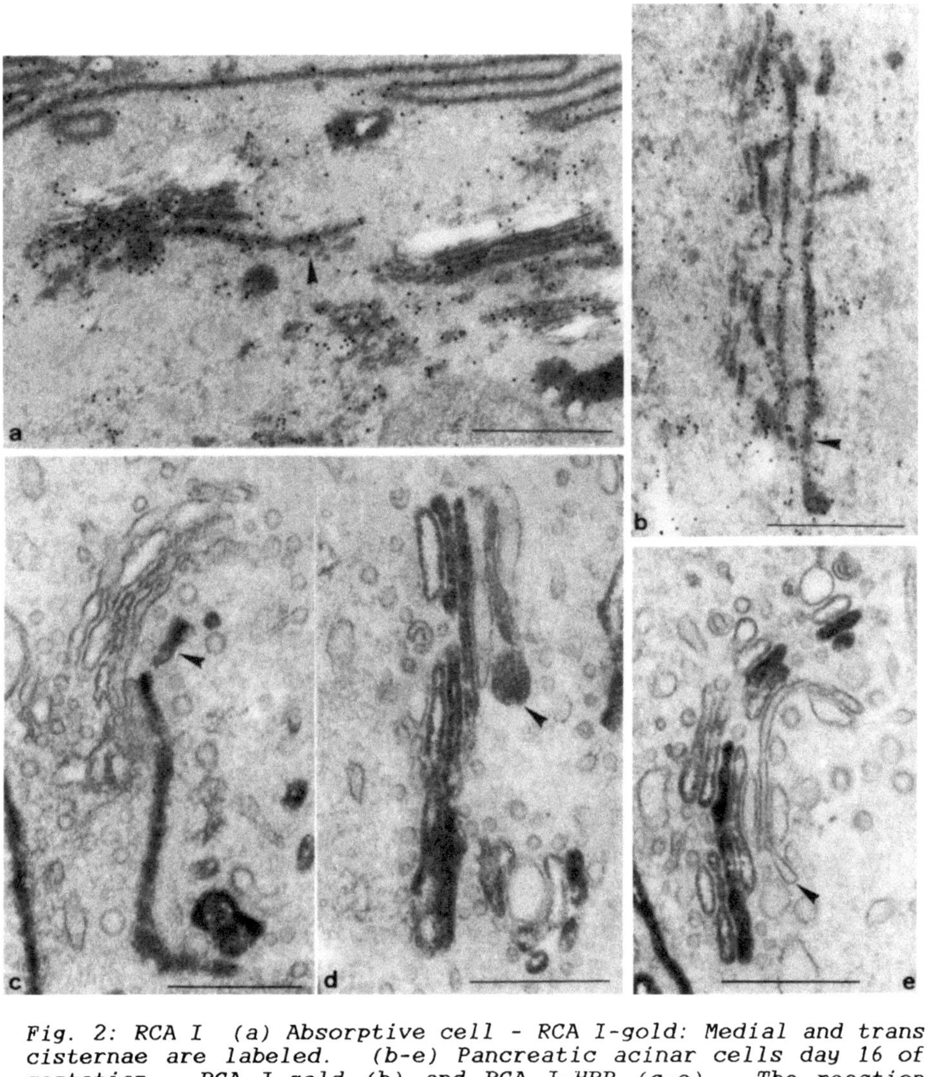

Fig. 2: RCA I (a) Absorptive cell - RCA I-gold: Medial and trans cisternae are labeled. (b-e) Pancreatic acinar cells day 16 of gestation - RCA I-gold (b) and RCA I-HRP (c-e). The reaction either is confined to medial and trans cisternae (b,c) or includes saccules of the cis side (d,e).
trans side ► *bars 0.5μm*

HPA

Goblet and absorptive enterocytes exhibited HPA label of cis and medial saccules, the cismost cisterna frequently being devoid of reaction. In the embryonic acinar cells HPA reaction either labeled cis and/or medial saccules or was extended across the entire battery of the stacked cisterns. In addition, HPA labeled several of the apical vesicles as well as lysosomes.

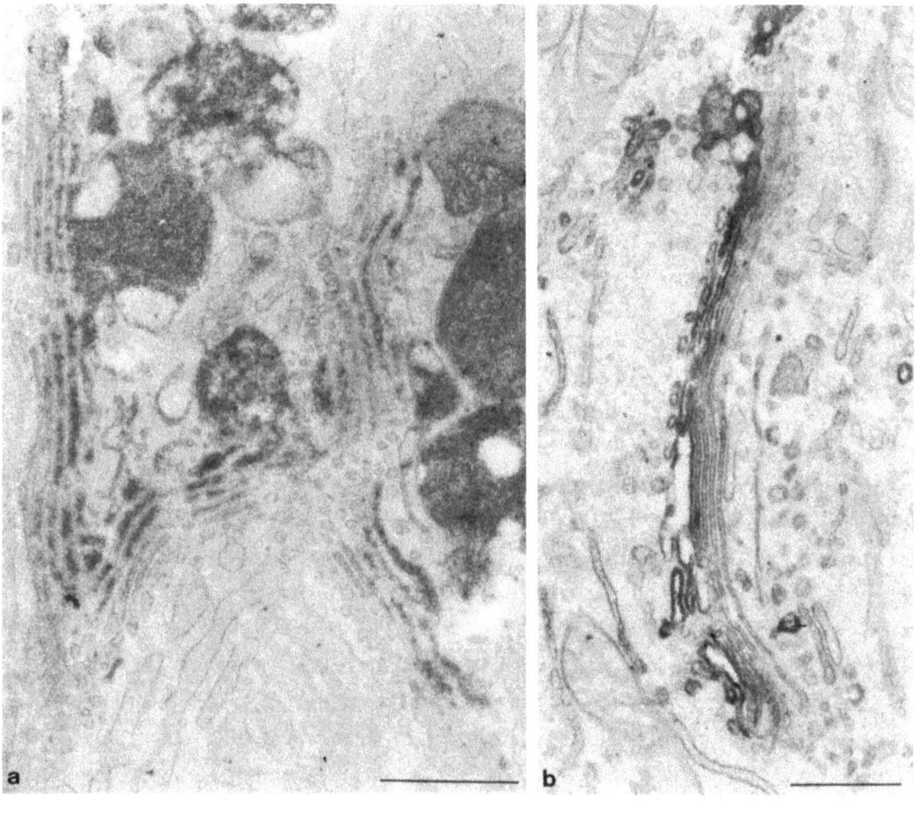

Fig. 3: PSA-HRP (a) Goblet cell - the reaction increases from the cis to the trans side of the stacks. (b) Absorptive cell - PSA-label is concentrated in cis cisternae. bars 0.5µm

PSA (Fig. 3)

In the goblet cells PSA reaction was apparent in the majority of the stacked cisternae, the reactions' intensities frequently increasing from the cis to the trans side (Fig. 3a); secretion granules were reactive. The absorptive cells were characterized by intense PSA staining of cis and/or medial saccules. On several stacks, faint reaction was observed in the transmost cisterna and Golgi-associated vesicles. Some of the cisterns of the endoplasmic reticulum showed weak staining; secondary lysosomes were intensely labeled.

In general, within the individual cisterns considerable variabilities of the reactions' intensities were noted with each of the lectins tested.

Control preparations were devoid of intracellular label.

Discussion

Our results showed variable lectin-binding patterns of the Golgi apparatus.

In the goblet cells and most of the absorptive cells the reactions obtained with ConA, RCA I and HPA were confined to distinct compartments of the stacks: ConA and HPA, recognizing mannose/glucose and N-acetyl-galactosamine respectively, stained cisternae of the cis Golgi side, whereas RCA I label, indicating presence of terminal ß-D-galactosyl residues, was concentrated in trans and/or medial saccules of the stacks. This "confined" reaction corresponds to that obtained in a number of other cell types (e.g. 6-9), and is in agreement with the conception of a functional subcompartmentalization of the Golgi stacks (recent reviews 3,13,14). The differentiated staining of clear-cut Golgi regions may be considered as mirroring Golgi-associated steps in the biosynthesis of carbohydrates, e.g. conversion of high-mannose oligosaccharides of glycoconjugates, demonstrated by the ConA reaction of cis saccules, to complex-type species, visualized by RCA I label of the trans Golgi side. The cis-confined HPA reaction may reflect initial steps in the synthesis of O-linked saccharides.

By contrast, the variable reactions obtained with ConA, RCA I, and HPA in the embryonic acinar cells, which in part include elements of all Golgi compartments, cannot readily be explained by a conception which assumes strict subdissection of the Golgi stacks. This is particularly obvious in the case of the RCA I reaction, since galactosyltransferase has immunocytochemically been shown confined to the trans section of the Golgi stacks

(15,16). The presence of terminal galactosyl residues in cisternae of each side of the stacks may be explained - (a) by assuming the galactose-transferring enzymes as being located in all of the Golgi compartments, such as has been suggested for 3 terminal glycosyltransferases from studies with hepatocytes' Golgi fractions (17), or - (b) by supposing galactose-terminated glycoconjugates as becoming transported from the site of their synthesis to other Golgi regions; in the case of trans-location of galactosyltransferase (15,16), this implies trans-to-cis directed traffic, which is reverse to the cis-to-trans transport usually assumed for newly synthesized molecules (1,3,6,18), or - (c) molecules retrieved from the cell surface or other cell compartments must be considered as possibly accounting for the lectin reactions as well.

By contrast to ConA, a fucose residue attached to the asparagine-linked N-acetyl-glucosamine is essential for high affinity binding to glycopeptides of the mannose/glucose - recognizing Pisum sativum lectin (11,12). Hence, the reactions obtained point to the presence of glycoconjugates containing core-fucosylated N-glycosidically linked complex oligosaccharides. In goblet cells, PSA-binding molecules, presumably representing secretory constituents, were demonstrable in most of the stacked cisterns as well as in secretory granules, the reactions increasing in intensity from the cis to the trans side; this pattern contrasts to that in the absorptive enterocytes, in which PSA label was concentrated in cisternae of the cis Golgi side. These patterns are particularly interesting in regard to questions as to functional subcompartments of the Golgi stacks, since the PSA-labeled Golgi sections possibly are related to sites at which fucose becomes inserted into oligosaccharides.

Acknowledgements

The authors gratefully acknowledge the excellent technical assistence of Mrs.Jutta Selbman, Mrs.Elfriede Scherzer, Mrs. Gerlinde Hartl and Mr. Helmut Oslansky.

This work was supported by the "Vermächtnis Josefine Hirtl".

References

1. Farquhar,M.G.,and Palade,G.E.: J.Cell Biol.91, 77s-107s(1981)
2. Goldfischer,S.: J.Histochem.Cytochem.30, 717-733 (1982)
3. Berger,E.G.: Cell Biol.Int.Rep.9, 407-417 (1985)
4. Goldstein,I.J., and Hayes,C.E.: Adv.Carbohydr.Chem.Biochem. 35, 127-340 (1978)
5. Sato,A., and Spicer,S.S.: Histochem.J.14, 125-138 (1982)
6. Griffiths,G., Brands,R., Burke,B., Louvard,D., and Warren,G.: J.Cell Biol.95, 781-792 (1982)
7. Tartakoff,A.M., and Vassalli,P.: J.Cell Biol.97, 1243-1248 (1983)
8. Roth,J.: J.Cell Biol.98, 399-406 (1984)
9. Lucocq,J.M., and Roth,J.: Electron Microscopy 1984, Vol.3, 1621-1628, (1984)
10. Pavelka,M., and Ellinger,A.: J.Histochem.Cytochem.33, 905-914 (1985)
11. Kornfeld,K., Reitman,M.L., and Kornfeld,R.: J.Biol.Chem. 256, 6633-6640 (1981)
12. Debray,H., Decout,D., Strecker,G., Spik,G., and Montreuil,J.: Eur.J.Biochem.117, 41-55 (1981)
13. Tartakoff,A.M.: Int.Rev.Cytol.85, 221-248 (1983)
14. Dunphy,W.G., and Rothman,J.E.: Cell 42, 13-21 (1985)
15. Roth,J., and Berger,E.: J.Cell Biol.92, 223-229 (1982)
16. Strous,G.J., Kerkhof,P.van, Willemsen,R., Geuze,H.J., and Berger,E.G.: J.Cell Biol.97, 723-727 (1983)
17. Bretz,R., Bretz,H., and Palade,G.E.: J.Cell Biol.84, 87-101 (1980)
18. Bergman,J.E., and Singer,S.J.: J.Cell Biol.97, 1777-1787 (1983)

BINDING OF PHA, WGA, AND SBA TO THE SURFACE OF RAT INTESTINAL EPITHELIAL CELLS *IN VITRO*

Margita Čuperlović, Gordana Cerović and Zdenka Milošević

Institute of Endocrinology, Immunology and Nutrition - INEP
11080 Zemun, Yugoslavia

Many plant lectins possess a strong affinity for the intestinal epithelium. Lectin histochemistry revealed that Phaseolus vulgaris lectin (PHA) binds to the glycocalix or surface coat of the small intestine and also to some of the subepithelial structures (7). The total binding pattern of wheat germ lectin (WGA) to the intestinal sections is somewhat different, but it also adheres to the brush border and surface coat of the intestinal villus (7, 8). The lectin of soybean seed (SBA), when used as a histochemical marker, has as affinity for certain types of goblet cells, but the staining of the surface coat is normally negative (22). It has been shown that some lectins are also bound to the intestinal epithelium when introduced into isolated intestinal segments or fed to intact animals either in purified form or as raw beans (4, 9, 13, 17, 18). The binding of lectin is then followed by further events, such as phytohemagglutinin induced diarrheal disease (2, 15), or penetration of macromolecules into the intestinal mucosa (1). It seems that such effects are not directly proportional to the amount of bound lectin. Light microscopy indicates very massive binding of PHA and almost negligible of SBA to the epithelial surface, but in both cases with definite biological effects.

The purpose of this study was to examine and quantitate the binding of PHA, WGA and SBA to epithelial cells isolated from the rat small intestine. The lectins were labelled by FITC conjugation for light microscopy and with ^{125}I for the binding studies.

Lectins, Vol. V

Material and Methods

Lectins. PHA was isolated from red-spotted kidney beans as previously described (6). SBA and WGA were extracted from soybean seeds and wheat germ, respectively, and purified by affinity chromatography (23).

For microscopy lectins were conjugated with fluorescein isothiocyanate (FITC) using the method of The and Feltkamp (21). The obtained lectin-FITC conjugates had a protein concentration of approximately 1 g/l and a F/P molar ratio of 2.3-2.5. Lectins were labelled with ^{125}I by the chloramine-T method based on Hunter and Greenwood (10). The reaction mixture contained: 20 μl 0.3 M PBS (pH 7.5), 50 μl 0.2% lectin solution, 3.5 μl of carrier-free ^{125}I (1 mCi) and 10 μl chloramine-T solution (3 mg/ml). After 40 sec of incubation the reaction was terminated by the addition of 50 μl sodium metabisulfite solution (1.8 mg/ml). Finally, 1 ml PBS and 20 μl of 1% bovine serum albumin were added. Unbound ^{125}I was removed by extensive dialysis against PBS. The obtained preparations of labelled PHA, WGA and SBA had specific activities of $2\text{-}4.10^6$ cpm/μg of lectin.

Cell preparation. Non-fasting female rats of the Wistar strain were killed by cervical dislocation. The small intestine was removed and cut into pieces approximately 10 cm long. Epithelial cells were isolated according to a modification of the method of Weiser (24). Intestinal segments were rinsed with saline and filled with solution containing: 1.5 mM KCl, 96 mM NaCl, 27 mM sodium citrate, 8 mM KH_2PO_4, 5.6 mM Na_2HPO_4 (pH 7.3). After 15 min of incubation at 37°C, the segments were refilled with PBS containing 1.5 mM EDTA and 0.5 mM dithiothreitol. Incubation was repeated for another 15 min and separated epithelial cells were collected by centrifugation (900 x g for 5 min). The cells were filtered through a nylon net and rinsed thrice with PBS. They were counted and the viability was tested using trypan blue dye. The viability amounted to 80-90%. The cells were finally resuspended in PBS to obtain an average count of 10^7 cells/ml.

Binding of lectin-FITC conjugates. The isolated epithelial cells (0.1 ml corresponding to 10^6 cells) were incubated in the lectin-FITC conjugate solution for 30 min at room temperature. The cells were rinsed three times with PBS and resuspended in 0.2 ml PBS containing 0.05% bovine serum albumin (BSA). They were viewed with a Zeiss fluorescence microscope (Fluoval 2). The specificity of lectin binding was tested by inhibition with appropriate sugars (N-acetyl-D-glycosamine for WGA and N-acetyl-D-galactosamine for PHA and SBA).

Binding of 125I-labelled lectins. Serial dilutions of tested lectins were prepared to cover the concentration range of 0.006 to 200 μg/ml, using PBS containing 0.05% BSA for dilutions. Ten μl of appropriately diluted ^{125}I-labelled lectin was added to 150 μl of these cold lectin solutions. The activity of the test tubes was counted. Then 50 μl of the cell suspension containing 5.10^5 cells was added to each tube. Thus, the final lectin concen-

tration during the binding assay amounted to 0.48-143 µg/ml of culture.
The tubes were then incubated for 1 h at room temperature with occasional shaking. The supernatant fluid was separated after brief centrifugation (5 min, 1000 rpm) and the cells rinsed with PBS. The activity was counted once more.
In order to test the specificity of lectin binding, 1 M solution of the inhibitory sugars was mixed with the same volume of lectin solution, and the whole procedure was repeated. The specifically bound lectin was calculated by subtracting the nonspecifically bound from the total bound lectin.

Results and Discussion

The fluorescence patterns obtained after treatment of isolated epithelial cells with lectin-FITC conjugates revealed that PHA and WGA were similarly distributed on the cell membrane, staining primarily brush border regions. Treatment with SBA-FITC conjugate did not result in visible fluorescence on the cell membranes.

Lectin histochemistry of the intestinal tissues gave a similar binding pattern for PHA and WGA-FITC conjugates (7). However, due to the very intensive fluorescence located at the surface of the epithelial cells, lectin binding to the brush border structure was not readily distinguishable from the staining of the adhering mucus. The method used for the isolation of intestinal cells, however, resulted in the removal of secretory glycoproteins by solubilization with dithiothreitol and by repeated washing of the cells. Therefore, it now appeares that intensive staining of the brush border membrane of intestinal epithelial cells is related not only to the glycocalix, but also to the carbohydrate-rich material which is an integral part of membrane structure.

The isolated epithelial cell preparations also permitted the quantitative study of the binding equilibrium of lectins at various lectin concentrations in the incubation media. The binding curves for the three investigated lectins are shown in Fig. 1. The lectin concentration (abscissa) is expressed in the logarithmic scale in order to emphasize the shape of the curve in the range of low concentrations of lectins in the culture media. All three binding curves have the same sigmoidal shape with the initial part characteristic for positive cooperativity of binding. For lectin concentrations higher than 75 µg/ml culture, the curves show a tendency of saturation. It has also been found that nonspecific binding increased with the increase of lectin concentrations being always highest for ^{125}I-PHA. The plotting of the data according to Scatchard showed more clearly the cooperativity of lectin binding. Fig. 2 shows these curves for PHA and SBA, while the shape of the WGA plot was similar to that for PHA and is therefore omitted. All three lectins gave biphasic Scatchard plots. The initial binding, at the lowest lectin concentration, appeared to be highly positively cooperative. The maximal values of B/F for PHA binding were reached at B values of approximately 0.5 μg PHA/10^7 cells or $2.5 \cdot 10^5$ molecules of lectin per cell. Maximal values for WGA and SBA binding were obtained for approximately 4.10^5 and 25.10^5 molecules per cell, respectively. In all three curves the maximal values were followed by concave sections, corresponding to negative cooperativity at higher occupancy of the binding sites.

Numerous studies of the interaction of lectins with eukaryotic cells claimed positive cooperative effects of lectin binding to lymphocytes, erythrocytes, fat cells, ovary cells and other systems (3, 5, 14, 16, 19). It was shown that cooperativity is a function determined by the lectin itself and also by the characteristics of the cell membrane. Polymerization of the lectin resulted in a change of the binding pattern of lectin to lymphocytes (14) and conformational changes of the cell membrane influenced the binding of lectins to cells or model membranes (11, 12, 16).

The effects of eventual polymerization of SBA or WGA during the binding procedure cannot be excluded in the interaction of isolated epithelial cells with the applied lectins. It remains to be investigated whether the aggregation of SBA can possibly be connected with the different binding of thes lectin to histological sections of the rat small intestine or enterocytes stained with SBA-FITC and to isolated cells incubated with ^{125}I-SBA. It is also possible that the applied process of cell isolation results in subtle changes of the cell membrane configuration, which then influences the binding of ^{125}I-SBA. Anyhow, at least some of the membrane glycoproteins of intestinal epithelial cells, presumbly with the highest affinity for PHA, WGA and SBA, are capable of positive cooperative binding of iodinated lectins.

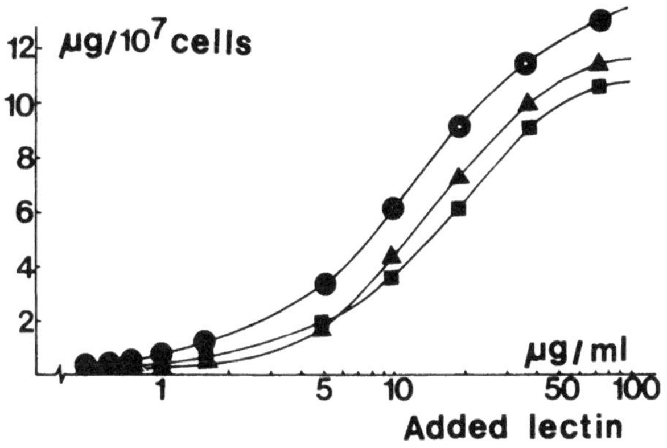

Fig. 1. Binding of iodinated lectins to isolated epithelial cells: ^{125}*I-PHA* ● ; ^{125}*I-SBA-* ▲ ; ^{125}*I-WGA* ■

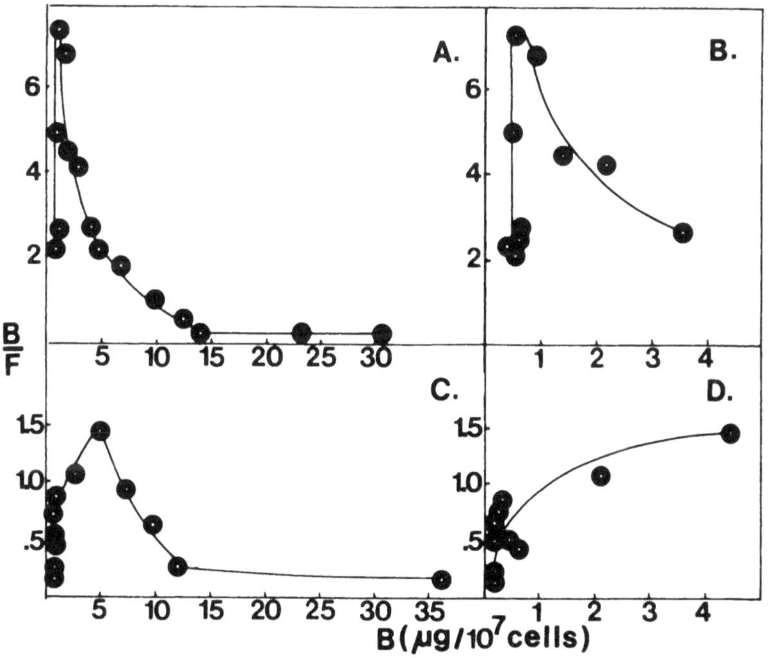

Fig. 2. Scatchard plots of specific binding of iodinated lectins to epithelial cells. Abscissa - bound lectin (μg/10^7 cells), ordinate - bound/free ratio. A -^{125}*I-PHA, whole plot; B -*^{125}*I-PHA initial part of the plot at an extended scale for B; C -*^{125}*I-SBA, whole plot; D -*^{125}*I-SBA initial part of the plot at an extended scale for B.*

It can be concluded that binding of PHA and WGA is localised mainly at the brush border region of the intestinal cell membrane. The binding of SBA was possible to detect only by using ^{125}J-lectin. All three investigated lectins, labelled by iodination, showed the characteristics of positive cooperativity of binding to the epithelial cells at low lectin concentrations, and negative cooperativity at higher occupancy of the binding sites.

References

1. Alvarez, Y. R., Torres-Pinedo, R.: Pediatr. Res. 16, 728-731 (1982).
2. Banwell, J. G., Abramowsky, C. R., Weber, F., Howard, R., Boldt, D. H.: Digestive Diseases Sciences 29, 921-929 (1984).
3. Bornens, M., Karsenti, E., Avrameas, S.: Eur. J. Biochem. 65, 61-69 (1976).
4. Bulajić, M., Čuperlović, M., Movsesijan, M., Borojević, D.: Periodicum Biolog. (in press).
5. Cuatrecasas, P.: Biochemistry 12, 1312-1323 (1973).
6. Čuperlović, M., Movsesijan, M., Djordjević, D.: Physiol. Pharmacol. Acta 17, 211-218 (1981).
7. Čuperlović, M., Movsesijan, M., Borojević, D., Bulajić, M.: Iugosl. Physiol. Pharmacol. Acta 19, 461-467 (1983).
8. Fischer, J., Klein, P. J., Vierbuchen, M., Skutta, B., Uhlenbruck, G., Fischer, R.: J. Histochem. Cytochem. 32, 681-689 (1984).
9. Higuchi, M., Suga, M., Iwai, K.: Agric. Biol. Chem. 47, 1879-1886 (1983).
10. Hunter, W. M., Greenwood, F. C.: Nature 197, 495-496 (1963).
11. Ketis, N. V., Girdlestone, J., Grant, C. W. M.: Proc. Nat. Acad. Sci. USA 77, 3788-3790 (1980).
12. Ketis, N. V., Grant, C. W. M.: Biochim. Biophys. Acta 685, 347-354 (1982).
13. King, T. P., Pusztai, A., Clarke, E. M. W.: Histochem. J. 12, 201-208 (1980).
14. Prujansky, A., Ravid, A., Sharon, N.: Biochim. Biophys. Acta 508, 137-146 (1978).
15. Pusztai, A., Clarke, E. M. W., King, T. P.: Proc. Nutr. Soc. 38, 115-120 (1979).
16. Reisner, Y., Lis, H., Sharon, N.: Exp. Cell Res. 97, 445-448 (1976).
17. Rossi, M. A., Filho, J. M., Lajolo, F. M.: J. Exp. Pathol. 65, 117-123 (1984).
18. Rouanet, J. M., Besancon, P.: Ann. Nutrit. Aliment. 33, 405-416 (1979).
19. Rutishauser, U., Yahara, J., Edelman, G. M.: Proc. Nat. Acad. Sci. USA 71, 1149-1153 (1974).
20. Stanley, P., Carver, J. P.: Proc. Nat. Acad. Sci. USA 74, 5056-5059 (1977).
21. The, T. H., Feltkamp, T. E. W.: Immunology 18, 865-873 (1970).
22. Tsuyama, S., Suzuki, S., Murata, F.: Acta Histochem. Cytochem. 16, 456-471 (1983).
23. Vretblad, P.: Biochim. Biophys. Acta 434, 169-176 (1976).
24. Weiser, M. M.: J. Biol. Chem. 248, 2536-2541 (1973).

THE DISTRIBUTION OF CARBOHYDRATE COMPONENTS AS REVEALED BY LECTIN HISTOCHEMISTRY IN THE ALIMENTARY TRACT OF TURTLES

Udo Schumacher, Ulrich Welsch
Anatomische Anstalt, Universität München
D-8000 München

Jürgen Barth
1. Med. Universitätsklinik, Universität Kiel
D-2300 Kiel

Comparative morphological studies have made clear that the vertebrate alimentary tract is endowed with a complex system of exocrine glandular cells which can constitute typical glands, surface lining epithelia, or be located as individual cells in the intestinal epithelium (7). In turtles mucus glands or glandular cells are integral parts of the esophagus, stomach and intestine. Luppa (2) studied the gastric mucus producing cells (surface lining epithelium, mucus cells of the fundic glands, pyloric glands) of *Testudo* with classical histochemical methods and observed that all of these cells contain neutral and in part acid mucopolysaccharides. Starting on the one hand from the above mentioned observations of Luppa (2) and on the other hand from the results obtained with lectins in the intestinal tract of mammals (1) we examined various sections of the intestinal tract of two species of turtles with different diets (carnivorous and herbivorous).

Material and Methods

Stomach and anterior midgut ("duodenum") of *Testudo horsefieldii* (3 adult animals) and of *Trionyx cartilagineus* (3 adult animals) were fixed in Bouin's fluid for two days and embedded in paraplast. Sections of 8 µm thickness were stained with Alcian Blue pH 0.5 (AB) and the periodic acid Schiff (PAS) reaction (3, 4). Furthermore such sections were incubated with various fluoresceinisothiocyanate (FITC) conjugated lectins, which possess different carbohydrate specificities: Concanavalin A (Con A), wheat germ agglutinin (WGA), soy bean agglutinin (SBA), leucoagglutinin (LA), Griffonia (=Bandeiraea) simplicifolia agglutinin I (GSA I), Dolichos biflorus agglutinin (DBA) (all obtained from medac Hamburg, FRG). The sections were incubated with lectins for 30 min. and rinsed afterwards in phosphate buffered saline (pH 7.4) for 2 h. Specimens not treated with FITC lectins served as a control to observe the autofluorescence of the tissue. The sections were studied under a Zeiss fluorescence microscope and photographed with an Ilford HP5-135 400 ASA black and white film.

Results

In the following text the main emphasis is put on the exocrine glandular cells.

A. *Testudo*

1. Stomach

a. Fundus. The fundic mucosa is composed of a columnar surface epithelium, tubular glands and a sparse connective tissue stroma. The surface epithelium lines rather narrow gastric pits. The apical half of the surface epithelium gives strong PAS (Fig. 1.1) and WGA (Fig. 1.3) reactions). This part of the cells, which contains the secretory granules, stains weakly with AB and DBA; Con A and SBA have binding sites at the plasma membrane and in the Golgi-area which forms a narrow zone above the nucleus; Con A outlines the entire cell membrane whereas SBA reacts only with the apical plasma membrane; Con A in addition binds intensively to fibrous components of the connective tissue, presumably to collagen. The staining pattern of GSA-I resembles in part that of SBA, the apical membranes and the junctional complex are outlined rather clearly, immediately below this membrane a non reactive zone can be recognized which is followed by a positive zone extending

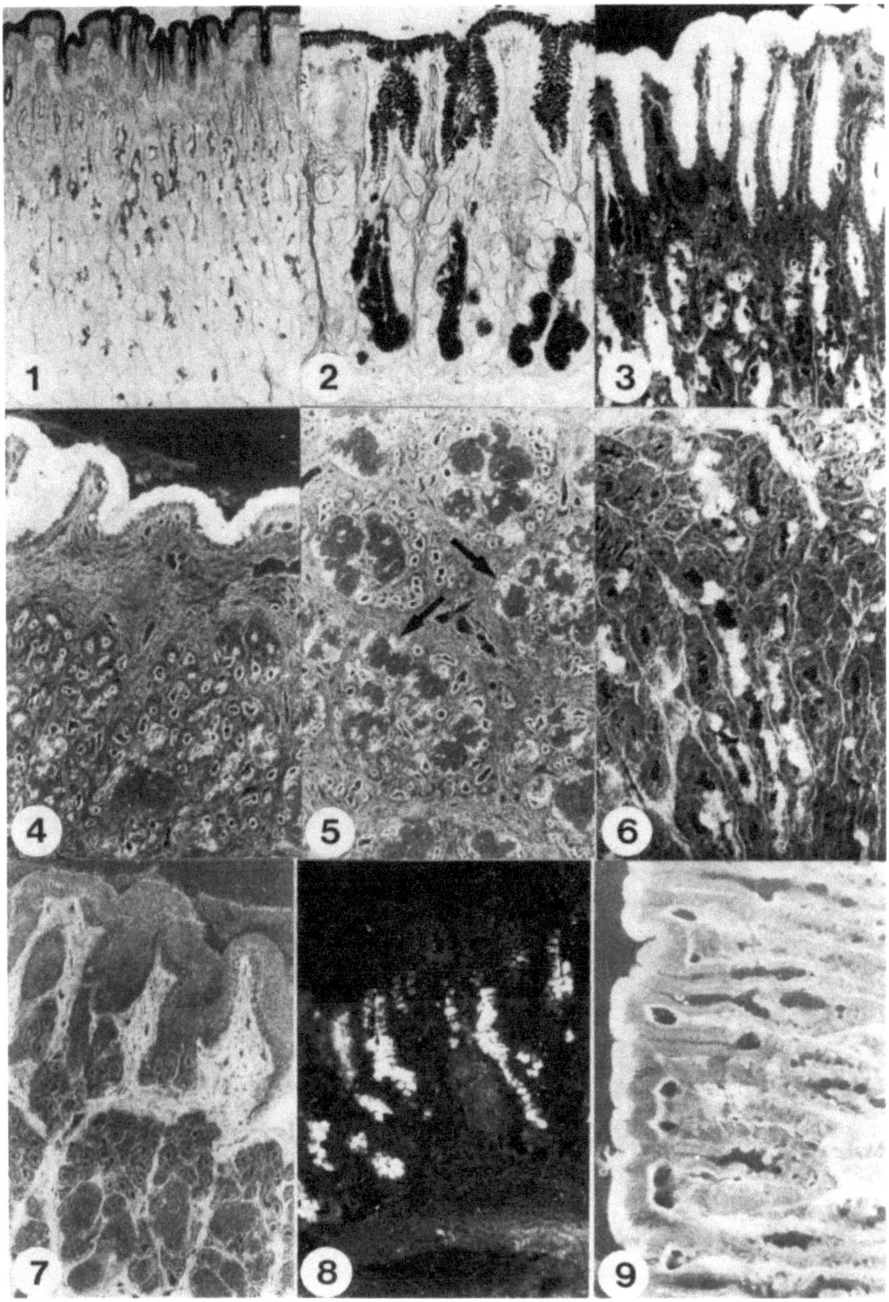

Fig. 1. Histochemistry of the fundic region of the stomach. 1.1: Testudo, PAS-reaction,x60. 1.2: Trionyx, PAS-reaction,x100. 1.3: Testudo, reaction with WGA,x165. 1.4: Trionyx, reaction with WGA,x100. 1.5: Trionyx, reaction with WGA, note that mucus cells (arrows), are stained only apically and basally,x165. 1.6: Testudo, reaction with Con A, x165. 1.7: Trionyx, reaction with Con A,x165. 1.8: Testudo, reaction with SBA,x100. 1.9: Testudo, reaction with LA,x100.

to the nucleus. LA stains the apical part of the cells as well as basement membranes and connective tissue material (Fig. 1.9).
In the fundic glands, consisting of mucus and main cells, a positive PAS reaction marks the presence of the mucous "light" cells (Fig. 1.1), Whereas AB gives negative results in the glands. All lectins tested bound to the mucous cells (Figs. 1.3, 1.8), WGA and GSA-I gave a strong reaction. The apical and basolateral membranes of the main "dark" glandular cells reacted positively with all lectins (Fig. 1).

b. Pylorus. In the pyloric surface epithelium the staining reactions did not differ from those of the fundic surface epithelium. The glands which are composed of one cell type only, give a positive reaction with PAS, GSA-I, WGA, Con A, SBA and DBA.

2. Anterior midgut
In the anterior midgut the goblet cells, the apex of the enterocytes as well as intraluminal strands of mucus stain positively with PAS (Fig. 2.1), AB, WGA (Fig. 2.4), Con A (Fig. 2.6). SBA only reacts faintly with the apical membrane of the epithelial cells.

B. *Trionyx*

1. Stomach
a. Fundus. The surface epithelium showed the same histochemical reactions as described for the gastric surface epithelium of *Testudo*. In the fundic glands of *Trionyx* a certain number of reactions are similar to those in *Testudo*. There are, however, also differences: The PAS reaction appears to be stronger in *Trionyx* (Fig. 1.2) than in *Testudo*. Con A preferentially binds to the apical membrane of the mucous cells (Fig. 1.7), whereas the intracellular mucus droplets were not stained as in *Testudo*. WGA combines with binding sites along the apical plasma membrane of all cells and with basally located material in the mucous cells (Figs. 1.4, 1.5). SBA reacts considerably weaker than in *Testudo*.

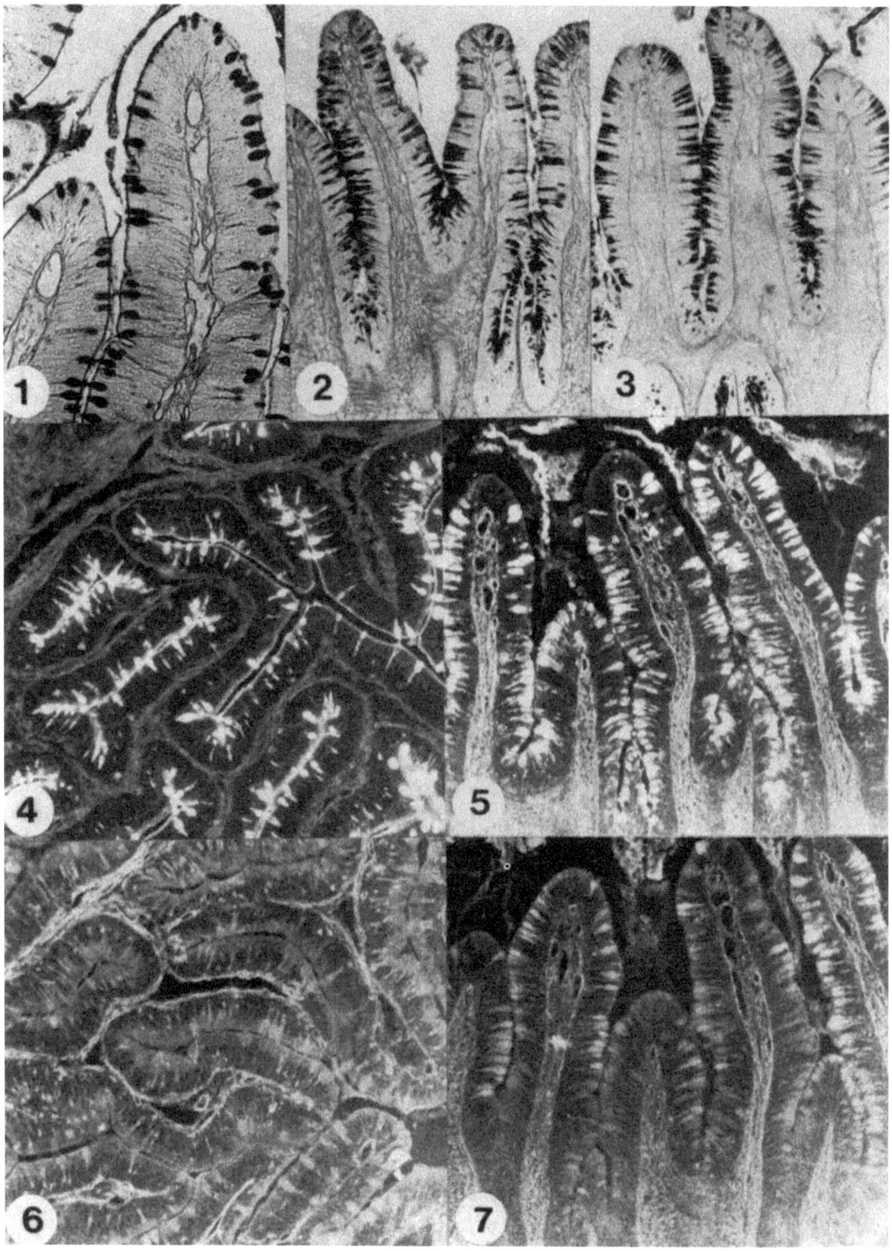

Fig. 2. Histochemistry of the anterior midgut.
2.1: Testudo, PAS reaction,x165. 2.2: Trionyx, PAS reaction,x100. 2.3: Trionyx, Alcian Blue,x100. 2.4: Testudo, reaction with WGA, x100. 2.5: Trionyx, reaction with WGA,x100. 2.6: Testudo, reaction with Con A,x100. 2.7: Trionyx, reaction with Con A,x100.

b. Pylorus. If compared to the pylorus of *Testudo*, the corresponding gastric part of *Trionyx* shows similar results as regards surface epithelium and PAS reaction and SBA in the glands. Differences concern AB which stained but faintly the basal parts of the glands, Con A which only reacted weakly with the membranes of the glandular cells and WGA staining not all of the cells.

2. Anterior midgut

The results of the *Trionyx* midgut are closely comparable to those of the same organ in *Testudo*: PAS reaction (Fig. 2.2), AB (Fig. 2.3), WGA (Fig. 2.5), Con A (Fig. 2.7), i.e., in principle, the goblet cells and the apex of the enterocytes gave positive results with most reagents tested.

Discussion

The present study has shown both, a remarkable quantity as well as diversity of carbohydrate containing cells and extracellular tissue components in the alimentary tract of *Testudo* and of *Trionyx* with a high degree of specialisation and compartmentalisation.

The PAS reaction indicates the presence of neutral glycoproteins in tissue sections, whereas Alcian Blue at the chosen pH demonstrates acidic mucopolysaccharides. Therefore this reaction and staining procedure can be used as screening methods for carbohydrate components in general. However, more detailed information on the distribution of specific monosaccharide units can be obtained by the use of lectins. For example, in *Testudo* the mucous cells of the fundic region react strongly with SBA, whereas the PAS reaction is weak. In the PAS reaction, however, the gastric surface epithelium is intensively stained; in contrast, SBA reacts only with the apical plasma membrane.

As regards the differences revealed by lectin histochemistry in the gastric mucins of the two turtle species, the question arises, whether these are due to a different composition of the gastric juice, e.g. different hydrogen ion concentrations, since ion diffusion depends on gastric mucus glycoprotein (5); and whether there are relations to dietary habits (*Testudo*, a terrestrial species of rather dry zones, being herbivorous, while *Trionyx*, an entirely aquatic species of tropical regions, is carnivorous). In addition it is still unknown whether blood group related intra-species variations as described in humans (6) also occur in turtles.

We consider that so far no definite explanation of the differences which have mainly been observed in the mucus composition of the stomach, can be offered. Nevertheless the carbohydrate histochemical results obtained from *Testudo* and *Trionyx* do call for further experimental work on the alimentary tract of turtles.

References

1) Fischer, J., P. J. Klein, M. Vierbuchen, R. Fischer, and G. Uhlenbruck: In: Lectins, ed. by T. C. Bøg Hansen. Walter de Gruyter, Berlin, 1983, 3, 167-178.

2) Luppa, H.: In: Biology of the Reptilia, ed. by C. Gans. Academic Press, New York, 1977, 6, 225-313.

3) Mowry, R. W.: J. Histochem. Cytochem. 4, 407, 1956.

4) Romeis, B.: In: Mikroskopische Technik. R. Oldenbourg, München Wien, 1969, 278.

5) Sarosiek, J., A. Slomiany, A. Takagai, and B. L. Slomiany: Biochem. Biophys. Res. Commun. 118, 523-531, 1984.

6) Schulte, B. A. and S. S. Spicer: J. Histochem. Cytochem. 31, 391-403, 1983.

7) Sheahan, D. S. and H. R. Jervis: Am. J. Anat. 146, 103-132, 1976.

BINDING OF DOLICHOS BIFLORUS AGGLUTININ (DBA) TO ENDOTHELIAL CELLS DURING THE EMBRYONAL PERIOD IS RESTRICTED TO THE NMRI-MOUSE.

Johanna Plendl and Wolfgang Schmahl

Gesellschaft für Strahlen- und Umweltforschung, München
Abteilung Nuklearbiologie
D - 8042 Neuherberg, Federal Republic of Germany

Originally the aim of the present study was to develop an histochemical method for an evaluation of the growing capillaries within the mouse brain during organogenesis (day 7 to day 12 of gestation). The visualization of the developing vessels, which should be possible even by low magnification, was intended to reveal aspect also of cellular connectivity between the primitive endothelial cells and the differentiating neuroblasts (1,2). Thus the relevant spectrum of cell surface glycoproteins was to be studied with the help of differential lectin binding. We used HRP-conjugates of a variety of lectins, including Dolichos biflorus agglutinin (DBA) recently reported to be useful for a highly specific demonstration of the neuro-muscular junction (3). As the latter represents an example of cell adhesion between neuroectodermal cells and a mesodermal derivative, similar to the neurovascular junctions (4) we examined its DBA-affinity pattern during organogenesis. The present study was initially restricted only to NMRI-mice, a strain we have used for many years for neurodevelopmental research (5). Ultimately, the unexpected findings of a rather specific DBA-affinity pattern, which was not reported in similar studies by other authors (6,7), led to a comparison with embryos of other strains of mice (DBA ,C57Bl6).

Lectins, Vol. V

Material and Methods

1. Material: HRP coupled Dolichos biflorus agglutinin, as well as neuraminidase from Clostridium perfringens (Type X) were provided by Sigma Chemie, Munich.
2. Animals: We used mice of the strains NMRI, C57Bl6, and CBA for this study. Femals were mated for two hours and subsequently examined for vaginal plugs. The morning of the next day was regarded as gestation day 1.
3. Animal experimentation: Embryos were removed from 3 pregnant females of the C57 and of the CBA strain on day 11. 5 pregnant NMRI-mice were anaesthetized on every other gestation day from day 7 until day 18. The embryos were removed either with the uterine wall in toto (days 7-9), or dissected (days 10-13). They were fixed in neutral buffered formalin and four embryos of each litter were embedded in paraffin. Serial sections of 5 µm thickness were cut on a microtome.
4. Analyses: The histological sections were deparaffinized and labeled with the HRP coupled lectin, either directly or after antecedent incubation of the slides (18h at 37°C) with neuraminidase (0.1 unit/ml). The labeling procedure was carried out as follows: The sections were treated with H_2O_2 to destroy endogenous peroxidase activity. Then the sections were overlayered with the peroxidase-coupled lectin (0.1 mg/ml) for 40 min at 20°C and subsequently washed in phosphate buffered saline for another 30 min. They were then incubated in 0.05% 3-3-diaminobenzidine in 0.01% H_2O_2 for 10 min at 20°C. After extensive washing in PBS the slides were counterstained with hämalaun, dehydrated in an alcohol series and xylol and finally covered with Eukitt. Controls were performed by adding the respective inhibitory sugar to the peroxidase-coupled lectin during the incubation period. This procedure resulted in only very faint labeling or no labeling at all.

Results

The following only concerns with the binding of DBA to the embryonal tissue of 3 different mouse strains. The results of these experiments on gestation day 11 are shown in Table 1.

NMRI-mice showed a rather narrow reaction pattern restricted only to the visceral yolk sac and the endothelium of all vessels, which were generally very intensely stained with the peroxidase reaction, thus indicating a binding of the HRP-DBA conjugate. The only exception was a lower intensity of the reaction at the liver sinusoids.

The gastrointestinal tract in NMRI-mice embryos was strongly positive on day 9 and day 10 but remained DBA-negative from day 11 onwards.

Table 1

Intensity of staining with DBA in embryonal tissues in different strains of mice on gestation day 11

	CBA	NMRI	C57B16
ECTODERM			
epidermis	++	-	+++
neural tube	-	-	-
MESODERM			
undiff. mesenchyme	(+)	-	p
chondrogenic blastema	-	-	-
vessel endothelium	-	+++	p[a]
endocard	-	-	-
nephrogenic tissue	-	-	-
liver sinusoids	-	+	-
ENTODERM			
visc. yolk sac	+++	+++	++
pharynx	-	-	-
stomach	-	-	-
intestine	++	-	-
hepaticduct	++	-	++
lung buds	-	-	-

The intensity of staining is given as: - = no reaction; (+) = weak reaction; + = positive reaction; ++ = strong reaction; +++ = very strong reaction.
p = patchy distribution of DBA-affine sites;
p[a] = some patches in aa. carotis.

DBA-labeling of the vessel endothelium: An unexpected finding observed only with NMRI-mice embryos was the strong labeling of all vessels starting from day 8. In principle there was no difference between the arteries and the venous system with the exception of day 9, when the cardinal vein was still weakly stained in contrast to all other vessels. A temporal gradient of the DBA-affinity to the endothelium occurred between day 9 and day 12 only in those vessels within or in close contact with the central nervous system. Until this time the DBA-affinity to vessels of the other body regions remained constantly high. In principle this was also true for all perineural vessels, however only as far as the neural tube was already closed. In areas with an incomplete closure the immediately adjacent vessels were DBA-negative.

The following description is restricted to those vessels closely connected with the developing central nervous system. Table 2 provides the changing DBA-affinities of endothelium within the ventricular zone. The most intense staining pattern occurred next to the pial regions. Invariably, the areas of neuronal migration showed a lower intensity of DBA-positive vessel staining, finally disappearing in areas of neuron settlement (former marginal zone).

Spinal cord: The vessels of the matrix zone were intensely labeled from day 10 to day 13. The intermediary zone revealed DBA-positive vessels only on days 11 and 12. The vessels of the spinal ganglia were DBA-positive from day 12 onwards.

Myelencephalon: There was a rather narrow period of strong DBA-affinity of the ventricular zone vessels only on day 10. This pattern regresses to an undetectable level within the next two gestation days. DBA-labeling of the vessels within the intermediary zone starts on day 12.

Meten- and mesencephalon: This region showed the latest appearance of DBA-labeling vessels within its ventricular zone on day 11. This pattern remained constant up to day 14.

Diencephalon: On gestation day 10 the vessels within the ventricular zone started with a weak DBA-affinity. This labeling was already intense on day 11 and continuously declined again between days 12 and 14.

Telencephalon: The vessel endothelium of the ventricular layer showed a marked DBA-affinity from gestation day 11 onwards. This further increased until day 12 but was rather inhomogenous from one specimen to another. The labeling intensity of these vessels was very strong on day 13 and sharply declines to weak staining the beginning of day 14. The intermediary zone vessels of the diencephalon and telencephalon generally showed a weaker DBA-affinity than the vessels of the ventricular zone, and also appeared about

Table 2

Intensity of staining the vessel endothelium within the ventricular zone with DBA during gestation in the NMRI-mouse.

Days of gestation	9	10	11	12	13
Spinal cord	-	+++	+++	+++	(+)
Myelencephalon	-	+++	++	++	-
Metencephalon	-	-	++	++	++
Mesencephalon	-	-	++	++	++
Diencephalon	-	+	+++	++	++
Telencephalon	-	-	++	*	+++

The intensity was given as in Table 1
* = rather inhomogeneous staining varying between different histological specimens

two days later than in the ventricular zone.

Beginning on day 13 the constant high affinity of DBA to the vessels of non-neuronal tissue also started to change. This implicated a general decrease of the DBA-affinity which however started at different gestation times in the various organs. An overview of this situation is provided in Table 3. The changing DBA-affinity of the vessel endothelium was not influenced by pre-incubation of the slides with neuraminidase. Thus a masking of the DBA-receptors by sialic acid could be ruled out.

DBA-labeling of neuronal cells: At the end of gestation day 12 and during the following day 13 a distinct DBA-labeling of the horizontal neurons of the superficial layer belonging to the frontal neocortical roof was observed only in NMRI-embryos. This affinity for the presumed Cajal-Retzius cells was strictly limited to the nuclear area. It occurred only at the beginning of the formation of the cortical plate and disappeared with its definite establishment.

A marked DBA-affinity was also seen within the cells forming the basis of the ganglionic eminences. This was a constant finding

Table 3

Intensity of staining with DBA of the endothelial cells in different organs of fetal NMRI-mice

Days of gestation	13	14	15	16
Endothelial cells within:				
CNS	++	(+)	-	-
perineural tissue	++	+	(+)	-
limb buds	+++	++	+	(+)
thorax walls	++	+	(+)	(+)
digestive system	+	(+)	(+)	-
liver	(+)	-	-	-
pancreas	(+)	(+)	-	-
kidneys	p+	p+	-	-
adrenals	++	+	-	-
genital tract	p(+)	p(+)	-	-

Abbreviations as in Table 1

between day 13 and day 18, but was restricted to only about 50% of the cells forming the basal brain parts. The DBA-positive cells were characterized by large, bright nuclei to which the labeling with HRP-DBA was restricted exclusively.

Discussion

Several authors have already used DBA in some animal species as a histochemical tool to characterize embryonal tissue glycoproteins. However, most of those studies reported a DBA-affinity to endodermal tissue (7-11). There are only marginal comments on a sparse and patchy staining of vessels by a marker-coupled DBA in neonatal Donryu rats (12) and in adult 129/slcp mice, as well as Balb/c mice where only brain capillary vessels were found to be DBA-positive (6). Two other reports are concerned with the patchy DBA-affinity pattern to vessel endothelium in adult DDK-mice and R3-mice (13,14). In these studies, DBA-positive endothelium was restricted to some body regions or to some organs, which neither could be related to any morphological pecularity nor to any distinct functional property.

The present study clearly shows that the affinity of the vessel wall and the affinity of certain neuronal cells to DBA is strictly specific for the NMRI-strain. Also it depends on the developmental stage of the embryos.

Biochemically, the affinity of cells to DBA can be deduced to an expression of a cell surface carbohydrate chain, which is either identical with or closely related to human blood group antigen A of the ABO system (15). Until now, it has been a generally accepted opinion that this relevant hapten structure is expressed only on human erythrocytes and human endothelial cells (16), the latter cells also being found to express the A antigen during fetal development (17). Some species of subprimates, dogs and rodents were also described to express the A antigen on erythrocytes, on intestinal mucosal cells, and on salivary gland cells (16). However, these findings were again restricted to adult animals only.

The present study shows that the expression of an A or A-like antigen occurs even in the developmental stage in NMRI-mice embryos. The biological role of this antigen (especially in its restriction to vessel endothelium and some brain cells) raises the question of a possible functional dependency. It may therefore be suggested that the earliest migrating neurons, the DBA-positive Cajal-Retzius cells are contact-guided by the radially orientated vessels which proved to possess similar DBA-positive determinants. After the establishment of this layer within the neocortical roof, however, the subsequent migration of cortical plate cells would appear to be no longer influenced by DBA-positive cells determinants.

Acknowledgements: We are particulary grateful to Ms. U Huffstadt, and Mrs. I Steege for their circumspect assistance and skillful collaboration.

References

1. Breipohl, W.: Acta Anat. 69, 393-408 (1968)
2. Conradi, N.G., and Sourander, P.: Acta neuropath. 50, 221-226 (1980)
3. Sanes, J.R., and Cheney, J.M.: Nature 300, 646-647 (1982)
4. Bär, T., and Wolff, J.R. In: The Cerebral Vessel Wall (1976) J. Cervos - Naverro, Ed., Raven Press, New York, pg 1-6
5. Schmahl, W.: Radiat. Enciron. Biophys. 21, 19-31 (1982) Schmahl, W.: ibid. 22, 95-115 (1983)
6. Watanabe, M., and Muramatsu, T., and Shirane, H.,and Ugai, K.: J. Histochem. Cytochem. 29, 779-790 (1981)
7. Valk, M.A. von der, and Hageman, P.: These proceedings Vol 2, 221-240 (1982)
8. Bee, J.A.: Differentiation 23, 128-140 (1982)
9. Noguchi, M., and Noguchi, T., and Watanabe, M., and Muramatsu,T. J. Embryol. Exp. Morphol. 72, 38-52 (1982)
10. Fischer, J., and Klein, P.J.: These Proceedings Vol 4, 95-100 (1985)
11. Duprat, A.M., and Rouge, P., and Flottes, S., and Bleys, M.: These Proceedings, Vol. 4, 193-201 (1985)
12. Miyake, T.: Acta Histochem. Cytochem. 29, 779-790 (1984)
13. Ponder, B.A.J., and Wilkinson, M.M.: Develop. Biol. 96, 869-877 (1983)
14. Uiterdijk, H.G., and Ponder, B.A.J.: These Proceedings, Vol 4, 161-168 (1985)
15. Etzler, M.E., and Kabat, E.A.: Biochemistry. 9, 869-877 (1970)
16. Oriol, R., and Cooper, J.E., and Davies, R.R., and Keeling, P.W.N.: Lab. Invest. 50, 514-518 (1984)
17. Szulman, A.E.: J. Exp. Med. 119, 503-515 (1964)

MATURATION OF CELLULAR SACCHARIDES IN FETAL MOUSE KIDNEYS

Harry Holthöfer[1] and Anita Nynäs-McCoy[2]

Department of [1]Bacteriology and Immunology and of [2]Pathology, University of Helsinki, Helsinki, Finland

The differentiation of metanephric kidneys includes profound changes in the organization and morphology of cells, to form the permanent functioning nephrons. During this process, cellular elements as well as extracellular matrix components develop synchronously, and at the same time various renal antigens appear in a determined order.

In the kidneys of newborn mice, some of the nephrons still undergo maturation. Thus all maturational stages of glomerular and tubular development can be found in these kidneys. The morphogenetic features of nephron differentiation are well characterized (1), but the molecular mechanisms leading to tissue maturation need clarification. Cellular glycoconjugates are thought to play an important role in the organization of various cell types of developing tissues (2), thus leading us to study the lectin histochemistry in the embryonic mouse kidney.

Lectins, Vol. V

Materials and methods

Embryos from CBA x $C57Bl_6$ mouse strain were obtained by cervical dislocation. The gestational day was determined from the appearance of the mucous plug.
The kidneys were dissected, snapfrozen in liquid nitrogen, and mounted in TissueTek.
Fluorochrome coupled lectins (RCA I, DBA, WGA, MPA, PNA) were obtained from EY laboratories (San Mateo, Ca).
The 3µm sections were fixed in 3.5% fresh paraformaldehyde, washed, and then incubated with the appropriate lectin solutions, as described earlier (3).
A Zeiss Standard microscope with a filter system for FITC-fluorescence was used for evaluation of the sections.

Results

The results showed that cellular glycoconjugates in fetal mouse kidneys appear after a defined maturational stage is reached. During further maturation, these cellular saccharides were expressed in a stage specific manner.

At the 11th gestational day, the uninduced mouse nephrogenic mesenchyme was positive for RCA I but not for the other lectins. At the 16th gestational day, glycoconjugates appeared to the collecting ducts and distal tubules but not to glomerules. At the 18th gestational day, the glomerular structures showed an advanced morphology, and could be identified with several of the lectins.

RCA I bound to the uninduced nephrogenic cells, and in addition, reacted with the basement membranes of the 18th day mouse embryonic kidneys (Fig. 1). At his stage, DBA failed to stain any glomerular elements, but instead, detected the cells of the collecting ducts (Fig. 2). WGA binding sites appeared in the podocytes and at the S-shaped stage of the glomerular development (Fig. 3). Interestingly, MPA binding sites could be observed transiently at the S-shaped body stage, but at later stages of the glomerular development these binding sites disappeared.
PNA proved to be a good marker for basement membranes in the developing mouse kidneys, starting from the S-shaped body stage (Fig. 4).

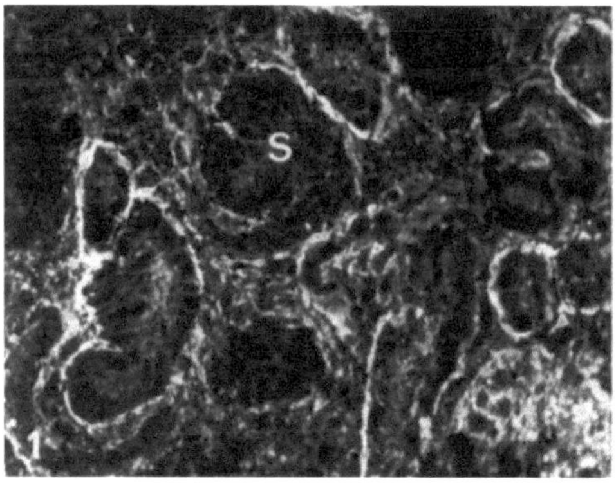

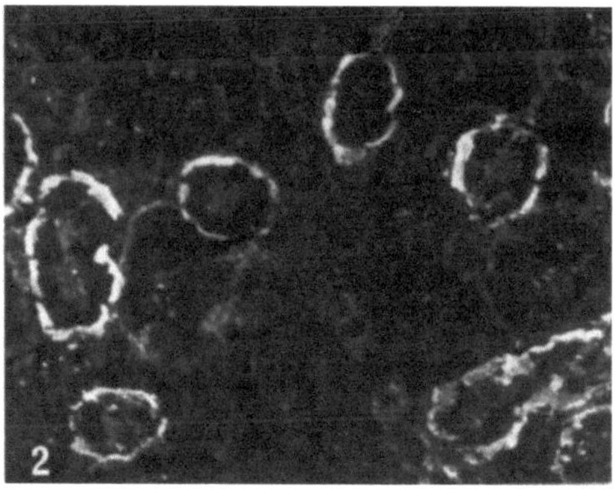

Fig. 1. 18th day mouse embryonic kidneys stained with FITC-RCA I. Fluorescence is confined to the uninduced cells of nephrogenic mesenchyme as well as to the basement membranes (S, s-shaped body).
Fig. 2. FITC-DBA stains the collecting ducts of developing mouse kidneys but no glomerular elements.

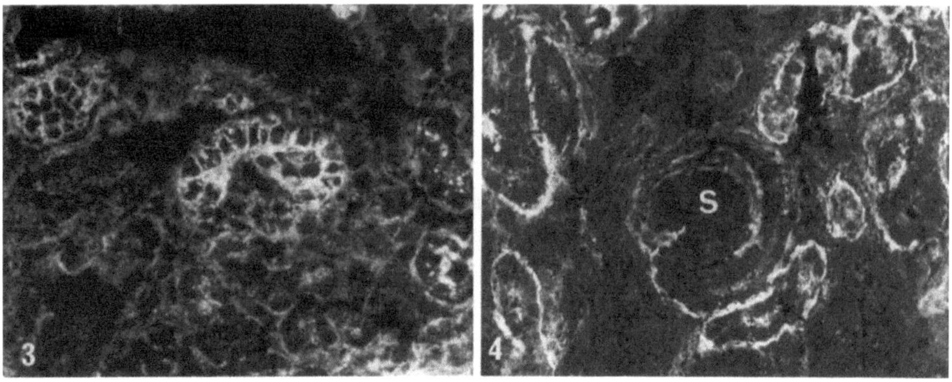

Fig. 3. FITC-WGA detects the developing glomerules in the 18th day mouse embryonic kidneys.
Fig. 4. The basement membranes of developing glomerules and tubules are visualized after staining with FITC-PNA.

The results are summarized in Table I.

TABLE 1

Lectin binding sites in developing mouse kidneys

Lectin	unind	s-shaped	cap loop	mature
RCA I	+	BM ++	BM ++	BM ++
DBA	-	-	-	-
WGA	-	POD +	POD ++	POD ++
MPA	-	BM +	BM +	-
PNA	-	BM ++	BM ++	BM ++

Abbreviations: unind, uninduced nephrogenic mesenchyme; s-shaped, s-shaped body stage; cap loop, capillary loop stage; BM, glomerular basement membrane; POD, podocytes.

Discussion

This study showed the developmental stage dependent expression of renal glycoconjugates in embryonic mouse kidneys as was detected with lectins.

Although immunohistochemical approaches have revealed that different renal antigens appear according to a strict schedule (4), little is known of the appearance of cellular saccharides during renal maturation. However, this might be important, as cell surface saccharides are proposed to function as recognizing structures in tissue maturation (2). On the other hand, the complex phenomenon of glomerular maturation is still poorly understood. Roth et al (5) have shown that HPA-binding sites appear in a specific polarized manner in developing rat kidneys, obviously suffecting the beginning of glomerular filtration.
Similarly, WGA binding sites appear in developing kidneys simultaneously with the sialic acids of podocytes in rat and human kidneys (3,4). The stage specific expression of lectin binding sites may reflect the functional activation of various renal structures and the lectins can thus be used as markers for these phenomena.

In studying the molecular mechanisms of tissue induction, RCA I and WGA may be useful probes. RCA I was found to detect the uninduced nephrogenic cells, and WGA detected the glomerular podocytes from the S-shaped body stage, when the changed elements of glomerules known to be important for glomerular function, appear.

MPA binding sites were expressed transiently during the glomerular development. This is obviously due to an addition of terminal saccharide residues to MPA receptor saccharides during further maturation, and thus leading to the covering of the binding sites for MPA. This kind of transient expression of cellular saccharides may be important in analyzing tumor tissues, as malignant tumors are known to express defectively organized glycoconjugates (6). Because the lectin binding sites were usually found in specific cell types, they could be used in analyzing the cellular origin of various cell types in renal tumors.

Our results indicate that lectins can be used as maturation stage specific markers for different cell lines in the mouse kidneys, and that they are potential tools in analyzing the cellular maturation and origin.

REFERENCES

1. Kazimierczak, J. (1971). Acta Pathol. Microbiol. Scand. (A) 218, 1.

2. Aplin, J.P., Hughes, R.C. (1982). Biochem. Biophys. Acta 694, 205.

3. Holthöfer, H., Virtanen, I., Pettersson, G., Törnroth, T., Alfthan, O., Linder, E., Miettinen, A. (1981). Lab. Invest. 45, 391-399.

4. Ekblom, P. (1981). J. Cell Biol. 91, 1.

5. Roth, J., Brown, D., Orci, L. (1983). J. Cell Biol. 96, 1189-

6. Hakomori, S., Kannagi, R. (1983). J. Natl. Cancer Inst. 71, 231.

A HISTOLOGICAL STUDY OF THE ALIMENTARY TRACT OF FOUR AFRICAN FISHES. FLUORESCEIN ISOTHIOCYANATE LABELLED (FITC) LECTINS.

André Danguy, Robert Kiss and Georges Lenglet.

Laboratoire de Biologie Animale et d'Histologie Comparée, Université Libre de Bruxelles, 50, avenue F.D. Roosevelt, B-1050 Bruxelles, Belgique.

Animal cells contain a number of glycoproteins and glycolipids, and lectins have successfully been used in characterization of these compounds because of their specificity for particular saccharide moieties. A wealth of information indicates the usefulness of lectins labelled with fluorescein isothiocyanate (FITC) or horseradish peroxidase (HRP) as histochemical reagents to study normal tissues (1-4), and various tumors (5-11). In contrast, the literature in lower vertebrates is meagre (12) and essentially focused on ova and roe (13, 14). As our laboratory is engaged in the histophysiology of the alimentary tract of African freshwater fishes (15) we kept this opportunity to evaluate in this tissue the binding of three FITC-labelled lectins. Indeed it has been taught that different digestive activities may be related to different lectin binding reactivities.

Lectins, Vol. V

Materials and Methods

Four African freshwater species were studied : *Schilbe mystus* (Linnaeus, 1758, omnivorous), *Stenopoma nigropannosum* (Reichenow, 1875, carnivorous), *Astronotus ocellatus* (Cuvier, 1829, carnivorous) and *Garra congoensis* (Poll; 1959, herbivorous, stomachless). The animals were killed with tricaine methane-sulfonate (MS 222). Oesophagus, stomach (if present), intestine and hepatopancreas were fixed in Bouin's fluid, embedded in paraffin and 5 µm thick sections were cut. After rehydratation in phosphate-buffered solution (PBS, pH = 7.3 ± 0.1), the sections were covered with conjugated lectin at the concentration of 100 µg/ml in PBS and left to incubate in a humidified box for 1 h at room temperature.

The following fluorescein isothiocyanate coupled (FITC) lectins (Vector Laboratories, Burlingame, CA) were used: Triticum vulgaris (WGA), Ulex europaeus I (UEA I) and Arachis hypogaea (PNA). After the incubation, the sections were washed in PBS, mounted with glycol-PBS and examined with a Leitz Orthoplan fluorescence microscope equiped for epiillumination and the appropriate filter combination for FITC. The specificity of the obtained staining was tested by preincubating the lectin conjugate in a 0.2 M solution of the appropriate inhibitory sugar: D-galactose for PNA, N-acetylglucosamine for WGA and L-fucose for UEA I.

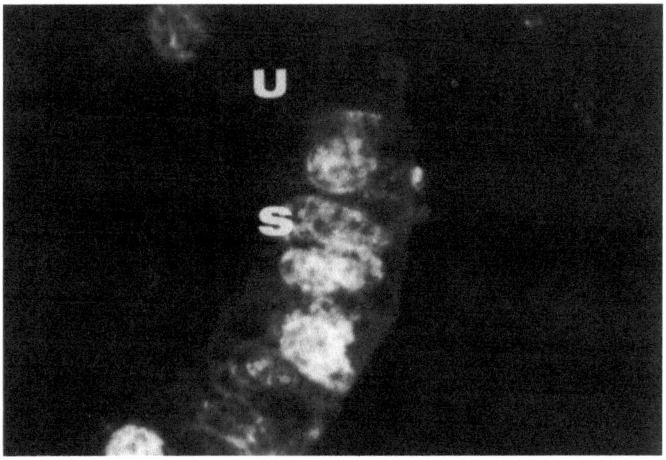

Fig. 1: Astronotus ocellatus. Oesophageal mucosa with PNA binding. Staining is evident in saccular mucus-secreting (S) cells. The undifferentiated (U) cells are negative.

Results

The binding patterns of three carbohydrate specific lectins in oesophagus, stomach and intestine of the four species are illustrated in Fig. 1-4 respectively. All the observations are summarized in Table 1.

Ulex europaeus I: this lectin stained some goblet cells in Garra and stomach glandular cells in Schilbe as well. The other cell types were unreactive. UEA I stained moderately the basal lamina and connective tissue of all four species suggesting fucose-rich residues in this tissue.

Wheat germ agglutinnin: this lectin presented selective bright staining of the oesophageal mucus-secreting cells except in Ctenopoma. Both gastric superficial epithelium and glands displayed the same kind of staining pattern except in Astronotus where these structures failed to react. Except in this last species, the goblet cells were faintly stained. Moreover, in Schilbe and Ctenopoma the apical membrane of the absorptive cells displayed a specific staining.

Peanut agglutinin: in the oesophagus, PNA stained brightly the mucus of the saccular mucous cells except in Schilbe. The gastric columnar superficial epithelium was faintly labeled. Regarding the gastric glands, the intensity of the fluorescence was higher than in the columnar epithelium. In Astronotus, these cell types were weakly stained. In the intestine, the goblet cells were unstained except in Garra.

With all lectins used, the binding could be inhibited by preincubating the lectin solution with the appropriate inhibitory sugar.

Discussion

For detection of carbohydrates at the light-microscope level, numerous staining methods exist but are limited in their specificity for particular sugar residues. Since the introduction of plant lectins in histochemistry, many works have demonstrated the feasibility of applying HRP and FITC lectins in this field. The present investigation emphasizes further the use of FITC lectins as histochemical tools for mapping the alimentary tract of lower vertebrates. Particularly three carbohydrate specific lectins have been chosen as probes in order to obtain new information about their respective binding patterns in fish digestive tract. In histological studies, various fixatives have been used. Recently some investigators claimed that the fixation of tissue in Bouin's fluid enhanced the staining of all the lectins compared to formalin

TABLE 1

Intensity of staining with lectins in various cells of the alimentary tract of four fishes.

Genus	Schilbe			Ctenopoma			Garra			Astronotus		
Locali-zation	PNA	WGA	UEA	PNA	WGA	UEA	PNA	WGA	UEA	PNA	WGA	UEA
OESOPHAGUS												
und. cells	-	-	-	-	-	-	-	-	-	-	-	-
muc. cells	-	+++	-	+	-	-	+++	+++	++	+++	+++	++
con. tis.	-	-	=	-	-	+	-	++	+	-	-	+
STOMACH												
s. ep.	+	+++	-	+	++	-	/	/	/	+	-	-
gl. ep.	+++	++	++	+++	+++	-	/	/	/	+	-	-
con. tis.	-	-	+	-	-	+	/	/	/	-	++	+
ANTERIOR INTESTINE												
g. cells	-	+	-	-	+	-	+	+	+	-	-	-
a. cells	-	+	-	-	++	-	-	-	-	++	-	-
con. tis.	-	-	+	-	-	+	-	+	++	-	++	+
MIDDLE INTESTINE												
g. cells	-	+	-	-	+	-	+	+	+	-	-	-
a. cells	-	+	-	-	++	-	-	-	-	++	-	-
con. tis.	-	-	+	-	-	+	-	+	++	-	++	+
POST INTESTINE												
g. cells	-	-	-	-	+	-	+	+	-	-	-	-
a. cells	-	+	-	-	++	-	-	-	-	++	-	-
con. tis.	-	-	+	-	-	+	-	+	++	-	-	+
HEPATO-PANCREAS	-	-	-	-	-	-	-	-	-	-	-	-

Intensity of reaction: - no; + weak; ++ moderately strong; +++ very strong.

Legend: con.tis.: connective tissue
und.cells: undifferentiated cells
muc.cells: saccular mucous cells
s. ep.: surface epithelium
gl. ep.: glandular epithelium
g. cells: goblet cells
a. cells: absorptive cells
/ structure not present (cyprinid fishes are stomachless)

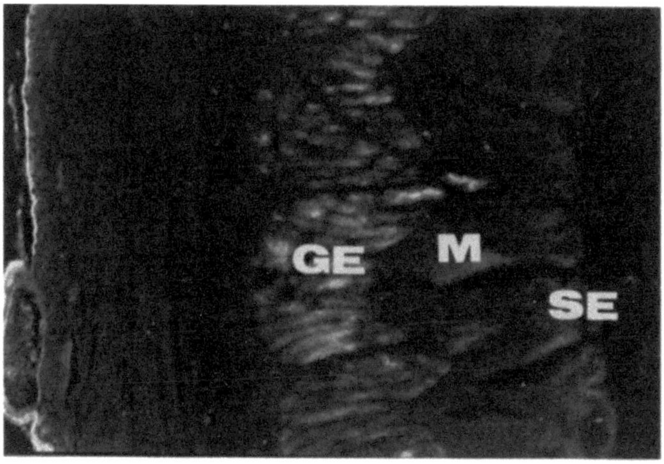

Fig. 2: Schilbe mystus. Longitudinal section of stomach. The glandular epithelium (GE) is stained after incubation with UEA I lectin. Surface epithelium (SE) remains unreactive whereas mucosa (M) is faintly stained.

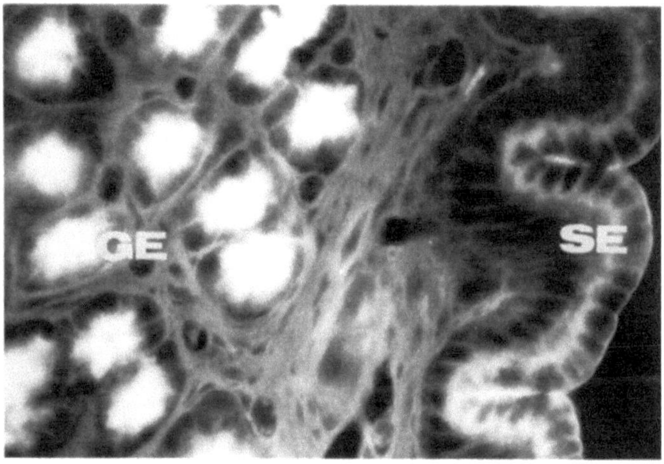

Fig. 3: Ctenopoma nigropannosum. Binding reaction of WGA lectin to the surface epithelium and glands of the stomach. The apical cytoplasm of glandular epithelial (GE) cells is strongly reactive. Surface epithelium (SE) is moderately stained.

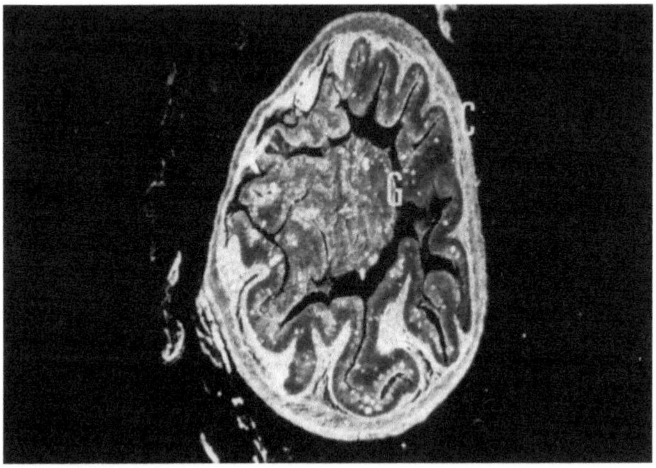

Fig. 4: Garra congoensis. Cross section of middle intestine. Connective tissue (C) is moderately stained after incubation with WGA lectin. Some reactive goblet cells (G) can be recognized.

fixation (16). Therefore in our work portions of oesophagus, stomach, intestine and hepatopancreas were fixed in the foregoing fluid. Careful examination of the Table 1 shows that the three lectins display different staining patterns. These results are somewhat different from those reported in the human species where gastric epithelial cells were described as being fucomucin positives (17, 18). This discrepancy is not surprising considering that the glandular structures of the fish stomach show a more primitive level of organization than the mammals; chief and parietal cells are lacking.

Comparison of our results with other publications in fish is quite difficult because information is lacking. Moreover such a comparison with lectin binding patterns in mammals is not easy because it must be born in mind that the histology of the digestive tract of Teleosts is simpler especially as the intestine is concerned, lined throughout by a simple columnar epithelium not elaborated into multicellular glands (19).

Whatever it can be, this paper indicates qualitative and quantitative differences in the composition of glycoconjugates in four African freshwater fishes. It is the opinion of the authors that these original observations may be due to some different digestive activities of these Teleosts which are related to the herbivorous, carnivorous habits of these fishes. The function of the sugar moieties accounting for the distinct characteristics of various cells in the digestive tract remains to be determined.

References

1. Cerović, G., Cuperlovic, M., Milosevic, Z.: These Proceedings, vol. 5

2. Holthofer, H., Nynäs-Mc Coy, A.: These Proceedings, vol. 5

3. Plendl, H., Schmahl, W.G.: These Proceedings, vol. 5

4. Schlepper-Schäfer, J., Hülsman, D., Kolb, H., Kolb-Bachofen, B.: These Proceedings, vol. 5

5. Danguy, A., Leclercq, G., De Vleeschouwer, N., Heuson, J.C.: Europ. J. Cell Biol. Suppl. 4, 5 (1983).

6. Breborowicz, J., Michalska, A., Breborowicz, D.: These Proceedings, vol. 5

7. Gabius, H.J., Engelhardt, R., Cramer, F.: These Proceedings, vol. 5

8. Gabius, H.J., Engelhardt, R., Cramer, F.: These Proceedings, vol. 5

9. Holthofer, H.: These Proceedings, vol. 5

10. Jacobovitz, D., Galand, P.: These Proceedings, vol. 5

11. Zippel, H.H., Hackenberg, R., Benzenberg, F., Hölzel, H., Schulz, K.D., Klein, P.J.: These Proceedings, vol. 5

12. Schumacker, U., Welsch, U., Barth, J.: These Proceedings, vol. 5

13. Nosck, J., Krajhanzl, A., Kocourek, J.: These Proceedings, vol. 3 (Bøg-Hansen, T.C., Spengler, G.A., eds) W. de Gruyter, Berlin-New York, 1983, pp 453-459.

14. Krajhanzl, A., Kocourek, J.: These Proceedings, vol. 5

15. Danguy, A., Lenglet, G., Kiss, R.B.: Cybium, 9, 193-201 (1985).

16. Soderstrom, K.O., Malmi, R., Karjaldinen, K.: Histochemistry, 80, 575-579 (1984).

17. Kuhlmann, W.D., Peschke, P., Wurster, K.: Virchows Arch. (Pathol. Anat.), 398, 319-328 (1983).

18. Peschke, P., Kulhmann, W.D., Wurster, K.: Experientia, 39, 286-287 (1983).

19. Barrington, E.J.W.: The Physiology of the Fishes, vol. 1 (M.E. Brown, ed.) Acad. Press, New York, pp. 109-161 (1957).

LECTIN HISTOCHEMISTRY OF THE CUTICLE OF THE ANTARCTIC KRILL *(EUPHAUSIA SUPERBA)*

Ulrich Welsch, Udo Schumacher
Anatomische Anstalt, Universität München
D-8000 München

Bettina Borisch
Anatomisches Institut, Universität Lübeck
D-2400 Lübeck

Introduction

The integument of Crustacea is composed of cuticle, epidermis and tegumental glands. The cuticle is a secretory product of the epidermal cells and its main function is that of an exoskeleton. According to different biological requirements this skeleton can be of variable chemical composition and it can display different physical properties (4).

The principal cuticular constituent in Crustacea and other arthropods is chitin, a polysaccharide in which the repeating unit is N-acetylglucosamine. According to Bouligand and Neville (1973) chitin is set in a protein matrix forming microfibrils (3). In the cuticle of crustacea acid mucopolysaccharides have been found by classical histochemical stains (4).

It is the aim of the present paper to report on lectin histochemical findings of the cuticle and tegumental glands of *Euphausia superba*, the Krill, a pelagic crustacean of considerable economic interest.

Lectins, Vol. V

Materials and methods

Seven µm thick deplasted and hydrated sections of Bouin fixed *Euphausia superba* were incubated for 30 min. in a moist chamber with the following FITC labeled lectins (E.Y.-products, San Mateo, California, obtained through medac Hamburg, dilutions used as indicated in brackets): DBA (1 : 40), LPA (1 : 1), GSA-I (1 : 1), GSA-B4 (1 : 20), PNA (1 : 80), SBA (1 : 10), PHA-L (1 : 40), HPA (1 : 20), Con A (1 : 80), RCA-I (1 : 40), UEA-I (1 : 40), WGA (1 : 80). After careful washing, the slides were mounted with a glycerol-PBS solution and photographed (for further methodological details see 5). Additionally a modified PAP technique for the demonstration of lectins was used (for details see 1). Furthermore sections were stained with Alcian Blue (pH 0.5 and 2.5) and the PAS reaction. Finally glutaraldehyde fixed Araldite embedded material was analysed in the electron microscope.

Results

Almost all staining and histochemical reactions showed a three layered structure of the *Euphausia* cuticle. The electron microscopical sections allowed to differentiate a broad homogenous pro-cuticle built up by chitin-microfibrils and an epicuticle which can be subdivided into a thin outer sublayer and a broad inner sublayer (Fig. 1).

1. PAS reaction. The outer sublayer of the epicuticle usually gave a strong positive reaction. In areas with a thick cuticle, e.g. the carapax, also a faint positive reaction could be observed in the procuticle; here the outer part of the epicuticle stained only weakly. The tegumental glands also stained faintly red.

2. Alcian Blue (pH 0.5 and 2.5): The procuticle, the outer sublayer of the epicuticle and the glandular cells of the tegumental glands in the legs stained blue at both pH values.

3. FITC lectins. Using FITC labeled lectins we always obtained a trilaminar staining pattern in the cuticle: the procuticle and the thin outer sublayer of the epicuticle reacted positively, whereas the broad inner sublayer of the epicuticle stained negatively (Fig. 2). Good to moderate degree of inhibition of the

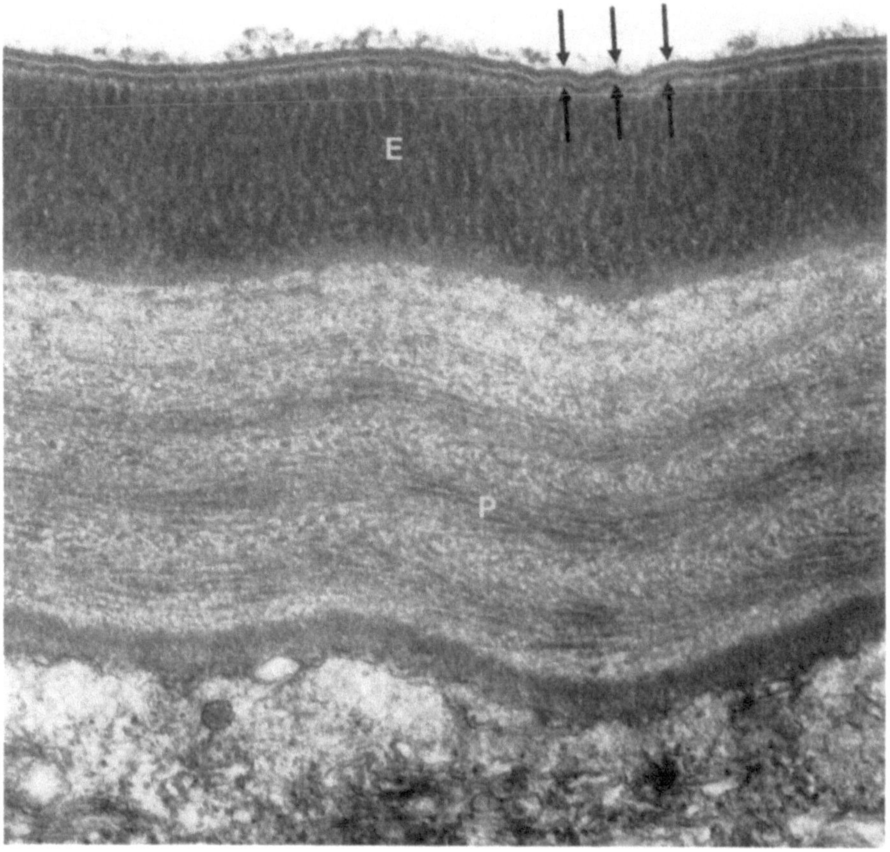

Fig. 1. Cuticle of Euphausia (EM x 25500). Note the procuticle (P) and the epicuticle (E). The outer sublayer is marked between 2 arrows.

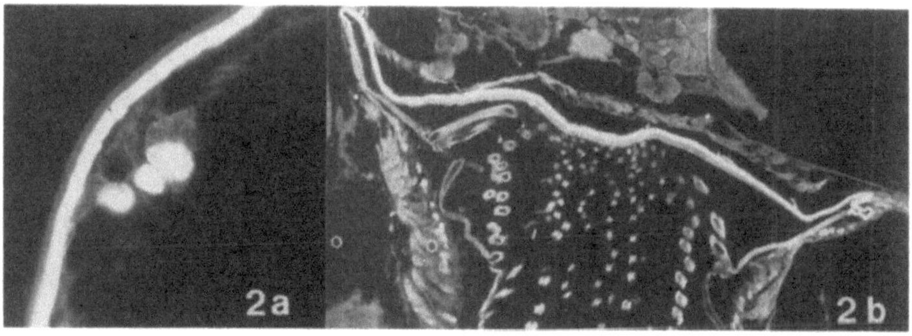

Fig. 2. Cuticula of Euphausia; 2a: LPA-FITC,x790; 2b: WGA-FITC,x165

fluorescence was obtained with D-GalNAc for DBA, SBA, HPA; with D-Gal for GSA-I, GSA-B4, PNA and RCA-I; D-Man and D-Gluc for Con A; D-GlucNAc for WGA; L-Fucose for UEA-I. No inhibition was observed with sialic acid for LPA and D-GalNAc for PHA-L.

The tegumental glands in the thoracic legs reacted strongly with HPA, moderately so with Con A, GSA-I, GSA-B4, SBA and weakly with WGA and PNA; the other lectins tested did not react with the glandular cells

4. PAP lectin technique. In the PAP lectin preparations the staining pattern of the cuticle was more variable than in the FITC lectin preparations; however, in principle the results obtained with both techniques are in agreement with each other. Often the procuticle stained positively in thick areas of the cuticle. Only the following lectins could be tested: SBA, UEA-I, WGA, PNA and GSA-I. All these lectins reacted positively in the procuticle of thick cuticular areas, in thin cuticles the positive reaction was to be seen only locally and often in basal (and apical : WGA) parts of the procuticle. The tegumental glands reacted strongly with GSA-I; a medium (SBA) or weak (WGA and PNA) reaction was observed with the other lectins (Fig. 3). Reaction with UEA-I was negative.

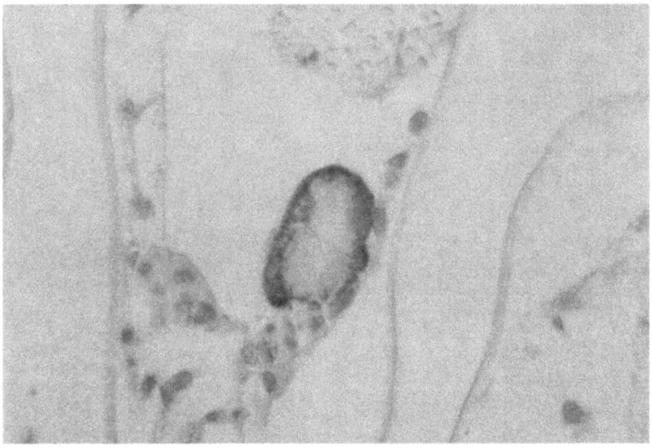

Fig. 3. Positive staining of a tegumental gland with SBA-PAP,x200

Discussion

The present study has shown that both cuticle and tegumental glands of *Euphausia* give positive results with various histochemical reactions for carbohydrates including a broad selection of different lectins. This fact points to the presence of various carbohydrate components in addition to the well known N-acetylglucosamine, the chemical unit of chitin. The probability that indeed such diverse carbohydrates are present is markedly strengthened by the fact that different methods gave principally similar results and that the lectin controls with inhibiting sugars resulted either in almost abolishing (PNA,SBA, RCA-I) or in moderately inhibiting (DBA, GSA-I, GSA-B4, Con A, UEA-I, WGA) the staining reaction. These carbohydrate components are predominantly localized in the procuticle in the sense of Stevenson (1985), i.e. the cuticular layer marked by chitin microfibrils. In this procuticle all tested FITC and PAP lectin preparations gave a positive result as well as Alcian Blue at pH 0.5 and 2.5, the latter stain being the classical method to demonstrate acid mucopolysaccharides, which had already been applied successfully to the cuticles of single crustacea, diplopoda and insects (3, 4). As to the question which components of the procuticle contain the different carbohydrates, one may speculate that they are localized in the protein matrix of the chitin molecules which according to chemical analysis also contain glycoproteins (3). It also seems possible that enzymes in the procuticle, the amount of which varies with different phases of the moulting cycle, may be in part responsible for this positive reaction. In this context it should be mentioned that *Euphausia* tends to moult relatively rapidly when handled or when it feels threatened.

The PAS reaction and the FITC labeled lectins also stained a narrow uppermost layer of the cuticle, which we interprete to correspond to the outer sublayer of the epicuticle. As can be seen in the electron microscope (Fig. 1) this layer is of complex structure and its irregular shaped filamentous surface may represent the positive light microscopical uppermost cuticular layer. Since tegumental glands are very few in the Krill and confined to certain body regions we think that their secretions do not contribute to this surface layer. The broad inner zone of the epicuticle obviously does not contain carbohydrates, according to chemical analysis it consists mainly of lipidic material. The differences observed between the PAP and FITC techniques may be due to the different resolving powers of these methods.

Acknowledgements. This study was supported by a grant from the Deutsche Forschungsgemeinschaft (We 360/2).

References

1) Borisch, B., P. Möller, D. Harms: Patholog. 4, 241-243,1983.

2) Bouligand, Y., A. C. Neville: J. Microsc. 17, 81a-82a, 1973.

3) Hackman, R. H.: In: Biology of the Integument I, ed. by J. Bereiter Hahn, A. G. Matoltsy, and K. S. Richards. Springer, Berlin, 1984, 583-610.

4) Stevenson, J. R.: In: The Biology of Crustacea, Vol 9; ed. by D. E. Bliss and L. H. Mantel. Academic Press, New York 1985, 1-24.

5) Welsch, U., U. Schumacher: Acta zool. (Stockholm), 65, 105-112, 1984.

PART 7

LECTINS AS TOOLS

HETEROGENEITY OF THE MONOCYTE/MACROPHAGE SYSTEM AS DEMONSTRATED BY THE USE OF LECTINS AND MONOCLONAL ANTIBODIES

H. Kreipe, H.J. Radzun, *U. Schumacher, M.R. Parwaresch
Department of Pathology, University of Kiel, D-23 Kiel

*Department of Anatomy, University of Munich, D-8000 Munich

It is well known that the monocyte/macrophage system constitutes a homogeneous cell system with regard to its origin from the bone marrow and the blood monocyte (1, 2). In contrast, applying different analytical methods such as morphology, enzymecytochemistry, and functional tests the monocyte/macrophage system displays a considerable heterogeneity (3, 4).
In the present study monocytes, peritoneal and alveolar macrophages as well as the monocyte/macrophage related permanent human cell line U-937 (5) were subjected to the analysis of their lectin binding patterns, reactivity with different monocyte/macrophage specific monoclonal antibodies, and of their surface glycoprotein patterns. Using these parameters it was the aim to find out discriminating properties for the different monocyte/macrophage subpopulations and to define certain differentiation stages within the monocyte/macrophage lineage.

Lectins, Vol. V

Materials and Methods

1) Separation of cell populations

Blood monocytes were separated from venous blood by density centrifugation and glass adherence. Peritoneal as well as alveolar macrophages were obtained by centrifugation (400 xg, 20 min, 4°C) from aspirated effusions of peritoneal cavity and bronchial lavages, respectively. Cells of the monocyte/macrophage related permanent human cell line U-937 (5) were harvested from optimally growing cultures using fetal calf serum (20 vol %) supplemented RPMI 1640 medium (Mediapharm, Aschaffenburg, West Germany) (for details refer to 6).

2) Analytical methods

Unfixed cytospin preparations were used for the analysis of lectin binding patterns and of reactivity patterns with monoclonal antibodies. Surface glycoproteins were analysed only in cell populations showing a viability and purity rate of over 80 % as demonstrated by trypan blue exclusion technique and morphological and enzymecytochemical methods.

a) Lectin binding assay

The following fluorescein isothiocynate-coupled lectins were applied and evaluated as described elsewhere (7):
Limulus polyphemus agglutinin, Ulex europaeus agglutinin, Soy bean agglutinin, Phaseolus vulgaris agglutinin, Concanavalin A, Wheatgerm agglutinin, and Helix pomatia agglutinin (Medac, Hamburg, West Germany).

b) Immunocytochemistry

A three step immunoperoxidase method was applied according to Stein et al (8) using the monoclonal antibodies Ki-M1, Ki-M6, Ki-M7, and Ki-M8 as primary agents. These antibodies were generated by fusion of the nonsecreting myeloma cell line X63-Ag 8.653 with splenic cells of Balb/c mice immunized with separated lysosomes of monocyte/macrophage related U-937 cell line and fresh minced lymph node specimens, respectively (9, 10). Ki-M1, Ki-M6, Ki-M7, and Ki-M8 proved to be entirely specific for the myelomonocytic cell line as judged by immunocytochemical analysis of various tissue specimens and blood cell populations.

c) Surface glycoprotein analysis

Surface glycoproteins were tritium labeled according to Andersson and Gahmberg (11) by oxidation of sialic acid residues with sodium metaperiodate followed by reduction with tritiated sodium borohydride. Electrophoresis was performed on sodium dodecyl sulfate containing polyacrylamide gels. Banding patterns were evaluated by fluorography using Kodak RPX-Omat X-ray film (12).

Results

1) Lectin binding pattern

Blood monocytes, peritoneal and alveolar macrophages showed a positive reaction with the majority of the different lectins applied in this study. Only Helix pomatia agglutinin was negative for these cell populations. Cells of the monocyte/macrophage cell line U-937 did not react with Limulus polyphemus agglutinin, Ulex europaeus agglutinin, and Soy bean agglutinin. The strongest reactivity for the different lectins was observed in case of alveolar macrophages (Fig. 1).

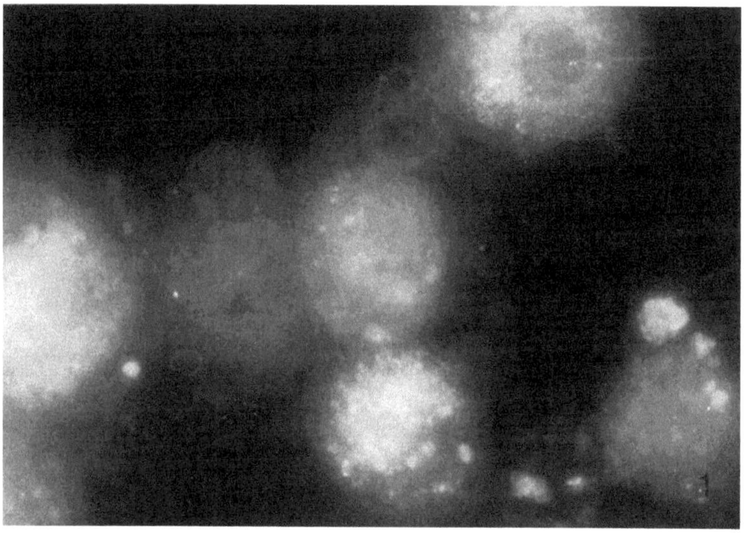

Figure 1)
Binding of Soy bean agglutinin to alveolar macrophages .
Cytocentrifuge preparation, x 1000.

2) Reactivity pattern with monoclonal antibodies

Cells of the permanent monocyte/macrophage related cell line U-937 showed only a positive reaction with the monoclonal antibody Ki-M7 specific for granulocytes, monocytes, and macrophages. For the other analysed cell populations an increasing reactivity with Ki-M1, Ki-M6, Ki-M7, and Ki-M8 could be observed from monocytes over peritoneal to alveolar macrophages. This is examplified in figure 2 for the monoclonal antibody Ki-M7 showing a dense diffuse immune reaction product within the cytoplasm of alveolar macrophages, whereas monocytes exhibit a granular immunoreactivity.

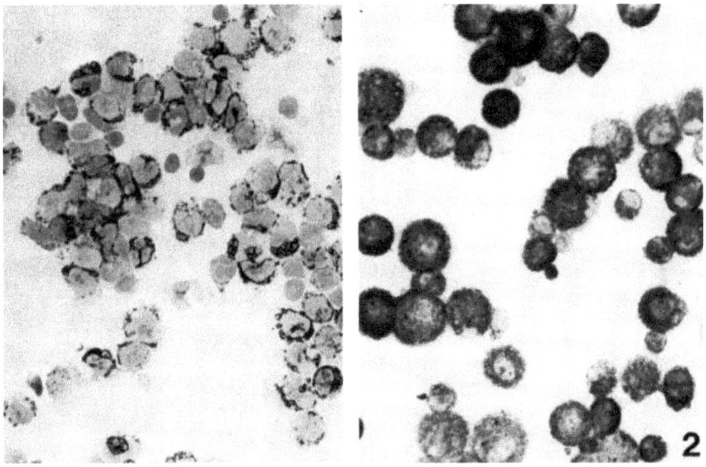

Figure 2)
Immunocytochemical reactivity of blood monocytes (left) and alveolar macrophages (right) with the monoclonal antibody Ki-M7 specific for granulocytes, monocytes, and macrophages. Cytocentrifuge preparations, immunoperoxidase method, x 350.

3) Surface glycoprotein pattern
Considering glycoprotein patterns a pronounced similarity could be observed between cells of the monocyte/macrophage related cell line U-937, monocytes, and peritoneal macrophages showing about seven strong bands between 250.000 and 88.000 daltons. In contrast alveolar macrophages displayed a broadened spectrum of glycoproteins with additional bands of lower molecular weights between 60.000 and 10.000 daltons (Fig. 3).

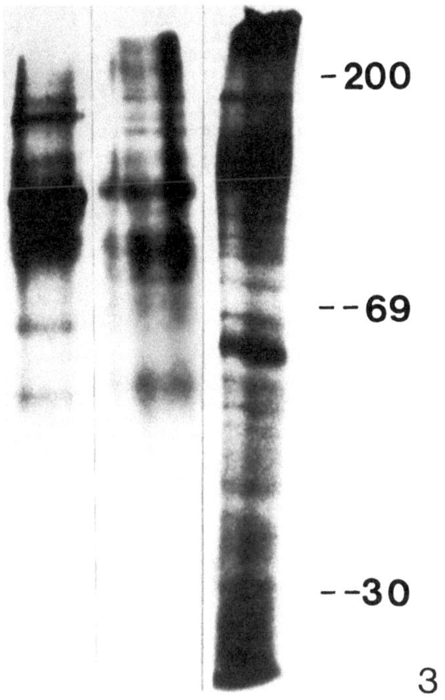

Figure 3)
Surface glycoprotein patterns of cells of the monocyte/macrophage related cell line U-937, blood monocytes, and alveolar macrophages (from left to right) with corresponding values of molecular weight (in kilodaltons).

Discussion

The heterogeneity of the monocyte/macrophage system is well established specially with regard to functional parameters: Besides phagocytosing macrophages the so-called immune accessories for B- and T lymphocyte immune response have been defined such as dendritic and interdigitating reticulum cells, respectively (13, 14). For the characterization of these subpopulations within the monocyte/macrophage system, however, clearly discriminating phenotypic marker profiles are so far available only to a limited extent (15).

Most of the properties analysed in this study showed a general tendency of being increasingly expressed during monocyte differentiation into peritoneal and alveolar macrophages (Fig. 1, 2). This issue was also true for the permanent monocyte/macrophage related cell line U-937, which is generally considered as a late monocyte precursor lacking only some of the typical monocytic properties (5, 6).

In contrast to the reactivity with lectins and monoclonal antibodies a clearly distinguishable phenotypic marker profile was observed in the glycoprotein pattern of alveolar macrophages in comparison to the other analysed cell populations showing additional glycoprotein bands of lower molecular weights between 10.000 and 60.000 daltons (Fig. 3).

Thus, glycoprotein patterns may be useful for the characterization of a certain activation state during monocyte/macrophage maturation, which is mirrored by alveolar macrophages (16). The other parameters such as lectin binding and reactivity with monoclonal antibodies applied in this study, did not seem to be appropriate to clearly discriminate between the analysed cell populations, which, however, all belong to the phagocytosing compartiment of the monocyte/macrophage system. It has to be shown, whether this finding holds also true for the immune accessories of this cell system.

Further studies will elucidate, to what extent lectin binding and immunocytochemical reactivity patterns as well as glycoprotein profiles can be correlated with differences of functional states of the monocyte/macrophage system measured in vitro (3, 4). Such experiments will probably elicit molecular structures being possibly controled by genetic mechanisms leading to the heterogeneous spectrum of monocyte/macrophage subpopulations.

References

1. Van Furth, G., Cohn, Z.A.: J. Exp. Med. 128, 415-433 (1968).

2. Parwaresch, M.R., Radzun, H.J., Dommes, M.: Am. J. Pathol. 102, 209-218 (1981).

3. Unanue, E.R.: Am. J. Pathol. 83, 396-417 (1976).

4. Nathan, C.F., Karnowsky, M.C., David, J.R.: J. Exp. Med. 133, 1356-1376 (1971).

5. Sundström, C., Nilsson, K.: Int. J. Cancer 17, 565-577 (1976).

6. Radzun, H.J., Parwaresch, M.R., Sundström, C., Nilsson, K., Eissner, M.: Int. J. Cancer 31, 181-186 (1983).

7. Welsch, U., Schumacher, U.: Virchows Arch. 44, 45-56 (1983).

8. Stein, H., Bonk, A., Tolksdorf, G., Lennert, K., Rodt, H., Gerdes, J.: J. Histochem. Cytochem. 28, 746-760 (1980).

9. Radzun, H.J., Parwaresch, M.R.: Cell Immunol. 82, 174-183 (1984).

10. Radzun, H.J., Parwaresch, M.R., Feller, A.C., Hansmann, M.-L.: Am. J. Pathol. 117, 441-450 (1984).

11. Andersson, L.C., Gahmberg, C.G.: Blood 52, 57-67 (1978).

12. Bonner, W.M., Laskey, R.A.: Eur. J. Biochem. 46, 83-88 (1974).

13. Unanue, E.R.: New Eng. J. Med. 303, 977-985 (1980).

14. Tew, J.G., Thorbecke, G.J., Steinman, R.M.: J. Reticuloendothel. Soc. 31, 371-380 (1982).

15. Ugolini, V., Nunez, G., Smith, R.G., Stastny, P., Capra, J.D.: Proc. Natl. Acad. Sci. 77, 6764-6768 (1980).

16. Radzun, H.J., Parwaresch, M.R., Kreipe, H.: J. Histochem. Cytochem. 31, 318-324 (1983).

DUAL PARAMETER FLOWCYTOMETRIC MEASUREMENT OF DNA-CONTENT AND LECTIN BINDING IN HUMAN BLADDER TUMORS.

T.F. Ørntoft, L. Bendtsen, H.S. Poulsen

The Institute of Cancer Research, Aarhus Kommunehospital, DK-8000 Aarhus C, Denmark

S.E. Petersen, H. Wolf

Dept. of Surgery, Aarhus Amtssygehus, DK-8000 Aarhus C

S. Pedersen, L. Bolund

Institute of Human Genetics, Aarhus University, DK-8000 Aarhus C

A large fraction of patients with low grade bladder tumors will suffer from local recurrence of their tumors after resection, but only a minor fraction of these will progress to the invasive stage (1). As current histopathological examination of the tumors do not provide a completely adequate picture of the biological potential of these, other risk parameters have been included to obtain better information on the indvidual disease course. A predictive parameter has been found in blood group antigen deletion, as retrospective immunohistological studies have shown a correlation between this phenomenon and development of invasive cancer (2). These results have been obtained by the use of anti-ABH blood group antisera and recently by the use of the anti-T lectin PNA (Peanut agglutinin)(3). The theory of PNA binding to bladder tumors is based on the hypothesis that it identifies blood group MN precursors, which are cryptic in normal tissue (4). A second predictive parameter has been found in the DNA content of the tumor (5). Tumors with close to diploid or tetraploid DNA amounts have a better prognosis than more aneuploid tumors (6). Microscopical morphology does not reflect the DNA content (5,6). To obtain information on the relationship between carbohydrate composition

Lectins, Vol. V

and DNA contents of the different tumor cell subpopulations we have developed a method for simultaneous measurement of lectin binding and DNA content on single cell suspensions from human bladder tumors.

Materials and Methods.

Thirteen human bladder tumors were collected in cooled PBS immediately after resection. They were physically disintegrated to single cell suspensions and stored frozen at -80°C, in Tris/PBS containing 7.5% DMSO. Prior to labelling, the cells were thawn and fixed in ice-cold absolute ethanol with vortexing. Parts of the tumors were fixed in formalin, paraffin embedded, and HE stained for light microscopic grading according to Bergkvist (7). The distribution of the tumors was: Normal: 1; grade I: 2; grade II: 4; grade III: 3; grade IV: 3. Invasive tumors were all grade III and grade IV. Labelling of surface glycoconjugates was done by incubation of ethanol-fixed cells with lectins (PNA, LCA, WGA, STA, VFA) conjugated to FITC (purchased from Kem-En-Tec, Hellerup Denmark) for one hour at 4°C. Lectins were used at a concentration of 10 ug/ml to avoid agglutination (8). Their binding specificity was controlled by absorption with appropriate simple sugars. After washing, RNA was removed by RNase, 10 ug/ml, and the DNA of the cells was stained by Propidium Iodide (PI), 75 ug/ /ml. The flow cytometric measurements were done in this PI-RNase solution, by the use of a FACS II (Beckton Dickinson). The FACS was calibrated by the use of green beads prior to each analysis Excitation was provided with a laser operating at 488 nm at 0.3 watt. The red and green signals were separated by a dichroic mirror, with a S525-nm short pass filter in front of the green photo detector, and a 620-nm long pass filter in front of the red photo detector. Scatter gating was used to exclude cellular debris, as confirmed by phase contrast microscopy of gated material. To obtain orthogonality in a red-green plot, a dual fluorescense compensation module was used if necessary.
Lectin binding was expressed as the ratio between the median of green fluorescence from the lectin bound to the cells concerned prior to and after absorption with lectin-specific sugars.

Results

By lack of correlation to forward-angle light scatter it was found that alterations in lectin binding was not simply caused by differences in cell size. A common finding was a wide variation in the lectin binding to cells with similar DNA contents from almost negative to strongly positive cells (Fig. 1A). In some tumors, such broad distributions could be subdivided into cells with low expression of lectin receptors and cells with a high expression of these receptors (Fig. 1B).

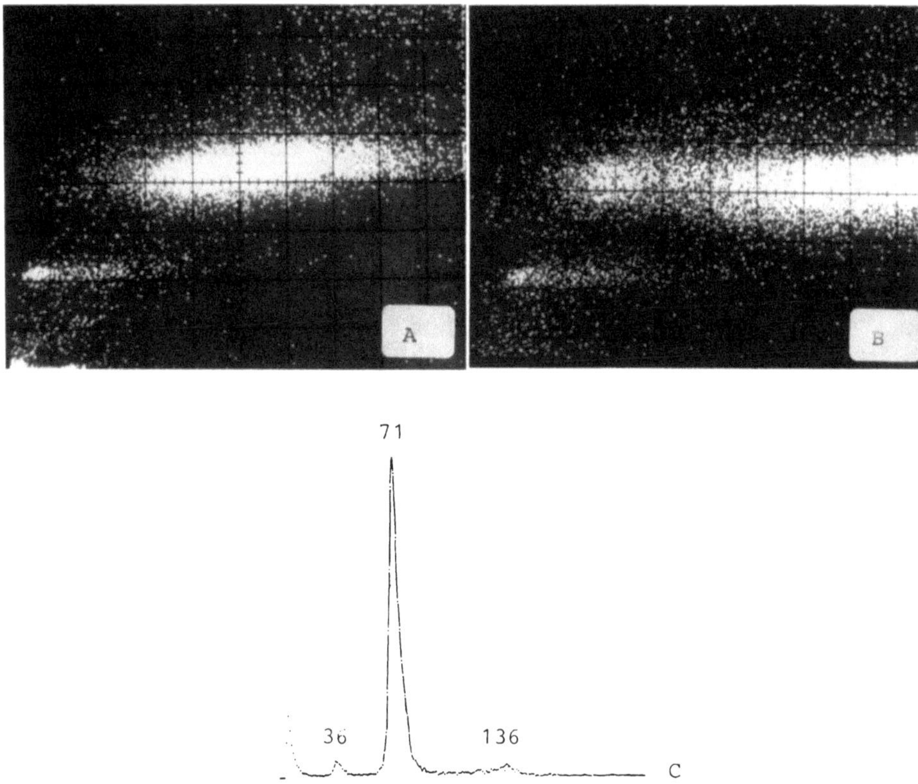

Fig. 1. Three results from a near tetraploid bladder tumor. Ordinate is red fluorescence (DNA), abscissa is green fluorescense (Lectin). A: PNA binding; B: LCA binding; C: DNA flowcytogram, ordinate is number of cells, figures on abscissa are amount of stained DNA. A small normal diploid population is present (36), the tumor cell population consists of cells in G_0/G_1 phase (71), and G_2/M phase (136).

Lectin binding to cells in G_1/G_0 and G_2/M phases of the cell cycle was different, but this difference seemed to be fully explained by an increased cell size in the G_2/M phase. Several tumors were examined both fresh from surgery and after months of storage at -80°C, without significant alteration of the DNA variation coefficient or the lectin binding. PNA which identifies Gal or Gal-GalNAc structures, did not react with diploid or near diploid cell populations (Table I). It bound to all abnormal DNA populations with the highest binding ratio (3.2 and 3.3) found in two hypotetraploid populations. LCA, which identifies Mannose structures, showed low level binding to near diploid populations, and remarkably, a similar degree of binding to near tetraploid populations.

In contrast, the examined hypotetraploid population was strongly positive. A few tumors were examined with WGA, STA, and VFA. WGA bound to near diploid cells to a surprisingly large extent, but seemed to be reduced in the more aneuploid cells, whereas the binding of STA and VFA (like PNA and LCA) was highest for the hypotetraploid cells. Treatment of PNA negative cells with neuraminidase from Vibrio Cholerae resulted in a change to PNA positivity.

TABLE I

Mean lectin binding (ratio positive/negative cells) in relation to ploidy of tumor subgroups.

	LECTIN BINDING			
DNA	PNA	LCA	WGA	VFA
Near diploid (ND)	1.0(n=7)	1.6(n=6)	3.3(n=1)	1.1(n=1)
Hypotetraploid (HoT)	2.6(n=3)	5.5(n=1)	1.8(n=1)	1.6(n=1)
Near tetraploid (NT)	2.2(n=3)	1.3(n=3)		
Hypertetraploid (HeT)	3.0(n=1)			

	VFA			
Near diploid (ND)	1.0(n=1)			
Hypotetraploid (HoT)	1.7(n=1)			
Near tetraploid (NT)	1.2(n=1)			
Hypertetraploid (HeT)				

TABLE II

DNA amount in relation to grade of atypia. For abbreviations, see Table I.

Grade of atypia	DNA
0-I	3ND
II	2 ND, 1 HoT, 1 NT
III-IV	1 ND, 2 HoT, 2 NT, 1 HeT.

Discussion

This paper describes the first simultaneous measurement of the two most well examined non-histological prognostic parameters in bladder tumors, carbohydrate composition, and DNA content. The ploidy alterations are related to the degree of atypia, so that 5 out of 6 grade III-IV tumors have abnormal hyperdiploid cell populations (Table II). The cellular lectin binding seems to be correlated to the degree of aneuploidy. Concerning PNA, this is evidenced by the acquisition of receptors for this lectin. These receptors seem to be present in largest amounts on hypotetraploid and hypertetraploid cell populations. Interestingly, populations with these DNA amounts have been shown to have a great potential for invasive growth (6). Therefore, a possible link between the presence of PNA receptors and an invasive behaviour might exist. This is in accordance with retrospective histological examinations which have shown PNA binding to be related to an invasive disease course (9).

Whether the expression of PNA receptors is a result of an invasive cell behaviour or a mediator of this is unknown. However, specific glycosylation patterns have been shown to be related to the ability to metastasize in other experimental systems (10). The effect of neuraminidase on PNA negative cells is in accordance with the idea that PNA receptors are incompletely sialylated M/N blood group antigens (4).

LCA showed no difference in binding to near diploid and near tetraploid cell populations, but was bound to a great extent to hypotetraploid populations as was PNA. Its receptor might therefore be involved in a similar way in invasive cell behaviour as proposed for the PNA receptor. The measured fluctuations in STA and VFA binding were modest. WGA binding might be inversely related to the degree of aneuploidy, but the material examined with these three lectins was too small to be conclusive.

References

1. Friedell GH, Bell JR, Burney SW, Soto EA, Tiltman AJ. 1976. Urol Clin North Am. 3:53-70

2. Decenzo JM, Howard P, Irish CE. 1975. J Urol 114:874-878.

3. Coon JS, Weinstein RS, Summers JL. 1982. Am J Clin Pathol 77:692-699.

4. Springer GF. 1984. Science 224:1198-1206.

5. Lamb D. 1967. Brit Med J 1:273-277.

6. Wijkström H, Granberg-Öhman I, Tribakait B. 1984. Cancer 53:1718-1723.

7. Bergkvist A, Ljungkvist A, Moberger G. 1965. Acta Chir Scand 130:371-378.

8. Reimann J, Ehman D, Miller RG. 1984. Cytometry 5:194-203.

9. Summers JL, Coon JS, Ward RM, Falov WH, Miller AW III, Weinstein RS. 1983. Cancer Res 43:934-939.

10. Yogeswaaran G, Stein BS, Sebastian H. 1978. Cancer Res. 38: 1336-1344.

A SUGAR-MEDIATED MECHANISM IN GLOMERULAR ENDOTHELIUM OF MICE

Bettina Borisch[1], Udo Schumacher[2], Kathy Hüniche[3], Wolfgang Kühnel[3]

[1]Pathologisches Institut der Universität Würzburg, FRG
[2]Anatomische Anstalt, Ludwig Maximilians-Universität, München, FRG
[3]Institut für Anatomie, Medizinische Universität zu Lübeck, FRG

Recent developments have implicated the cell-surface glycoproteins, with their endowement of highly variable carbohydrate moieties, as potential mediators of information exchange and have prompted hypotheses as to the manner in which such recognition phenomena are evoked. The classical studies of Hirst, Burnet and their collegues (1) have demonstrated the role of sialic acid in the successfull attachment of influenza virus to the erythrocyte membrane. It was recently shown by Williams et al. (2) that the endothelia of microvessels, isolated from rat epididymal fat pads, take up glycosylated albumin in preference to albumin. Freshly isolated endothelial cells are known to take up macromolecules by "constitutive micropinocytosis" as an expression of vascular transport (3).

Up to now lectin-conjugates have been used extensively for mapping oligosaccharide moities on the surface of endothelium of different localisation in vitro (3, 4). Normal human kidney has been investigated by the use of peroxidase-labeled lectins at the light and electron microscopic level (5). Lectin binding reactions have been carried out on normal human kidney (6) as well as on diseased kidney (7). Recently, Holthöfer (8) has shown that lectins can also be used as maturation stage specific markers for different cell lines in the mouse kidney.

Very little is known on the in vivo binding of lectins in the kidney, a knowledge that could aid in the understanding of basic mechanisms of the kidney.

In the present study we wanted to examine the in vivo binding of lectins by different domains of the renal endothelium. This would

Lectins, Vol. V

detect the glycoconjugate pattern in the living animal and then lead to the question if there is only an attachment or also a consecutive uptake of material.

Material and methods

Animals. Six to eight weeks old NMRI-mice of either sexes were used. The mice were given free access to water and a standard diet.

Lectins. Ferritin conjugates of red kidney bean lectin (PHA-L) were purchased from E.Y. laboratories (San Mateo, CA), as well as all fluorescein isothiocyanate (FITC) or Texas Red coupled lectins and the monosaccharides used as hapten inhibitors of lectin binding.

Animal experiments. The basic experimental protocol involved the following steps: Fluorochrome coupled lectins (GonA, DBA, GSA I, LTA, LPA, PHA-L, PNA, SBA, UEA I, WGA) were each injected into the caudal vein of an experimental animal. The mice received 0.3 mg labeled lectin each. Animals were sacrified after 10 minutes, 1 hour and 12 hours, respectively. They were anesthetized with ether and then perfused via the left ventricle with phosphate buffered saline (PBS). The perfusion was carried out for 5 minutes at a rate of 6-8 ml/min.
Control animals were perfused with the appropriate inhibitory sugar of the lectin prior to the perfusion with the lectin itself. Sugar concentrations ranged from 0.2 M for UEA I i.e. to 0.4 M for PNA i.e.
The kidneys, liver and spleen were taken and directly snap-frozen in liquid nitrogen. Sections of the organs were cut (7 µm thick) on an American Optical system cryostat. Cryostat sections were immediately mounted with a glycerol-PBS-buffer and then viewed under a ZEISS standard microscope. Photographs were taken with an ILFORD HP 5 135 (400 ASA) film.

A second group of animals was treated with the above mentioned lectins which had been coupled to ferritin as an electron dense marker. This group of animals was proccessed following the same protocol as the FITC-group, with the exception that perfusion with PBS was followed by a fixation by perfusion with 2.5 % glutaraldehyde in 0.1 M NaCacodylate-HCl. Kidneys, liver and spleen were further processed for electron microscopy. Blocks of tissue were postfixed in 1 % OsO_4. After dehydration in a graded series of ethanol the tissue was embedded in Araldit M. Ultrathin sections were stained with lead citrate and uranyl acetate (LKB 2168 Ultrastainer) and viewed under a Philipps EM 400. In the second group as well, control animals were used to test the specificity of lectin binding by competitive inhibition with the appropriate monosaccharide.

Results

With the help of FITC-conjugated lectins in vivo binding sites could be shown. From all the lectins used only PHA-L showed a strong binding to glomerular structures (Fig. 1). The other lec-

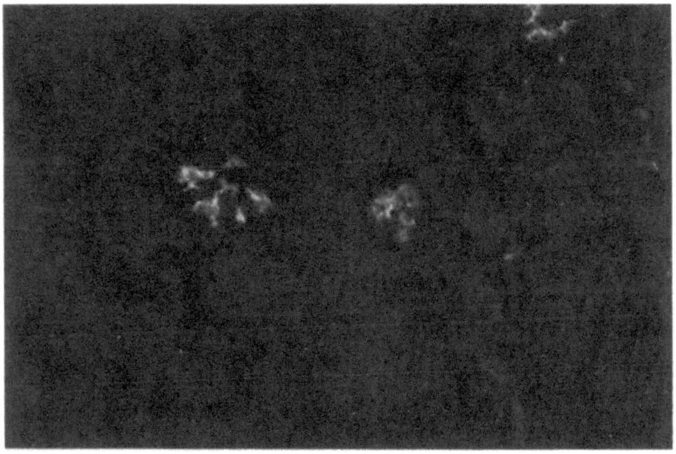

Fig. 1: PHA-L decorates glomerular structures after 1 hour in vivo in mice. FITC-PHA-L, x 90.

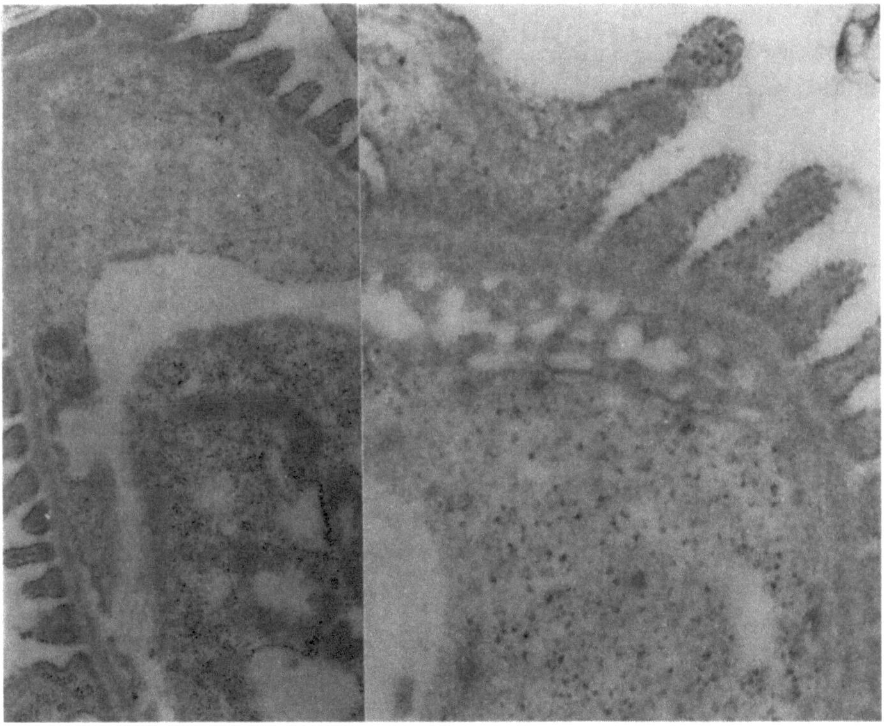

Fig. 2: Glomerular capillary wall. Endothelial and epithelial cells are labeled with PHA-L-ferritin. Lectin was injected 1 h before. Left x 21.000, right x 48.750.

tins under investigation did not demonstrate any in vivo binding to glomerular and other vascular structures. All lectins showed a coarse binding pattern in the liver and the spleen.

The binding of PHA-L to glomerular structures could be nearly completely abolished by the pretreatment with N-acetyl-D-Galactosamine whereas the binding in the liver and the spleen was unaffected by inhibitory sugars.

Electron microscopy revealed the following distribution of ferritin-labeled PHA-L: Ferritin particles are found not only on endothelial cells of the glomerulum but also on epithelial cells and their pedicles. Macrophages showed abundant PHA-L in their cytoplasm (Fig. 2).

Both, the FITC-labeled as well as the ferritin-labeled PHA-L showed nearly no difference in binding sites and binding intensity by comparison of the 10 min and the 1 h animals. After 12 h the fluorescence labeling was weaker. Furthermore, the glomerular capillaries seemed to be slightly altered.

Discussion

This study showed the binding of a certain lectin, PHA-L, to distinct parts of the vascular system, to the renal glomerulum under in vivo conditions.

Although conventional approaches have revealed that lectins bind to glomerular structures under in vitro conditions (that means on tissue sections) (6,8), little is known on the in vivo interaction between lectins and the vascular system of the kidney. However, this might be of special interest, as cell surface glycoproteins may play a role in information exchanges and recognition phenomena (1). On the other hand such complex phenomena are poorly understood in human diseased glomerulum.

We could demonstrate interactions between the rodent glomerulus and circulating macromolecules, i.e. lectins. From a panel of 10 tested lectins only one was shown to interact with a distinct part of the vascular system. Our experiment provides no explanation for this phenomenon, but suggestions on steric inhendrance, and accessibility of the accepting molecules are possible.

In our experiments, PHA-L was not only found in endothelial but also in the epithelial layer of the glomerulus. Small solutes such as electrolytes, glucose and urea pass freely across capillary endothelia, whereas the walls of these capillaries restrict the passage of macromolecules, such as PHA-L. Probably the PHA-L-ferritin molecule has not been filtered freely through the glomerular filtration barrier. So explanations possible are: Either there has been a transport of PHA-L or the lectin has been creating alterations in the filtration barrier, so that a leakage of macromolecules becomes possible.

It is well known that glycoproteins contribute to the anionic charge of the glomerulus and that polycutions induce alterations of the glomerular epithelium (10). The glomerular filtration barrier is both size and charge selective (11). PHA-L could have interfered with this mechanism.

Our results suggest that the binding of PHA-L to the glomerulum is a recognition phenomenon between a lectin and an appropriate cell-surface acceptor and that presumably the binding enhances alterations in the filtration barrier, so that PHA-L can reach the epithelial cells and the urinary space. This interaction between PHA-L and the renal glomerulum could serve as a modell for the selected uptake of macromolecules in the kidney.

Acknowledgements. The skillful help of Miss D. Schroll in preparing the manuscript is kindly acknowledged.

REFERENCES

1. Burnet, F.M.: Physiol. Rev. 31, 131 (1951)

2. Williams, S.K., Devenny, J.D., Bitensky, M.W.: Proc. Natl. Acad. Sci. U.S.A. 78, 2393-2397 (1981)

3. Simionescu, M., Simionescu, N.: Microvasc. Res. 16, 426-452 (1978)

4. Borisch, B., Möller, P., Harms, D.: Pathologe 4, 241-243 (1983)

5. Farraggiana, T., Malchiodi, F., Prado, A., Churg, J.: J. Histochem. Cytochem. 30, 451-458 (1982)

6. Holthöfer, H., Virtanen, I., Petterson, G., Törnroth, T., Alfthan, O., Linder, E., Miettinen, A.: Lab. Invest 45, 391-399 (1981)

7. Holthöfer, H., Rapola, J.: Lectins, Biology, Clinical Biochemistry. Vol. 4, 83 ff. (1985)

8. Holthöfer, H., Nynüs-McCoy, A.: Vol. 5, these proceedings.

9. Klein, P.J., Vierbuchen, M., Fischer, J., Ortmann, M., Uhlenbruck, G., Fischer, R.: Immun. Infekt. 10, 151-158 (1982)

10. Seiler, M.W., Remcke, H.G., Venkatachalam, M.A., Cotran, R.S.: Lab. Invest. 36, 48-61 (1977)

11. Bliss, D.J., Brewer, D.B.: Virchows. Arch. (Cell. Pathol.): 48, 351-359 (1985)

PHENOTYPE OF SBA$^+$ T CELLS IN HUMAN BONE MARROW

A. Raedler, E. Hachmann[+], M. Buhl, H. Greten, H.-G. Thiele
Med. Dept., [+]Dept. Anaesth., Univ. Hamburg, FRG

SBA$^+$ T cells can be demonstrated in the subcortical area of human as well as murine thymus (1,2). It has been proposed that this area is populated with immigrating bone marrow-derived T cells that give rise to cortical thymocytes (3,4). Furthermore it has been observed that agglutination of bone marrow cells with SBA prior to transplantation successfully removes T cells or T cell precursors thus preventing graft-versus-host reactions (5,6). The experiments presented here are undertaken to characterise the population of SBA$^+$ T cells in the human bone marrow to give answer to the question whether SBA$^+$ bone marrow-derived T cells resemble SBA$^+$ T cells in the thymic subcortex in regard to their phenotype and functional properties.

Lectins, Vol. V

Material and Methods

Preparation of bone marrow cells. Aspiration of bone marrow was performed on patients (n=6) that were suspected to suffer from iron deficiency. 1-2 ml of bone marrow cells were added to 4 ml 0.9 saline supplemented with 1000 U heparin and given into EDTA coated tubes. 2.5 ml of 'Lymphoprep' (Nyegaard) were overlayered with 5.0 ml of diluted blood and centrifugated for 30 min (20C, 560 g). The lymphocytes were recovered from the interphase and washed twice with phosphate-buffered saline (Gibco). The pellet was suspended in 15 ml of cell culture medium (RPMI 1640, Gibco) supplemented with 10% fetal calf serum (Medac) and cultured for 2 h at 37°C for removal of monocytes. Non-adherent cells were then collected and layered on Petri dishes (0.5.5 cm, $5x10^6$ cells per plate) coated with goat anti-human immunoglobulin (Tago, 4 ml, 10 µl/ml). After incubation for 60 min at room temperature non-adherent (non-surface immunoglobulin bearing) cells were recovered.

Controls of this separation step were performed by labelling non-adherent cells with goat anti-human Ig $(Fab)_2$ antibodies (1:20, Tago) and anti-Leu 12 (1:20, Becton-Dickinson) and adherent ones with T3 (5 µl/10^6 cells).

Enumeration of SBA^+ cells. $1x10^6$ cells were incubated in SBA/FITC (1 mg/ml, diluted 1:10, Medac). Pellets were then suspended in 50 µl of rhodamine coupled goat $(Fab)_2$ antimouse IgG (Tago) diluted 1:60 with PBS (30 min, 4°C) and subsequently washed three times with culture medium. Cells were then resuspended, examined and counted (2-3 x 100 cells) under a fluorescence microscope (Zeiss).
Controls were performed by supplementation with 0.2 mmol solution containing the inhibitory sugar N-acetyl-galactosamine.

Separation of SBA^+ cells by lectin-mediated panning. $5x10^6$ non-adherent, sIg^- cells were layered on petri dishes (Ø 5.5 cm) coated with SBA (4 ml, 0.001%, 12h, 4°C). After incubation for 30 min at room temperature non-adherent cells were recovered. Adherent cells were harvested after supplementing N-acetyl-galactosamine (0.2 mmol, 10 min, 4°C). Effectivity of panning was controled by labelling of adherent cells and non-adherent cells with SBA/FITC.

Phenotypic characterization of SBA^+ cells. $1x10^6$ non-adherent sIg^- cells were incubated in TRITC coupled OKT3, OKT4, OKT8, OKT9, OKT 11 (ortho-mune, 5 ul/ml, diluted 1:10), PNA/TRITC (Medac, 1mg/ml, diluted 1:10) anti-HLA/A,B,C and DR (common epitope, Becton, Dickinson, 1:15) for 30 min, 4°C. In case of anti-HLA antibodies pellets were suspended in 50 µl of TRITC-coupled goat $(Fab)_2$-antimouse IgG (Tago, diluted 1:60 with phosphate-buffered saline (30 min, 4°C) and subsequently washed three times in culture medium.

Autologeous pokeweed mitogen (PWM) stimulated B cell assay. Surface immunoglobulin bearing cells were separated using direct panning through goat antihuman immunoglobulin coated Petri dishes from the pool of non-glass adherent lymphocytes. One million cells were given in culture medium (RPMI 1640,

Gibco, plus 1o% fetal calf serum and 50 µg/ml PWM, Gibco and antibiotics) and cultured for 5 days. One million SBA^+ T cells (prepared by panning as described above) no T cells or SBA^+ cells plus total peripheral T cells were added to the B cell cultures. After 5 days the supernatants were collected and analysed for Ig by an enzyme linked immunoabsorbent assay using goat anti-human Ig (Medac) and an alkaline phosphatase coupled antibody (Medac), respectively.

Results

Separation of SBA^+ non-adherent, non sIg^+ mononuclear cells (MNC). After preparation of mononuclear cells from bone marrow via density centrifugation and removal of monocytes and B cells the remaining cells were separated into SBA^+ and SBA^- cells through lectin-mediated panning. It was found that 53% (S.D. 14.4) of the cells were SBA^+.

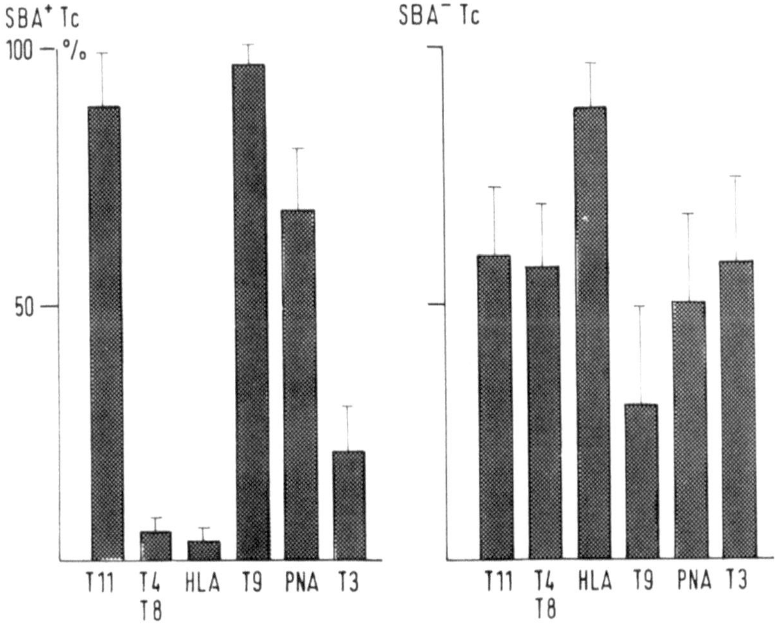

Fig.1: (left) Characterization of phenotype of separated non-adherent sIg^-, SBA^+ bone marrow cells revealing a predominant subset of SBA^+, $T11^+$, PNA^+, $T4^-$, $T8^-$, HLA^- cells.

Fig.2: (right) Characterization of phenotype of separated non-adherent, sIg^-, SBA^- bone marrow cells.

Characterization of cell surface determinants on SBA^+ and SBA^- bone marrow-derived non-adherent, non-sIg^+ MNC. As shown in Fig.1 SBA^+ cells express the T 11 antigen in 88% (S.D. 11.9) and are thus referred to as T cells. T3 antigens (21%, S.D. 8.4) are only weakly detectable on these cells. Moreover most SBA^+ T cells are negative for T4 or T8 (94.3%) and are negative or only weakly positive for HLA/DR, A, B, C, (96%) but express T9 (97%) on their cell surface. SBA non-adherent, sIg^- lymphocytes are heterogenous concerning these cell surface structures as is shown in Fig.2.

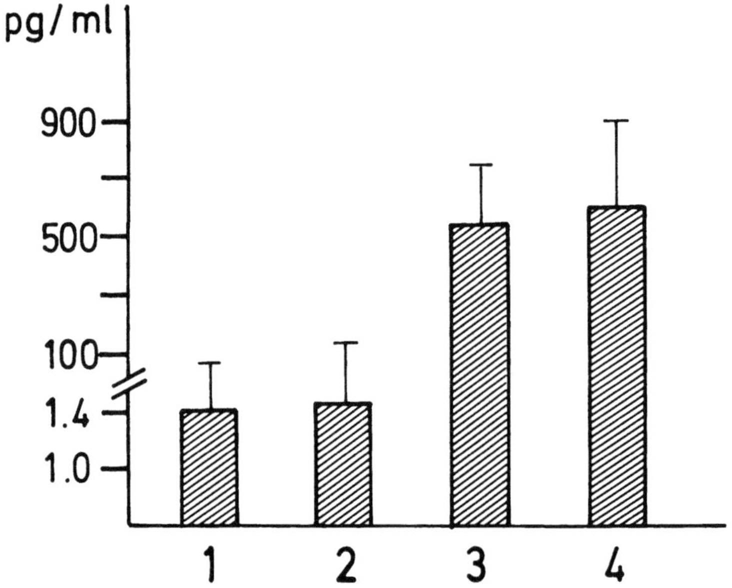

Fig.3: Influence of SBA cells on Ig-secretion of autologeous B cells. Adding of SBA⁺ T cells to B cells alone do not increase Ig production (left columns). Adding of SBA⁺ T cells to B cells plus total T cells do not suppress help of T cells for B cells (right columns).

Functional activity of SBA^+ T cells in pokeweed mitogen driven autologeous B cell assays. The presence of SBA^+ T cells in autologeous B cell cultures does not lead to an increase of Ig in the supernatants (Fig.3). Moreover SBA^+ T cells do not influence the increase of Ig in the supernatants of autologeous B cell cultures that are supplemented with normal peripheral T cells.

Discussion

The first successful approach in separating functionally different T cells within the murine as well as human thymus was accomplished by the use of PNA (7-10). Thymocytes bearing receptors for this lectin are found to reside in the thymic cortex and exhibit functional immaturity in contrast to medullary PNA^- thymocytes. Moreover it was found that in the thymic subcortex PNA staining was far less pronounced. However thymocytes in this area show a strong binding of SBA under defined experimental conditions in contrast to cortical thymocytes.

Whereas PNA exhibits a strong affinity to germinal centre cells in stimulated lymph nodes, SBA seems not to react with lymph node lymphocytes (1o-12).
The thymic subcortex, populated with SBA^+ cells, is thought to be the area where thymocyte progenitors arrive from the bone marrow. These cells appear to be blasts and are obviously involved in intensive mitotic activities (13,14). From autoradiographic experiments it was further suggested that these cells represent the precursors of cortical thymocytes (4). SBA^+ cells

additionally reside in the bone marrow. This has been deduced from experiments showing that removal of SBA^+ cells together with sheep erythrocyte receptor bearing cells from bone marrow grafts can prevent graft-versus-host reaction (5,6). Bearing in the mind that the receptor for sheep erythrocytes is identical with the T 11 antigen (which has been proved in competitive binding experiments), and furthermore that the majority of SBA^+ cells is shown to expres the T 11 antigen, it is unclear why both agglutionation with SBA and rosetting with sheep erythrocytes is necessary to abolish anti-host activity. This might, however, be explained by a higher sensitivity of monoclonal antibodies compared to rosetting in detecting this T cell associated determinant, so that SBA^+ but weakly T 11^+ T cells are removed only if both separation procedures are applied. Another explanation is based on the assumption that among the few SBA^+ T cells are such that can induce reaction against the host. The majority of SBA^+ T 11^+ T cells detected in the human bone marrow are found to be T4, T8 negative, do not significantly bind antibodies that recognize MHC gene products or T3 but react with PNA and with antibodies directed against the T9 antigen. In autologeous B cell assay these SBA^+ T cells do neither increase nor suppress immunoglobulin secretion. These data support the hypothesis that SBA^+ T cells derived from the bone marrow are candidates for precursors of subcortical thymocytes which themselves are supposed to give rise to the majority of cortical thymocytes.

Summary

Starting from the observation that SBA^+ T cells can be found in the thymic subcortex which is supposed to be the entry of bone marrow-derived pre-thymocytes. SBA^+, non-adherent, sIg^- bone marrow cells were analysed for their functional and phenotypic features.

It was found that the vast majority of non-adherent, sIg^- SBA^+ cells are T 11^+ and are thus regarded as T cells. These cells prove to be $T9^+$ and PLA^+ but express only scarsely T3 antigens and MHC gene products on their cell surface. T4, T8 antigens are not demonstrable on the majority of SBA^+ T cells. Furthermore these cells do not help or suppress Ig secretion of autologeous B cells. It is thus concluded that SBA^+ T cells derived from human bone marrow are immature in regard to their functional and phenotypic properties and may be looked upon as candidates for progenitors of cortical thymocytes.

References

1. Raedler A, Raedler E, Becker WM, Arndt R, Thiele HG (1982), Immunology 46: 321

2. Raedler A, Raedler E, Scholz K, Arndt R, Thiele HG (1983), Thymus 5: 311-326

3. Bryant BJ (1971), Adv exp Med Biol 12: lo3

4. Weissmann IL, Masuda T, Olive C, Friedberg SH (1975), Isr J Med Sci 11:1267

5. Reisner Y, Kirkpatrick D, Dupont B, Kapoor N, Pollack M, Good RA, O_Reilly RJ (1981), Lancet, August 15: 327-331

6. Reisner Y (1983), Biology, Biochemnistry, Clinical Biochemnistry Vol. 3, Eds T.C. Bøg-Hansen and G.A. Spengler, Walter de Gryter, Berlin-New York

7. Reisner Y, Linker-Israeli M, Sharon N (1976), Cell Immunol 25: 129

8. London J, Berrih S, Bach JF (1978), J Immunol 121: 438

9. Irlê D, Piguet PF, Vassalli P (1978), J exp Med 148: 32

lo. Raedler A, Raedler E, Arndt R, Thiele HG (1981a) Cell Tiss Res 218: 219

11. Rose ML, Birbeck MSC, Wallis VJ, Forrester JA, Davies AJS (198o), Nature (Lond.) 284: 364

12. Raedler A, Raedler E, Arndt R, Thiele HG (1981), Immunology Letters 2: 335-338

13. Gorgollón P, Ottone-Anaya M (1978), Acta Anat loo: 136

14. Darbrowski MP, Peel WE, Thomson AER (198o), Europ J Immunol. lo: 821

INITIAL STUDIES ON A NOVEL LECTIN-AFFINITY METHOD FOR THE PURIFICATION OF PLASMAMEMBRANE FROM SOYBEAN

S. Bassarab, R.B. Mellor, D. Werner

Department of Botany, University of Marburg,
D-3550 Marburg/L, F.R.G.

It is increasingly apparent that the plasmamembrane, in addition to its role as cell delimiter, acts as a transmission station between the cell and its environment, including the apportioning of cellular membrane budget (1). The difficulties encountered in the isolation of pure plasmamembrane have, however, hindered its analysis. Plasmamembrane purification, especially in plants, is hampered by the large membrane size, its heterogeneity, and the lack of absolutely reliable marker activities. Recently a new two-phase method for plasmamembrane isolation has been published which utilizes the charge density differences of different membrane type surfaces (2). The two-phase system was originally calibrated using the differential glucan synthetase I and II activities (3 and 4). Glucan synthetase I und II activities are not found in all tissues (5) and thus, in order to provide stricter controls, we have started to devise a totally independent method of isolating small quantities of plasmamembrane with which fractions from two-phase extractions may be compared.

Lectins, Vol. V

Materials and Methods

Preparation of protoplasts. *Glycine max* cv Maple Arrow (K. Behm GmbH, Hamburg) was grown and infected with *Rhizobium japonicum* 61-A-101 (Nitragin Co, Milwaukee) as described before (6). Nodules were harvested 20-30 days after infection, peeled of non-infected rind cells, and used immediately. Tissue culture, originally from seedling roots of similar plants, was grown and maintained exactly as before (5). Protoplasts were prepared and isolated by enzymic digestion exactly as previously (5) which yielded 4 ml protoplast suspension (0,5 x 10^6 cells/ml) from 5 g fresh weight tissue.

^{125}I Labelling. Protoplasts were surface radioiodonated using chloramine T as before (5).

Preparation of Affinity Media. Blotting media was cut into 0.5 x 0.5 cm squares and incubated at 28°C for the appropriate length of time in 100 µl 1 mg/ml lectin solution in 0.025 M $NaKPO_4$ blotting buffer, pH 8.3. After binding of the lectin, the affinity leaflets were washed in blotting buffer and transferred for 45 min into a 5 mg/ml solution of cold-iodonated BSA in blotting buffer to block excess binding sites, washed once again, and dried. Such leaflets were used immediately.

Affinity Binding. 100 µl labelled protoplast solution was incubated at 28°C with one leaflet of lectin-coated blotting medium. After the appropriate length of time the leaflets were washed in 100 ml osmotically-adjusted blotting buffer and then in 3x 100 µl the same buffer for 15 min each wash after which the ^{125}I content was at a low, stable level. The protoplasts were then osmotically broken in 100 ml blotting buffer only, and subsequently with 3 x 100 µl either (a) blotting buffer or (b) blotting buffer containing the appropriate monosaccharide (normally 50 mM). The radioactive content of those leaflets processed by method (a) were designated "total binding" (the last 3 washes containing negligable activity) or, with leaflets processed by method (b), "remaining activity". The ^{125}I content of the "remaining activity" leaflets, when added to the ^{125}I content of the monosaccharide washes ("specifically removed activity") were always approximately equal to that of "total binding".

SDS-PAGE. Proteins were separated in 10 % polyacrylamide gels by previous methods (6) using TCA-precipitated samples from 24x scaled-up lectin-affinity isolations.

Materials. Lectins were from Sigma München (RCA_{60} = L8508, LCA = L5880, LTA = L9254). Zeta Probe was from BioRad, München, nitrocellulose from Millipore, Massachusetts, and ^{125}I (Na salt) (607 MBq/µg I) was from Amersham, Braunschweig.

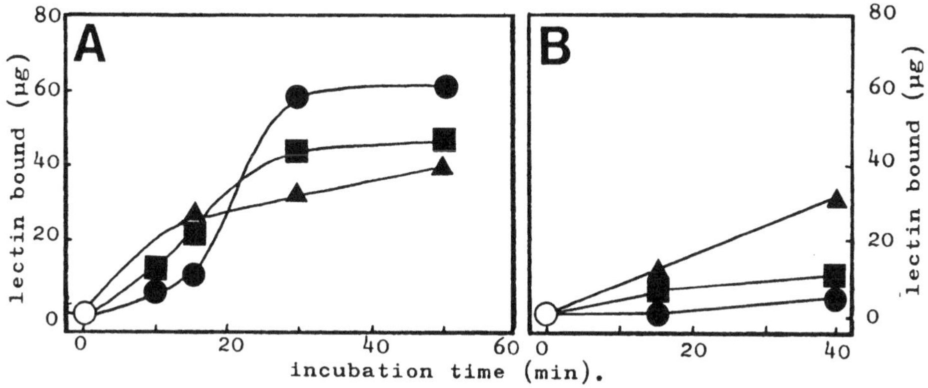

Figure 1. The binding of lectins RCA (▲), LTA (■), and LCA (●) onto 0.5 x 0.5 cm leaflets of blotting media. Starting conditions were 100 µl 100 µg/ml lectin.
A = Nitrocellulose; B = Zeta probe.

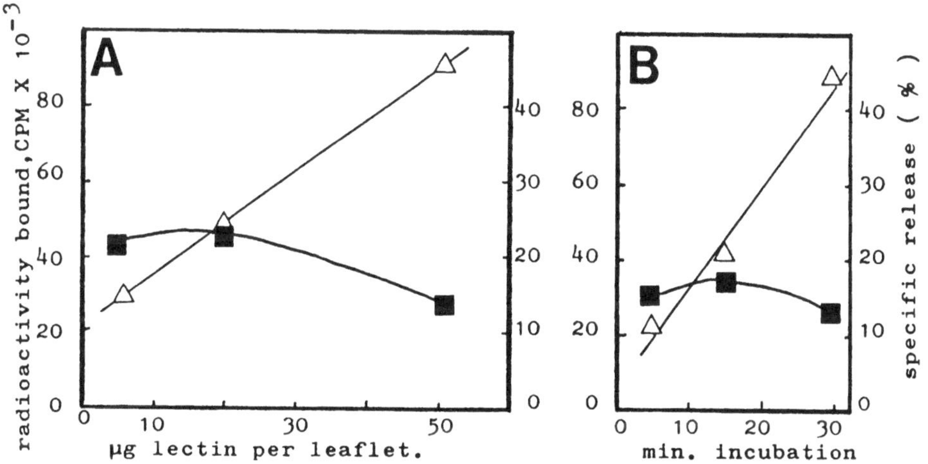

Figure 2. The binding of ^{125}I-protoplasts from tissue culture (Δ) to RCA-coated leaflets either (a) for 30 min with differing RCA concentrations or (b) with differing incubation times (52 µg/leaflet RCA). Specific release (■) is the amount of ^{125}I recovered in 3 x 15 min washes with 50 mM galactose.

Results and Discussion

Figure 1 shows that all the 3 lectins tested bound to blotting media. For unknown reasons the lectins bound to nitrocellulose in greater quantities than to Zeta Probe. Thus we routinely used nitrocellulose. ^{125}I protoplasts bound to lectin-leaflets in a time dependent manner (Fig. 2). Protoplasts bound also to nitrocellulose leaflets without attached protein and also to those with lectin but without BSA blocking, at high efficiencies (data not shown). Because such binding was not sugar-reversable we conclude that the cells bound covalently directly onto unoccupied sites on the paper. Thus BSA-blocking was routinely performed.

Although protoplast release from high-lectin-concentration leaflets was greater than that from low-lectin leaflets, Fig. 2 shows that this increase was small as a percent of total binding, which we take as evidence for binding by several lectin "bridges" per protoplast. This hypothesis was not contradicted by experiments where the effect of different sugar concentrations upon cell release was studied. Fig. 3 shows that cell release was only very slightly influenced by increases in sugar concentration over 50 mM.

Different lectins bound ^{125}I protoplasts at different efficiencies. Fig. 4 supports the general theory that tighter binding leads to less specific release, although, as would be expected from their larger size (25 nm diameter), nodule protoplasts bound to leaflets better that did tissue culture cells (18 nm diameter). They were also released in higher quantities (Fig. 4). The reason for this is unknown, but may be due to a range of factors, including having modified cell-surface glycoconjugates (See 7, 8). We investigated the effect of different lectin mixes on protoplast retention, and our results are in line with the above hypothesis, namely that mixes of two lectins bound more cells, but that washes with the appropriate monosaccharides (both at 50 mM) reduced the specific release to 10 - 12 %. Less than 10 % protoplasts were released with mixes of all 3 lectins (data not shown).

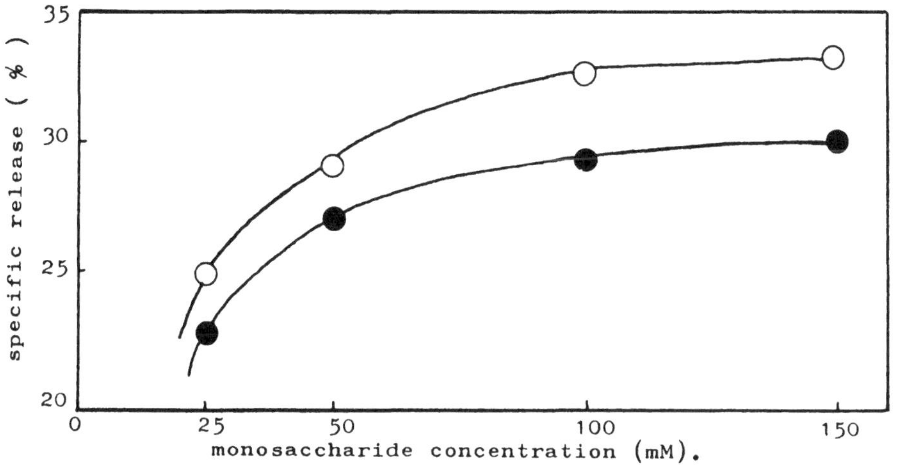

Figure 3. Effect of different monosaccharide concentrations on the release of ^{125}I-protoplasts from either tissue culture (○) or nodules (●) from RCA (45 µg/leaflet) coated leaflets.

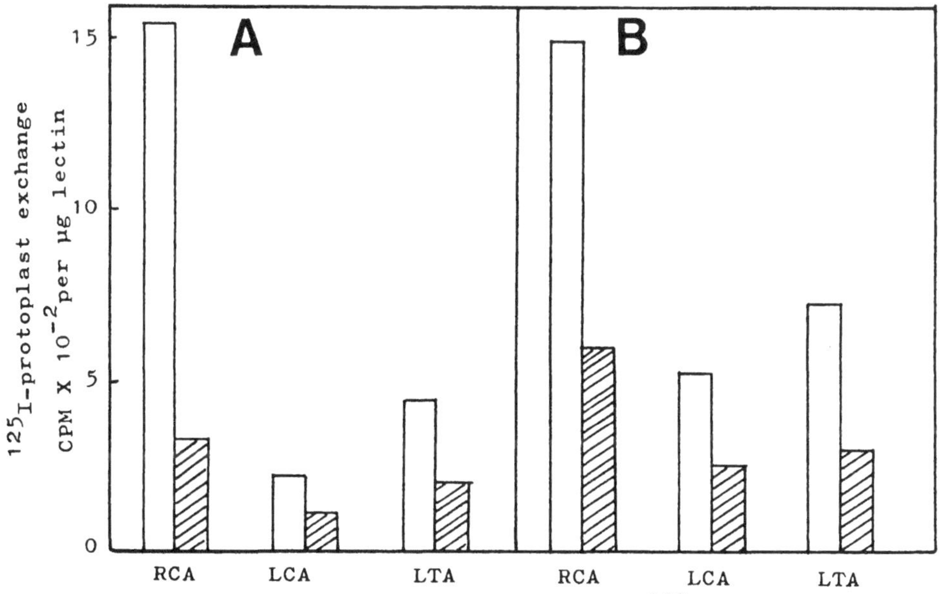

Figure 4. Specific binding and release of ^{125}I-protoplasts from (a) tissue culture or (b) nodules using different leaflet-bound lectins (50 mM monosaccharide).
□ total bound; ▨ specific release

Evidence that protoplasts bound to the leaflets (presumably by the plasmamembrane) and not other organelles released from broken cells, was forthcoming from SDS-PAGE. Comparison of polypeptide profiles of mitochondria, ER and peribacteroid membranes isolated by sucrose density gradient centrifugation (9) were compared to those obtained from putative plasmamembranes from 2-phase extraction and lectin affinity leaflet monosaccharide washes.

Fig. 5 shows that the mitochondrial, ER and peribacteroid membrane fractions from nodules cells were not significantly cross-contaminated and that these organelles were not present in detectable quantities in either the 2-phase system or lectin affinity method derived plasmamembrane fractions. The four major bands derived from the 2-phase separation of tissue culture cells (open arrows) were also in the lectin affinity method-derived plasmamembrane fraction, showing that the bands in the latter fraction were not derived from con-

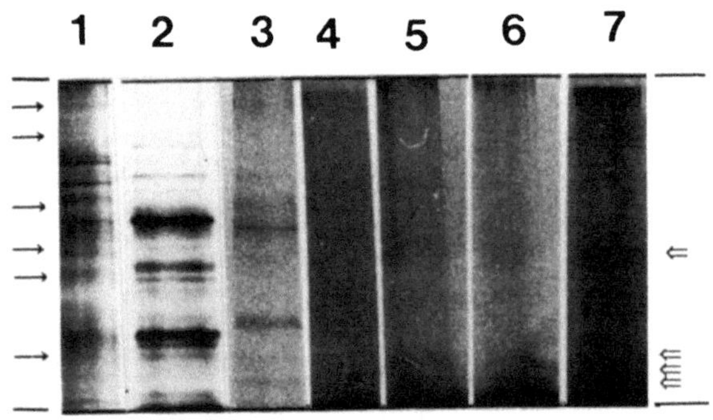

Figure 5. SDS-PAGE profiles of 1) ER, 2) mitochondria and 3) peribacteroid membranes from 0.4 g nodules. 4) plasmamembrane from 10 g 2-phase partitioned nodules. 5) plasmamembrane, lectin affinity isolated from 5 g nodules. 6) as 5, from tissue culture. 7) as 4, from tissue culture. Open arrows indicate the major bands from lane 7. Closed arrows = MW standards 132, 97, 66, 45, 35 and 12 Kd in descending order.

taminants, dislodged lectin etc. The four major bands from tissue culture cell putative plasmamembrane were also found in the lectin-affinity monosaccharide wash from nodule cells. That no discernable bands were found in the plasmamembrane fraction from the 2-phase system reflected the basic technical difficulty of the method, that even from 10 g tissue (from about 100 plants) the end-material was only a few micrograms protein.

These initial results are encouraging and the lectin-affinity methods warrants further research.

Acknowledgements

The Deutsche Forschungsgemeinschaft financially supported this work through the SFB 305 "Ökophysiologie, Verarbeitung von Umweltsignalen" project.

References

1. Figura, K.v., U. Klein, A. Hasilik. 1980. In: Biological Chemistry of Organelle Formation (Th. Bücher, W. Sebald and H. Weiss, eds.). Springer Verlag, p. 119.

2. Kjellborm, P., C. Larsson. 1984. Physiol.Plant 62, 501.

3. Ray, P.M. 1977. Plant Physiol. 59, 594.

4. Widell, S., T. Lundborg, C. Larsson. 1982. Plant Physiol. 70, 1429.

5. Ostrowski, E.D., R.B. Mellor, D. Werner. 1985. Physiol.Plant (in press)

6. Mellor, R.B., T.M.I.E. Christensen, D. Werner. 1985. Proc.Natl.Acad.Sci.USA (in press)

7. Mellor, R.B., W. Dittrich, D. Werner. 1984. Physiol. Plant Pathol. 24, 61.

8. Mellor, R.B., D. Werner. 1985. In: Lectins, Biology, Biochemistry and Chemical Biochemistry (T.C. Bog Hansen and J. Brebowitz, eds.). 4. W. de Gruyter, p. 267.

9. Mellor, R.B., T.M.I.E. Christensen, D. Werner. 1984. Z.Naturforsch. 40c, 73.

MICROELECTROPHORETIC MOBILITY TEST FOR LECTIN BINDING TO PURPLE MEMBRANE FRAGMENTS

M.R. Kantcheva, D.E. Kovatchev, N.G. Podimitorva, S.P. Stoylov

Department of Biophysics and Radiobiology, Biological Faculty, Sofia University, Sofia; Department of Biology, Higher Medical Institute, Varna; and Department of Physics and Biophysics, Medico-Biological Institute, Sofia, Bulgaria.

For most bioparticles there is evidence for a negative outer biomembrane surface charge of about one electronic charge per 2000 $Å^2$ (1). This surface charge is responsible for redistribution of ions in the surrounding medium and for formation of an electric double layer at the membrane surface. There are different mechanisms for changing the surface charge and its corresponding electric double layer. The most common of them are binding, screening and modification with chemical agents (2).

The microelectrophoretic method provides a way to estimate the average charge density over the entire cell membrane. Since lectin binding to biological membranes results in some electrostatic surface change, using a microelectrophoretic technique, valuable biophysical information can be obtained. In this test, a quantitative measurement of the surface charge density is related to the bioparticle stability, in addition to aggregation and functional activity. In an electrophoretic mobility test using lectins with different carbohydrate specificity, the membrane morphology, the chemical composition as well as some topological properties could be investigated.

The negative charge on the two surfaces of the pruple membrane of *Halobacterium halobium* consists of the surface charge of the protein component bacteriorhodopsion and polar lipid headgroups (3). The present paper sets out to demonstrate the capability of the of the elecetrophoretic mobility test as a probe for purple membrane surface organization.

Lectins, Vol. V

MATERIAL AND METHODS

Purple membrane fragments isolated from *H. halobium* in a concentration of 57 μg/ml were used (4). Lectin-membrane interaction was tested for PHA-P and PHA-M (Difco) in concentrations between 10 and 150 μg/ml.

A cytopherometer "Opton" was used for electrophoretic mobility measurements (5). Mobilities of the membrane fragments were calculated in m^2/V.s (+ S.E.M.); each independent value was obtained by timing the movement of 15 to 20 fragments pooled from three membrane preparations. The standard deviation for the measurements of purple membrane mobility was between 3% and 6%. Purple membrane fragments were incubated 30 min with and without lectin at 22°C before resuspending in electrophoresis buffer. For monitoring the effectivity of PHA-P and PHA-M an incubation time of 30 min as well as 24 have been used. PHA treatment (50 μg/ml) of purple membrane fragments was monitored in the range of pH=2 to pH=8. Purple membrane fragments resuspended in buffer-saline medium (0.05 M-NaCl-0.001 M-$NaHCO_3$ pH adjusted with 0.05 M HCl or NaOH) in a concentration o 5.7 μg/ml were used for electrophoretic mobility measurements. The ionic strength of the suspending medium was adjusted by different amounts of NaCl. The results are expressed as percent of, electrophoretic mobility change

RESULTS

Interaction of lectin with the purple membrane results in a decrease in the elelctrophoretic mobility. Fig. 1 presents the effect of PHA binding to purple membrane fragments at low ionic strength (1.4 mM NaCl). The electrophoretic mobility decreases by increasing the PHA concentration. The electrokinetic alteration is specific for membrane surface carbohydrate components inasmuch as the effect is abrogated in the presence of N-acetyl-galactosamine.

The pH dependence on the effects of PHA on the purple membrane electrophoretic mobility is presented at Fig. 2. There is a significant increase in the lectin induced electrophoretic mobility change starting from the isoelectric point of the membranes, pH=2, with a maximum value at pH=3.

Our goal was also to test the efficiency of PHA-P and PHA-M (Table 1). A striking decrease in the electrophoretic mobility of the purple membrane could be related to the protein content of PHA-P. The effect is most pronounced at a lectin dose of 150 μg/ml. The increase of the incubation time to 24 hours did not change signifi-

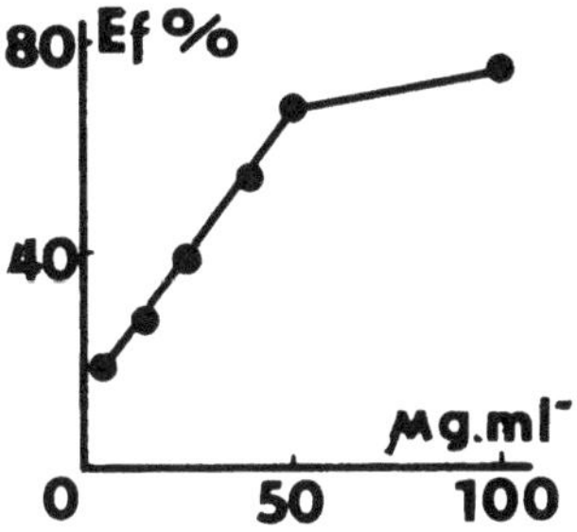

Fig. 1. The mobility decrease of purple membrane fragments with PHA. PHA was bound to purple membrane fragments (57 μg/ml) at low ionic strength (1.4 mM NaCl). The result is expressed as the mobility decrease Ef in % of the mobility of the untreated membrane fragments.

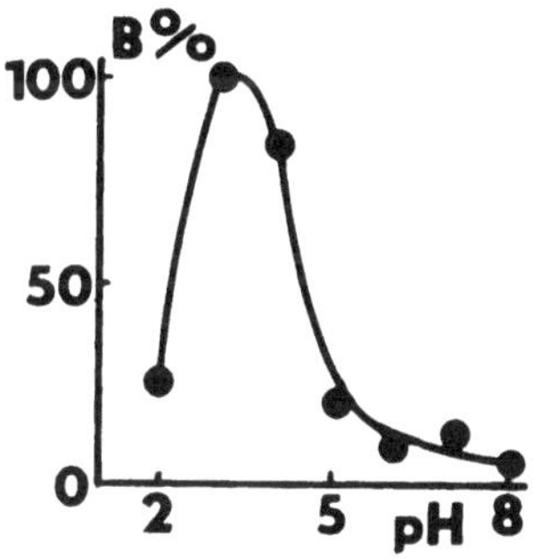

Fig. 2. The mobility decrease of purple membrane fragments as a funtion of pH. PHA (50 μg/ml) was bound to purple membranes (57 μg/ ml). The mobility decrease is given in % of the maximal mobility decrease.

cantly the effect of PHA-P binding to purple membrane fragments. However, increasing the time of incubation to 24 hours resulted in clearly manifested decrease in the electrophoretic mobility of the PHA-M treated samples.

DISCUSSION

There is a specific trend in the pH-dependent electrophoretic mobility under PHA treatment. It has to be noted that the lectin recognition of membrane saccharides involves matching pairs of surfaces endowed with specific electrostatic properties which are modified by the interactions. The electrostatic properties determine the stability of the resulting aggregate. The effect of PHA bindig was found to have the greatest effet on the mobility between the isoelectric point of purple membrane fragments, pH=2, and the isoelectric point of PHA, pH=6. The electrophoretic mobility change is a result of the electrostatic interaction and the specific lectin membrane affinity.

Previous experiments have shown that when the ionic strength was changed from 1.4 mM NaCl to 140 mM NaCl, the effect was similar to a PHA dose of 15 µg/ml (5). At low ionic strength (1.4 mM NaCl) a dose of 50 µg/ml is sufficient for about 70% decrease in the electrophoretic mobility (see Fig. 1). The same concentration produced a negligable effect at physiological ionic strength. It is obvious that optimizing the experimental conditions, a low ionic strength is recommended for the test. By using PHA products with different degree of purification and activity the incubation time plays also an important role (see Table 1).

The lectin affinity to purple membranes could be related to the galactolipid membrane components as a single carbohydrate participant. Having in mind the spacial organization of the purple membrane surface at different ionic strengths of the suspendig medium, the following model for lectin membrane interaction could be suggested. At low ionic strength the bacteriorhodopsin terminals are expected to be more extended and so they could enhance lectin binding to surface oriented galactolipids. The high ionic strength could be responsible for another structure of the surface protein components which colapse at the membrane surface and may produce an effect of screening of the specific binding sites; so galactolipid molecules become less accessible for the lectin molecules. The electrostatically determined topology of the membrane surface is also essential for lectin membrane interaction.

Estimation of ionic strength effects on the electrophoretic mobility after membrane modification represents a biophysical approach to an investigation of the topology of surface components (6). The evidence of ionic strength effects by lectin binding could be explained with the deeper position of galactolipid components in the purple membrane if compared to bacteriorhodopsin terminals. So, making changes in the ionic strength, som details in the three-dimensional

TABLE 1

The electrophoretic mobility of pruple membrane fragments at different concentrations of PHA-P and PHA-M.

Lectin Concentration µg/ml	Incubation time hours	Electrophoretic mobility X $10^{-8} m^2/V.s$ PHA-P	PHA-M
0	0.5	-2.188 ± 0.041	-2.188 ± 0.041
	24	-2.123 ± 0.035	-2.123 ± 0.035
15	0.5	-1.884 ± 0.057	-2.245 ± 0.047
	24	-2.002 ± 0.059	-1.985 ± 0.038
50	0.5	-1.695 ± 0.064	-2.099 ± 0.041
	24	-1.584 ± 0.046	-1.720 ± 0.030
150	0.5	-1.048 ± 0.061	-1.817 ± 0.096
	24	-1.061 ± 0.037	-1.351 ± 0.045

surface topology could be elucidated. The model suggested above does not consider ionic strength induced changes in the PHA molecule related to molecular weight, shape and secondary structure, which could also modify the lectin membrane affinity. In contrast to PHA, a low binding activity of Con A for purple membrane fragments was observed (7). This effect indicates that mannose residues are not present in the purple membrane surface except in the form of isolation impurities, or it could be the result of nonspecific adsorption. Summarizing the experimental results, two main conclusions could be assumed:

1. The suspending medium properties - pH and ionic strength as well as the time of incubation and lectin probe activity - are valuable means for control and optimization of lectin membrane interaction.

2. The electrophoretic mobility test by lectin binding as a real support for other methodological approaches may be useful for the study of carbohydrate structure and topology of the membrane surface.

Acknowledgement: Mr. Hjalmar Johansen is thanked for typing the manuscript.

REFERENCES

1. Gilbert, D.L., Ehrenstein, D.: Current topics in membranes and transport, *22*, 407-421, (1984).

2. Durocher, J.R., Glader, B.E., Conrad, M.E.: Proc. Soc. Exp. Biol. Med., *144*, 249-251, (1973).

3. Neugebruer, D. CH., Oesterhelt, D., Zingsheim, H.: J. Mol. Biol., *125*, 123-135, (1978).

4. Oesterhelt, D., Stoeckenius, W.: Methods Enzymol. (Packer, L. ed.) Vol. *31 A*, 667-678, (1974).

5. Kantcheva, M.R., Podimitrova, N., Stoylov, S., Kovatchev, D.: Electrophoresis, *5*, 160-161, (1984).

6. Sato, Ch., Kojima, K., Nishizawa, K.: Rad. Research, *69*, 367-374, (1977).

7. Popdimitrova, N., Kantxheva, M., Stoylov, S., Kovatchev, D.: Proceedings of the Fourth Meeting of the International Electrophoresis Society (Neuhoff, V. ed.), 499-500, Verlag Chemie Weinheim (1984).

LIMAX FLAVUS AGGLUTININ: A NEW TAXONOLECTIN FOR THE IDENTIFICATION OF TRYPANOSOMA CRUZI, THE AGENT OF CHAGAS'DISEASE

J.Schottelius

Department of Protozoology, Bernhard-Nocht-Institute for Nautical and Tropical Medicine, Bernhard-Nocht-Str.74, D-2000 Hamburg 4, Germany

Lectins are a group of proteins and glycoproteins of plant and animal origin which interact specifically with glycoconjugates by binding of specific carbohydrate residues. In spite of the fact that in the meantime thousands of seeds and animal products have been tested with regard to the existence of lectins (5, 11, 16) only a few lectins have been found specific for N-acetylneuraminic acid: Triticum vulgaris (7), Limulus polyphemus (12), Carcinoscorpus rotunda cauda (2) and Homarus americanus (8). The slug lectin of Limax flavus reacts specifically with N-acetylneuraminic acid and N-glycolylneuraminic acid (13). Both carbohydrates have been found on the surface of Trypanosoma cruzi, the agent of Chagas'disease (17). The Aaptos papillata II lectin (4) was the first lectin which was used to distinguish between T.cruzi, T.rangeli and T.conorhini (3, 14). But the disadvantage of this lectin is that it is not available on the market. As this lectin reacts with N-acetylneuraminic acid on the cell surface of the epimastigote culture forms of T. cruzi and T.cruzi-like stocks (23) the slug lectin of Limax flavus was tested to find out whether this agglutinin can be used instead of the Aaptos lectin for interspecific differentiation tests to distinguish between stocks of T. cruzi, T.cruzi-like (9), T.ranegli, T.rangeli-like and T.conorhini. As T. conorhini has pseudocrypt antigens of the 1st order (24) which react with the lectin Aaptos papillata II (14) it was tested whether these receptors react also with the slug lectin of Limax flavus.

Lectins, Vol. V

Materials and Methods

Four-days old epimastigote culture forms of 53 T.cruzi and T.cruzi-like stocks from different Latin American countries, 7 T.rangeli and T.rangeli-like stocks as well as 1 stock of T.conorhini were used in this study. The flagellates were cultured at 26° C in Brain Heart Infusion Agar (Difco) supplemented by defibrinated undiluted rabbit blood. In order to obtain a homogeneous parasite culture three passages were made for four days each. Cells were centrifuged at 1500 x g for 5 min and washed once in 0.9 % NaCl.

The culture forms were resuspended in 0.) % NaCl at a concentration of 10^{8} cells / ml. The trypanosomes were tested with the lectin of Limax flavus (1000 µg/ml, 100 µg/ml)(Calbiochem - Behring, American Hoechst Corp., USA) and Aaptos papillata II (Dr.H.Bretting, Univ. Hamburg)(1 : 256 against 5 % solution of human erythrocytes, blood group 0). A parasite suspension (50 µl) and 50 µl of the lectins were incubated together for 15 min at 26° C. The agglutination tests were controlled by microscope using slides and cover slips.

Thiobarbituric acid test(Aminoff) for sialic acid: The measurement of N-acetylneuraminic acid on the cell surface was carried out according to the method of Aminoff (1). The culture forms were washed three times in 0.9 % NaCl and centrifuged (1400 x g for 5 min). The cell pellet of T.conorhini stock Hawai (100 µl, 50 µl) was incubated with 200 µl and 100 µl neuraminidase from Vibrio comma (Behringwerke, Marburg, FRG)(1 U/ml) for one hour in a water bath of 37° C. The cells were centrifuged (1400 x g for 5 min) and the supernatant (100 µl) mixed with 125 µl of solution A (535 g $NaJO_4$, 12,5 ml 1 N H_2SO_4, 87,5 ml H_2O) and incubated for 30 min in a water bath of 37° C. The excess of periodate was reduced with 100 µl solution B (2 g $NaAsO_2$, 50 ml 1 N HCl, 50 ml H_2O). Immediately after the yellow colour had disappeared 1 ml of thiobarbituric acid reagent (8.1 g thiobarbituric acid to 500 ml H_2O, pH 9.0 with 5 N NaOH) was added. The test tube was covered and heated in a boiling water bath for 7.5 min, then cooled in water and shaken with butanol-HCl (95 ml butanol plus 5 ml 37 % HCl). The two layers were separated by centrifugation (2000 rpm for 10 min). The coloured top layer was taken off and examined at 549 nm in a spectrophotometer. as standard was used N-acetylneuraminic acid (Serva, Heidelberg, FRG).

Pronase test: A pellet of 10^{8} cells was incubated in pronase (0.1 mg / ml in 0.9 % NaCl, Calbiochem, Calif., USA) for 30 min at 37° C. The cells were washed in 0.9 % NaCl and resuspended at a concentration of 10^{8} cells / ml in 0.) % NaCl. These cells were tested with the lectin of Aaptos papillata II (1 : 256) and Limax flavus (100 µg / ml).

Control reactions: The agglutination tests of the T.cruzi stocks Molino 1, Y and the T.cruzi-like stock Morcego 1354 (representative for the other stocks of the T.cruzi complex) were controlled by inhibition tests using 0.2 M N-acetyl-D-glucosamine (Serva, Heidelberg, FRG) and N-acetylneuraminic acid (Sigma, München, FRG). Furthermore, 50 µl cell pellets of the same stocks were incubated in 100 µl neuraminidase (1 U/ml, Vibrio comma, Behringwerke, Marburg, FRG) for one hour at 37° C before they were tested with the lectins. The cells were also incubated for autoagglutination. After the pronase test the agglutination reactions of T.conorhini with the lectins of Aaptos papillata II and Limax flavus were controlled by inhibition tests with the above mentioned N-acetyl-D-glucosamine and N-acetylneuraminic acid.

Results

Agglutination test: The stocks of T.cruzi and T.cruzi-like (Table 1) were agglutinated by the lectins of Limax flavus and Aaptos papillata II. These reactions could be inhibited by 0.2 M N-acetylneuraminic acid and 0.2 M N-acetyl-D-glucosamine and by treatment of these cells with neuraminidase. In contrast, all stocks of T.rangeli and T.rangeli-like and T.conorhini were not agglutinated by these lectins (Table 1, 2).

Periodic acid / Thiobarbituric acid test for sialic acid: After treatment of T.conorhini with neuraminidase and the periodic acid; thiobarbituric acid method, the supernatant showed only a pale yellow-green colour. No N-acetylneuraminic acid was detectable on the surface of this flagellate.

Pronase test: After incubation in pronase the culture forms of T.conorhini were agglutinated by the lectins of Aaptos papillata II and Limax flavus. These agglutination reactions could also be inhibited by both mentioned carbohydrates. Autoagglutinations were not observed (Table 2).

Differentiation of T.cruzi, T.rangeli and T.conorhini: The agglutination tests show that the lectin of Limax flavus can be used for the identification of the agent of Chagas'disease. Differences between T.cruzi and T.cruzi-like stocks were not observed (Table 1).

Discussion

The results demonstrate clearly that the slug lectin of Limax flavus can substitute the sponge lectin of Aaptos papillata II. Selective agglutination reactions with certain strains of T.cruzi and T.cruzi-like, which were described for the lectins of Triticum vulgaris and Arachis hypogaea (18, 22, 23), were not observed. The inability of the T.rangeli and T.rangeli-like stocks and T.conorhini to react with the anti-neuraminyl specific lectins of Aaptos papillata II and Limax flavus is in agreement with the result that neither the stocks of T.rangeli and T.rangeli-like (19, 20) nor T.conorhini have N-ace-

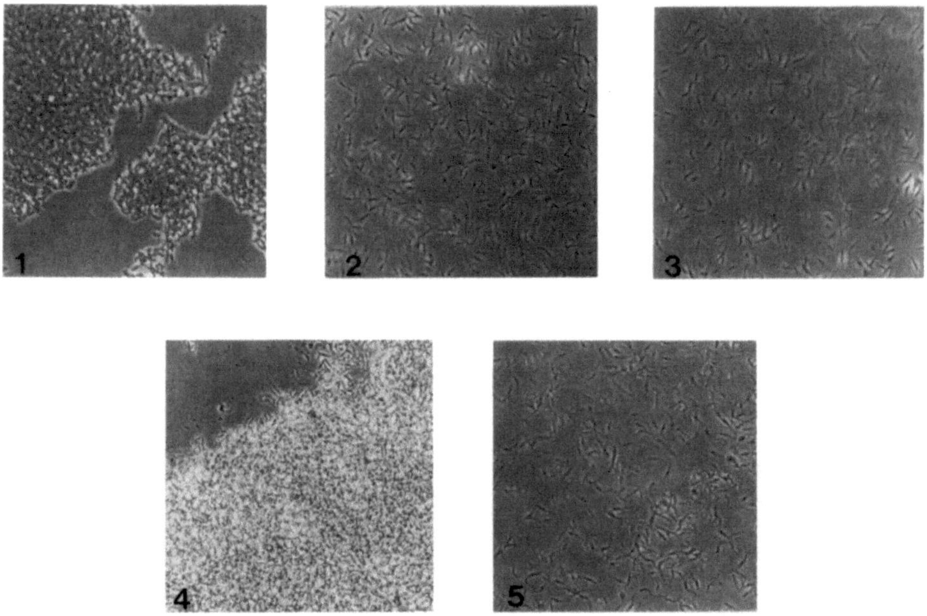

Figure 1. Agglutination reactions of Trypanosoma cruzi (1), Trypanosoma rangeli (2), and Trypanosoma conorhini (3), with the lectin of Limax flavus and reactions of Trypanosoma conorhini (4) and Trypanosoma rangeli with the lectin of Soja hispida.

TABLE 1

Agglutination reactions of different stocks of Trypanosoma

Lectin	Group 1	Group 2	Group 3
Limax flavus	+	Ø	Ø
Aaptos papillata II	+	Ø	Ø

+ = agglutination reaction ; Ø = no agglutination reaction

Group 1: Trypanosoma cruzi Complex (T.cruzi human stocks from Brazil: Y, PF, 436/2; Colombia: Molino 1, Catatumbo II; Costa Rica: CR 2-1, CR 2-2, CR 3; Guatemala: G; Peru: P-60; Venezuela: OPS-21, OPS-31, OPS-32, OPS-42, OPS-49; T.cruzi-like wild animal stocks from Brazil: WA 301/130, Morcego 1354; Venezuela: OPS-4, OPS-6, OPS-7, OPS-13, OPS-17, OPS-29, OPS-39, OPS-65, OPS-69, OPS-87; T.cruzi-like domestic animal stock from Venezuela: OPS-12; T.cruzi-like bug stocks from Brazil: MR, FL; Chile: Chile 10; Colombia: Ubaque 4, Ubaque 6; Costa Rica: CR-4, CR 5-1, CR 5-2, CR 6; Mexico: Tehuantepec; Venezuela: OPS-1, OPS-9, OPS-22, OPS-53; OPS-59, OPS-60, OPS-63; T.cruzi stocks from Venezuela of unknown origin: MA, DS, ES, FL, 57 A, A 35, EP, JAN) Group 2: Trypanosoma rangeli Complex (T.rangeli human stocks from Colombia: San Agustin; Venezuela: Venezuela; T.rangeli-like wild animal stock from Colombia: DA-3412; of unknown locality: ITMAP-1140; T.rangeli-like domestic animal stock from Venezuela: RGB; T.rangeli-like bug stocks from Colombia: Choachi 4, Choachi 6) Group 3: Trypanosoma conorhini (T.conorhini bug stock from Hawai: Hawai).

TABLE 2

Control reactions for agglutination of different stocks of Trypanosoma

Lectin	Carbohydrate	T.cruzi T.cruzi-like	T.conorhini after pronase Treatment
Limax flavus	0.2 M GluNAc	Ø	Ø
Limax flavus	0.2 M Neu5Ac	Ø	Ø
Aaptos papillata II	0.2 M GluNAc	Ø	Ø
Aaptos papillata II	0.2 M Neu5Ac	Ø	Ø

GluNAc = N-acetyl-D-glucosamine Neu5Ac = N-acetylneuraminic acid tylneuraminic acid on their cell surfaces. This substance was only found on the surface of T.cruzi and T.cruzi-like stocks (17, 19, 20, 23). In contrast to these parasites, the stock of T.conorhini was only agglutinated by the Aaptos papillata II lectin and also by the agglutinin of Limax flavus after treatment with pronase. This reaction demonstrates that the Aaptos and Limax receptor exist as a pseudocrypt antigen of the first order (24). The Aaptos papillata II lectin can be inhibited by the carbohydrates N-acetyl-D-glucosamine, chitobiose, chitotriose, chitotetraose (10). The reaction of the Aaptos II lectin with the flagellates of the T.cruzi complex is based on the presence of N-acetylneuraminic acid on the surface of the trypanosomes (23). Therefore we tested whether T.conorhini also has this substance linked to the pseudocrypt antigens. Cat erythrocytes have neuraminic acid as pseudocrypt antigens which are detectable after pronase treatment (24). Therefore, it would be desireable to test whether this is also the case with T.conorhini. However, it was not possible to demonstrate N-acetylneuraminic acid on the surface of T.conorhini, and , it was concluded that the receptor for the Aaptos papillata II and Limax flavus lectins is not a sialoglycoconjugate. The positive reaction of T.conorhini after treatment with pronase with the lectins of Limax flavus and Aaptos papillata II demonstrates that both agglutinins may react with the same receptors. This reaction can be inhibited by N-acetylneuraminic acid as well as by N-acetyl-D-glucosamine, both of which have the same configuration on the C_2 and C_3 atom. This configuration is essential for the reaction of these lectins (15). The lectins of Aaptos papillata II and Triticum vulgaris are very similar concerning their inhibiting carbohydrates (4, 6). As the lectin of Limax flavus is also specific for N-acetylneuraminic acid (13) the carbohydrates which inhibit the anti-neuraminyl lectins of Triticum vulgaris and Aaptos papillata II may also partly have an inhibitory effect on the Limax lectin. The stocks of T.rangeli and T.rangeli-like have no receptor for the Aaptos II and Limax lectin. Therefore, the commercially available Limax flavus lectin can be used for unequivocal identification of the culture forms of the agent of Chagas'disease also in the presence of T.rangeli and T.rangeli-like flagellates and T.conorhini. The slug lectin reacts with a species-specific receptor on the cell surface of the T.cruzi and T.cruzi-like stocks.

To distinguish between the trypanosomes of the T.cruzi complex, T.rangeli complex and T.conorhini the complement lysis test can be used (21). The lectins of Ricinus communis-120 and Soja hispida can distinguish between T.cruzi and T.rangeli because the culture forms of the trypanosomes of the T.cruzi complex only have receptors for both lectins. Therefore, in the case of a double infection both trypanosome complexes can be separated by these lectins or by the Limax lectin (Fig. 1). If there is an additional infection with T.conorhini it is possible to take only the Limax lectin for the separation of T.cruzi and T.cruzi-like flagellates. The lectins of Ricinus communis-120 and Soja hispida cannot be used, as both T.cruzi, T.cruzi-like and T.conorhini have receptors for these agglutinins (21). After elimination of the flagellates of the T.cruzi complex with the Limax lectin T.rangeli and T.rangeli-like flagellates and T.conorhini can be differentiated with the lectins of Ricinus communis-120 or Soja hispida, because the trypanosomes of the T.rangeli complex cannot react with these agglutinins.

Conclusions

The experiments demonstrate that the lectin of Limax flavus can be used to distinguish between T.cruzi, T.rangeli and T.conorhini, because only T.cruzi was agglutinated by this lectin. In contrast, T.rangeli has no receptors for this agglutinin while T.conorhini has these receptors as pseudocrypt antigens of the 1st order. Furthermore could be shown that the reactinns of T.cruzi with Limax flavus lectin can be eliminated by treatment with neuraminidase.

Acknowledgements: Supported by the Commission of the European Communities and the Bernhard-Nocht-Institute for Nautical and Tropical Medicine, Department of Protozoology, Hamburg, No. TSD-005-D(B)

References

1. Aminoff, D.: Biochem. J. 81, 384-392 (1961)
2. Bishayee, S., and Dorai, D.T.: Biochim. Biophys. Acta 623, 89-97 (1980)
3. Bretting, H., and Schottelius, J.: Z. Parasitenkd. 57, 213-219 (1978)
4. Bretting, H., Kabat, E.A., Liao, J., Pereira, M.E.A.: Biochemistry 15, 5029-5038 (1976)
5. Gold, E.R., and Balding, P.: Receptor specific proteins. Excerpta Medica, Amsterdam, pp. 1-440 (1975)
6. Goldstein I.J., Hammrström, St., Sundblad, G.: Biochim. Biophys. Acta 405, 53-61 (1975)
7. Greenaway, P.Y., and Le Vine, D.: Nature New Biol. 241, 191-192 (1973)
8. Hall, J.L., and Rowlands, D.T.: Biochemistry 13, 828-832 (1974)
9. Hoare, C.A.: The Trypanosomes of Mammals. Oxford and Edinburgh, Blackwell Sci. Publ., pp. 3-749 (1972)
10. Kabat, E.A.: J. of Supramol. Structure 8, 79-88 (1978)
11. Krüpe, M.: Die Blutgruppen des Menschen. Eine Übersicht. Selbstverlag Martin Krüpe, Fulda (1970)
12. Marchalonis, J.J., and Edelman, G.M.: J. Mol. Biol. 32, 453-465 (1968)
13. Miller, R.L., Collawn, J.F., Fish, W.W.: Biol. Chem. 257, 7574-7580 (1982)
14. Mühlpfordt, H., and Schottelius, J.: Tropenmed. Parasit. 28, 1-7 (1977)
15. Peters, B.P., Ebisu, S., Goldstein, I.J., Flashner, M.: Biochemistry 13, 5505-5511 (1979)
16. Prokop, O., and Uhlenbruck, G.: Lehrbuch der menschlichen Blut- und serumgruppen. VEB Georg Thieme Leipzig, pp. 1-924 (1966)
17. Schauer, R., Reuter, G., Mühlpfordt, H., Andrade, F.B., Pereira, M.E.A.: Hoppe Seyler's Z. Physiol. Chem. 364, 1053-1057 (1983)
18. Schottelius, J.: Z. Parasitenkd. 68, 147-154 (1982)
19. Schottelius, J.: Europ. J. Cell Biol. Supple. 4, p.18 (1983)
20. Schottelius, J.: Tropenmed. Parasit. 35, 160-162 (1984)
21. Schottelius, J., and Müller, V.: Acta Tropica 41, 29-38 (1984)
22. Schottelius, J., and Raabe, W.: Pesquisa Basica em Doenca de Chagas, V. Reuniao Anual, Caxambu, Minas Gerais, Brazil 6 a 9 de novembro, Resumos B-49 (1978)
23. Schottelius, J., and Uhlenbruck, G.: Z. Parasitenkd. 69, 727-736 (1983)
24. Uhlenbruck, G.: Immunbiologie. Eine Einführung. Wilhelm Goldmann Verlag, München, pp. 1-407 (1971)

USE OF LECTINS FOR THE CHARACTERIZATION OF UNTRANSFORMED AND TRANSFORMED AMPHIBIAN CELL LINES

Van Vyve Ph., P.J. Querinjean and J.J. Picard

University of Louvain, Laboratory of Developmental Genet

Place Croix du Sud, 5(Bte 3), B-1348 Louvain-la-Neuve, I

Glycoproteins and glycosphingolipids alterations represent a constant modification observed in most transformed and tumor cells obtained after either viral or chemical carcinogenesis. These changes result in either the reduction or the elongation of oligosaccharides from the normal cell glycoconjugates and, in some instances, in the loss or the appearance of new glycoproteins or glycolipids. Different technical approaches using several lectins were used to identify the various cell surface modifications between untransformed and transformed cells (1-4).

In vitro mutagenesis with N-methyl-N'-nitro-N-nitrosoguanidine of an epithelioid cell line (XB 693 C) initiated from stage 35 tadpoles of *Xenopus borealis* allowed us to obtain a tumorogenic cell line (XB 693 T) (5) inducing transplantable metastasizing adenocarcinomas in a histocompatible family (6).

Untransformed (XB 693 C) and transformed (XB 693 T) cell lines of *Xenopus borealis* were studied by agglutination in the presence of three different lectins of various specificities, namely, concanavalin A (con A), wheat germ agglutinin (WGA) and *Ulex europaeus* agglutinin 1 (UEA 1) and by revelation with peroxidase of the con A-binding glycoconjugates from the cell extracts separated by SDS-polyacrylamide gel electrophoresis (SDS-PAGE).

Lectins, Vol. V

Materials and Methods

1. *Chemicals* : The lectins, con A(Serva), UEA 1(Serva) and WGA(USB) and the inhibiting sugars(Serva) were dissolved at the appropriate concentrations in Holfreter's solution (7) or in PBS. Molecular weight standard proteins were purchased from Pharmacia or Boehringer. Horseradish peroxidase, grade II, 100 U.I./mg, was obtained from Boehringer.

2. *Cells* : Untransformed (XB 693 C) and transformed (XB 693 T) cell lines were obtained from *Xenopus borealis* species as described (5). Cells were cultured in Falcon flasks with modified Minimum Essential Medium(MEM) containing 75% MEM, 15% distilled water, 10% fetal calf serum(Gibco) and 0.15% gentamycine. Cells were recovered at confluence after 1h incubation in a 1.88mM EDTA solution, pH 7.4 (8).

3. *Agglutination* : Agglutination tests of living cells collected in a Holfreter's solution, pH 7.2 were performed at 25°C in microtitration plates. Each V-shaped well contained 5.10^5 cells in 100µl and 100µl of lectin solution at different final concentrations ranging from 5.10^{-5} to 50µg/ml for con A and WGA and from 5.10^{-5} to 250µg/ml for UEA 1 in a physiological Holfreter's solution. Cell agglutination by con A, WGA and UEA 1 were also tested in presence of their corresponding inhibiting sugars : α-methyl-D-mannopyranoside , β-D-N-acetylglucosamine (GlcNAc) and L-fucose (Fuc), respectively.

4. *SDS-PAGE* : Cell extracts were prepared by mixing one volume of the cell pellet in Barth's solution, pH 8.1 (9) and 5 volumes of a lysis buffer containing 9M urea, 2% β-mercaptoethanol (v/v) and 4% Triton (v/v). Samples were stored at -20°C until use. Samples were submitted to electrophoretic separation in 0.7mm thick slab gels according to a modification of the Laemmli procedure (10). After the electrophoresis, SDS was removed from the gels by several rinses (11); overnight in water/propanol/acetic acid : 67.5/25/7.5 (v/v/v), 10h in water/propanol/acetic acid : 91/10/9 and overnight in 10% acetic acid. Proteins were stained with 0.25% Coomassie Brillant Blue R250 in methanol/water/acetic acid : 5/4/1 (v/v/v). Mannose-containing glycoconjugates were revealed by con A. After equilibration in PBS, the gels were cut in two strips. The first one was treated with 3ml of 100µg/ml con A solution in PBS supplemented with $CaCl_2$ for 10h. Unbound con A was rinsed overnight in PBS. The second strip, used as a control, was incubated with a similar con A solution containing 0.2M α-methyl-D-mannopyranoside as inhibiting sugar. The gels were then incubated for 2h with a 100µg/ml peroxidase solution in PBS. Excess peroxidase was removed by three successive 2h rinses in PBS. Bound peroxidase was revealed by a solution containing 0.06% 4-chloro-1-naphtol (w/v) and 0.01% hydrogen peroxyde (v/v) in PBS according to Hawkes (12). The reaction was stopped after a few minutes with 7% acetic acid. Dried gels were analyzed with a Chromoscan 3 (Joyce Loebl) densitometer equipped with a 590nm filter.

Results

1. Agglutination

Agglutination tests of living cells were performed in parallel with untransformed (XB 693 C) and transformed (XB 693 T) cells. Minimal agglutination titres read after 30 min were the lowest concentrations of lectin giving a cell agglutination and they are presented in Table 1 for the three lectins tested.

A clear inhibition of cell agglutination was obtained for both cell lines with 25mM α-methyl-D-mannopyranoside for con A, with 25mM GlcNAc for WGA and with 2.5mM fucose for UEA 1.

2. SDS-PAGE of untransformed and transformed cell extracts

SDS-PAGE separation of untransformed (C) and transformed (T) cell extracts from XB 693 C and XB 693 T *Xenopus borealis* cell lines are shown in Fig. 1. Subtle differences are visible between both cell types after Coomassie blue staining (Fig. 1a). In contrast, several characteristic bands for each cell type were identified after con A binding and peroxidase revelation (Fig. 1b). For the untransformed cell line, most glycoprotein bands were revealed in the high molecular weight range (Mr > 97kd) whilst several con A-binding glycoproteins were identified between 90 and 20kd for the transformed cell extract. The specificity of the staining for the con A-binding components was confirmed by the negative control strip for which 0.2M α-methyl-D-mannopyranoside was added to the con A solution (Fig. 1c). These results were consistently observed with different extracts, both in 7.5 and 10% acrylamide gels.

Table 1
Agglutination titres of two Xenopus borealis cell lines for con A, WGA and UEA 1 in μg lectin/ml

Lectins	Agglutination titres	
	XB 693 C	XB 693 T
Con A	5.0	2.5
WGA	1.6	1.8
UEA	250.0	3.9

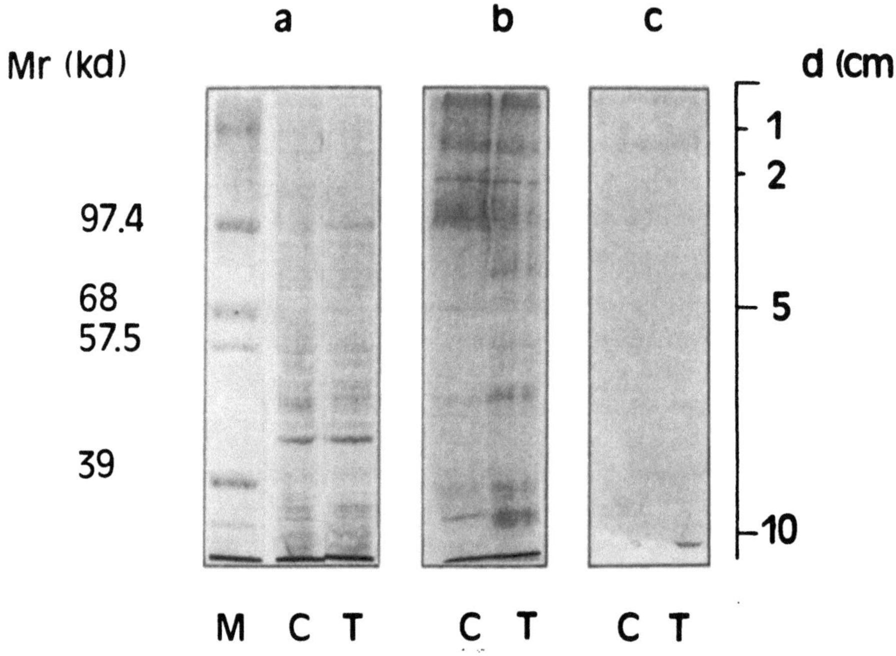

Fig. 1. SDS-PAGE (10%) of Mr standard proteins (M), XB 693 C (C), and XB 693 T (T). Cell extracts were revealed with Coomassie blue (a), with con A (b), and with con A mixed with 0.2M α-methyl-D-mannopyranoside (c).

These differences were further confirmed by densitometric analysis of both con A-stained tracks (Fig. 2). Some con A-binding glycoproteins were more abundant in untransformed cells than in transformed ones, i.e. the 130, 114 and 98kd bands. Conversely, a few bands were more prominent in transformed cells, i.e. the 50, 34 and 32kd glycoproteins.

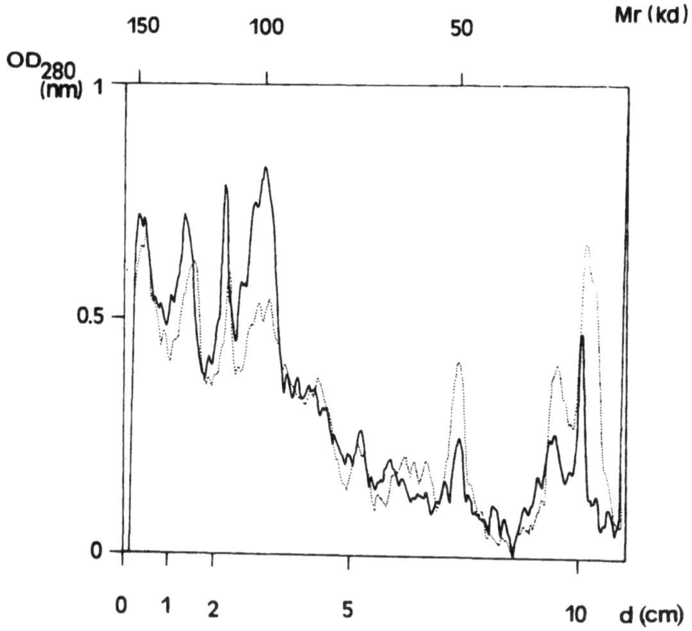

Fig. 2. Densitometric scans of the SDS-PAGE (10%) revealed with con A (Fig. 1b). (——) XB 693 C and (....) XB 693 T cell extracts.

Discussion

The agglutinability of transformed or tumor cells is usually slightly higher or identical to that of their normal counterparts in Mammalian models (3). The Amphibian cell lines studied here exhibit identical agglutination titres when tested with WGA, but a higher agglutinability was observed for transformed cells with con A and UEA 1. The selectivity factor of 64 observed with UEA 1 is worth to be noticed and reflects a significant fucose increase in the composition of the cell surface glycoproteins as already mentioned by Yogeeswaran (4).

Con A-binding of glycoproteins separated by SDS-PAGE of XB 693 C and T cell extracts allows the detection of several bands in both cell types. High molecular weight components are present in higher amounts in untransformed cells. Three striking differences show up in the low Mr range. In transformed cell extracts, there is a marked increase of the 50 and 34kd components as well as the appearance of a new 32kd glycoprotein.

These data extend to Amphibian cell lines current observations collected with lectins on several Mammalian transformed and tumor cells.

Aknowledgments

We thank Mrs S. Denis for expert technical assistance. This work was supported by grants 3.4154.79 and 9.4503.83 from the Fonds de la Recherche Scientifique Médicale, Brussels, Belgium, and by a grant from Boehringer-Ingelheim, Wien, Austria.
P.Q. is a Research Associate from the Fonds National de la Recherche Scientifique, Brussels, Belgium.

References

1. Lis, H., and Sharon N. : Ann. Rev. Biochem. *42*, 541-547 (1973).
2. Goldstein, I.J., and Hayes C.E. : Adv. Carbohydr. Chem. Biochem. *35* : 128-340 (1978).
3. Nicolson, G.L. : Cancer Markers (Snell, S.,ed.) The Humana Press, Clifton, New Jersey, pp. 403-443 (1980).
4. Yogeeswaran, G. : Adv. Cancer Res. *38*, 289-349 (1983).
5. Picard, J.J., Afifi, A. and Pays A. : Carcinogenesis *4*, 739-743 (1983).
6. Afifi, A., Querinjean, P. and Picard J.J. : Lab. Anim. Sci. *35*, 139-141 (1985).
7. Holfreter, J. : Roux Arch. Entwicklungsmekanik *124*, 404-406 (1931).
8. Freed, J.J., Mezger-Fred, L. and Schatz S.A. : Biology of Amphibian Tumors (Mizell, M.,ed.), Springer-Verlag, Berlin, pp. 101-111 (1969).
9. Barth, L.G. and Barth L.J. : J. Embryol. Exp. Morphol. *7*, 210-222 (1959).
10. Laemmli, U.K. : Nature *227*, 280-285 (1970).
11. Fairbanks, G., Steck, T.L. and Wallach D.F.H. : Biochemistry *10*, 2606-2617 (1971).
12. Hawkes, R. : Anal. Biochem. *123*, 143-146 (1982).

MYELIN-ASSOCIATED GLYCOPROTEIN OF THE DEVELOPING RABBIT OPTIC NERVE. SEPARATION BY MICRO-POLYACRYLAMIDE GEL ELECTROPHORESIS AND IDENTIFICATION BY CONCANAVALIN A BINDING.

B. Galas-Zgorzalewicz

Department of Developmental Neurology, University School of Medicine, 60-355 Poznan, Poland

V. Neuhoff, H.G. Zimmer

Max-Planck-Institut für experimentelle Medizin, Forschungsstelle Neurochemie, 3400 Göttingen, F.R.G.

M. Dambska

Medical Research Center Polish Academy of Sciences, 02-093 Warsaw, Poland

The glycoprotein components of the myelin sheath have been studied using periodic acid - Schiff staining (1), incorporation of radioactive sugar precursors (2, 3, 4) or lectin binding techniques (5, 6, 7). These investigations demonstrated that in CNS myelin one glycoprotein of molecular weight about 110 000 is particularly prominent. This major myelin glycoprotein, described first by Quarles et al. (2) was called myelin associated glycoprotein (MAG). Light microscopic immunochemical studies have shown that MAG is present in periaxonal region of the developing myelin sheath (8). On the contrary Webster et al.(9) proved on the basis of electron microscopic and immunochemical studies that MAG is localized within the compact CNS myelin. It was also shown that MAG increases in the whole brain during development (3, 4). The interpretation of the last findings is rather complicated for two reasons:

firstly each brain structure is myelinated at different times during ontogeny and secondly there are regional differences in myelin composition of the CNS. Consequently, the rabbit optic nerve was chosen as a structure particularly suitable for the biochemical studies during postembrional ontogenesis (10). Since the available material in early stages of development was limited micromethods were applied.

Materials and methods

The studies were carried out on optic nerves obtained from Chinchilla rabbits of both sexes aged 6, 8, 10, 12, 16, 32, 42, and 180 days of life. The sequence of myelination was determined by means of quantitative morphometry as described by Dambska and Zgorzalewicz (11). The myelin fraction was prepared essentially by the method of Norton and Poduslo (12) with modification to the microscale as outlined by Fagg et al. (13). The total protein content was determined by means of a micromethod of Neuhoff et al. (14) and the individual protein fractions were separated using micro-SDS-polyacrylamide gradient gel electrophoresis according to Rüchel et al. (15).

Fluorescein isothiocyanate (FITC) labelled concanavalin A (Con A) was used to identify lectin binding to MAG by the Cahill and Morris (16) method. The microgels were scanned using a Zeiss scanning fluorescence microscope equipped with filters for fluorescein excitation and emission wavelengths. A PDP-12 computer linked to the microscope served both to drive the scanner and to record the output as described by Lane et al. (17). Glucose oxidase (D-glucose:oxygen 1-oxidoreductase EC 1.1.3.4) from Sigma was used as standard. FITC labelled Con A was obtained from Serva. All other chemicals were from either Biomol or from Merck.

Results

Direct staining of glycoproteins in polyacrylamide microgels with FITC labelled Con A have shown in all developmental stages the presence of one predominant major glycoprotein of molecular weight 110 000, besides small amounts of minor Con A binding glycoproteins localized in the low molecular region of the gel. (Fig. 1).

The content of MAG in the rabbit optic nerve increases significantly between the 6th and 10th day of postnatal development and then drops rapidly till the 16th day with a moderate decreasing tendency through subsequent period. The Con A binding to MAG

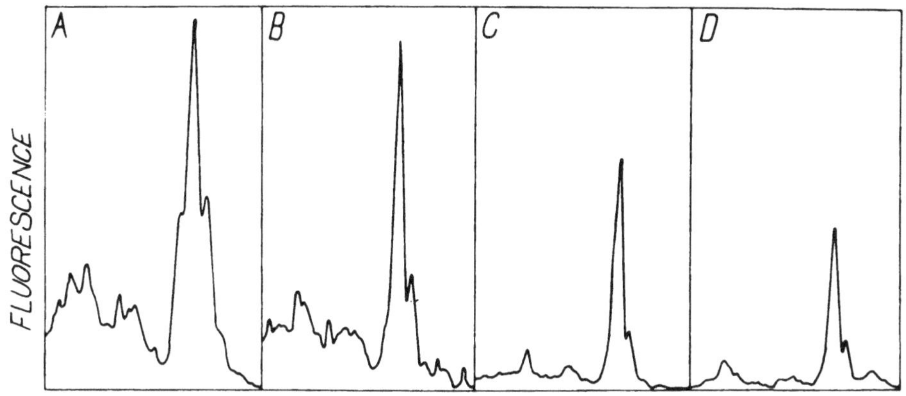

Fig. 1. Developmental pattern of Con A binding myelin glycoproteins. Myelin proteins were separated by SDS polyacrylamide gradient microgel electrophoresis at 10 (A), 12 (B), 32 (C) and 180 (D) days of postembrional ontogenesis of the rabbit optic nerve. After fixation of the micro - gels they were stained with FITC labelled Con A and scanned with computer controlled fluorescence microscope.

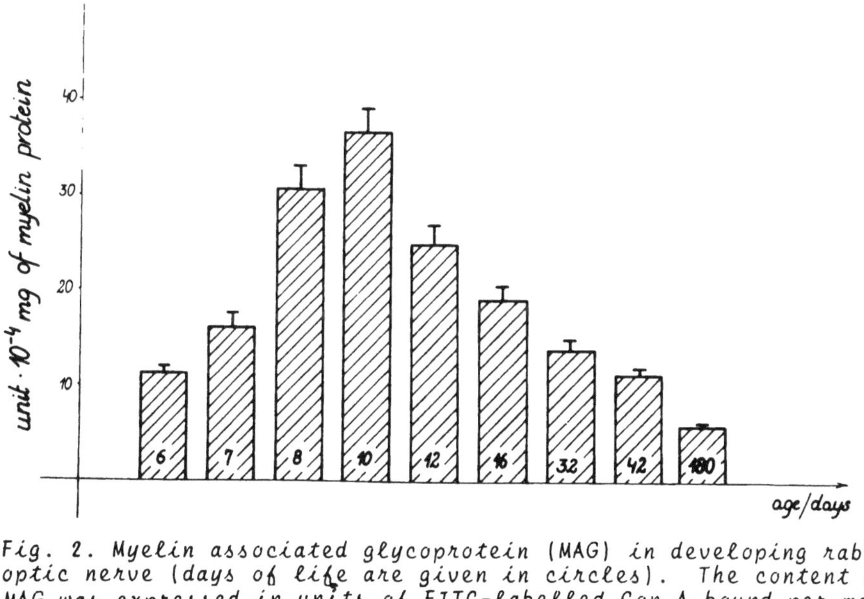

Fig. 2. Myelin associated glycoprotein (MAG) in developing rabbit optic nerve (days of life are given in circles). The content of MAG was expressed in units of FITC-labelled Con A bound per mg of myelin protein. One unit was defined as the amount of MAG which was stained with the same fluorescent intensity as 1 µg of glucose oxidase. Values are means of 6 determinations. Bars represent S.D.

isolated from adult optic nerve (180 days of postnatal age) was 5 times smaller than in the 10th day of life (Fig. 2).

The ultrastructural investigations performed in the time span studied revealed that myelination of the rabbit optic nerve starts about the 6th day after birth. In the 8th day 30 % of nerve fibres are already surrounded by, albeit few, myelin lamellae. The process of myelination runs very fast until the 16th day and on the 32nd day nearly all nerve fibres have myelin sheaths, but of various thickness. Then this process continues until adulthood of the animals, consisting morphologically of increasing numbers of myelin lamellae around the nerve fibres (Fig. 3).

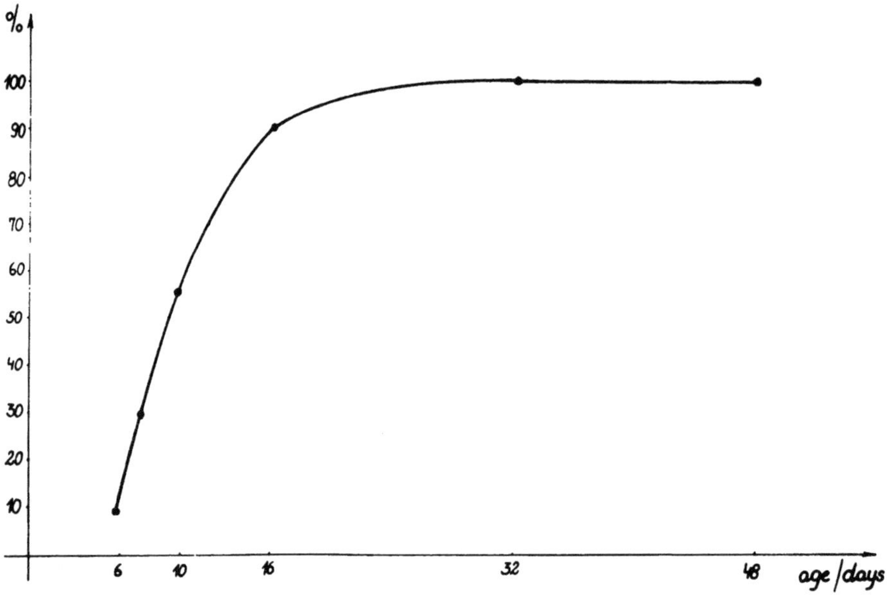

Fig. 3. Percentage of myelinated axons of the rabbit optic nerve as a function of age.

Discussion

A technique which combines the specific binding properties of Con A with high resolution of polyacrylamide microgel electrophoresis was applied in the studies on myelin glycoproteins. This faciliates studies from the 6th day extrauterine life, when the myelination of the optic nerve starts in rabbits. Both myelin proteins of this nerve separated by microgel electrophoresis, as it was shown by Zgorzalewicz et al. (10), and MAG detected on microgel with FITC labelled Con A showed marked changes especially between 6th and 10th day after birth. Therefore, the applied microtechniques have proved to be very efficient methods for research concerning myelin sheath formation when material is limited.

In the early stages of development of the optic nerve the increase of MAG content runs paralell to the degree of myelination of its axons, which indicates the function of this component in formation of myelin sheath. It can be therefore suggested that the external surface membrane components of myelin sheaths to which the glycoproteins belong may be involved in specific recognitation roles during the process of myelination. The presence of glycoproteins in purified myelin isolated from mature optic nerve, proves that they are related to the function in the maintenance of the compact lamellar structure of myelin sheath. The studies conducted by Lane and Fagg (6) on the glycoprotein composition of myelin fractions of the rat optic nerve also showed considerable changes in glycoprotein patterns during ontogenetic development. In addition to MAG, ten other glycoproteins were found localized in the low molecular region of the gel. Both Zanetta et al. (5) and Poduslo et al. (7) showed the presence of several minor glycoproteins associated with purified myelin preparations. Which of these are intrinsic to the myelin sheath and which are derived from membranes contaminating myelin is not yet clear. Therefore the complex pattern of Con A - reactive glycoproteins found in purified myelin preparations at various stages of development of the optic nerve raises several important questions concerning their exact localization in the myelin sheath as well their functions in the CNS.

Acknowledgement

The work was partially supported by a research grant of the Alexander von Humboldt Foundation F.R.G.

ACKNOWLEDGEMENT: The work was partially supported by a research grant of the Alexander von Humboldt Foundation F.R.G.

REFERENCES

1. Zimmermann, A.W., Quarles, R.H., Webster, H., Matthieu, J.M., Brady, R.O.: J. Neurochem. 25, 749-757 (1975).
2. Quarles, R.H., Everly, J.L., Brady, R.O.: Biochem. biophys. Res. Commun., 47, 491-497 (1972).
3. Quarles, R.H., Everly, J.L., Brady, R.O.: Brain Res. 58, 506-509 (1973).
4. Matthieu, J.M., Brady, R.O., Quarles, R.H.: Brain Res. 86, 55-65 (1975).
5. Zanetta, J.P., Sarlieve, L.L., Madel, P., Vincendon, G., Gombos, G.: J. Neurochem. 29, 827-838 (1977).
6. Lane, J.D., Fagg, G.E.: J. Neurochem. 34, 163-171 (1980).
7. Poduslo, J.F., Harman, J.L., McFarlin, D.E.: J. Neurochem. 34, 1733-1744 (1980).
8. Sternberger, N.H., Quarles, R.H., Itoyama, Y., Webster, H.: Proc. Natl. Acad. Sci. USA 76, 1510-1514 (1979).
9. Webster, H., Palkovits, C.G., Stoner, G.L., Favilla, T., Frail, D.E., Braun, E.P.: J. Neurochem. 41, 1469-1479 (1983).
10. Zgorzalewicz, B., Neuhoff, V., Zimmer, H.G.: Neuropat. Pol. 21, 161-168 (1983).
11. Dambska, M,m Zgorzalewicz, B.: Neuropat. Pol. 21, 28-33 (1983).
12. Norton, W.T., Poduslo, S.E.: J. Neurochem. 21, 749-757 (1973).
13. Fagg, G., Waehneldt, T., Neuhoff, V.: Adv. Exp. Med. Biol. 100, 135-145 (1978).
14. Neuhoff, V., Philipp, K., Zimmer, H.G., Mesecke, S.: Hoppe-Seyler's Z. Physiol. Chem. 360, 1657-1670 (1979).
15. Rüchel, R., Mesecke, S., Wolfrum, D.I., Neuhoff, V.: Hoppe-Seyler's Z. Physiol. Chem. 354, 1351-1368 (1973).
16. Cahill, A.L., Morris, S.J.: J. Neurochem, 32, 855-867 (1979).
17. Lane, J.D., Zimmer, H.G., Neuhoff, V.: Hoppe-Seyler's Z. Physiol. Chem. 360, 1405-1408 (1979).

PRECIPITATION OF HUMAN SERUM GLYCOPROTEIN WITH LECTINS

Renate Dörner and Volkmar Sachs

Abteilung Transfusionsmedizin-Immunhaematologie des Klinikums der Christian-Albrechts-Universität Kiel, Klaus-Groth-Platz 2, D 2300 Kiel, BRD

Several studies have shown that lectin can precipitate human serum glycoproteins, e.g. anti-galactans and several single lectins(1-5). We studied the precipitating activity of 39 crude extracts and 7 purified lectins of plant seeds, mushrooms, snails and eels. The glycoproteins we used are of human origin except ferritin, which was prepared from horse spleen and c-reactive-protein (CRP) from Limulus polyphemus.
The aim of the investigation was to obtain more information about the reactivity of lectins and to pick out lectins, which could be of interest as a tool for clinical diagnostics.

Lectins, Vol. V

MATERIAL AND METHODS:

1. Isolation of the lectin crude extracts and purified lectins:
Plant material: Seeds were obtained of 17 cultivars and variants of 9 species of lupinus (L.albus,L.angustifolius Schleswig-Holstein, L.angustifolius Steb, L.angustifolius Stevens,L.cosentinii Chile, L.cosentinii Erregulla, L.cosentinii Guss., L.cosentinii Murr., L.luteus black, L.luteus yellow Chile, L.luteus yellow Schleswig-Holstein, L.microcarpus, L.mutabilis, L.polyphyllus, L. pubescens, L. subcarnosus, L.succulentus) as well as Abrus precatorius, Erythrina cristagalli, Iberis amara, Laburnum alpinum, Laburnum angyroides, Lotus corniculatus, Salvia sclarea, Robinia pseudoaccacia, Euonymus europaeus, Lathyrus pratensis, Lathyrus sativus and Dolichos lablab.
Extracts: We prepared crude extracts by sodiumchloride extraction, ammoniumsulfate precipitation (60%), desalting by Sephadex G-25 (PD-10,Pharmacia,Freiburg) and concentration by Minicon B 15 or Amicon cells (Amicon Corp.,Danvers,MA) leading to a final protein concentration of appr. 5g/l. Lectins were purified from some crude extracts by affinity chromatography on GalNac-, Galacto-, Manno- and Sialo-Gel (EY Laboratories, Medac,Hamburg).

2. Lectins: Bauhinia purpura, Macula pomifera, Helix aspera, Soy bean, Perseau americana, Limax flavus, Griffonia simplicifolia I (pure), Griffonia simplicifolia II (pure), Lotus tetragonolobus, Magnifera indica, Anguilla anguilla and Datura stramonium were delivered by EY Laboratories, Medac, Hamburg.

3. Glycoproteins: Ferritin (horse spleen), transferrin, α_1-antitrypsin, orosomucoid, CRP(Limulus polyphemus), fibronectin, collagen, fibrinogen, lactoferrin are from Sigma (München), α_2-macroglobulin, haptoglobin, β_2-glycoprotein I, IgG, IgA, IgM from Behringwerke (Marburg). The lipoprotein fractions VLDL, LDL,HDL were prepared by differential ultracentrifugation. The glycoprotein concentrations were 5mg/ml.

4. Double affinity diffusion tests were performed by standard method (Ouchterlony) in 1% agarose ME (Sea Kem) or 1.2% agar noble (Difco) in diethylbarbiturate-acetate buffer pH 8.4. The plates were washed in 0.15 sodiumchloride solution during the night,dried and stained in 0.1% amido black.

Table 1

List of lectins and abbreviations in alphabetic order:

Agaricus bisporus	ABA	Lathyrus pratensis	LPA
Anguilla anguilla	AAA	Lathyrus sativus	LSA
Abrus precatorius	APA	Limax flavus	LFA
Bauhinia purpura	BPA	Lotus corniculatus	LCA
Cytisus sessilifolius	CSA	Lotus tetragonolobus	LTA
Datura stramonium	DSTA	Macula pomifera	MPA
Dolichos lablab	DLA	Magnifera indica	MIA
Erythrina cristagalli	ECA	Perseau americana	PAA
Euonymus europaeus	EEA	Robinia pseudoaccacia	RPA
Griffonia simplicifoliaI	GSA	Salvia sclarea	SSA
Griffonia simplicifoliaII	GSA	Sophora japonica	SJA
Helix aspera	HASA	Sophora tetraptera	STA
Homerus americanus	HAMA	Soy bean	SBA
Iberis amara	IAA		
Laburnum alpinum	LALA		
Laburnum angyroides	LANA		

Table 2

List of lectin crude extracts of cultivars and variants of Lupinus (L.) and abbreviations as numbers:

L. albus	1	L.luteus yellow Schleswig-Holstein	11
L. angustifolius Schleswig-Holstein	2	L. microcarpus	12
L. angustifolius Steb	3	L. mutabilis	13
L. angustifolius Stevens	4	L. polyphyllus	14
L. cosentinii Chile	5	L. pubescens	15
L. cosentinii Erregulla	6	L. subcarnosus	16
L. cosentinii Guss.	7	L. succulentus	17
L. Cosentinii Murr.	8		
L. luteus black	9		
L. luteus yellow Chile	10		

RESULTS:

We performed double affinity diffusion of lectins and glycoproteins. Fig. 1 shows a typical example of precipitation pattern

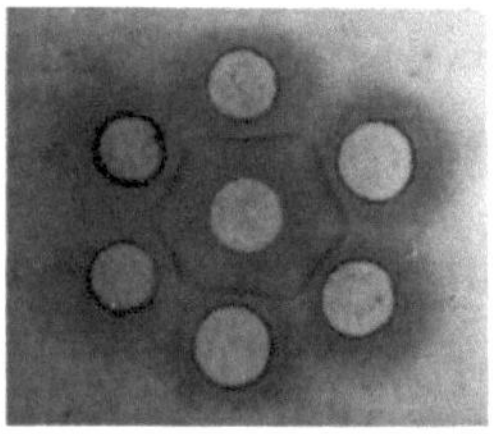

Figure 1: Double affinity diffusion test of Limax flavus crude extract (25µl) and CRP of human source (25µl).

The results demonstrate that D-galactose and N-acetylgalactosamine specific lectins precipitate a broader spectrum of glycoproteins than other lectins we tested, especially α_1-antitrypsin, α_2- macroglobulin, haptoglobin, orosomucoid and fibrinogen. Table 3 shows a summary of our experiments. Only sophora japonica and griffonia simplicifolia I don't show precipitation and haemagglutination in our system (Table 3).

The D-mannose specific lectins from Dolichos lablab, Lathyrus pratensis and Lathyrus sativus show precipitation with α_2-macroglobulin and lactoferrin. The lectin from Dolichos lablab reacts also with IgA and IgM (Table 4). All other crude extracts and purified lectins precipitate only single glycoproteins except Magnifera indica, which precipitates β_2-glycoprotein I, ferritin, IgG, IgA, IgM as well as fibrinogen and the lectin from Perseau americana which precipitates ferritin, IgG and fibrinogen (Tables 5 and 6).

Table 3

Precipitation of D-galactose and/or N-acetylgalactosamine specific lectins with glycoproteins in double affinity diffusion tests.

Lectin	APA	RPA	BPA	EEA	MPA	SBA	ECA	SJA	GSA I	HASA	ABA
Ferritin	-	-	-	+	-	-	+	-	-	-	-
Transferrin	-	-	-	-	-	-	-	-	-	-	-
α_1-Antitrypsin	+	+	-	+	-	-	-	-	-	-	-
α_2-Macroglobulin	+	+	-	+	-	-	+	-	-	-	-
Orosomucoid	-	+	-	-	+	-	-	-	-	-	-
β_2-Glycoprotein I	+	-	-	-	-	-	-	-	-	-	-
Haptoglobin	+	+	-	+	-	+	+	-	-	-	-
Ceruloplasmin	-	+	-	+	-	-	-	-	-	-	-
IgG	+	-	-	-	-	-	-	-	-	-	-
IgA	+	-	-	-	+	-	-	-	-	-	-
IgM	+	-	-	-	-	-	-	-	-	-	-
VLDL	-	-	-	-	-	-	-	-	-	-	-
LDL	+	-	-	-	-	-	-	-	-	-	+
HDL	+	-	-	-	-	-	-	-	-	+	-
CRP	-	-	-	-	-	-	-	-	-	+	-
Fibronectin	-	-	-	-	-	-	-	-	-	-	-
Collagen	-	-	-	-	-	-	-	-	-	-	-
Fibrinogen	+	+	-	+	-	-	-	-	-	-	-
Lactoferrin	+	-	+	-	-	-	-	-	-	-	-
Haemagglutination	+	+	+	+	+	+	+	-	-	+	+

The precipitation is scored: + precipitation, - no precipitation; APA...... etc. are abbreviations for lectins and crude extracts (table1)

15 crude extracts precipitate the lipoproteinfraction LDL, 7 of them are cultivars and variants of Lupinus (Tables 3-6).

The reaction of the lectin from Lotus corniculatus and α_2-macroglobulin, of the lectin from Lupinus pubescens and IgG as well as of the lectin from Laburnum alpinum and IgG were observed only once and could not be reproduced.

Table 4

Precipitation of D-mannose (1,2,3), N-acetylglucosamine (4,5,6), L-fucose (7,8,9) and sialic acid (10) specific lectins with glycoproteins in double affinity diffusion tests.

Lectin	DLA_1	LPA_2	LSA_3	$LALA_4$	GSA II_5	$DSTA_6$	AAA_7	LTA_8	CSA_9	LFA_{10}
Ferritin	-	-	-	-	-	-	-	-	-	-
Transferrin	-	-	-	-	-	-	-	-	-	-
α_1-Antitrypsin	-	-	-	-	-	-	-	-	-	-
α_2-Macroglobulin	+	+	+	-	-	-	-	-	-	-
Orosomucoid	-	-	-	-	-	-	-	+	-	-
β_2-Glycoprotein I	-	-	-	-	-	-	-	-	-	-
Haptoglobin	+	-	-	-	-	+	+	-	-	-
Ceruloplasmin	-	-	-	-	-	-	-	-	-	-
IgG	-	-	-	(+)	-	-	-	-	-	-
IgA	+	-	-	-	-	-	-	-	-	-
IgM	+	-	-	-	-	-	-	-	-	-
VLDL	+	-	-	-	-	-	-	-	-	-
LDL	+	-	-	+	+	+	+	-	-	-
HDL	-	-	-	-	-	-	-	-	-	-
CRP	-	-	-	-	-	-	-	-	-	+
Fibronectin	-	-	-	-	-	-	-	-	-	-
Collagen	-	-	-	-	-	-	-	-	-	-
Fibrinogen	-	-	-	-	+	-	-	-	-	-
Lactoferrin	+	+	+	-	-	-	-	-	-	-
Haemagglutination	+	+	+	+	+	o	o	-	-	+

The precipitation is scored: + precipitation, - no precipitation, o not tested;
DLA ..... etc. are abbreviations for lectins and crude extracts (Table 1) (+) result could not be reproduced with a different IgG preparation.

Table 5

Precipitation of lectins with glycoproteins in the double affinity diffusion tests:

Lectin	MIA	PAA	JAA	LANA	LCA	STA	SSA	HAMA
Ferritin	+	+	+	-	-	-	-	-
Transferrin	-	-	-	-	-	-	-	-
α_1-Antitrypsin	-	-	-	-	-	-	-	-
α_2-Macroglobulin	-	-	-	+	(+)	-	-	-
Orosomucoid	-	-	-	-	-	-	-	-
β_2-Glycoprotein I	+	-	-	-	-	-	-	-
Haptoglobin	-	-	-	-	-	-	-	-
Ceruloplasmin	-	-	-	-	-	-	-	-
IgG	+	+	-	-	-	-	-	-
IgA	+	-	-	-	-	-	-	-
IgM	+	-	-	-	-	-	-	-
VLDL	-	-	-	-	-	-	-	-
LDL	-	-	-	-	-	-	-	+
HDL	-	-	+	-	-	-	-	-
CRP	-	-	-	-	-	-	-	-
Fibronectin	-	-	-	-	-	-	-	-
Collagen	-	-	-	-	-	-	-	-
Fibrinogen	+	+	-	-	-	-	-	-
Lactoferrin	-	-	-	-	-	-	-	-
Haemagglutination	+	+	+	+	+	-	-	+

The precipitation is scored: + precipitation, - no precipitation; MIA .... etc. are abbreviations for lectin crude extracts (Table 1) (+) result could not be reproduced.

Table 6

Precipitation of lectins from cultivars and variants of Lupinus with glycoproteins in double affinity diffusion tests:

Lectins	1	2	3	4	5	6	7	8	9	10	11	12	13	14	15	16	17
Ferritin	-	-	-	-	-	-	-	-	-	-	-	-	-	-	-	-	-
Transferrin	-	-	-	-	-	-	-	-	-	-	-	-	-	-	-	-	-
α_1-Antitrypsin	-	-	-	-	-	-	-	-	-	-	-	-	-	-	-	-	-
α_2-Macroglobulin	-	-	-	-	-	-	-	-	-	-	-	-	-	-	-	-	-
Orosomucoid	-	-	-	-	-	-	-	-	-	-	-	-	-	-	-	-	-
β_2-Glycoprotein I	-	+	-	+	-	-	-	-	-	-	-	-	-	-	-	-	-
Haptoglobin	-	-	-	-	-	-	-	-	-	-	-	-	-	-	-	-	-
Ceruloplasmin	-	-	-	-	-	-	-	-	-	-	-	-	-	-	-	-	-
IgG	-	-	-	-	-	-	-	-	-	-	-	-	-	+	+	(+)	-
IgA	-	-	-	-	-	-	-	-	-	-	-	-	-	-	-	-	-
IgM	-	-	-	-	-	-	-	-	-	-	-	-	-	-	-	-	-
VLDL	-	-	-	-	-	-	-	-	-	-	-	-	-	-	-	-	-
LDL	-	+	-	+	-	+	+	-	-	-	+	-	+	-	-	-	+
CRP	-	-	-	-	-	-	-	-	-	-	-	-	-	-	-	-	-
Fibronectin	-	-	-	-	-	-	-	-	-	-	-	-	-	-	-	-	-
Collagen	-	-	-	-	-	-	-	-	-	-	-	-	-	-	-	-	-
Fibrinogen	-	-	-	-	-	-	-	-	-	-	-	-	-	-	-	-	-
Lactoferrin	-	-	-	-	-	-	-	-	-	-	-	-	-	-	-	-	-
Haemagglutination	+	+	-	+	+	+	+	-	o	+	+	-	-	-	+	-	+

The precipitation is scored: + precipitation, - no precipitation, o not tested; 1,2,3 ... etc. are abbreviations for crude extracts (Table 2);
(+) the result could not be reproduced with a different IgG preparation.

The double affinity diffusion test of different cultivars and variants of the same species of Lupinus angustifolius, Lupinus cosentinii and Lupinus luteus have shown different results with our glycoprotein panel (Tables 3 and 6).

DISCUSSION:

We have shown here that lectins prepared from seeds of different cultivars and variants of the same species of plants can show a difference in the precipitating activity (Table 3,5 and 6). This may be caused by changes of protein composition either qualitative or quantitative (6). This indicates the importance of standardisation of lectin preparation procedure which must start with well defined cultivars or variants of plant seeds as primary material.

Furthermore our results show, that many human serum glycoproteins carry ß-D-galactosyl or N-acetylgalactosyl residues at the terminal or subterminal end of the carbohydrate chain, which is the part recognized by several lectins but not by all (Table 3). The reason for this might be sterical hindrance of binding.

It appears that the limax flavus lectin might be of interest for clinical diagnostics, as it precipitates CRP prepared from Limulus polyphemus and as we could show now CRP from human source.

The D-mannose specific lectin Dolichos lablab, the N-acetylglucosamine specific lectins of Laburnum alpinum, Griffonia simplicifolia II, Datura stramonium as well as of some variants of Lupinus (Table 4) show great affinity to the LDL fraction. This confirms the observation that ß-lipoprotein has terminal α-D-glucosidic and α-D-mannosidic groups which were available for lectins of Con A (7), Dolichos lablab, Laburnum alpinum, Griffonia simplicifolia II and Datura stramonium. It seems that variants of Lupinus (Table 2: lectin 2,4,6,7,11,13,17) bind to the same carbohydrate groups.

The lectins from Lupinus angustifolius precipitate β_2-glycoprotein I, the lectins from Lupinus cosentinii and luteus a protein of the β_2-γ-region and the lectins from Lupinus polyphyllus, pubescens and subcarnosus IgG. These results will be a field for further investigation.

ACKNOWLEDGEMENTS: We thank Miss B. Grupe for the excellent technical assistance and Mrs. A. Hartmann for typing the manuscript.

REFERENCES:

1. Uhlenbruck, G., P. Vaith, D. Karduck, H. Haupt and W.E.G. Müller: Protides Biol. Fluids 27, 595-598 (1979).
2. Uhlenbruck, G., B.A. Baldo, G. Steinhausen, H.G. Schwick, B.P. Chatterjee, V. Horejisi, A. Kranjhansl and J. Kocourek: J. Clin. Chem. Clin. Biochem. 16, 19-23 (1978).
3. Spengler, G.A., R.-M. Weber and F. Skvaril: Schweiz. Med. Wschr. 102, 1618-1619 (1972).
4. Jaquet, H., W. Wojnarowski, G.I. Pardoe and G.A. Spengler: Z. Immun. Forch. 150, 212-213 (1975).
5. Osunkoya, B.O. and A.I.O. Williams: Clin. exp. Immunol. 8, 205-212 (1971).
6. Rougé, P., M. Richardson, C. Chatelain, A. Yarwood, B. Sousa-Cavada: these proceedings.
7. Bøg-Hansen, T.C.: Anal. Biochem. 56, 480-488 (1973).

HUMAN TEAR GLYCOPROTEINS.
SDS-GEL ELECTROPHORESIS, BLOTTING AND LECTIN BINDING.

P.Halken[1,2], T.C.Bøg-Hansen[1] and J.U.Prause[2]

[1]The Protein Laboratory, Sigurdsgade 34, DK-2200, Copenhagen N, Denmark, and [2]The Eye Pathology Institute, Frederik V's Vej 11, DK-2100, Copenhagen Ø, Denmark.

The normal human cornea does not contain blood vessels, and a major part of the metabolic substances needed thus crosses the precorneal tear film.

Disturbances in the tear film reduce the stability of the film resulting in an inhibited metabolic activity of the cornea (1). Accordingly, tear film stability is essential for maintenance of the corneal function. Reduced stability may cause corneal drying, ulceration, and impaired visual acuity.

The normal human tear film is a three layered structure. The outermost layer consists of lipids, the main, middle layer is aqueous, containing ions, proteins, and glycoproteins. The inner mucous layer contains glycoproteins which coat the surface of the corneal cells. Disturbances in each of the layers may influence the tear film stability, and especially changes in the mucous layer reduce the stability.
Tear glycoproteins are produced mainly by conjunctival goblet cells, but the tear glands also secrete soluble glycoproteins (2).
Various diseases can induce changes in production and composition of tear film mucous glycoproteins (1), but until now only a few studies have attempted a characterization of the tear glycoproteins and their changes.
With the use of lectins as probes for glycoproteins, we present a simple and sensitive method for characterization of soluble tear fluid proteins by SDS-polyacrylamide gel electrophoresis, electroblotting and binding of lectins.

Lectins, Vol. V

MATERIALS:

Sample collection: Tear samples were collected in 50µl glass capillary tubes (type 50 P. Vitrex micropipettes purchased from KEBO Laboratorieudstyr Aps, Virum, Denmark) pooled, and stored at -4°C until use.

SDS-PAGE and lectin-blotting: Electrophoresis was performed in a BioRad Protean slab gel cell, the glass plates (for gel casting) being 16x18 cm, separated by 1.5 mm spacers (BioRad Laboratories, Richmond, Ca. U.S.A.). Sample wells (10 per gel) were 9 mm wide. Blotting was performed in a semi-dry blotting apparatus (JKA-Biotech, Vanløse, Denmark). Acrylamide, bisacrylamide (BIS), TEMED (N,N,N,N-tetramethylethylene diamine), and ammonium persulphate were products of BioRad. Na-riboflavine-5-phosphatedihydrate was produced by BDH, W.Germany. Sodium-dodecyl-sulphate (SDS) was obtained from Serva, Heidelberg, W.Germany. Pyronin G, 2-mercaptoethanol, H_2O_2, and Tween 20 were purchased from Merck, Darmstadt, W.Germany. 3-amino-9-ethyl-carbazole and α-methyl-D-glucoside were products of Sigma, St.Louis, Mo.U.S.A. N-acetyl-α-D-glucosamine was produced by Janssen Chimica, Beerse, Belgium. D-galactose was from Fluka AG, Switzerland. Nitrocellulose membrane (type BA 85, 0.45 µm) was purchased from Schleicher and Schuell, Dassel, W. Germany. Biotinylated lectins and avidin-conjugated horse-radish peroxidase (avidin-HRP) were obtained from Kem-En-Tec, Hellerup, Denmark.
All other reagents were of analytical grade.

METHODS:

SDS-PAGE: Electrophoresis was performed using a modified Laemmli system (3), with minor alterations: the separation gel (0.8 M Tris HCl, pH 8.8) was prepared as a linear gradient gel with T = 5-25% and C = 0.5-1.5%. The gel was photopolymerized overnight at 4°C using 10 µM riboflavine, 4.5 mM TEMED and 26 mM ammonium persulphate as catalysts. The stacking gel (T = 3% and C = 5%) contained 0.14 M Tris HCl, pH 6.5, 0.1 M glycine, 10 µM riboflavine, 7 mM TEMED, 26 mM ammoniumpersulphate, and was photopolymerized for 30 minutes. In both gels BIS acted as crosslinker. The electrode buffer was 0.14 M Tris, 0.1 M glycine and 0.1% (w/v) SDS.
Aliquots of pooled tears were mixed with equal volumes of sample buffer, giving a final concentration of 0.14 M Tris HCl, pH 6.5, 5% (v/v) mercaptoethanol, 5% (w/v) sucrose, 1% (w/v) SDS, and 0.01% (w/v) pyronin G.
The sample solutions were heated at 100°C for 3 minutes and applied immediately to the gel (50 µl per well). Electrophoresis was carried out at 15°C with 5 mA/cm^2 during stacking and 50 mA/cm^2 during separation, total running time: 5 hours.

Blotting: The proteins were transferred electrophoretically on to a nitrocellulose membrane (4). After transfer, lanes - corresponding to the lanes in the gel - were cut out of the nitrocellulosemembrane. These were quenched for 2 minutes with 10 ml 2% (v/v) Tween in 10 mM Tris HCl, pH 7.2, 1 M NaCl and subsequently washed for 5 minutes in a corresponding buffer containing 0.05% Tween and 1 mM Ca^{++}, Mg^{++}, and Mn^{++} (= incubation buffer).

Overlay: 20 µg biotinylated lectin in 10 ml incubation buffer was used as overlay for 20-24 hours, followed by 3x5 minutes wash with

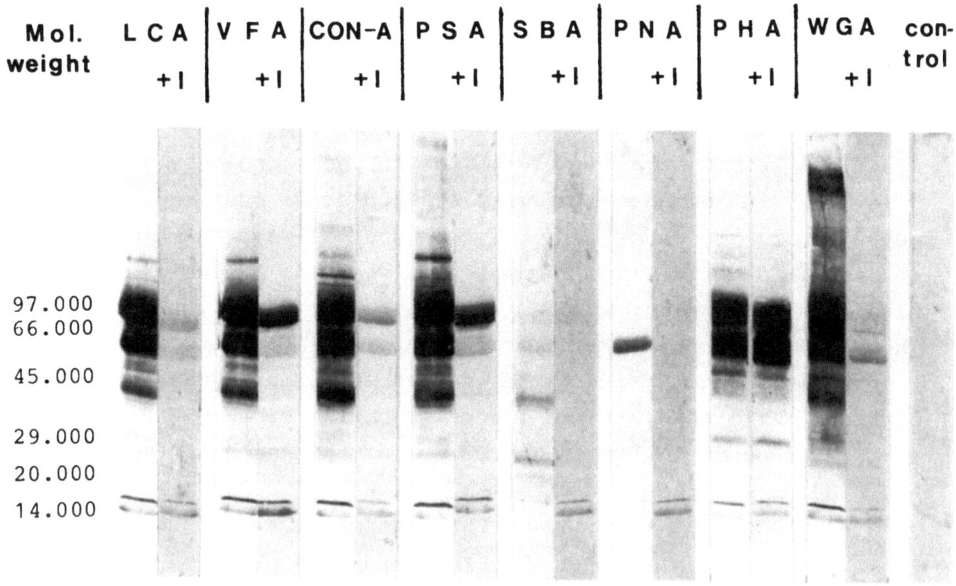

Fig. 1. Tear glycoproteins visualized after SDS-gel electrophoresis and electroblotting with biotinylated lectins and avidin-HRP.
More than 30 bands with varying lectin binding capacity could be demonstrated.
Binding specificity of lectins was ensured by disappearance of bands upon incubation with the corresponding inhibitory monosaccharides (+I, 10% w/v): LCA, VFA, Con-A, PSA: α-methyl-D-glucoside; SBA, PNA, PHA: D-galactose; and WGA: N-acetyl-α-D-glucosamine.
Specificity of avidin-HRP binding was demonstrated by excluding the lectin (control). Uninhibited and inhibited blots were transferred from different gels, which is the cause of the slight differences in position and appearance of corresponding bands.

the incubation buffer. Specific monosaccharide inhibition was performed concurrently; avidin-HRP (4 µg/10 ml incubation buffer) was overlayed for 3 hours, followed by 3x5 minutes wash in incubation buffer. Incubations and washes were performed under agitation and at room temperature.

Staining: Lectin bound avidin-HRP was visualized using 0.05 M sodium acetate - glacial acetic acid pH 5.5 containing 0.04% (w/v) carbazole and 0.015% (w/v) H_2O_2 mixed immediately before use. Blots were stained for 30 minutes and subsequently washed in H_2O.

RESULTS:

We performed SDS-polyacrylamide gel electrophoresis of human tear proteins with subsequent blotting and lectin staining. With the described method more than 30 different bands could be visualized (Fig. 1). Two broad and heavily stained bands could be seen in the 60-100 kD region on all blots except the PNA- and SBA-blots. These bands did not disappear completely when inhibiting the lectin binding with monosaccharides. Likewise, two heavily stained bands in the 15-20 kD region persisted faintly on all blots after inhibition. Blots incubated with LCA showed 16 bands. VFA yielded 20 bands of which the staining intensity of the broad 95 kD band and of the two bands in the 15-20 kD area diminished slightly with inhibition as was the case with PSA, which produced 25 bands. Con-A gave 22 bands, SBA 9 bands and PNA 2 bands. Paradoxically, the 15-20 kD bands on the SBA and PNA blots stained more intensely when lectin binding was inhibited. WGA produced 14 bands and a slight diffuse staining of the entire blot from the 25 kD area up. The PHA blot showed 16 bands. Only the heaviest of these disappeared due to inhibition.

In order to optimize the assay, various parameters were investigated. Three modifications of the Laemmli buffer system were tried (3,5,6) and (3) was found to be best suited for this assay. The denaturizing agents dithiothreitol (DTT) and mercaptoethanol were found to be equally effective. Mercaptoethanol was chosen as it is reported to be more stable in storage. Using various amounts of denaturizing agent in the sample buffer the conclusion was reached, that 5% (v/v, final concentration in the sample solution) would reduce sufficiently the amount of protein present in the tear samples. The effect of heat denaturation was investigated: Sample solutions were subject to heat denaturation by immersion in boiling water (100°C) for 2-5 minutes and in hot water (90°C) for 10-20 minutes. Since no change could be seen with prolonged denaturation, a heat denaturation time of 3 minutes at 100°C was chosen.

The acrylamide and BIS concentration (T and C) in both stacking and separation gel (continuous and gradient)was varied: T of the stacking gel was varied between 3 and 6%. In the continuous separation gels T was varied between 9 and 20% and C was varied from 0.5-1.5%. To expand the molecular weight range, gradient gels with T from 9-12% up to 5-25% were cast, using a C gradient (0.5-1.5%) in the latter gels. Gradient gels with T = 5-25% and C = 0.5-1.5% were found to give the most satisfying results. During these tentative steps of T and C variation the gels were stained in Coomassie Brilliant Blue R-250 (Serva).

A stacking gel height of 0.5-1 cm was chosen, enabling the samples to form a sharp band before entering the separation gel.

Lectin-incubation: Comparing blots incubated with biotinylated lectin for periods from 2 hours to 7 days, it appeared that the minimum incubation time should be 4 hours. Incubation periods of more than 24 hours did not reveal additional bands. For convenience, an incubation period of 20-24 hours was chosen. Then 10-40 μg biotinylated lectin per 10 ml incubation buffer was overlayed for 24 hours and it was shown that concentrations higher than 20 μg/10 ml incubation buffer did not give further details in the blots.

Avidin-HRP incubation: Incubation time and concentration of avidin-HRP were also examined, and our studies indicated that variations in incubation time could be used with appropriate changes in the avidin-HRP concentration. An incubation time of 3 hours was chosen and it was found that the highest concentration of avidin-HRP without non-specific binding was 4 μg/10 ml. Consequently, this concentration was chosen.

DISCUSSION:

Because of the high content of salts in tears it could be expected that ionic interference with the ionic zones of the SDS-PAGE system would occur, thus reducing the resolving ability of the system, but apparently the described method overcomes this problem. The presented method visualized more than 30 different lectin binding components. Inhibition of lectin-glycoprotein binding with monosaccharides resulted in a total or partial disappearance of all bands. The residual binding could be non-specific, but differences in binding kinetics and affinity of the lectins for the glycoproteins and the monosaccharides could also be the reason. The latter explanation is supported by the observation that the bands only partly inhibited were originally the most heavily stained. These bands probably represent very large amounts of glycoprotein (see below). It is a well-known fact that inhibition of PHA is hard to obtain. Therefore, persisting bands on the PHA blot should not automatically be regarded as representing non-specific lectin binding.

The broad, diffuse, heavily stained bands seen in most blots are caused by large amounts of glycoprotein: When adding smaller volumes of tear sample to the assay, the two broad 60-100 kD bands could be seen to consist of several narrow bands, thus giving an improved resolution in this area. This can be seen in some of the inhibited blots in Fig. 1. However, reducing the tear volume caused a disappearance of the weakest bands of the blots and as a compromise, a tear sample volume of 20 µl per lane, giving the highest number of distinct bands, was chosen.

Tear samples are obtained easily, and with a somewhat shorter incubation time, results may be available within 24 hours. This indicates that the method might be useful when diagnosing external eye disorders. Furthermore, other biological fluids could also be examined by the presented method, thus expanding the diagnostic range.

ACKNOWLEDGEMENTS: This study was supported by The Danish Medical Research Council, and with grants gratefully accepted from Jacob Madsen og hustru Olga Madsens Fond, Mimi og Victor Larsens Fond, Carla Cornelia Storch Møllers Legat. The Harboe Foundation is thanked for a generous grant to the Protein Laboratory.

REFERENCES:

1. Holly, F.I., Lemp, N.A. (1977). Survey of Ophthalmol. 22, 69-87.

2. Jensen, O.A., Falbe-Hansen, I., Jacobsen, T., Michelsen, A. 81969). Acta Ophthalmol. 47, 605-619.

3. Schäfer-Nielsen, C., Rose, C. (192). Biochem. Biophys. Acta 696, 323-331.

4. Kyhse-Andersen, J. (1984). J. Biochem. Biophys. Methods 10, 203-209.

5. Janssen, P.T., Van Bijsterveld, O.P. (1981). Clin. Chim. Acta 114, 207-218.

6. Internal laboratory manual. The Protein Laboratory, University of Copenhagen, Denmark, unpublished.

ACID PHOSPHATASE FROM GRASS TISSUES:
II. CON A-BINDING ROOT ACID PHOSPHATASE ISOENZYMES AND THEIR IMMUNOLOGICAL RELATIONSHIP

Irena Lorenc-Kubis and Bronislawa Morawiecka
Institute of Biochemistry, University of Wroclaw, Poland

The lectins with their narrow specificity for carbohydrates became a valuable tool for the purification of glycoproteins (1), for the study of carbohydrate structure (2) and for the identification of glycoconjugates in tissues (3-5). In our studies we demonstrated the distinct differences between glycoprotein acid phosphatases of grass seeds, coleoptiles and leaves in their capacity to bind to Con A which may reflect changes in the sugar component during development as a results of postsynthetic modifications(6).Moreover the enzymes were activated or inhibited by binding to con A or LCA (7, 8). Little is known about the root acid phosphatases, their interaction with lectins and the immunological relationship. The aim of this paper was to isolate the acid phosphatases from barley and wheat roots, to study their affinity to lectins and their immunological relationship.

Lectins, Vol. V

Material and Methods

Roots from 7 day seedings of barley (Hordeum sativum) and wheat (Triticum vulgare) were harvested, frozen, homogenized with 0.1 M acetate buffer pH 5.1 and centrifuged. The supernatant was used for purification of acid phosphatases by affinity chromatography on con A-Sepharose (Pharmacia Fine Chemicals) as described in (7, 8). The bound proteins were displaced from the column with 15 mM (acid phosphatase Ba) and 500 mM alpha-methyl-D-mannopyranoside (acid phosphatase Bb) and further purified by chromatography on SP-Sephadex C-50 or QAE-Sephadex A-50 (Pharmacia Fine Chemicals). The active fractions were eluted from the columns with 0.2 M and 0.5 M NaCl in 0.05 M acetate buffer pH 4.6 or in 0.05 M buffer Tris-HCl pH 8.0, respectively.

Acid phosphatase activity was measured spectrophotometrically (9) and developed histochemically as described in (7). Protein was determined by the method of Lowry et al. (10). Double affinity diffusion with free lectins: concanavalin (con A), Lentil lectin (LCA), soy bean agglutinin (SBA) and wheat germ agglutinin (WGA) was performed as described in (6). Rocked affinity electrophoresis was performed according to (11) and immunodiffusion according to Quchterlony (12).

Antibodies were purified from rabbit antiserum (13) raised against purified acid phosphatase B, from Poa pratensis caryopses.

Results

The acid phosphatases from roots of 7 day seedings of barley and wheat were separated by affinity chromatography on con A-sepharose into two fractions: the non-binding (A) and the con A-binding (B). The con A-binding acid phosphatases were eluted from the column with 15 mM alpha-methyl-D-mannopyranoside (Ba) and with 500 mM solution of this sugar (Bb). Acid phosphatase Ba from wheat consituted the majority of the activity of the enzyme bound to con A.

The enzyme was further fractionated by chromatography on SP-Sephadex C-50. Two isoenzymes (Ba_1 and Ba_2) were obtained by elution from the column with 0.2 M NaCl and 0.5 M NaCl in 0.05 M acetate buffer pH 4.6. Barley acid phosphatases Ba and Bb were not adsorbed on SP-Sephadex. These enzymes were further purified by chromatography on QAE-Sephadex A-50. It was found that barley acid phosphatase Ba was eluted from the column with 0.2 M NaCl (Ba), while in the case of Bb two isoenzymes (Bb_1 and Bb_2) were obtained. The isoenzymes were eluted from the column with 0.2 M and 0.5 M NaCl, respectively.

The all purified isoenzymes from barley (Ba_1, Bb_1, Bb_2) and two isoenzymes of wheat (Ba_1, Ba_2) cross-reacted with antibodies raised against isoenzyme B_1 from Poa pratensis caryopses (Fig. 1).

To show other molecular property of the acid phosphatase, the affinity of the isoenzymes to lectins was tested. It has been shown that all isoenzymes of barley and wheat roots interact with free con A and form an affinity precipitate. In the presence of LCA the affinity precipitate of the barley enzyme exhibited a much weaker staining than with con A (fig. 2), and wheat isoenzymes formed only a trace affinity precipitate with LCA, Binding to WGA and SBA was not observed, altogether indicating that the carbohydrate of the root acid phosphatases contain exposed D-mannose units but is devoid of exposed N-acetylglucosamine and N-acetylglactosamine units.

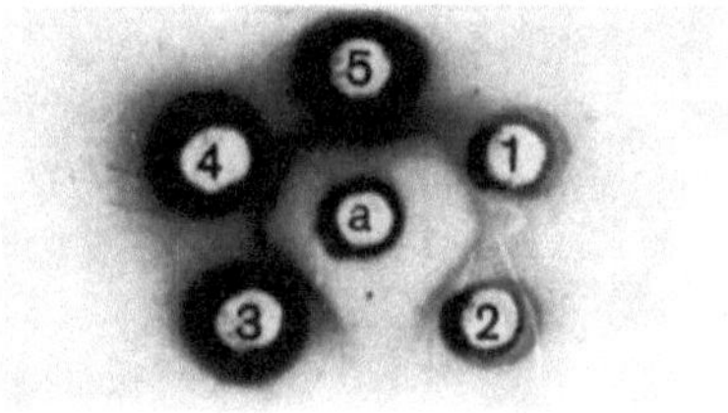

Fig. 1. Ouchterlony immunodiffusion of roots AcPases from wheat (1, 2) and barley (3, 4, 5) against anti-seeds AcPase B_1 of Poa pratensis antiserum (a). AcPase Ba_1 (1), AcPase Ba_2 (2), AcPase Ba_1 (3), AcPase Bb_1 (4) and AcPase Bb_2 (5).

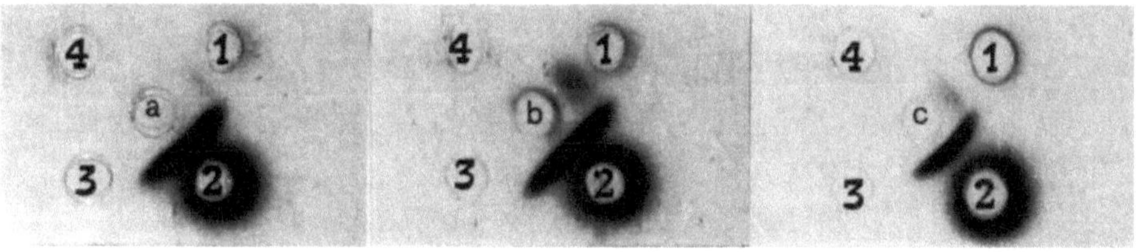

Fig. 2. Double diffusion in agarose gel. Lectins are directed against barley roots AcPases. 10 µg of AcPase Ba_1 (a), AcPase Bb_1 (b) and AcPase Bb_2 (c) are placed in the centre well. Lectins (100 µg) in outside wells: 1. LCA, 2. Con A, 3. WGA, 4. SBA.

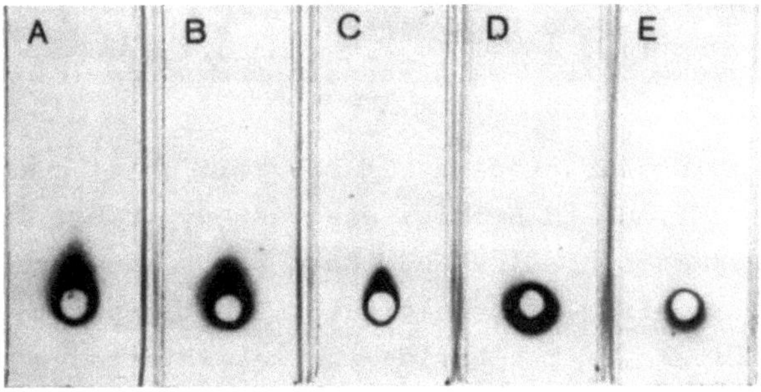

Fig. 3. Rocket-affinity electrophoresis of 8 µg roots AcPases of barley (A, B, C) and wheat (D, E) into agarose gel containing con A (25 µg/cm²). AcPase Ba_1 (A), AcPase Bb_1 (B), AcPase Bb_2 (C), AcPase Ba_1 (D) and AcPase Ba_2 (E).

Rocket affinity electrophoresis with free con A (Fig. 3) showed different affinity precipitates of barley and wheat isoenzymes with respect to in morphology and staining intensity. Isoenzymes Ba_1 and Bb_1 from barley form a distinct precipitate with free con A while the precipitate of Bb_2 and wheat acid phosphatase Ba_1 was significantly weaker.

We could also show that con A and LCA activate the barley and wheat acid phosphatases. The increase of the enzyme activity ranged from 20 to 50%. The activation of the enzymes was completely reversed by addition of alpha-D-mannose what indicates that the carbohydrate part of the enzymes is an important structural element.

Discussion

The acid phosphatase isoenzymes from barley and wheat roots are glycoproteins like acid phosphatases from other grass tissues (6-8). The enzymes interact with immobilized and free con A. The affinity of the root acid phosphatases to free con A differed considerably from that of enzymes of leaves and coleoptiles from barley which did not form any precipitate with free con A (6). The root acid phosphatases are similar to the enzymes of seeds from all grasses which also form an affinity precipitate with free con A. This indicates that also the root acid phosphatases posses two or more binding sites with expossed D- mannose units. The differences in the morphology of rocket affinity precipitates may indicate a subtle structural distinction between the barley and the wheat acid phosphatases in their sugar moiety (11). Differences of the sugar component could be the number and position of carbohydrate chains and the exposition of D-mannose units. The cross reaction of root acid phosphatases from barley and wheat with antibodies raised against the enzyme from Poa pratensis seeds indicates a close structural relationship among the enzymes. An immunological relationship among polymorphic forms of acid phosphatases has also been shown in the case of the enzyme isolated from seeds, coleoptiles and leaves from various grass species (6, 14, 15). It seems that the grass acid phosphatases in spite of the differences in sugar moiety are family-related proteins. Presumably the polypeptide part of the enzymes has been evolutionary conserved for retention of some important physiological function.

Acknowledgements: The authors are grateful to Ms Renata Herba for technical assistance and to Ms Caroline Nielsen for typing the manuscript. The study was supported by a grant from the Polish Academy of Science, number MR.II.1.2.6.

References

1. Dunaley, J.T.: Mol. Cell. Biochem. 21, 43-63 (1978)

2. Sharon, N., Lis, H.: Mol. Cell. Biochem. 42, 167-187 (1982)

3. Freeman, H.J., Lotan, R., Kim, Y.S.: Lab. Invest. 42, 405-412. (1980).

4. Maylie-Pfenninger, M.F., Jamieson, J.D.: J. Cell. Biol. 86, 96-103 (1981).

5. Pena, S.D.J., Gordon, B.B., Karpati, G., Carpenter, S.: J. Histochem. Cytochem. 29, 542-546 (1981)

6. Lorenc-Kubis, I., Morawiecka, B. in: Lectins, Biology, Biochemistry, Clinical Biochemistry, T.C. Bøg-Hansen and J. Breborowicz (Eds.) Vol. 4, W. de Gruyter, Berlin, 1985 pp. 259-266.

7. Lorenc-Kubis, I., Bøg-Hansen, T.C. in: Lectins, Biology, Biochemistry, Clinical Biochemistry, T.C. Bøg-Hansen (Ed.) Vol. 1. W. de Gruyter, Berlin, New York 1981 pp. 157-166.

8. Lorenc-Kubis, I., Morawiecka, B., Wieczorek, E., Wisniowska, J., Wierzba, E., Ferens, M., Bøg-Hansen, T.C. in: Lectins, Biology, Biochemistry, Clinical Biochemistry, T.C. Bøg-Hansen (Ed.) Vol. 1. W. de Gruyter, Berlin, New York 1981 pp. 169-178.

9. Yoshida, H., Tamlya, N.: J. Biochem. 69, 525-534 (1971).

10. Lowry, O.H., Rosenbrough, N.J., Farr, A.L., Randall, R.J.: J. Biol. Chem. 193, 265-275 (1951).

11. Bøg-Hansen, T.C., Bjerrum, O.J., Brogren, C.-H.: Anal. Biochem. 81, 78-87 (1977).

12. Ouchterlony, O. in: Handbook of Experimental Immunology, Weir, D.M. (Ed.) Blackwell Sci. Publ. Oxford and Edingburgh, pp. 655-706 (1974).

13. Harboe, N., Ingild, A.: Scand. J. Immunol. 2, Suppl. 1, 161-164 (1973).

14. Lorenc-Kubis, I. in: Lectins, Biology, Biochemistry, Clinical Biochemistry, T.C. Bøg-Hansen and G.A. Spengler (Eds.) Vol. 3. W. de Gruyter, Berlin, New York 1983 pp. 379-386.

15. Lorenc-Kubis, I., Morawiecka, B., Bøg-Hansen, T.C. in: Lectins, Biology, Biochemistry, Clinical Biochemistry, T.C. Bøg-Hansen (Ed.) Vol. 2. 509-515 (1982)

MICROHETEROGENEITY OF ALPHA$_1$-ACID GLYCOPROTEIN, ALPHA$_1$-ANTICHYMOTRYPSIN AND ALPHA$_1$-ANTITRYPSIN IN RHEUMATOID ARTHRITIS.

A. Mackiewicz, T. Pawlowski, K. Wiktorowicz, S. Mackiewicz

Department of Immunology and Rheumatology, University School of Medicine, Szkolna 8/12, 60967 Poznan, Poland.

Alpha$_1$-acid glycoprotein (AGP), Alpha$_1$-antichymotrypsin (ACHT) and Alpha$_1$-antitrypsin (AAT) are members of the so-called acute phase proteins. Their concentration increases up to four times in patients' sera after infection, tissue injury, tissue necrosis or malignant neoplasia. All the three proteins contain carbohydrate moieties which represent 45%, 26% and 12% of the molecular weight of AGP, ACHT and AAT respectively. Glycoproteins are microheterogeneous (1-3) and show various microheterogeneity patters in different clinical states and pregnancy (4-6, 9-13).

Rheumatoid arthritis is a chronic inflammatory disease with periods of remissions and exacerbations. The evaluation of the patient with rheumatoid arthritis assessment of the disease activity, and the effectivness of the therapy should be judged on its ability to control this.

In the present work we decided to study the microheterogeneity of AGP, ACHT and AAT in various grades of rheumatoid arthritis activity and evaluate the clinical usefulness of microheterogeneity determination in the assessment of disease activity.

Lectins, Vol. V

Materials and methods

Sera from 48 patients fulfilling the criteria of the American Rheumatism Association for rheumatoid arthritis and sera from 12 healthy individuals were studies. For each patient the disease activity was assessed according to the modified multivariate analysis of Mallya and Mace (7) which is comprised of morning stiffness, pain scale, grip strength, articular index, hemoglobin concentration and erythrocyte sedimentation rate. On the basis of the global score in the analysis the patients were assigned to one of four grades of disease activity I - inactive (4 patients), II - mildly active (15 patients), III - moderately active (13 patients), and IV - severe (16 patients).

APG, ACHT, AAT and C-reactive protein (CRP) levels were measured by electroimmunoassay.

Affinity electrophoresis with Concanavalin A as a ligand was carried out according to Bøg-Hansen (8), as described in detail (9). Con A (Sigma, type IV, lot 97C-7620) in concentration of 40 μmol/l was incorporated in the first dimension gel. In order to dissolve the affinity precipitate an intermediate gel containing 7.5% of alpha-methyl-D-mannoside (Sigma) was included.

The relative concentration of AGP, ACHT and AAT variants was determined by planimetry and expressed as percentages.

The results of the AGP, ACHT, AAT and CRP serum levels as well as the relative amounts of AGP, ACHT and AAT variants observed in the various grades of disease activity, were compared statistically by a student t-test or a Cochran-Cox test (all values were normally distributed). Correlations between the relative amounts of the AGP variants and AGP as well as CRP serum levels, and correlations of serum CRP and AGP-variant 0 with rheumatoid arthritis activity grade were sought using Spearman's rank correlation coefficient.

Results

The levels of AGP, ACHT, AAT as well as CRP in the sera of the healthy individuals and in the sera of rheumatoid arthritis patients are shown in Table 1.

Alpha$_1$-acid glycoprotein: Con A affinity electrophoresis revealed three microheterogeneity variants of AGP (Fig. 1): AGP-variant 0 (non-reactive with Con A). AGP-variant 1 (weakly reactive with Con A), and AGP-variant 2 (strongly reactive with Con A). The relative amounts of the AGP variants in the healthy individuals and the rheumatoid arthritis patients are presented in Table 2.

Alpha$_1$-antichymotrypsin: In Con A affinity electrophoresis four variants were observed: ACHT-variant 0 (non-reactive with Con A),

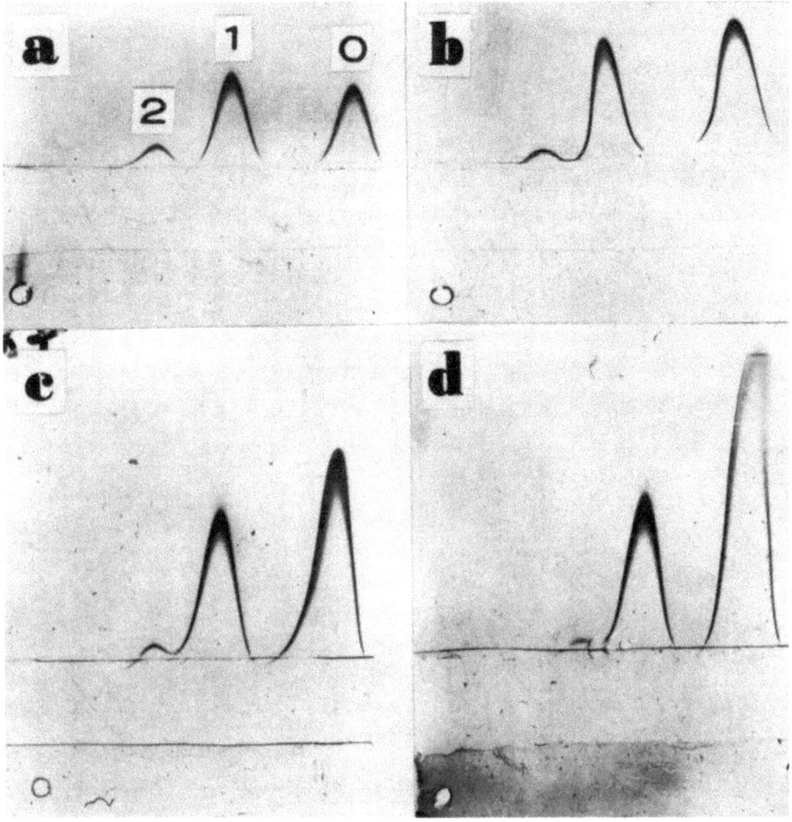

Fig. 1. Affinity electrophoresis with con A in the first dimension gel. The microheterogeneity pattern of alpha$_1$-acid glycoprotein observed in: a = healthy individuals, b = rheumatoid arthritis patients with activity grade II, c = grade III and d = grade IV. For experimental details see (8-11).

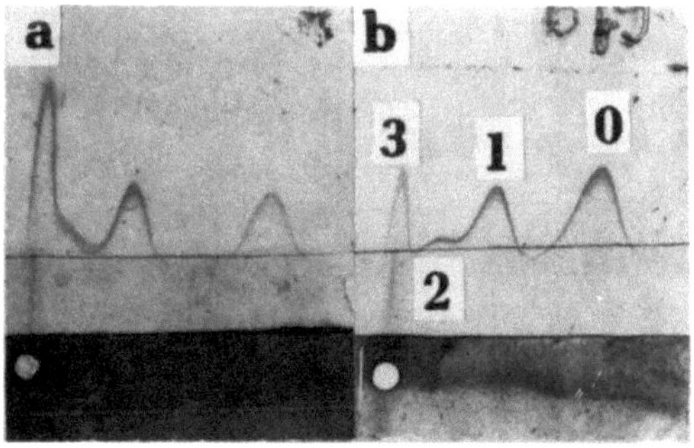

Fig. 2. The microheterogeneity pattern of alpha$_1$-antichymotrypsin observed in : a = healthy individuals and b = rheumatoid arthritis patients grade IV. Affinity electrophoresis with con A, cf. Fig.1.

TABLE 1

Alpha$_1$-acid glycoprotein, alpha$_1$-antichymotrypsin, Alpha$_1$-antitrypsin and C-reactive protein levels in sera of the healthy individuals and the rheumatoid arthritis patients.

	n	AGP	ACHT	AAT	CRP
Normal	12	425±135[b]	400±113	2361±578	1 - 1
Grade I	4	850±145	770±247	2785±445	4.8±4.7
Grade II	15	1120±342	775±238	2955±967	15 ± 14
Grade III	13	1498±382	1212±827	3440±800	42 ± 33
Grade IV	16	1652±43o	1500±970	4343±1141	64 ± 40

The protein level is given as mg/l ± SD; n = number of cases studied.

TABLE 2

The relative amounts of alpha$_1$-acid glycoprotein variants in sera of healthy individuals and rheumatoid arthritis patients

	n	Variant 0 %	Variant 1 %	Variant 2 %
Normal	12	43±4[x]	46±3	11±2
Grade I	4	44±4 a NS	47±2 a NS	9±3 a NS
Grade II	15	47±5 a p<0.05 b NS	43±4 a p<0.05 b p<0.05	10±4 a NS b NS
Grade III	13	52±5 a p<0.001 b p<0.001 c p<0.01	41±2 a p<0.05 b p<0.05 c NS	7±3 a p<0.05 b NS c p<0.05
Grade IV	16	59±8 a p<0.001 b p<0.001 c p<0.001 d p<0.05	36±7 a p<0.001 b p<0.001 c p<0.01 d p<0.05	5±3 a p<0.001 b p<0.01 c p<0.001 d NS

The relative amount is given in % ± SD with significance level, p. a = compared to normal; b = compared to rheumatoid arthritis grade I; c = to grade II; d = to grade III; NS = not significant = p>0.05.

ACHT-variant 1 (weakly reactive with Con A), ACHT-variant 2 (strongly reactive with Con A), and ACHT-variant 3 (precipitated in the first dimension gel, observed as an immunoprecipitate (Fig. 2) when sugar was included in the intermediate gel). The relative amounts of the variants specified above in the healthy individuals and the rheumatoid arthritis patients are shown in Table 3.

Alpha$_1$-antitrypsin: Affinity electrophoresis with Con A divided AAT into four variants (Fig. 3): two weakly reactive with Con A (AAT-variant 1a and 1b), one variant strongly reactive with Con A (ATT-variant 2), and one variant precipitating in the first dimension gel (AAT-variant 3). The relative amounts of the ATT-variants observed in the healthy individuals and the rheumatoid arthritis patients are presented in Table 3.

The relative amounts of the AGP-variants 0 and 1 showed significant correlation with the AGP serum level ($p<0.01$), while variant 2 did not. The coefficients of correlation for both variants were low (0.420 and 0.413 respectively). The correlation between the AGP-variant 0 and the CRP serum level was also significant ($p<0.001$ with a coefficient of 0.445). Comparison of the AGP-variant 0 and the CRP serum level with each rheumatoid arthritis activity grade showed a highly significant correlation (p 0.001 for both) with coefficients $r = 0.75$ and $r = 0.69$ respectively).

Discussion

The results of AGP, ACHT and AAT measurement in the rheumatoid arthritis patients sera indicated that the mean levels of these proteins increased in parallel to the increases in the rheumatoid arthritis activity. However, much overlap of the results was observed, and differences between the activity grades were not always statistically significant. The determination of serum levels of the inflammation markers is therefore not very useful for the physician in the assesment of rheumatoid arthritis activity.

Affinity electrophoresis with Con A as a ligand showed, for all the three glycoproteins, the same microheterogeneity pattern in normal serum as in serum from patients with inactive rheumatoid arthritis. The proportion of microheterogeneity variants of AGP, ACHT and AAT changed with the rheumatoid arthritis activity. The differences were most significant for AGP. For ACHT variations were less significant. The most stable microheterogeneity pattern was found for AAT, as only AAT-variant 2 and to some extend AAT-variant 3 changed significantly. The similarity of the microheterogeneity changes in the three glycoproteins studied suggest either the existence of the same glycosylation precessess in the liver, or else cellular uptake of certain microheterogeneity variants of AGP, ACHT and AAT in the course of rheumatoid arthritis.

Microheterogeneity composition of AGP observed in the sera of all the healthy individuals are almost identical to the normal patterns described by Hansen et al. (4). Moreover, our AGP patterns of rheumatoid arthritis patients are similar to those observed by the same authors in patients with inflammatory lung diseases. In contrast, the prevalance of glycosylated variants was observed by Nicollet and coworkers in patients with sepsis (5) and by the present authors in lupus patients with intercurrent infection (not published). The conflicting data suggest that different mechanisms of AGP glycosylation or cellular uptake may occur in various pathological states.

It is difficult to compare the ACHT microheterogeneity observed here with the results obtained by Bowen et al. (3). As they did not use sugar to dissolve the affinity precipitate. However, the ACHT pattern shown by Bowen et al. in rheumatoid arthritis patients is very similar to that observed in our study in rheumatoid arthritis activity grade IV, but is different from the normal pattern.

As the determination of serum CRP level is an accepted biochemical test for the assessment of the rheumatoid arthritis activity, the CRP concentrations were measured also in the sera of the patients studied here, in order to compare them with the relative amounts of AGP variants. The comparison showed highly significant correlation between the serum CRP level and the AGP-variant 0, and further comparison with the rheumatoid arthritis activity grades showed that the AGP-microheterogeneity variants are better indicators of the rheumatoid arthritis activity than CRP levels.

Table 3

The relative amount of alpha$_1$-antichymotrypsin and alpha$_1$-antitrypsin variants in sera of healthy individuals and rheumatoid arthritis patients.

	n	ACHT 0	ACHT 1	ACHT 2	ACHT 3	ATT 1a	ATT 1b	ATT 2	ATT 3
Normal	12	24±4	27±3	15±4	37±6	9±2	37±4	32∓4	22±3
Grade I	4	26±1	27±1	9±1	40±2	9±2	38±4	32±4	24±3
Grade II	15	26±6	31±5	17±6	26±8	9±2	38±9	29±6	24±10
Grade III	13	26±4	32±4	14±3	28±4	11±3	44±7	24±3	21±10
Grade IV	16	31±5	30±5	17±6	22±10	11±3	39±3	22±6	28±7

The relative amount is given in % ± SD.

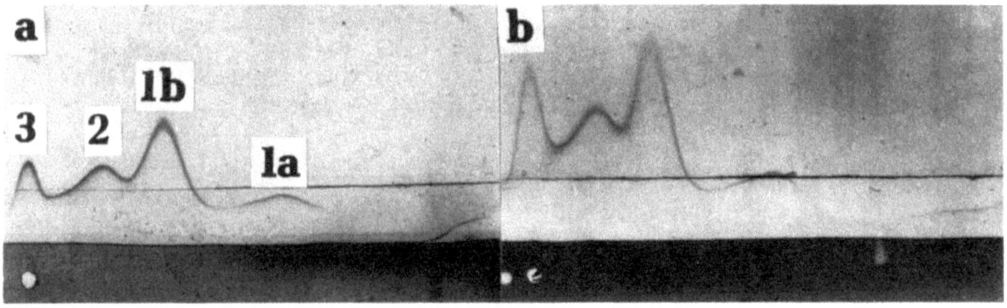

Fig. 3. The microheterogeneity pattern of alpha$_1$-antitrypsin observed in: a = healthy individuals and b = patients with rheumatoid arthritis grade IV (as Fig. 2).

Acknowledgement: This work was supported by a grant from the Polish Academy of Sciences (No 10.5). The authors are tankful to Ms Caroline Nielsen for typing the manuscript.

References

1. Mackiewicz, A., Wiktorowicz, K., Mackiewicz, S.: Arch. Immunol. Ther. Exp., *33*, 703-713 (1985)
2. Bayard, B., Kerchaert, J.-P., Laine, A., Hayem, A.: Eur. J. Biochem., *124*, 371-376 (1982)
3. Bowen, M., Raynes, J.G., Cooper, E.H.: In: Lectins, Biology, Biochemistry, Clinical Biochemistry Vol. 2 (T.C. Bøg-Hansen, ed.). Walter de Gruyter, Berlin, Nwe York, pp. 403-411 (1982)
4. Hansen, J.E., Larsen, V.A., Bøg-Hansen, T.C.: Clin. Chim. Acta *138*, 41-47 (1984)
5. Nicollet, J., Leberton, J.P., Fontaine, M., Hiron, M.: In: Lectin - Biology, Biochemistry, Clinical Biochemistry (T.C. Bøg-Hansen, ed.). Walter de Gruyter, Berlin, New York, pp. 413-421 (1982)
6. Wells, Ch., Bøg-Hansen, T.C., Cooper, E.H., Glass, M.R.: Clin. Chim. Acta *109*, 59-67 (1981)
7. Mallaya, R.K., Mace, B.E.W.: Rheumatol. Rehab. *20*, 14-20 (1981)
8. Bøg-Hansen, T.C.: Scand. J. Immunol. 4, Suppl. 2, 89-99 (1975)
9. Mackiewicz, A., Jakubek, P., Sajdak, S., Breborowicz, J.: Placenta *5*, 373-380 (1984).
10. Kerckaert, J.P., Bayard, B.: In: Lectins, Biology, Biochemistry, Clinical Biochemistry, Vol. 1 (T.C. Bøg-Hansen, ed.). W. de Gruyter, Berlin, New York, pp. 271-281 (1981)
11. Breborowicz, J., Mackiewicz, A.: ibid pp. 303-314 (1981)
12. Wells, C., Cooper, E.H., Glass, M.R., Bøg-Hansen, T.C.: ibid pp. 339-346 (1981)
13. Hansen, J.E.S., Larsen, V.A., Bøg-Hansen, T.C.: Clin. Chim. Acta *138*, 41-47 (1984)

DETERMINATION OF THE DISSOCIATION CONSTANT FOR THE HEPARIN FIBRONECTIN COMPLEX

M. Caron, R. Joubert

Laboratoire de Biochimie Fondamentale, U.F.R. Santé, Médecine, Biologie Humaine
93012 Bobigny Cedex , France

T.C. Bøg-Hansen

The Protein Laboratory, University of Copenhagen, DK-2200-Copenhagen, Denmark

A wide range of biological activities are now widely shown to depend on the interactions occuring between lectins and glycoconjugates (1-3). Although the knowledge about human lectins is still limited, such molecules are now obvious in mammalian body fluids : CRP, SAP, Cl_q, vitronectin, thrombospondin, Von Willebrand factor and especially fibronectin (3-6).

Fibronectin is a large, adhesive glycoprotein which is found in a number of locations, most notably on cell surfaces, in extracellular matrixes, and in blood. This glycoprotein is present in substantial amounts, i.e. up to 300 μg/ml, in human plasma. Fibronectin is a dimer with a polypeptide chain subunit about 220 000 dalton, covalently linked by a single disulfide bond near the carboxyl termini ; it contains distinct domains to mediate specific interactions with many cellular and molecular structures including heparin.

In the present study, we tested the reactivity of human plasma fibronectin with free heparin by Crossed Immuno-Affinoelectrophoresis (7), and we obtained informations concerning dissociation constant and state of heterogeneity of the complexes.

Lectins, Vol. V

Materials and Methods

Crossed immuno-affinoelectrophoresis in 1 % agarose (A 37, IBF, France) in 0.02 M Tris-barbital buffer (pH 8.6) containing 0.1 to 2 mg/ml of soluble heparin in the first dimension were performed according to the method of Bøg-Hansen (7). In the first dimension, 10 µl of normal human plasma was run at 10 V/cm, 15°C, for 60 min. Then, electrophoresis in the second dimension was performed at 5 V/cm for 16 hr in an antibody-containing gel. The concentration of anti-plasma fibronectin (Behring) was of 5 µl per ml of agarose gel. After electrophoresis the gels were washed several times with 0.15 M NaCl and distilled water, dried and stained. Determination of the dissociation constant was performed according to the method of Bøg-Hansen and Takeo (8).

Results

Dissociation constant

The protein precipitates obtained using heparin affinity crossed immuno-electrophoresis (Fig. 1) were different to those described commonly, when lectins are used in the first dimension. With increasing concentrations of heparin, plasma fibronectin was moving faster. It can be suggested that heparin transfered a negative charge to fibronectin by the complex formation. The complex was in equilibrium at low concentrations of heparin and we could measure the dissociation constant (Fig. 2), considering that fibronectin was moving faster according to the affinity between the glycosaminoglycan and the protein, from a general Takeo plot (8). We determined the dissociation constant to be : $6.10^4\ M^{-1}$. The comparison of this result with the high affinity determined by Scatchard analysis (9) (approximately 10^7 to $10^8\ M^{-1}$) suggest an influence of the electric field on the apparent affinity measured by affinity electrophoresis (10).

Heterogeneity

At low concentrations of heparin, there was some cross linking of fibronectin molecules by heparin molecules as a slight precipitate is seen in the first dimension already in Fig. 1 b,c. With a certain concentration of heparin (Fig. 1d) there was a distinct precipitate in the first dimension and the immunoprecipitate of fibronectin (or of complexes involving plasma fibronectin)

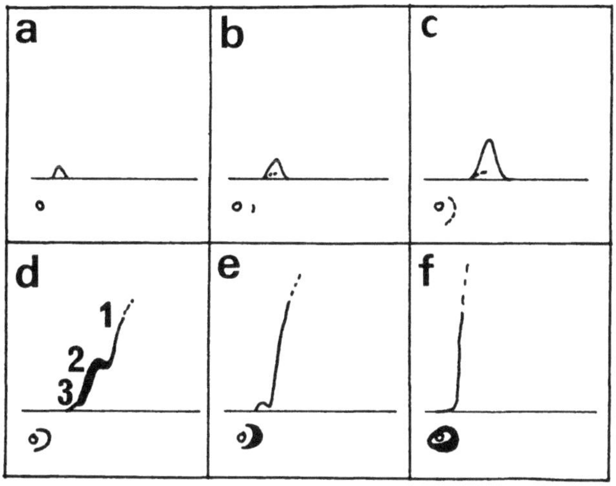

Fig. 1. Crossed immunoelectrophoresis of normal human plasma (10 µl) against antiserum to fibronectin in the second dimension. The first dimension gel contained (a) no heparin ; (b) 100, (c) 200, (d) 500, (e) 1000, (f) 2000 µg/ml heparin

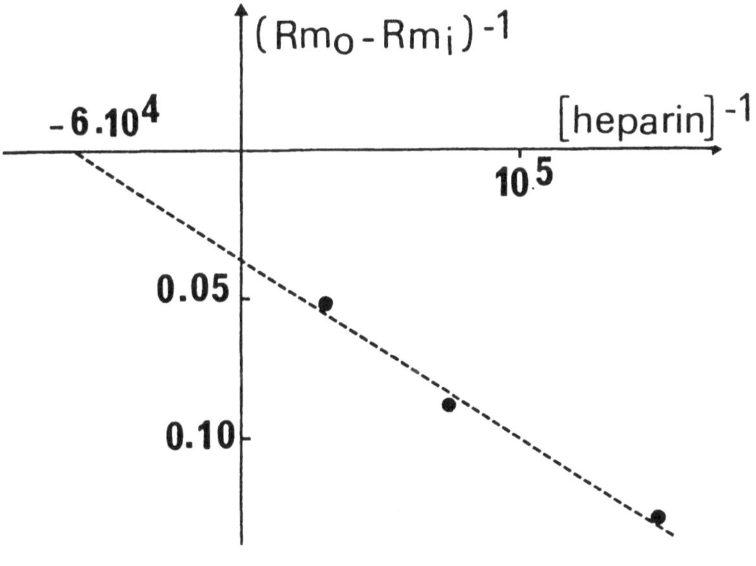

Fig. 2 Plots for determination of dissociation constant of heparin-fibronectin complex. Plotted according to equation

$$\frac{1}{R_{mo}-R_{mi}} = \frac{k}{R_{mo}-R_{mc}} + \frac{1}{C} + \frac{1}{R_{mo}-R_{mc}}$$

changed character completely. There appeared a multipeak precipitate with a very high fast peak 1 and a very pronounced, strongly stained, intermediate peak 2 and a small peak 3. This result indicates a heterogeneity of plasma fibronectin : some fibronectin reacted strongly with heparin (peak 1), some fibronectin reacted less strongly with heparin (peak 2), whereas some fibronectin (peak 3) did not complex with heparin at all. The comparison of the position of fibronectin without heparin (Fig. 1a) with the position of peak 3 suggests that this fibronectin did not react with heparin. With increasing amounts of heparin in the first dimension gel (Fig. 1e,f) there was an increased formation of fast moving complexes and of the precipitate in the first dimension gel. The complex was so stable that it was carried into the second dimension and it was clearly detectable before staining.

Discussion

Human plasma fibronectin was shown to be homogeneous on a normal immunoelectrophoretic system, but displayed a degree of heterogeneity when heparin was incorporated into the first dimension. With increasing concentrations of heparin the immunoprecipitate of fibronectin is changed by the transfer of negative charges and the formation of insoluble complexes. The change in the pattern of fibronectin observed with high concentrations of heparin, strongly suggest a heterogeneity of fibronectin. Since we did not use pooled serum but an individual sample the heterogeneity cannot be related to differences between individual plasma. It may be induced by participation of others plasma proteins, as for example fibrinogen. Although the data presented in this report do not allow to fully understand the implications of the heterogeneity of plasma fibronectin detected by Crossed Immuno-Affinoelectrophoresis, observations concerning the precipitation of this protein with free heparin (11) suggest that it is the main constituants of the observed precipitates. However, since a given fibronectin molecule can bind to different saccharides and glyconjugates molecules, a large number of reactions may interfere with its electrophoretic analysis. Finally, a full understanding of the multi-peak precipitates will require identification of all the plasma molecules with which plasma fibronectin actually associates.

References

1. Neufeld, E.F., and Ashwell; G. : The Biochemistry of Glycoproteins and Proteoglycans (Lennarz, W.J., ed.), pp. 242-266 (1980)

2. Barondes, S.H. : Sciences 223, 1259-1264 (1984)

3. Caron, M., Rougé, P., Deugnier, M.A. and Joubert, R. : Marker Proteins in Inflammation 2 (P. Arnaud, J. Bienvenu, P. Laurent, eds.). W. de Gruyter, p. 641-650 (1984)

4. Uhlenbruck, G., Karduck, D., Haupt, H. and Schwick, H.G. : Z. Immun. Forsch. 155, 262-266 (1979)

5. Hayman, E.G., Pierschbacher, M.D., Ohgren, Y. and Ruoslahti, E. : Proc. Natl. Acad. Sci. USA 80, 4003-4009 (1983)

6. Caron, M. and Joubert, R. : Marker Proteins in Inflammation 3, in press

7. Bøg-Hansen, T.C. : Anal. Biochem. 56, 480-488 (1973)

8. Bøg-Hansen, T.C. and Takeo, K. : Electrophoresis 1, 67-71 (1980)

9. Yamada, K.M., Kennedy, D.W., Kimata, K. and Pratt, R.M. : J. Biol. Chem. 255, 6055-6063 (1980)

10. Caron, M. Faure, A., Keros, R. and Cornillot, P. : Biochim. Biophys. Acta 491, 558-565 (1977)

11. Faure, A., Tepenier, D. and Caron, M. : Technologies de Purification des Protéines (A. Faure, C. Doinel, J.F. Stoltz, eds.), in press

A NEW QUANTITATIVE AND HIGHLY SPECIFIC ASSAY FOR LECTINBINDING ACTIVITY.

O. Vang, K. Pii Larsen and T.C. Bøg-Hansen

The Protein Laboratory, University of Copenhagne

Determination of lectin may be performed in several ways, the most widely used being the rapid semi quantitative haemagglutination test, see for instance (1). Other systems, allow for a highly quantitative lectin determination based on antibody recognition of lectin such as in single radial immunodiffusion (2), radio immuno assay (RIA) (3), and enzyme-linked immunosorbent assay (ELISA) (4).

Another quantitative assay for lectin makes use of sugar immobilized to agarose beads (5). The amount of lectin bound to the agarose beads may be estimated with radioactively labelled antilectin antibodies. In this case only protein with lectin activity is detected and recognized by the anti-lectin. Therefore this lectin assay has a high degree of specificity. We have developed further a lectin assay that also makes use of simultaneous lectin and antibody activity in a new design to combine the double specificity and enhance the sensitivity. Our assay (ELLA) which was original developed to quantitate glycoproteins with enzyme-conjugated lectins (6, 7).

For our studies we used the misteltoe (Viscum album) lectins. At least two lectins are found in the extract from misteltoe with galactose and N-acetyl-galactosamine specificity (8, 9). For measuring the lectins we have combinde a modification of the ELLA technique (6, 7) with the immobilized sugar method (5) in a microtiter plate coated with glycoprotein (asialoorosomucoid and asialofetuin). When lectin is added as the second layer, active lectin is bound to the immobolized glycoconjugate and then visualised with anti-misteltoe lectin and a secondary antibody conjugated to peroxidase. In this way we base our measurements on the carbodhydrate specific binding of the misteltoe lectins.

MATERIALS AND METHODS

Enzyme-linked lectin assay (ELLA):
As the 1st layer, the microtiterplate (Nunc-Immuno Plate II, Code: 4-42404 from TekNunc, Roskilde, Denmark) was coated with glycoprotein (asialoorosomucoid or asialofetuin) in PBS in concentrations of 0.1 mg/ml, 0.05 mg/ml, 0.025 mg/ml, 0.0125 mg/ml. 100 µl was added to each well and the plate was incubated at room temperature overnight.

The next day, the plate was washed three times in PBS containing 0.1% Tween 20 (PBS-T). Then remaining binding sites on the plate were blocked with PBS-T and 1% bovine serum albumin (PBS-T-BSA) for 1 h at 37°C. Following the blocking a second washing procedure was performed as above.

The second layer consisted of the crude extract or the purified lectins (100 µl) in dilutions from 0.0095 µg/ml to 20 µg/ml in PBS-T-BSA. After 2 h incubation at 37°C washing was done as above.

The antibody incubation buffer used in 3rd layer was PBS-T-BSA with 1% v/v Triton X-100 (PBS-T-BSA-TX) to prevent unspecific binding. The antibody incubation was performed at 37°C for 2 h. In inhibition experiments anti-lectin was mixed with lactose (to final concentrations of 6.25 mM, 12.5 mM, 50 mM and 200 mM) in PBS-T-BSA-TX.

After one more washing step, the 4th layer, Goat-anti-rabbit-IgG antibodies conjugated to horse radish peroxidase (Kem-En-Tec, Hellerup, Denmark) were added diluted 1:2500 v/v in PBS-T. The incubation was performed for 1 h at 37°C. The the plate was washed three times with PBS-T.

Following the last wash, 100 µl of a freshly prepared substrate solution was added to each well and the plate incubated in the dark for 15 min or until a suitable colour was developed. The substrate solution was 8 mg of 1.2-phenylenediamine dihyrochloride (OPD tablets, Kem-En-Tec, Denmark) dissolved in 20 ml 0.1 M citric acid-phosphate pH 5.0 to which was added 5 µl 30% v/v H_2O_2 immediately before use. The colour reaction was stopped by adding 150 µl 1M H_2SO_4 to each well, and the colour at 490 nm was measured in an ELISA-reader (Minireader Mr 590, Dynatech Microelisa).

Sialidase (C. perfingens, type V, Sigma) was conjugated to Mini-Leak-gel (Kem-En-Tec, Hellerup, Denmark) by incubtaion of 15 mg sialidase in 8 ml 1 M potassium phosphate at pH 7.3 with 1.5 g Mini-Leak. After coupling at room temperature for 48 h, 82% of protein was bound and the gel was washed with 1 M NaCl (3 times 10 ml) to remove the unbound enzyme. The gel was then blocked with 1 M Tris-HCl (pH 8.6) for 2 h. The sialidase gel was used after washing with 1 M NaCl (3 times 10 ml).

Orosomocuid (α_1-acid glycoprotein, AGP) was a gift from Plasma Derivatives Laboratory (American Red Cross, Bethesda) and Fetuin type III was from Sigma. 8 ml glycoprotein (1.0 mg/ml) in PBS (pH 7.2) was mixed with the sialidase gel and incubated for 1 h at 37°C. The gel was then separated from the glycoprotein by centrifugation. The efficiency of the sialidase treatment was tested by crossed immunoelectrophoresis.

Misteltoe was harvested on apple trees, november 1984 in France. Steams and leaves were dried and broken in a blender (for safety reasons performed in a fume cupboard) and extracted in PBS (5 mg/g plant material) overnight at 4 °C. The extract was filtered through cheese cloth and centrifuged at 10.000 g for 10 minutes (crude extract). The lectin fraction was precipitated between 30 % and 70 % of $(NH_4)_2SO_4$, dialysed for 3 days with 3 changes of PBS buffer. Lectins were then affinity-purified on lactose Mini-Leak gel: The dialysed $(NH_4)_2SO_4$ precipitate was applied on the column and after a slight wash with PBS, lectins were eluted with 4 % w/v lactose in PBS. This eluate (unseparated lectins) contained two lectins and were separated futher on an affinity-column containing N-acetyl-galactosamine-Mini-Leak. From this column one lectin was eluted by 4% lactose (the Gal-lectin) and the other with 0.1 M N-acetyl-galactoseamine (the GalNAc-lectin). After removing the sugar by dialyses the lectins were used for ELLA.

Antibodies against useparated lectins were raised in rabbits and purified by salting out and ionexchange (10).

RESULTS

Optimizing conditions: Several system parameters were tested to obtain the highest activity with the lowest possible background: The concentration of each layer and the incubation time.

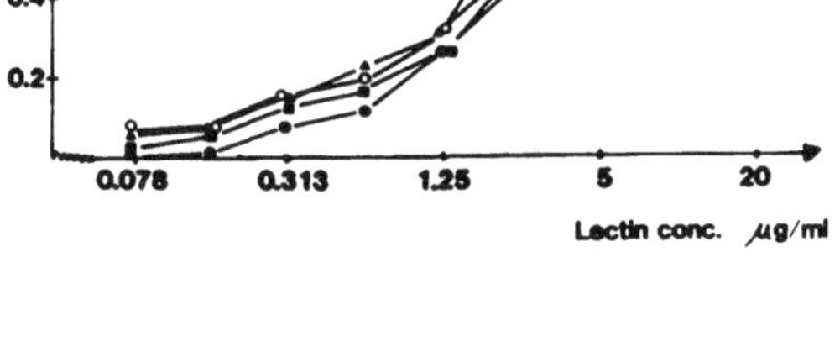

Fig. 1: Enzyme Linked Lectin Assay (ELLA): The binding of the lectins to asialoorosomucoid as a function of the concentration of the unseparated lectins in 2nd layer for different concentrations of glycoprotein in 1st layer. Wells were coated with 100µl of asialoorosomucoid in PBS: ▲= 0.0125 mg/ml, ●= 0.025 mg/ml, ■= 0.05 mg/ml, ○= 0.1 mg/ml. The dilution of anti-lectin and HRP-secondary antibody were 1:20,000 and 1:25,000 respectively.

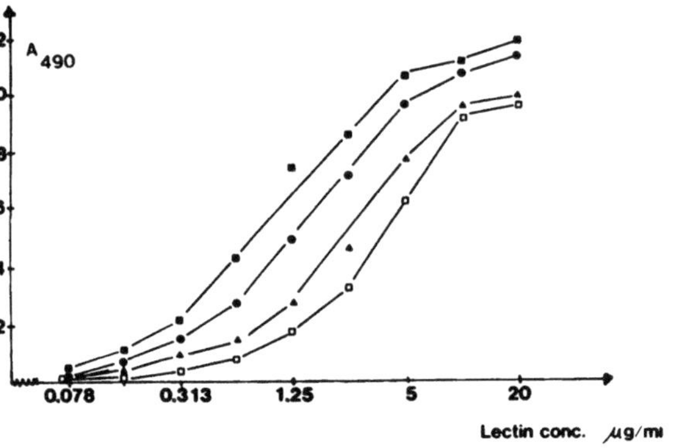

Fig. 2: Enzyme Linked Lectin Assay (ELLA): Binding of the unseparated misteltoe lectins as a function of the lectin concentration at different times in 2nd layer. Wells were coated with 100µl of 0.1 mg/ml asialoorosomucoid in PBS and incubated with lectin for different times at 37°C: □ = 0.5 h, ▲ = 1 h, ● = 2 h, ■ = 3 h. The dilution of anti-lectin and HRP-secondary antibody were 1:20,000 and 1:2,500 respectively.

Fig. 1 shows the binding of the unseparated lectins to asialoorosomucoid as a function of the lectin concentration at different glycoprotein concentration in 1st layer. The binding activity of the lectin does not increase with increased amount of glycoprotein indicating that the system is saturated with 0.025 mg/ml asialoorosomucoid in the first layer. To prevent background staining, blocking with 1 % BSA after first layer was found to be critical at low glycoprotein concentration. To prevent background staining, the wells were coated first with 0.1 mg/ml of the glycoprotein in the first layer and then the blocking step with 1 % BSA was performed.

The time dependent binding of unseparated misteltoe lectins to asialoorosomucoid in the second layer is shown in Fig. 2. When the incubation time was increased from 0.5 to 3 h the amount of bound lectin increased 2.6 times (at the lectin concentration 2.5 g/ml) and the sensitivity increased ten-fold, while the background staining did not increase appreciable.

The optimally balanced concentration of the third layer was found by applying the rabbit anti-lectin in different dilution (1:5.000-1:50.000) and 1:20.000 was found to be best.

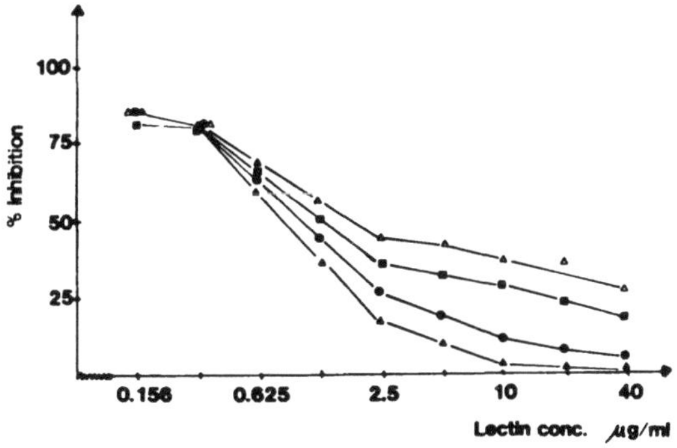

Fig. 3: Enzyme Linked Lectin Assay (ELLA): Lactose inhibition of the lectin binding to asialoorosomucoid. Wells were coated with 100µl of 0.1 mg/ml asialoorosomucoid in PBS and incubated with unseparated lectins as 2nd layer. Anti-lectin in 3rd layer (1:20,000) was mixed with different amounts of lactose: ▲ = 6.25 mM, ● = 12.5 mM, ■ = 50 mM, △ = 200mM.

Inhibition with sugar: The inhibitory effect of lactose on the binding to asialoorosomucoid is shown in Fig. 3, where the percent inhibition is calculated as (11):

$$\% \text{ inhibtion} = \left(1 - \frac{\text{A490 (lectin with inhibitor)}}{\text{A490 (lectin without inhibitor)}}\right) \text{X } 100$$

Complete inhibition was not seen at a lectin concentration of 0.156 µg/ml with 200 mM lactose.

The measurement of lectin concentration in a crude extract: The amount of lectin in crude extracts was measured by adding the purified unseparated lectin in increasing amounts as an internal standard (Fig. 4).

Binding of two misteltoe lectins by two different glycoproteins: The unseparated lectins and the two separated lectins were tested with asialoorosomucoid or asialofetuin in the 1st layer. Fig. 5. shows that asialofetuin binds both lectins better than asialoorosomucoid and that the Gal-lectin binds both the asialoorsomucoid and aialofetuin better than the GalNAc-lectin. The unseparated lectin samples shows an intermediate degree of binding.

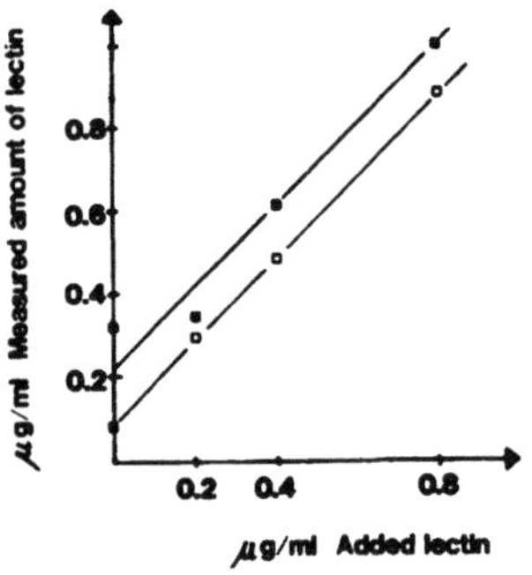

Fig. 4: Enzyme Linked Lectin Assay (ELLA): Measurement of the lectin concentration in two crude extracts. Wells were coated with 100µl of 0.1 mg/ml asialoorosomucoid in PBS. As 2nd layer 50µl of two crude extracts were mixed with 50µl of purified unseparated lectins as an internal standard. The total lectin concentration was calculated from an standard curve with purified lectins.

DISCUSSION

The enzyme-linked immunosorbent assay (ELISA) has been a standard method for quantitating lectins (12). In the enzyme-linked lectin assay (ELLA) described here, only proteins with glycoprotein binding activity are detected, whereas the conventional ELISA systems do not discriminate between active and inactive lectins, but measure all molecules. Our system works with polyspecific polyclonal antibodies and shows that it is not necessary neither to use monoclonal nor monospecific polyclonal antibodies as the third layer. It has been shown that the misteltoe lectin binds some IgA and IgG molecules (13), but we have not found any contribution to the background due to this binding at the low anti-lectin antibody concentrations used here.

The incomplete inhibition by lactose of the binding to asialoorosomucoid may be caused by the high affinity constant of the lectin-glycoprotein system. Another explanation is that the lectin may be bound with more than one binding site to asialoorosomucoid. In Fig. 4 the two sets of points are distributed around two straight lines with slope = 1. This is expected after addition of increasing amounts of the internal standard, the purified **lectins**. Therefore

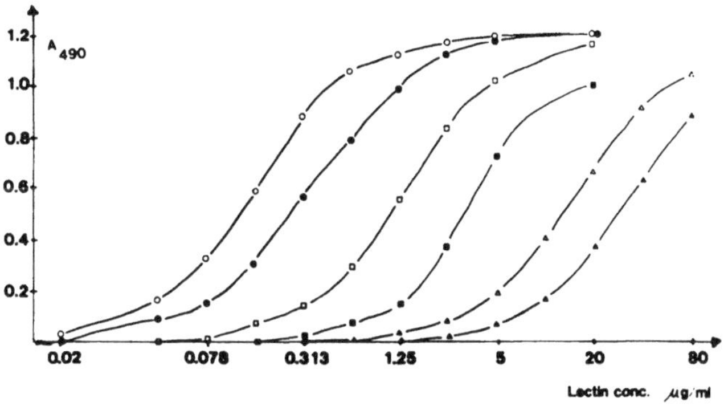

Fig. 5: Enzyme Linked Lectin Assay (ELLA): Binding intensities of the unseparated lectins and the two separated lectins on asialofetuin and asialoorosomucoid as a function of the lectin concentration. The wells were coated with 100µl asialofetuin (open symbol) and asialoorosomucoid (0.1 mg/ml) (filled symbol). ● = Gal-lectin, ▲ = GalNAc-lectin, ■ = unseparated lectins.

we have established that it is possible to measure accurately the content of active lectin in crude misteltoe extract by the ELLA method. Thus, this method constitutes a good assay for the lectin during purification procedures.

Compared with ELISA our test as presented here is less sensitive. In the experiment with the Gal-lectin bound to asialofetuin (Fig. 5) we found a sensitivity as low as 50 ng/ml or 50 times less sensitive than reported with conventional ELISA measuring misteltoe lectin (12). There is no reason to believe, however, that the ELLA system has generally a lower sensitivity than the ELISA system, and further experimentation will show wether our test can be developed to a higher paragraf level of sensitivity.

We observed a difference in binding to two different glycoproteins. The main qualitative difference in the carbohydrate parts of fetuin and orosomucoid is the O-linked carbohydrate structure present in fetuin (14, 15). This difference may be responsible for the differences observed. Consequently the ELLA-technique may allow detection of small structural differences of glycans.

This assay was developed using specific glycoproteins and lectins, but in priciple, the assay could be used for any carbohydrate-binding protein.

REFERENCES

1. Sandhu, R.S., Arora, J.S. Chopra, S.K. and Kamboj, S.S.: (These procedings) and Lis, H. and Sharon, N.: Ann. Rev. Biochem. 42, 541-574.
2. Rouge, P. and Pere, D.: Lectins - Biology, Biochemistry, Clinical Biochemistry (Bøg-Hansen, T.C. ed.) Vol 2, pp137-150, Walter de Gruyter, Berlin, New York 1982.
3. Talbot, C.F. and Etzler, M.: Plant. Physiol. 61, 847-850, 1978.
4. Causse, H. Lemoine, A. and Rouge, P.: These procedings, Vol 5.
5. Howard, J. and Shannon, L.: Anal. Biochem. 79, 234-239, 1977.
6. McCoy, J.P., Varani, J. and Goldstein, I.J.: Anal. Biochem. 130, 437-444, 1983.
7. McCoy, J.P., Varani, J. and Goldstein, I.J.: Exp. Cell Res. 151, 96-103, 1984.
8. Samtleben, R., Kiefer, M. and Luther, P.: Lectins - Biology, Biochemistry, Clinical Biochemistry (Bøg-Hansen, T.C.,ed.) Vol. 4, pp 617-626, Walter de Gruyter, Berlin, New tork 1985.
9. Franz, H., Ziska, P. and Kindt, A.: Biochem. J. 195, 481-484, 1981.
10. Harboe, N. and Ingild, A.: A Manual of Quantitative Immunoelectrophoresis (Axelsen, N.H., Krøll, J. and Weeke, B., ed.), pp 161-164, Universitetsforlaget, Oslo, 1973.
11. Engwall, E. and Perlmann, P.: J. Immunol. 109, 129-135, 1972.
12. Ziska, P. and Franz, H.: Lectins - Biology, Biochemistry and Clinical Biochemistry (Bøg-Hansen, T.C., ed.) Vol 4 pp 473-480, Walter de Gruyter, Berlin, New York, 1985.
13. Ziska, P. and Franz, H.: Experimenta 35, 161, 1979.
14. Katagiri, Y.U. and Ochiai, H.: Anal. Biochem. 147, 222-229, 1985
15. Fournet, B., Montreuil, J., Strecker, G., Dorland, L., Haverkamp, J., Vliegenhart, J.F.G., Binette, J. and Schmid, K.: Biochemistry 17, 5206-5214.

APPENDIX

A REPORT ON THE STANDARDIZATION OF LECTINS

M. Caron

Laboratoire de Biochimie, U.F.R. Santé, Médecine, Biologie Humaine, F-93012 Bobigny Cedex, France

A. Faure

Centre National de Transfusion Sanguine, F-91943 Les Ulis Cedex, France

Introduction

Results obtained with Con A samples distributed by the IUIS/WHO Subcommittee on Standardization of Lectins are summarized

More informations about standardization are available in references 1,2,3 and in the present volume.

Experimental

1. Weights and solubilities of lyophilized samples were determined. Most samples were soluble, but sample C could not be dissolved completely and samples E and O were poorly soluble.
2. Protein contents have been determined by measuring the absorption at 280 nm assuming an ε of II.O and compared with the theorical concentration (Table 1)
3. Ultraviolet absorbance spectra were recorded between 260 and 300 nm using a temperature-controlled Cary 219 spectrophotometer (4). The samples were filtered on Falcon sterile 0.2 µm in order to remove aggregates and insoluble sample was ommited. No significant differences were noted between the spectra.
4. Electrophoresis in polyacrylamide gradient gel were performed in an exponential 4-30 % gradient in 0.09 M Tris, o.08 M boric acid, 0.003 M Na_2EDTA, pH 8.4 (5). Differences were shown on the densitograms of electrophoresis patterns (Fig. 1) : one or serveral peaks with a molecular weight corresponding to the theorical molecular weight of Con A appear for several samples. The presence of high molecular complexes of this lectin was apparent. Samples 3, 5, 8, E, K, N, O contained mostly such complexes.
5. Hemagglutination activity was determined using A^+ red blood cells. A Technicon Autonanalyser was used for the evaluation of agglutination after contact with Con A (6). The concentration of hemoglobin after hemolysis of the non-agglutinated erythrocytes was calculated by densitometry at 440 nm. The comparison of the optical density (expressed as the hight of the peak measured on the printer) enabled us to compare the activity of the different Con A samples in the same conditions (Table 2).

Lectins, Vol. V

TABLE 1

Protein content

Sample	Theorical mg/ml	Measured* mg/ml	Measured/Theorical
A	4.0	2.1	0.52
	2.0	1.4	0.70
C	4.0	0.6	0.15
	2.0	0.4	0.20
E	0.72	0.79	1.09
K	4.0	3.4	0.85
	2.0	1.6	0.80
N	4.0	4.3	1.07
	2.0	2.0	1.00
O	4.0	2.4	0.60
	2.0	2.2	1.10
1	2.5	2.8	1.12
	1.25	1.28	1.02
3	1.8	2.5	1.13
	0.9	1.2	1.33
4	2.0	1.7	0.85
5**		0.48***	
6	4.0	2.9	0.73
	2.0	1.6	0.80
8	1.3	1.4	1.07
9	5.7	1.0	0.18
10	4.0	3.9	0.98
	1.0	0.9	0.90

* Measured by spectroscopy at 280 nm
** precipited
***dilution 1/10

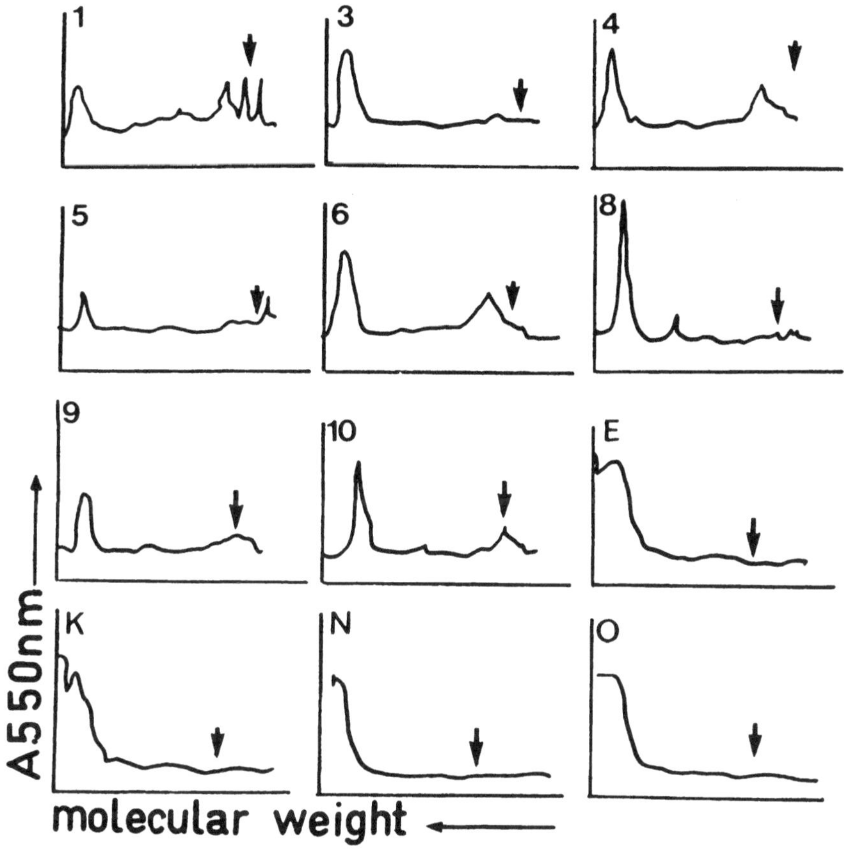

Fig. 1: Densitograms of gradient PAGE electrophoresis patterns of samples of Con A. Arrows indicate the theorical molecular weight of Con A.

In conclusion, these analysis are in accordance with those previously reported (3). The two main reasons of the differences detected between the samples seem to be (i) the contaminating impurities of several samples and (ii) the denaturation of Con A molecules which may be responsible for the presence of high molecular weights aggregates. As these aggregates may have unwanted biological or chemical effects in various experimental or diagnostic systems, our results support the need for more exactly defined lectin preparations.

TABLE 2

Evaluation of haemagglutination activity

Sample*	Hight of the peak (nm)
1	153
2	113
3	148
4	148
5	112
6	141
7	136
8	144
9	84
10	145

* 8.10^{-3} mg/ml in NaCl 0.15 M, BSA 0.5 %

References

1. T.C. Bøg-Hansen, Breborowicz, J. and Franz, H. : Lectins, Biology, Biochemistry, Clinical Biochemistry (T.C. Bøg-Hansen, ed.) Vol. 2. pp. 791-793, W. de Gruyter, Berlin 1982

2. T.C. Bøg-Hansen : Lectins, Biology, Biochemistry, Clinical Biochemistry (T.C. Bøg-Hansen and G.A. Spengler, eds.) Vol. 3, pp. 699-700, W. de Gruyter Berlin 1983

3. J. Breborowicz, Franz, H., Caron, M., Faure, A., Gryska, K., Kint, J.A., Sabbe, L.M.J., Kohn, J., Lorenc-Kubis, I., Rudiger, H., Takeo, K., Toftager-Larsen, K., Wu, M. and Bøg-Hansen, T.C. : Lectins, Biology, Biochemistry, Clinical Biochemistry (T.C. Bøg-Hansen and J. Breborowicz, eds.) vol. 4, pp. 637-647, W. de Gruyter, Berlin 1985

4. M. Caron, Ohanessian, J., Becquart, J. and Gillier-Pandraud, H. : Biochim. Biophys. Acta 717, 432-438 (1982)

5. M. Caron and Faure, A. : Electrophoresis 4, 257-258 (1983)

6. M. Caron, Faure, A., Ohanessian, J., Delrieu, J. : Arch. Int. Physiol. Biochim. 93, 1-4 (1985)

CHARACTERIZATION OF DIFFERENT CONA PREPARATIONS BY MEANS OF THEIR HISTAMINE RELEASING ACTIVITY FROM HUMAN BASOPHILIC GRANULOCYTES.

I. Sehrt and P. Luther

Research Institute of Lung Diseases and Tuberculosis, Berlin, GDR

The lectin Concanavalin A (ConA) from jack bean (Canavalia ensiformis) has been shown to induce histamine release from the basophils or mast cells of several mammalian species, including human basophilic leukocytes (1 - 4). There are a great deal of similarities between immunologically and ConA induced histamine release: both reactions are noncytotoxic, require Ca^{2+}, are temperature dependent and are inhibited by the same compounds as β adrenergic agents and methylxanthines (2). Therefore, the histamine releasing property of ConA is very attractive for immunologists for a variety of studies including testing of potential inhibitors for pharmacologic mediators or for investigations on receptor-receptor-interactions on cell membranes.

We chose the histamine release method for the characterization of different ConA preparations provided by commercial companies and research laboratories to the International Working Party on Standardization of Lectins for Diagnosis.

Lectins, Vol. V

MATERIALS AND METHODS

The experimental procedure performed was described in detail previously (5). In brief, blood was drawn by venipuncture and allowed to sediment in a mixture of dextran, glucose and heparin. The leukocyte rich plasma was withdrawn after 45 - 60 min. The cells were washed twice with tris (hydroxymethylaminomethan) buffer (0.025 M), containing (0.12 M) NaCl, (0.005 M) KCl, and human serum albumin (0.03 mg/ml), pH 7.6. The washed cells were finally resuspended in tris buffer containing the above mentioned constituents together with calcium and magnesium ions (0.0006 and 0.001 M, respectively). Histamine release investigations were carried out in a total volume of 2.4 ml. The cells were incubated for 60 min at 37°C following addition of ConA (final concentration of the 14 different samples: 4.2 ug/ml in 16 series and 42 ug/ml in 8 series of trials, respectively). After separation of the cells by centrifugation histamine released into the supernatant was assayed spectrofluorometrically.

The amount of histamine released (HR) was expressed as percent of total histamine content of the cells after correction for spontaneous release (recorded in the absence of lectin). A positive response was defined as a release of more than 10 %.

RESULTS

ConA induced histamine release (HR) from basophilic leukocytes was examined in a total of 24 individuals, including 7 donors with known allergy (Table 1).

The first series of experiments the ConA preparations to be tested were carried out with cells from 16 individuals. A final lectin concentrations of 4 ug/ml was used as it was shown to give maximum histamine release in previous investigations. Since there are conflicting reports in the literature concerning the amount of histamine secreted in normal and allergic individuals, cells from 9 normal and 7 allergic donors were included (Table 2).

TABLE 1.

Number, age, sex and state of the donors.

	Number	Age (years)		Sex		Non allergic	Allergic
		mean	range	m	f		
Series 1	16	41	18-59	3	13	9	7
Series 2	8	38	22-47	2	6	8	-
Total	24	40	18-59	5	19	17	7

TABLE 2

Degree of the histamine release, HR, of leukocytes from 16 individuals induced by 14 different ConA preparations.

Donor		ConA preparation													
		1	3	4	6	7	8	9	10	A	C	E	K	N	O
		% HR													
Hi	N	0	1	0	4	9	12	2	0	1	9	8	5	1	3
Si	N	2	3	2	5	0	11	1	10	2	6	5	9	5	2
La	N	7	6	18	7	14	7	5	9	10	6	13	4	11	16
Ne	A	3	-	7	6	5	-	10	6	9	5	-	2	1	3
Uh	A	7	-	11	5	9	-	6	7	10	5	-	5	5	8
Re	N	18	21	20	8	21	21	14	8	16	13	11	18	20	23
De	N	13	23	12	17	21	13	11	20	15	13	16	16	21	24
Ma	N	10	16	12	8	22	10	19	26	8	21	11	18	21	22
Gr	A	23	-	28	20	15	-	9	25	10	8	-	19	26	22
Sl	A	29	-	26	25	13	-	12	25	8	9	-	13	24	17
Wi	N	21	34	15	29	41	24	27	44	17	29	15	34	30	23
We	N	22	-	12	26	35	-	22	33	10	22	-	21	29	22
Pr	N	-	39	-	26	28	-	20	30	17	23	37	-	32	40
Se	A	21	34	33	28	22	18	27	41	26	15	38	24	34	34
Sv	A	26	-	66	55	26	-	34	57	37	19	-	23	52	-
Pa	A	36	39	28	26	37	-	20	41	39	24	37	-	32	36

NA = normal individuals. A = allergic individuals. Final lectin concentration: 4.2 µg/ml. The differences in mean histamine release, HR, between the three donors groups are signigicant (U-test, $p<0.01$).

As seen in Table 2 significant donor dependent differences in the magnitude of histamine secretion were observed. Five out of the 16 individuals did not respond significantly to the ConA samples tested at a concentration of 4 ug/ml. There was only in some cases a very weak reaction, but mostly no release at all. The mean value of histamine release was less than 10 % (6.0 ± 4.0 %).

The remaining 11 individuals (70%) did respond to ConA challenge. They seem to segregate into two groups: those which respond moderately with an average release of 17.1 ± 5.9 % (3 nonallergic and 2 allergic donors) and those predominantly showing a good responsiveness with a mean release value of 28.9 ± 9.1 % (3 normals and 3 allergics). There was a tendency for higher values with cells from allergic donors, but this difference was statistically not significant.

Extremely high values for histamine release (more than 50 %) were measured in an allergic individual with the ConA samples 4, 6, 10 and N.

The average release rates induced by the various ConA preparations are listed in Table 3, including the experiments with the two groups of donors, whose cells responded to ConA challenge. As to be seen all ConA samples tested are potent releasing agents. The mean values varied between 17 and 32 %.

Based on the measured activities the ConA samples can be divided into 3 groups:

Group 1 with the lowest activity (17 - 20 %): 8, C, A, 9

Group 2 with intermediate activity (20 - 26 %): K, 1, E, 6, 4, 7

Group 3 with the highest activity (> 26%): O, N, 3, 10

The differences in releasing activity between the various ConA preparations were tested by U-test and found to be significant between most of the samples belonging to group 1 and 3 (Table 3).

In an further series of 8 experiments we examined the response of basophilic leukocytes to supraoptimum amounts of ConA, using a final lectin concentration of 42 ug/ml (Table 4). Contrary to the results of the investigations with maximum ConA concentration for histamine release the basophils of all individuals responded similarly and no significant donor dependent differences were observed. With the majority of ConA samples (1, 4, 8, A, E, K, N) no activation of the cells happened (mean value: 5.4 ± 3.4): with the ConA

TABLE 3

Classification of ConA samples by histamine release (HR). Average histamine releasing activity of 14 different ConA preparations using maximum lectin concentration.

ConA sample	Number of individuals tested	HR (%) ($\bar{x} \pm s$)	U-test
1	10	21.9± 7.4	nsd
3	7	29.4± 9.3	>9;>C p<0.05
4	10	25.2±16.3	nsd
6	11	24.4±12.6	nsd
7	11	25.5± 8.9	nsd
8	5	17.2± 5.7	<10; <N p<0.05
9	11	19,5± 7.7	<3; <10; <N p<0.05
10	11	31.8±13.4	>8; >9; >A; >K p<0.05; >C p<0.01
A	11	18.4±11.0	<N p<0.05; <10 p<0.05
C	11	17.8± 6.7	<3; <0 p<0.05 <10; <N p<0.01
E	7	23.6±13.0	nsd
K	9	20.7± 6.0	<10; <N p<0.05
N	11	29.2± 9.0	>8; >9; >A; >K p<0.05 >C p<0.01
O	10	26.3± 7.5	>C p<0.05

The conA concentration was 4 μg/ml. nsd = no significant difference.

preparations 3, 6, 10 and 0 only a very poor response was registered (mean value: 13.0 ± 4.4 %). However the samples 7, 9 and C did provoke a significant response in the basophils of all individuals tested. The mean values were 30.2 ± 5.0 for sample 7, 27.0 ± 10.4 for sample 9 and 28.5 ± 9.4 for sample C (Table 4).

DISCUSSION

The present studies confirm that human basophilic leukocytes activated to release histamine by challenging with Con A may differ in their releasing activity (3, 6, 7). In our investigations 30 % of the donors did not respond significantly to ConA at the concentration used (4 ug/ml), and a considerable difference in the magnitude of secretion between the responding cells of the remaining 70 % of individuals were observed. The reason is not clear. It has been suggested that ConA acts by binding to pyranosyl type sugar moieties of the Fc part of the IgE molecule located close to the epsilon-one determinant within the C 2 domain in basophilic leukocytes (8), thereby inducing crosslinking of IgE. Differences in the accessibility of this ConA binding site could be one possible explanation for the wide variation in the response of cells to ConA. But in addition to the IgE-mediated mode of action a crosslinking of sites other than cell-fixed IgE cannot be excluded as a direct interaction with the basophil cell surface.

Siraganian found a better response to ConA with cells from allergic donors (2), other authors could not confirm this finding (6, 9). Only one of our allergic patients showed a very high histamine release of over 50 %. The other allergic as well as nonallergic individuals showed no responsiveness to ConA or a moderate one. On the whole significant difference between the cells of nonallergic and allergic donors were not observed.

The concentration for maximum histamine release (about 4 ug/ml) was in the same range as found by Bergstand et al. (7) (3.3 to 10 ug/ml) and was only slightly higher than that registered by Siraganian and Siraganian (2) (1 - 3 ug/ml). All the various ConA samples studied - 1, 3, 4, 6, 7, 8, 9, 10, A, C, E, K, N, O - revealed histamine releasing activity using this ConA concentration. But a variation in the amount of histamine released was observed. There may be responsible intrinsic properties of the ConA molecule as denaturation, loss at metal ions, aggregation and fragmentation (see the 3. Report of the International Working Party on Standardization of Lectins for Diganosis), although effects of contaminating impurities and additives cannot completely be excluded. The variation in the average response of cells to each ConA sample ranged from 17 to 32 %. The highest activities (mean values: > 26 %) were registered with the samples 10, N. 3 and O, the lowest (means values 17-20 %) with the samples 8, C, A, and 9. There were statistically significant differences between most samples of these two groups, using the U-test

TABLE 4

Degree of histamine release, HR, of leukocytes from 8 individuals induced by 14 different ConA preparation.

Donor	1	4	8	A	E	K	N	3	6	10	0	7	9	0
Da	-	3	-	6	9	8	12	-	15	22	26	-	32	31
Wo	-	4	-	6	6	9	11	-	8	15	13	-	34	40
Pa	3	-	2	15	2	8	8	12	-	-	7	38	-	42
Hi	10	-	0	6	1	9	6	11	-	-	9	31	-	27
Ju	-	1	-	4	-	4	9	-	10	14	17	-	23	20
Wi	5	-	2	-	3	3	6	16	-	-	14	30	41	31
We	6	4	-	3	6	2	9	13	10	11	11	27	19	21
Po	1	6	0	2	9	-	-	11	8	12	-	25	13	16
$\bar{x}$	5,0	3,6	1,0	6,0	5,1	6,1	8,7	12,6	10,2	14,8	13,9	30,2	27,0	28,5
s	3,4	1,8	1,2	4,3	3,2	3,0	2,3	2,1	2,9	4,3	6,3	5,0	10,4	9,4

Final lectin concentration: 42 µg/ml

for significance. The remaining ConA preparations K, 1, E, 6, 4, and 7 had intermediate activities.

In the supraoptimum range of lectin concentration an inhibitory effect of ConA in accordance with results of other authors (2, 8) was registered with the majority of the ConA samples tested, possibly by a primary membrane effect causing stabilization of basophilic cell membrane as discussed for the inhibitory action of viscum lectin (10, 11). However, the preparations 7, 9, and C failed to inhibit release at the concentration used (42 ug/ml), but provoked a significant secretion in all individuals with an average activity of 28.4 $\pm$ 8.4 %.

Thus estimation of histamine releasing activity is a suitable parameter for the characterization of ConA preparations used for lectin-cell-interaction studies.

Acknowledgement: Mr. Hjalmar Johansen is thanked for typing the manuscript.

REFERENCES

1. Keller, R.: Clin. Exper. Immunol. 13, 139-147 (1973).
2. Siraganian, P.A., Siraganian, R.P.: J. Immunol. 112, 2117-2125 (1974)
3. Siraganian, R.P., Siraganian, P.A.: J. Immunol. 114, 886-893 (1975)
4. Sullivan, T.J., Greene, W.C., Parker, Ch.W.: J. Immunol. 115, 278-282 (1975)
5. Sehrt, I., Bergmann, K.-Ch., Dehnert, I., Christ, R.: Dtsch. Gesund.-Wesen 32, 303-307 (1977)
6. Morita, Y., Matsumoto, Y., Miyamoto, T., Horiuchi, Y.: Ann. Allergy 40, 47-50 (1978)
7. Bergstrand, H., Björnsson, A., Frick, I.M., Lundquist, B., Nilsson, A.: Agents Actions 14, 602-605 (1984)
8. Magro, A.M., Bennich, H.: Immunology 33, 51-58 (1977)
9. Hook, W.A., Brown, H., Oppenheim, J.J.: Proc. Soc. Exper. Biol. Med. 147, 659-663 (1974)
10. Luther, P., Sehrt, I., Bergmann, K.-Ch., Reutgen, H.: Acta biol. med. germ. 33, 1623-1628 (1978)
11. Sehrt, I., Luther, P.: Lectins - Biology, Biochemistry, Clinical Biochemistry, ed. Bøg-Hansen, T.C., Vol. II, Walter de Gruyter, Berlin, New York, 1982, 45-56

HYDRAZIDATED HISTOCHEMICAL MARKERS - ALTERNATIVES TO LECTINS FOR DETECTION OF SIALIC ACID IN NEOPLASTIC TISSUES

Zoia Muresan, Rodica Dutu, Nicolae Voiculetz and Virgil Muresan

Institute of Oncology, Bd. I Mai 11 and Institute of Cellular Biology and Pathology, 8 Hasdeu St., Bucharest, Romania.

Glycoconjugates are integral components of the extracellular matrix and cell surface, with a major role in cellular recognition and other interaction phenomena. Among them, sialic acid-containing molecules are particularly interesting due to the terminal position of this sugar residue on the carbohydrate chain of glycoproteins and glycolipids. Malignant cells often exhibit an altered pattern of cell surface glycoconjugates and they show a characteristic spectrum of glycosylated molecules in the extracellular matrix and in secretions (1, 2).

Lectins have been used frequently to detect abnormal glycoconjugates as markers of malignancy with a diagnostic relevance for tumor biochemsitry and cancer therapy (3-7). In recent years, several studies have proven the utility of hydrazide-derivatized molecules for the detection of cell surface carbohydrates (8, 9). The use of hydrazidated markers resides in their capacity to interact with free aldehyde groups generated at the cell surface by chemical or enzymatic means. Thus, sialic acid-containing molecules may be histochemically revealed by a mild oxidation with sodium m-periodate which leads to formation of highly reactive aldehyde groups detectable by tracers containing amino or hydrazide groups (10, 11)

This study presents the synthesis of two new hydrazide derivatives, peroxidase-hydrazide (hydrazide-HRP) and fluorescent albumin-hydrazide (hydrazide-BSA-FITC), and their use for the detection of sialic acid-bearing molecules by light and fluorescence microscopy. We also present results obtained by using the electron dense tracer ferritin-hydrazide to detect cell and tissue sialoglycoconjugates by light microscopy.

Lectins, Vol. V

Materials and Methods

Synthesis of hydrazidated tracers. Bovine serum albumin (GIBCO, Grand Island, NY) was labeled with fluoroscein isothiocyanate and then coupled with adipic acid dihydrazide via carbodiimide (11) to obtain hydrazide-BSA-FITC. Briefly, one mg FITC was added to 40 mg BSA in sodium phosphate buffer pH 9.5; the solution was stirred 30 min at 23°C and then overnight at 4°c. Unreacted FITC was removed by gel filtration on Sephadex G-25; 4 mg adipic acid dihydrazide and 3 mg 1-ethyl-3-(-dimethylaminopropyl) carbodiimide HCl were added per mg fluororescent conjugate. The reaction was continued for 16 h at 23°C. Excess reagents were removed by gel filtration and the conjugate was stored at 4°C at a concentration of 5 mg/ml.

Hydrazidated horseradish peroxidase (HRP type VI, Sigma Chemical Co., St. Louis, MO) was obtained by a modification of the method of Nakane and Kawoi (12). The carbohydrate residues of peroxidase (2 mg) were oxidized with 35 mM sodium m- periodate (30 min, 23°C). Free aldehyde groups were then reacted with adipic acid dihydrazide (40 mg/ml) for 3 h at 23°C and quenched with 0.3 mg/ml sodium borohydride. The conjugated hydrazide-HRP was dialysed against phosphate buffer and stored at 4°C at a concentration 1 mg/ml.

Ferritin-hydrazide was synthesized as described elsewhere (11).

Characterization of the tracers. The enzymatic actitity of hydrazide-HRP was spectrophotometrically measured at neutral pH by using 3,3'-diaminobenzidine as H_2 donor and compared to that of nonderivatized peroxidase

The ability of hydrazide derivatives to interact with free aldehyde groups generated on cell surface was measured by an agglutination assay using sodium m-periodate-oxidized human red blood cells (11).

Wheat germ agglutinin-FITC (WGA-FITC) was a gift from Kem-En-Tec Biotechnology Corporation, Hellerup, Denmark.

Tissue preparation for light microsxopy. Fresh-frozen sections through rat kidney and mouse mammary tumor tissue were oxidized with 1 mM sodium m-periodate for 15 min at 4°C washed with phosphate buffer and incubated for 30-60 min at 23°C with one of the following tracers: hydrazide-BSA-FITC, hydrazide-HRP or ferritin hydrazide. In controls, tissue was not oxidized or was pretreated with a neuraminidase solution (0.5 mg/ml) and then incubated with wheat germ agglutinin-FITC or one of the above mentioned tracers. Sections treated with hydrazide-HRP were incubated with 0.15% 3,3'-diaminobenzidine and 0.01% H_2O_2 (15 min, 23°C in dark). Ferritin-hydrazide was rendered visible by light microscopy after performing the reaction with potassium ferrocyanide (13). Ehrlich ascitic cells and mechanically dissociated mouse thymocytes were oxidized with sodium m-periodate and react with hydrazide-derivatized tracers as described above.

Results

Characterization of the tracers. Spectrophotometric measurements of the peroxidatic activity of the hydrazide-HRP conjugate indicated that the enzyme preserved its native activity to a high degree. For the hydrazide BSA-FITC conjugate, the hydrazidation procedure did not affect the fluoroscence properties of FITC and an average of 1-2 moles of fluorochrome were contained in one mole hydrazide-BSA-FITC. This tracer was highly agglutinating for oxidized human erythrocytes, indicating a good accessibility to cell surface aldehyde groups (Fig. 1a). Hydrazide-HRP agglutinated to a lesser extent the oxidized erythrocytes, probably due to a lower degree of hydrazidation or to a less favorable exposure of its hydrazide residues. Neuraminidase-treated, oxidized erythrocytes were not agglutinated by any of the tracers, proving that the interaction was mediated by oxidized sialyl residues.

Detection of sialoglycoconjugates on the surface of dissociated cells. Both hydrazide-BSA-FITC and hydrazide-HRP were able to reveal sialic acid-bearing molecules in periodate-oxidized human erythrocytes, mouse thymocytes and Ehrlich ascitic cells in suspension. The labeling of glutaraldehyde-fixed, sodium borohydride-quenched, sodium m-periodate-oxidized cells was uniformly distributed on the cell surface (Fig. 1b). In freshly dissociated, nonfixed, oxidized thymocytes and ascites cells, the incubation with hydrazide-BSA-FITC induced the formation of caps (not showm). Capping was also observed when using ferritin-hydrazide, which appeared as a blue deposit at one pole of the cells after fixation and the Prussian blue reaction (not shown).

Use of hydrazidated markers as histochemical probes for sialoglycoconjugates. When applied on nonfixed frozen, periodate-oxidized sections of rat kidney, hydrazide-BSA-FITC heavily labeled the glomeruli (Fig. 2). A similar increased staining of glomeruli was obtained with ferritin-hydrazide after performing the histochemical reaction for iron (Fig.3). Additionally, the hydrazide-HRP reaction product revealed sialyl residues on the luminal aspect of proximal tubules (Fig. 4). A conjugate of wheat germ agglutinin-FITC also stained heavily the glomeruli (Fig. 5). The specificity for sialic acid of hydrazidated derivatives, used in conjunction with the mild periodate oxidation, was proven by the significant reduction of labelling after neruaminidase treatment or in nonoxidized specimens. Hydrazide-BSA-FITC was also used revealing sialic acid conjugates in frozen sections of spontaneous mouse (R III stain) mammary tumors (Fig. 6).

Discussion

In this paper we describe the synthesis of two new hydrazide-derivatized tracers and demonstrate their usefulness in investigation of tissue and cellular sialoglycoconjugates by light microscopy. The tracers have been used to detect sodium m-periodate-oxidizable sialyl residues. m-Periodate oxidizes sialic acids not O-acetylated at carbon atoms C_8 and C_9. The technique can be extended for detection of O-acetylated sialic acids by using the periodate oxidation after de-O-acetylation with potassium hydroxide (10).

The two hydrazidated tracers were synthesized using starting molecules (BSA, HRP) which are well characterized proteins and suit most of the experimental conditions in histochemistry and cytochemistry. The nonspecific binding of the tracers to tissue and cells is relatively low and the sugar specificity can be proven by adequate controls, e.g., enzymatic removal of the specific sugar, absence of oxidation.

Since both hydrazide-BSA-FITC and hydrazide HRP are large molecules and do not penetrate plasma membranes they can be used as cell surface markers. Providing that the mild periodate oxidation and the covalent binding of the tracers to activated sialyl residues do not perturb significantly the physiology of the membrane, the hydrazidated derivatives may represent valuable tools for studying the membrane related events in living cells.

Our tracers are easily detectable on cell surfaces or on tissue sections, preserving the fluorescence or enzymatic properties of the initial molecules. Hydrazide-HRP can be also used in ultrastructural cytochemistry since its enzymatic reaction product is electron opaque. On the other hand, ferritin-hydrazide which is a

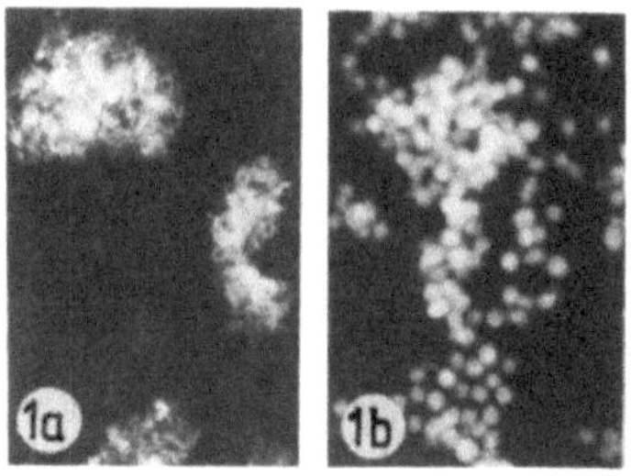

Fig. 1. Labeling of cells with hydrazide-BSA-FITC. (a) Periodate oxidized human erythrocytes. (b) Glutaraldehyde-fixed, periodate-oxidized mouse thymocytes.

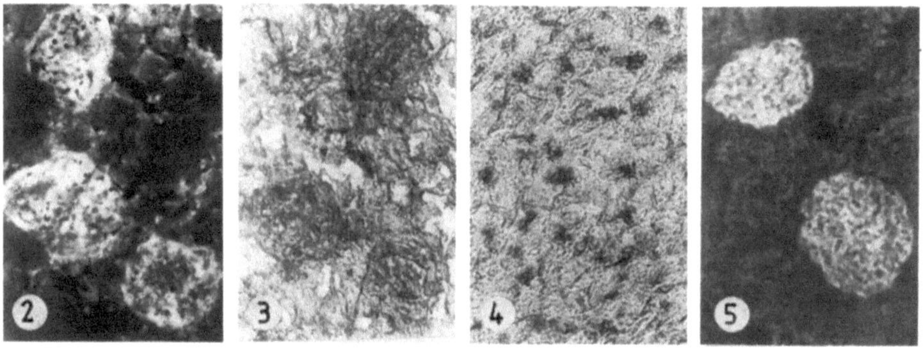

Figs. 2-5. Labeling of periodate-oxidized frozen sections through rat kidney by hydrazide-BSA-FITC (2), ferritin-hydrazide and Prussian blue reaction (3), hydrazide-HRP (4), and wheat germ agglutinin-FITC (5).

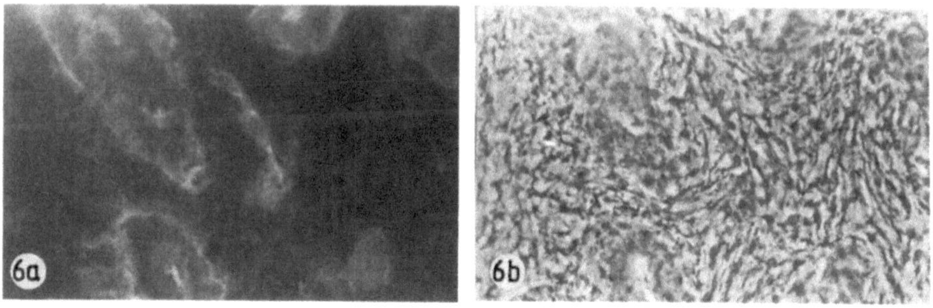

Fig. 6. Labeling of periodate-oxidized frozen sections of mouse (R III strain) mannary tumor with hydrazide-BSA-FITC (a), and the corresponding phase micrograph (b).

tracer for electron microscopy (11, 14) proved to be useful also for detection by light microscopy of sialoglycoconjugates in cryostat sections and on dissociated cells. Such tracers, detectable by both light and electron microscopy, are very useful when extending histochemical studies to the ultrastructural level.

The tracers used here are multivalent ligands, having more than one aldehyde-reactive hydrazide group on each molecule. Therefore they are able to bind simultaneously to several oxidized sialyl residues on the cell surface and to induce cell agglutination and redistribution of binding sites (patching, capping) in nonfixed cells. Such a phenomenon is characteristic for other multivalent ligands too (lectins, polycations, antibodies), and may be overcome by prior fixation of cells.

Several studies have shown that malignant transformation leads to the synthesis of abnormal cell surface and tissue glycoconjugates, including those which are sialylated (1, 15). More than that, by using a sialic acid-specific lectin, limulin, we have previously shown that tumor pancreatic acinar cells display differences in the amount and distribution of sialylated molecules, as compared to their normal counterpart (2). Preliminary results indicate that breast tumor tissue exhibits characteristic distribution of sialyl residues, as detected with hydrazide-derivatized tracers after mild periodate oxidation. It is therefore concluded that hydrazidated histochemical markers can be used as alternatives to sialic acid-specific lectins and in conjunction with lectins to detect histochemical modification in the glycoconjugate content and distribution as induced by a malignant transformation.

Acknowledgements. The authors acknowledge Ms. Mariana Banita for the skilful technical assistance and Mr. Al. Bifu for the help in preparing the illustrations. Kem-En-Tec Biotechnology Corporation, Hellerup, Denmark is acknowledged for kindly providing the wheat germ agglutinin-fluorescein isothiocynate conjugate. Ms. Caroline Nielsen is acknowledged for retyping the manuscript

References

1. Hakomori, S.: Cancer Res. 45, 2405 (1985).

2. Muresan, Z., Muresan, V., Iwanij, V., Jamieson, J.D.: J. Histochem. Cytochem. In press (1986).

3. Calafat, J., Jansen, H.: Lectins. Biology, Biochemistry, Clinical Biochemistry, Volume 3, T.C. Bøg-Hansen and G.A. Spengler, eds. Walter de Gruyter, 1983, Berlin, New York, pp 119.

4. Vierbuchen, M., Klein, P.J., Rösel, S., Fischer, J., Uhlenbruck G.: Lectins. Biology, Biochemistry, Clinical Biochemistry, Volume 3, T.C. Bøg-Hansen and G.A. Spengler, eds. Walter de Gruyter, 1983, Berlin, New York, pp 145.

5. Klein, P.J., Vierbuchen, M., Fischer, J., Farrar, G., Uhlenbruck, G.: Lectins. Biology, Biochemistry, Clinical Biochemistry, Volume 3, T.C. Bøg-Hansen and G.A. Spengler, eds. Walter de Gruyter, 1983, Berlin, New York, pp 157.

6 Klein, P.J., Vierbuchen, M., Fischer, J., Schulz, K.-D., Farrar, G., Uhlenbruck, G.: J. Steroid Biochem. 19, 839 (1983)

7. Louis, C.J., Sztynda, T., Cheng , Z.-M., Wyllie, R.G.: Cancer 52, 1244 (1983).

8. Skutelsky, E., Bayer, E.A.: J. Cell Biol. 96, 184 (1983).

9. Muresan, V., Constantinescu, M.C.: J. Histochem. Cytochem. 33, 474 (1985).

10. Heppelmann, B., Rahmann, H.: Histochem. 70, 199 (1981).

11. Muresan, V., Constantinescu, M.C.: Rev. Roum. Biochim. 21, 189 (1984).

12. Nakane, P., Kawaoi, A.: J. Histochem. Cytochem. 22, 1084 (1974)

13. Gabe, M.: Hostological Techniques, Masson and Springer Verlag, 1976, New York, pp 311.

14. Roffman, E., Spiegel, Y., Wilcbech, M.: Biochem, Biophys. Res. Commun. 97, 1192 (1980).

15. Altevogt, P., Fogel, M., Cheingsong-Popov, R., Dennis, J., Robinson, P., Schirrmacher, V.: Cancer Res. 43, 5138 (1983).

QUANTIFICATION OF SOYBEAN SEED LECTIN IN SOYBEAN TISSUES DURING THE LIFE-CYCLE OF THE PLANT, BY AN ENZYME LINKED IMMUNOSORBENT ASSAY (ELISA)

Henri Causse, Alain Lemoine and Pierre Rougé

Laboratoire de Biologie cellulaire, Faculté des Sciences Pharmaceutiques, Université Paul Sabatier, 35 chemin des Maraîchers, 31 062 Toulouse, France

Until now, a number of hypothetical roles for lectins in plants have been advanced but none of them has ever been demonstrated conclusively. However, most of the putative functions suggested for plant lectins imply the localization of these highly reactive molecules in various organs of the plant at well defined developmental stages. Thus, it is hoped that studies of the distribution and localization of lectins throughout the life-cycle of the plants will lead to a better understanding of their biological function(s). Qualitative immunodiffusion rather than classical hemagglutination techniques was first used by Howard et al. (1) to detect lentil lectin into the different organs of the plant. Subsequently, radial immunodiffusion (RID) has been widely used to detect and quantify lectins in various plant extracts (2). Recently, more sensitive radioimmunoassays (RIA) have been developed to measure *Dolichos biflorus* lectin (3) and soybean lectin (SBL) from *Glycine max* (4,5). Although specific and very sensitive, RIA techniques require sophisticated material and could be replaced by equally sensitive enzyme immunoassays (6,7). We have developed a sensitive ELISA (enzyme-linked immunosorbent assay) technique and we report here on the use of this technique to quantify the amounts of soybean seed lectin (SBL) in various tissues of soybean seeds and seedlings of different ages, throughout the life-cycle of the plant.

Lectins, Vol. V

MATERIALS AND METHODS

1. Plant material: The seeds of soybean (*Glycine max* (L.) Merr., cv. Hodgson), pea (*Pisum sativum* L., var. petit provençal nain) and lentil (*Lens culinaris* Medik., var. verte du Puy) were used. *Canavalia ensiformis* DC. seed meal was purchased from Serva Finebiochemica (Heidelberg), for concanavalin-A isolation.

2. Plant growth conditions: Soybean seeds were grown on Vermiculite, irrigated with Hoagland-Arnon I nutrient solution, under controlled illumination (about 6 500 lux, 16h photoperiod), humidity (60% day, 80% night) and temperature (25°C day, 18°C night) conditions. Lots of seedlings were harvested on various days after planting and were carefully dissected into cotyledons and embryo axis and, where possible, into cotyledons, roots, epicotyls and hypocotyls. The samples were homogenized in 50 mM Tris (pH 7.6), 0.15M NaCl with mortar and pestle and analysed for SBL and protein. Subsequently, various developmental stages of the immature seeds were defined by using several parameters of the seeds such as diameter, fresh weight and dry matter content. At each stage, a lot of seeds was homogenized as previously described and SBL and protein were studied.

3. Lectin and protein isolation: SBL was isolated according to the procedure of Allen and Neuberger (8). Lectins from pea, lentil and concanavalin-A were isolated by affinity chromatography on Sephadex G-100 (9). The purity of lectins was checked by disc polyacrylamide gel electrophoresis at basic pH (10). Protein from mature soybean seeds, cotyledons, embryo axis, seed coats and seedlings were isolated from meal extracts with 0.1M phosphate (pH 7.2) buffer.

4. Preparation of antibodies: Antibodies against SBL were raised in rabbits using Freund's complete adjuvant, according to an immunization schedule described previously (11). Specific antibodies to SBL were purified by passing antisera through a column of SBL covalently coupled to CNBr-Sepharose 4B (Pharmacia) and subsequently eluting at acidic pH.

5. Immunocytochemical localization of SBL: The immunocytochemical localization of SBL in seeds and seedlings was performed by indirect immunofluorescence. Dry seeds were imbibed in distilled water for 40h at 4°C and subsequently dissected into cotyledons and embryo axis. The embryo axis were integrally fixed whereas the cotyledons were cut in small pieces of about 1 mm^3, before fixation. Small blocks were similarly cut off from the cotyledons of immature seeds and young seedlings during seed maturation and seed germination, respectively. The tissues were fixed for 1h at 4°C, under vacuum, in 40 mM cacodylate (pH 7.2), 2.0M $MgCl_2$ containing 4% paraformaldehyde in order to reduce autofluorescence. After dehydration in an ethanol gradient, the tissues were embedded in prepolymerized glycolmethacrylate (12) and polymerization was carried out for 15h under UV light, at 4°C. Semi-thin sections (0.5 µm thick) were treated as follows : 1) incubated with anti-SBL antibodies (0.5 mg IgG/ml) during 12h at 4°C, 2) washed three times (10 min each) with phosphate buffered saline (30 mM phosphate (pH 7.2) buffered saline) containing 1% Tween 20, 3) incubated for 3h at 37°C with fluoresceine isothiocyanate-labelled goat anti-rabbit IgG (Nordic Immunological Laboratories, Tilburg, the Netherlands) 1/10 diluted in buffer, 4) washed three times (10 min reach) with phosphate buffered saline containing 1% Tween 20 and 5) covered with two drops of a solution of Mowiol (Hoecht Laboratories, Frankfurt, West Germany). Sections were examined with a Leitz Laborlux 12 microscope equiped for immunofluorescence. Control staining was done by replacing anti-SBL IgG by IgG from non-immunized rabbits.

6. Enzyme-linked immunosorbent assay (ELISA): The ELISA technique used was a competitive technique adapted from Engvall and Perlman (13), based on the ability of SBL to inhibit the specific interaction between SBL labelled with alkaline phosphatase (EC 3.1.3.1.) and anti-SBL antibodies. The controls included replacement of anti-SBL antibodies by a non-immune rabbit immunoglobulin preparation and replacement of SBL by various lectins such as concanavalin A, pea and lentil lectins. Another control was made by replacing plant extracts containing SBL by SBL-free plant extracts.
To 50 µl of PBS (phosphate buffered saline) containing 1.5 mg SBL and 1.5 mg alkaline phosphatase (Sigma), glutaraldehyde was added to a final concentration of 0.2% and the mixture was allowed to stand for 2h, in the dark, at room temperature. The mixture was dialysed against several changes of PBS and 5 ml of 50 mM Tris (pH 8.0) containing 5% bovine serum albumine (Sigma) and 0.02% NaN_3 were added. The resulting conjugated SBL was stored frozen at -30°C until used. Microtiter polystyrene plates (Limbro, SI MVC 96, Flow Laboratories) were used and incubation was performed in polystyrene boxes containing wet filter paper. Anti-SBL antibodies, conveniently diluted in 0.1M Na_2CO_3 (pH 9.6) containing 0.02% NaN_3, were placed in wells and allowed to adsorb at 4°C during 12h. The wells were washed four times by flooding with 0.15M NaCl containing 0.05% Tween 20 and 0.02% NaN_3. A mixture of 25 µl of properly diluted conjugated SBL and 25 µl of plant extract was then added to each well and SBL (conjugated and unconjugated) and anti-SBL antibodies were allowed to react for 3h at 37°C and for 12h at room temperature. The wells were rinsed as above and 150 µl of 50mM Na_2CO_3 (pH 9.8), 1.0M $MgCl_2$ containing 1 mg/ml 4-nitrophenyl phosphate (Sigma) were added. The reaction was stopped after a 30 min incubation period at 37°C by adding 50 µl 4.0M NaOH and the absorbance at 410 nm was recorded using a Multiskan Titertek spectrophotometer (Flow Laboratories).

7. Hemagglutination: Hemagglutinating activities of seed extracts were measured with untreated rabbit erythrocytes (2% suspension in PBS) by the serial dilution method in microtitration plates. The reciprocal of the greatest dilution which caused full agglutination was given as the titer.

8. Protein determination: Total soluble protein from seeds and seedlings extracts was determined colorimetrically by the micro-biuret method of Goa (14) with bovine serum albumin (Sigma) as standard.

RESULTS

Experiments were performed to determine the appropriate dilutions of conjugated SBL and anti-SBL antibodies to be used in order to achieve maximum sensitivity. Dilutions of 1/512 and 1/5 000 for conjugated SBL and anti-SBL antibodies respectively, allowed the accurate estimation of lectin concentrations as little as 50 ng/ml. The high specificity of the technique was attested by the fact that standard absorbance at 410 nm was not affected in the presence of other lectins (concanavalin A, pea or lentil lectins), even at concentrations as high as 100 µg/ml. In addition, no changes in absorbance were detected when SBL-free plant extracts were used instead of lectin-containing plant extracts.
The amounts of SBL in the cotyledons, embryo axis and seed coats from soybean

seeds, as measured by ELISA, were reported in Table 1. SBL appeared as mainly located in cotyledons since embryo axis contained very low amounts of SBL while seed coats were almost devoid of lectin. These results were in agreement with the hemagglutinating activities of the different parts of the seed. The immunocytochemical localization of SBL showed that the lectin is mainly located within the protein bodies of both the cotyledons and embryo axis of mature seeds (Fig. 1), as previously reported by Horisberger and Vonlanthen (15).

Table 1

Distribution of SBL and protein in soybean seed (SBL and protein are expressed as mg/g fresh weight and each value is the average of three separate determinations); hemagglutinating activities are expressed as titers.

Tissue	SBL content measured by ELISA	Protein content	Hemagglutinating activity
Cotyledon	1.96	92.4	131 072
Embryo axis	0.026	178.5	4 096
Seed coats	0.0007	3.7	8
Whole seed	1.93	97.6	131 072

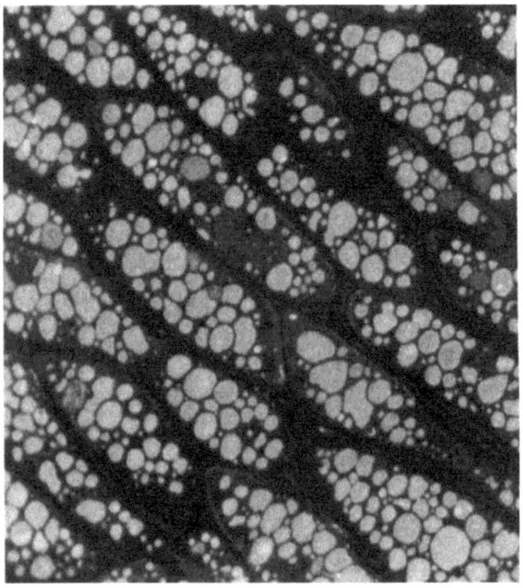

Figure 1. Immunocytochemical localization of SBL in the protein bodies of the cotyledonary parenchyma cells from soybean seed, with rabbit antibodies against SBL followed by fluoresceine isothiocyanate-labelled goat anti-rabbit immunoglobulins. The cell walls were lightly marked, probably in a non-specific manner. No fluorescence was observed in control ecperiments (results not shown), were antibodies against SBL were replaced by immunoglobulins from non-immunized rabbit.

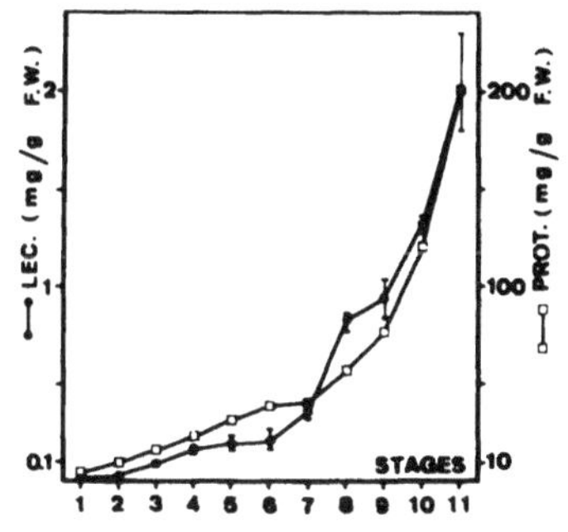

Figure 2. Changes in the lectin and protein content (expressed as mg/g fresh weight) of the cotyledons during seed germination. Lectin content was measured by ELISA and each data point is the average of three separate determinations.

In germinating seeds, SBL content of cotyledons decreased gradually until abscission and by day 18 following the onset of seed germination negligible quantities of SBL were still present in cotyledons (Fig. 2). Total soluble protein exhibited a similar pattern but showed a transient increase just after planting. Similarly, the SBL content of the embryo axis decreased (Fig. 3) but rather quickly since only one week after the beginning of the germination process, the total amount of SBL in the entire plant was less than 5% of the original amount of SBL in the embryo axis. By day 16, the young plant contained only 70 ng SBL/

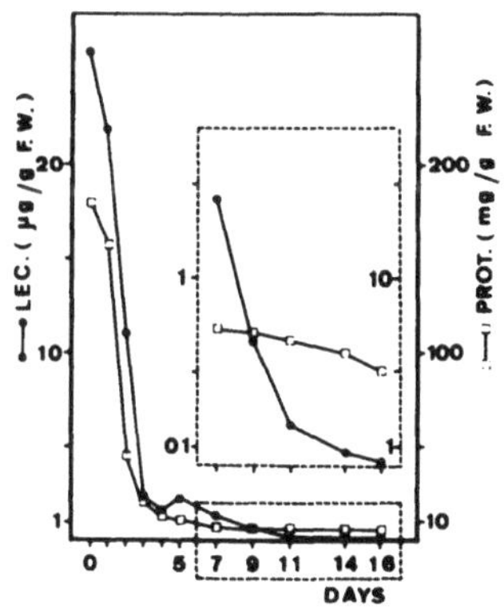

Figure 3. Changes in the lectin and protein content (expressed as mg/g fresh weight) of the embryo axis and the young plant during seed germination. Lectin content was measured by ELISA and each data point is the average of three separate determinations.

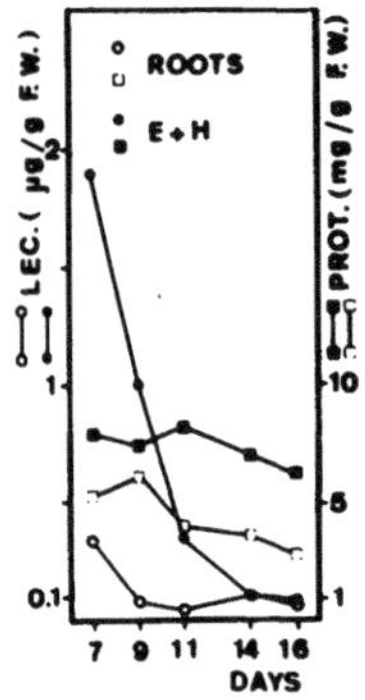

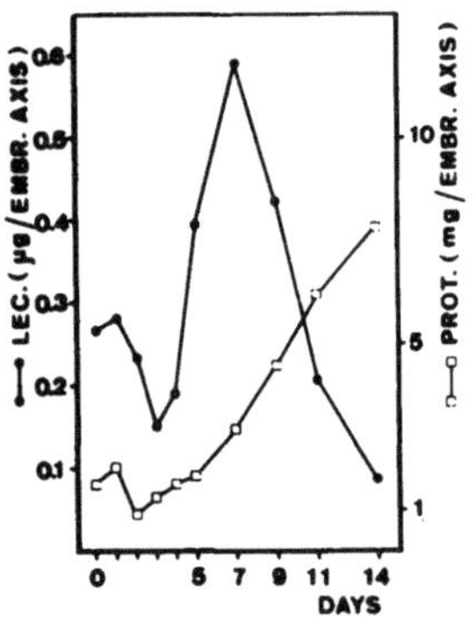

Figure 4. Changes in the lectin and protein content (expressed in ug and mg/g fresh weight, respectively) of roots, epicotyls (E) + hypocotyls (H) during seed germination. Lectin content was measured by ELISA and each data point is the average of three separate determinations.

Figure 5. Changes in the lectin and protein content (expressed in ug and mg/embryo axis, respectively) of the embryo axis and the young plant during seed germination. Lectin content was measured by ELISA and each data point is the average of three separate determinations.

g fresh weight. Such a rapid decrease also occurred in roots, epicotyls and hypocotyls (Fig. 4) but by day 16 the SBL content of roots still remained easily detectable (about 100 ng/g fresh weight). Again, soluble protein exhibited a very similar pattern. Interestingly, when SBL amounts were expressed as µg/embryo axis or young plant (Fig. 5), data showed a rapid and transient increase in lectin content in the young plant tissues between the 5th and the 11th days of germination. Throughout the first stages of the germination process, SBL remained located within the protein bodies, as shown by indirect immunofluorescence.

During seed maturation, SBL gradually increased and accumulated at about the same rate as soluble protein to reach a maximum level in the mature seed (Fig. 6). Although lectin appeared during the early stages of the seed ripening process, it could not be detected by indirect immunofluorescence within the protein bodies until these organelles were sufficiently dehydrated and enriched in proteins.

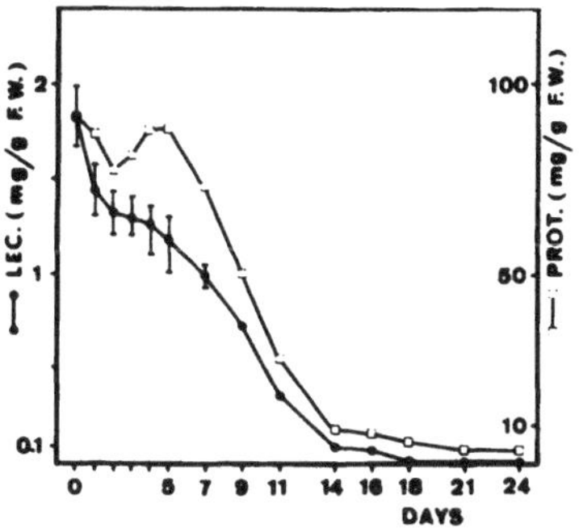

Figure 6. Changes in the lectin and protein content (expressed in mg/g fresh weight) of seeds during seed maturation. Lectin was measured by ELISA and each data point is the average of three separate determinations.

DISCUSSION

The ELISA technique we have developed appears as a very sensitive and specific tool since it can easily and accurately detect lectin concentrations of as little as 50 ng/ml. Such a sensitivity is similar to that exhibited by RIA, as previously reported in (3,4).
Using this ELISA technique, we determined that SBL in the seeds is essentially confined to the cotyledons, while the embryo axis and seed coats contain very low amounts of lectin. The measured SBL levels agree with those reported for other soybean cultivars by Gibson et al. (5) and with the hemagglutinating activities estimated for cotyledons, embryo axis and seed coats (see Table 1). However, using RIA, Pueppke et al. (4) have measured higher SBL amounts in the seed coat of *Glycine max* var. Beeson. In addition, SBL has been found to be located within the protein bodies by indirect immunofluorescence but the cell walls were often slightly labelled, probably in a non-specific manner, as already reported by Horisberger and Vonlanthen (15). In this respect, recent evidences (16,17) have shown that SBL was in part membrane-associated.
During seed germination SBL, primarily located in the cotyledons, decreases gradually until they are shed and the lectin behaviour does not markedly differ from that of the soluble protein bulk. In the young plant, issued from the embryo axis, SBL sharply decreases but a transient increase of the SBL content occurs during the first days following the onset of germination (see Fig. 5). The total amount of SBL in the embryo axis is thus not sufficient to account for the

SBL levels found in the different parts of the young plant and such an increase could be interpretated on the basis of a SBL transportation from the cotyledons or by a *de novo* synthesis of lectin. However, attempts to demonstrate *de novo* synthesis of a soybean root-lectin whose properties are very similar to SBL (18) have been unsuccessful (19). The recent discovery in soybean of another locus gene designated le2 which has some homology with the seed lectin gene and is expressed at low levels in the roots, could explain such an observed increase (20) Two weeks after sowing, the young plant is almost devoid of SBL although an appreciable amount of lectin still remains in roots. SBL reappears in the very young seed and then accumulates slowly during the seed ripening process.
These results indicate that SBL is mainly associated with the seed and quickly disappears during the development of the plant. Thus, among the hypothesis proposed for the role(s) of lectins in plants, they favour those in which seeds or seedlings could be involved (21).

REFERENCES

1. Howard, I.K., Sage, H.J., and Horton, C.B.: Arch. Biochem. Biophys. 149, 323-326 (1972).
2. Rougé, P., and Père, D.: in Lectins, Biology, Biochemistry, Clinical Biochemistry, (Bøg-Hansen, T.C. ed.), De Gruyter, Berlin, 2, 137-150 (1982).
3. Talbot, C.F., and Etzler, M.E.: Plant Physiol. 61, 847-850 (1978).
4. Pueppke, S.G., Bauer, W.D., Keegstra, K., and Ferguson, A.L.: Plant Physiol. 61, 779-784 (1978).
5. Gibson, D.M., Stack, S., Krell, K., and Huose, J.: Plant Physiol. 70, 560-566 (1982).
6. Borrebaeck, C.A.K., and Mattiason, B.: Physiol. Plant. 58, 29-32 (1983).
7. Borrebaeck, C.A.K., and Mattiason, B.: in Abstracts of the 12 th Internatl. Congress of Biochemistry, Parth, Australia, p. 196 (1982).
8. Allen, A.K., and Neuberger, A.: FEBS Lett. 50, 362-364 (1975).
9. Agrawal, B.B.L., and Goldstein, I.J.: Biochim. Biophys. Acta 147, 262-271, (1967).
10. Davis, B.J.: Ann. N. Y. Acad. Sci. 121, 404-427 (1964).
11. Rougé, P.: Biologie des hémagglutinines chez le Pois (*Pisum sativum* L.), Thèse Doct. ès-Sci., n° 746, Toulouse (1977).
12. Ledux, E.H., and Bernhard, W.: J. Ultrastruc. Res. 19, 196-201 (1967).
13. Engvall, E., and Perlmann, P.: Immunochemistry 8, 871-884 (1971).
14. Goa, J.: Scand. J. Clin. Lab. Invest. 5, 218-222 (1953).
15. Horisberger, M., and Vonlanthen, M.: Histochemistry 65, 181-186 (1980).
16. Bowles, D.J., Lis, H., and Sharon, N.: Planta 145, 193-198 (1979).

17. Pueppke, S.G., Friedman, H.P., and Su, L.C.: Plant Physiol. 68, 905-909, (1981).

18. Gade, W., Jack, M.A., Dahl, J.B., Schmidt, E.L., and Wold, F.: J. Biol. Chem. 256, 12905-12910 (1981).

19. Gade, W., Schmidt, E.L., and Wold, F.: Planta 158, 108-110 (1983).

20. Vodkin, L.O.: in Chemical Taxonomy, Molecular Biology, and Function of Plant Lectins, (Goldstein, I.J., and Etzler, M.E. ed.), Alan R. Liss Inc., New York, 87-98 (1983).

21. Lis, H., and Sharon, N.: in The Biochemistry of Plants, (Stumpf, P.K., and Conn, E.E. ed.), Academic Press, New York, 6, 371-447 (1981).

SERUM AND HEMOCYTE-ASSOCIATED LECTINS OF THE OYSTER *CRASSOSTREA VIRGINICA*

Gerardo R. Vasta
Department of Biochemistry, Medical University of South Carolina, Charleston, South Carolina 29425, USA

Introduction

The recognition of non-self as part of a defense mechanism has been postulated to be the role of some lectins from invertebrates. It was suggested initially for the lectins present in the albumin gland of gastropods, which secretes the fluid that coats the oocytes. These lectins were named "protectins" under the idea that their function was to protect the oocyte from bacterial infection (1). Although it seems almost certain that the invertebrates, as a broad group, lack the typical adaptive vertebrate immune response characterized by immunoglobulins and lymphocytes, invertebrates are able both to differentiate self from non-self and to respond against non-self substances (2).

The invertebrate response to non-self is primarily cellular; that is, non-self substances are either phagocytosed by individual hemocytes, or, if too large to be endocytosed, are surrounded by a multilayered capsule of hemocytes (3). Although the cellular components of the invertebrate internal defense system have been described, the mechanism by which this system recognizes non-self is not well understood. In addition to general surface phenomena, such as wettability or hydrophobicity (4), and the presence of non-specific cell surface receptors such as glass-adherent proteins (5) it is likely that specific cell surface receptors also mediate non-self recognition mechanisms in invertebrates. The presence of cell surface carbohydrate-binding specific receptors has been described on a variety of cells including slime molds, hepatocytes, lymphoid cells and macrophages from vertebrates (6). However, at present, lectins on the surfaces of phagocytic cells have been described only in a few invertebrate species (7,8). Occurrence of true lectins on the surface membranes of invertebrate phagocytes would be of interest in understanding the potential evolutionary relationships between the immune system of vertebrates and the internal defense system of invertebrates. Therefore, lectin molecules on the surfaces of phagocytic cells may comprise a primitive non-self recognition mechanism common to both invertebrates and vertebrates.

In addition to cellular responses to non-self, humoral factors have also been implicated both as recognition molecules and effector molecules in invertebrates; levels of lysosomal enzymes released into the serum from hemocytes may serve to partially degrade foreign substances. Moreover, bacteriocidal, hemolytic, and parasite-immobilizing substances have been described in invertebrates. Since invertebrates do not synthesize antibodies comparable to those of vertebrates, other molecules must be responsible for the specific recognition of non-self. Humoral lectins, which are ubiquitous among invertebrates and are known to be present in most body fluids, are likely candidates for the role of specific recognition molecules. Because lectins bind specifically to polysaccharides and glycoconjugates, humoral lectins may attach to or agglutinate, and thus immobilize, microbial or metazoan parasites which come in contact with hemolymph and mediate their subsequent phagocytosis, melanization, or encapsulation. Opsonic properties ascribed to lectins have been the subject of controversy because hemocytes can recognize foreign particles, cells, or macromolecules independent of the presence of serum lectins, although in most cases previous exposure of the foreign material to serum lectins enhances the rate of phagocytosis. This indicates that although humoral lectins might facilitate this recognition and enhance the response, hemocytes display non-self recognition molecules, probably located on the cell membrane, which interact with the foreign material and may be essential for cellular recognition.

Serum and hemocyte-associated lectins of the oyster Crassostrea virginica

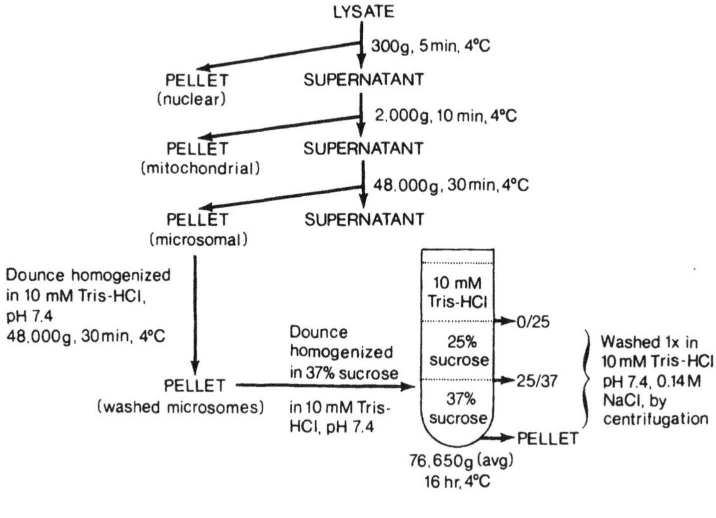

Figure 1: Buoyant density centrifugation of oyster hemocyte membranes. Lysate obtained by disrupting approximately 10^9 hemocytes in 200 ml HBS with a Stansted cell disruptor at 15 psi back pressure.

Table 1: Agglutinating Activity of Cell Fractions from Disrupted Oyster Hemocytes

	Recovered AU[a]	Percentage total recovery	Specific activity (titer/protein)
Lysate	1600	100	40
300*g* pellet	250	15.6	46.0
Supernatant I	800	50	20.0
2000*g* pellet	250	15.6	46.0
Supernatant II	550	34.37	14.0
48,000*g* pellet	600	37.5	141.0
Supernatant III	0	0	0

[a] Figures are in agglutination units (AU), defined as the amount of lectin present in the highest dilution of serum capable of agglutinating RBC with a score of +2 in a scale from 0 to +4.

In our initial publication (9), we reported the detection of lectin activity in oyster hemocyte microsomal fractions (Figure 1, Table 1). Lectin activity in associated oyster hemocytes was demonstrated by a high agglutination titer for certain erythrocytes in the lysate produced by rupture of a suspension of hemocytes by a cell disruption (Figure 1, Table 1). The membrane association of a significant proportion of the lectin was shown by the sedimentability of agglutinating activity, especially in the microsomal (48,000g) pellet. As shown in Table 1, this pellet contained approximately 37% of the total lysate activity and an increase of 3.7-fold in specific activity. The plasma membrane association of the lectin was initially suggested by the results of subjecting the microsomal pellet to buoyant density centrifugation (Figure 1). The microsomal fraction banding at the 25/37% sucrose interface contained approximately 50% of the recoverable activity. Microsomal lectin activity was expressed as high agglutination titers for vertebrate erythrocytes (RBC), in particular from rat and horse in a pattern similar to that of a fraction of the serum lectin. The binding of both serum and microsomal preparations to RBC could be inhibited specifically by some mono- and oligosaccharides (such as galactose, GalNAc, lactose and stachyose) and glycoproteins (such as bovine submaxillary mucin, fetuin and porcine stomach mucin). Intact hemocytes bound to the same radiolabeled glycoprotein (bovine submaxillary mucin) which inhibited the hemagglutination by hemocyte microsomes, while glycoproteins which did not inhibit binding of microsomes to RBC such as ovomucoid, failed to bind to the intact hemocytes (Table 2). Our studies showed that there are at least two distinct divalent cation-dependent serological specificities in the serum lectin while the hemocyte membrane lectin shows only one of the two serum patterns of serological reactivity although the membrane lectin apparently possesses no unique agglutination activity. The plasma membrane association of this lectin is shown by its copurification with the plasma membrane fraction of disrupted hemocytes, using sucrose density gradient centrifugation (Figure 1).

Table 2: Radioactivity Associated with 4 x 10^6 Oyster Hemocytes after Incubation with ^{125}I-Labeled Glycoproteins

	cpm × 10^{-3}			
Labeled glycoprotein	Initial	After 1st wash	After 2nd wash	After 3rd wash
BSM (4°C)	*a*	110.74	60.48	50.20
BSM (25°C)	955.29	270.47	207.30	176.83
Ovomucoid (4°C)	*a*	54.34	12.91	9.26
Ovomucoid (25°C)	767.80	62.52	31.72	23.54

[a] 4°C initial reading not performed.

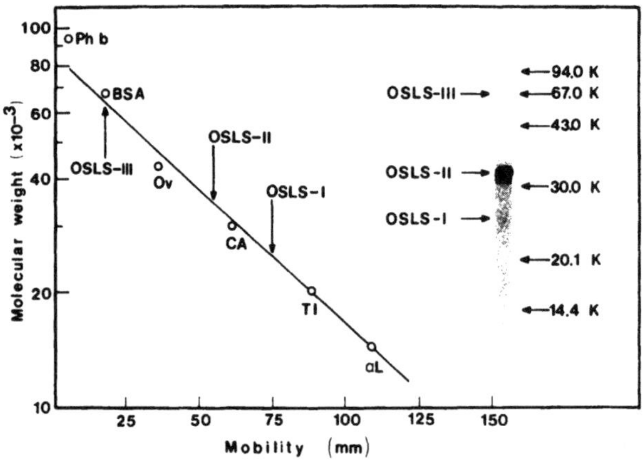

Figure 2: SDS-PAGE under reducing conditions of P-treated formaldehyde-fixed rat RBC-binding lectins from oyster serum: Estimation of approximate subunit molecular weights from the plot of log MW vs mobility. The inset shows Coomassie blue-stained bands indicated by arrows on the left. Positions of the molecular weight standards are indicated on the right.

Subsequently (10) we purified, by affinity chromatography, a major lectin (CVL-II) of subunit Mr = 34 Kd (Figure 2, 3) from serum (lectins of subunit Mr = 67 Kd and 26 Kd are also present) that is different from the one isolated by Acton et al (11). We performed the amino acid analysis of the purified serum lectins and compared it to the amino acid compositions of 250 different proteins from our data bank through the parameter SΔQ. The results obtained suggest that although the SΔQ values are relatively high CVL-II exhibits an amino acid composition that is similar to several tunicate, merostome and other molluscan lectins and to human C-reactive protein. Subsequently we examined the serological cross-reactivity of the purified oyster lectin CVL-II with antibodies made against elasmobranch immunoglobulins and mammalian acute phase proteins performing ELISA assays. The oyster lectin CVL-II did not cross-react with the antiserum made against the 19S IgM pentamer of the shark C. galapagensis (Vasta, Marchalonis and Rosenshein, unpublished), but cross-reactivity was observed when the antiserum sheep anti-C-reactive protein (human) was tested, supporting data from the comparison of amino acid composition by SΔQ. In addition, we have identified 24 amino acid residues from the N-terminus sequence of the oyster lectin CVL-II (Figure 4) and efforts are being made in our laboratory in order to obtain the complete amino acid sequence of this molecule.

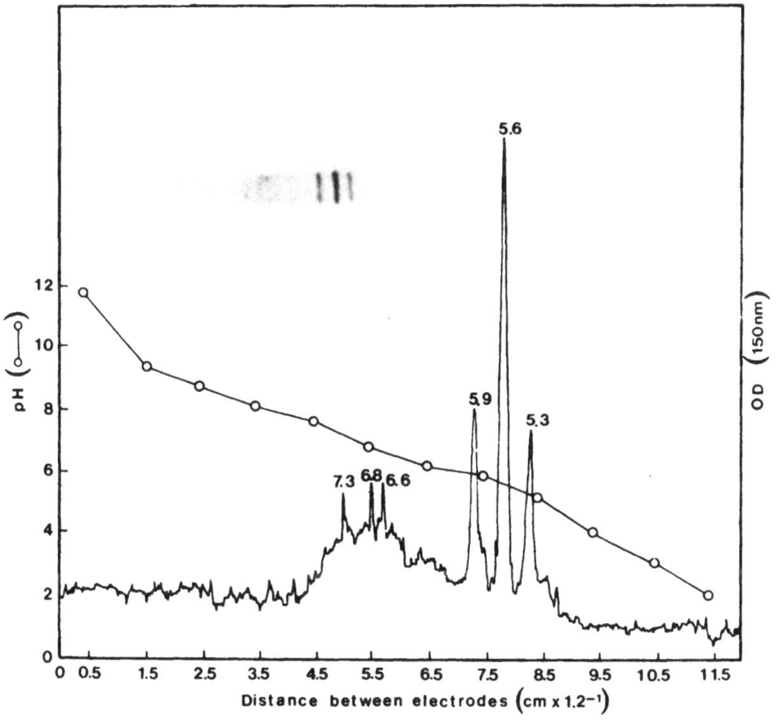

Figure 3: Isoelectric focusing of rat RBC-binding lectins from oyster serum: Densitometric scan (150 mM) of the gel shown in the inset. The pI indicated on each peak was extrapolated from the pH gradient (O----O).

Figure 4

Partial N-Terminus Amino Acid Sequence of Crassostrea Major Serum Lectin CVL-II

1				5					10
Ser -	Ala -	Arg -	Asn -	Glu -	Ala -	Asn -	Val -	Asn -	Ileu -
11				15					20
- Tyr -	Leu -	His -	Leu -	Cys -	Asp -	Asp -	Glu -	Ala -	Val -
21				25					
- Asn -	Tyr -	Ala -	Asn -	-					

We raised antibodies in rats (10) against the purified major oyster serum divalent cation-dependent lectin CVL-II and showed that those antibodies [Rt anti-OSL (oyster serum lectin CVL-II] will bind to the intact oyster hemocyte cell surface and hemocyte microsomal fractions, blocking their hemagglutinating ability (Figure 5). Antibodies raised in rabbits against the total oyster serum proteins (Rb anti-OSP) exerted a similar effect even at much greater dilutions. Immunodiffusion and immunoelectrophoresis tests confirmed that sera of immunized rats and rabbits contained antibodies against the oyster purified lectins. Our immunofluorescence studies performed on the intact oyster hemocytes showed that rat antibodies against the purified oyster serum lectin CVL-II bound to the hemocyte surface at 0°C and the label was not endocytosed at that temperature (Figure 6). Direct observation with fluorescence microscopy of labeled hemocytes with both rat or rabbit antisera showed fluorescence patterns at the cell surface that resembled "patching-" and "capping-like" phenomena, suggesting some surface redistribution of the label may have occurred at that temperature (Figure 6).

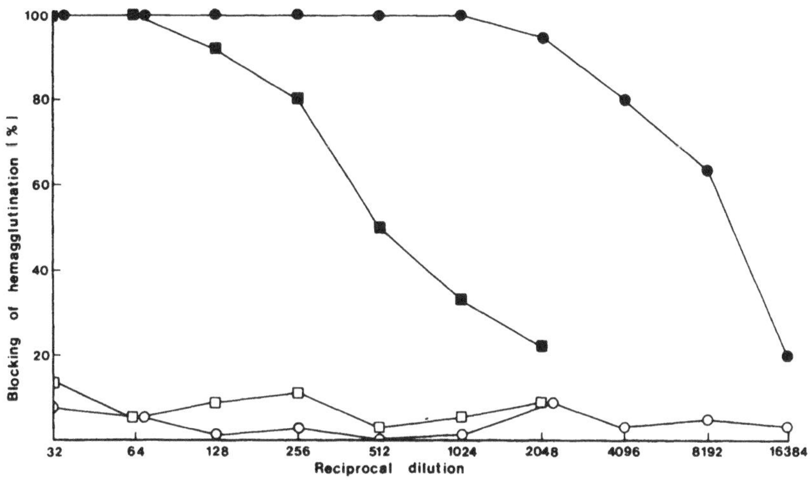

Figure 5: Blocking of the binding of hemocyte microsomes to P-treated rat and horse RBC by rat and rabbit antisera: hemocyte microsomes were incubated with serial dilutions of (●) rabbit anti-oyster serum proteins; (O) normal rabbit serum; (■) rat anti-oyster serum lectin; and (□) normal rat serum; and tested for agglutination of P-treated rat RBC.

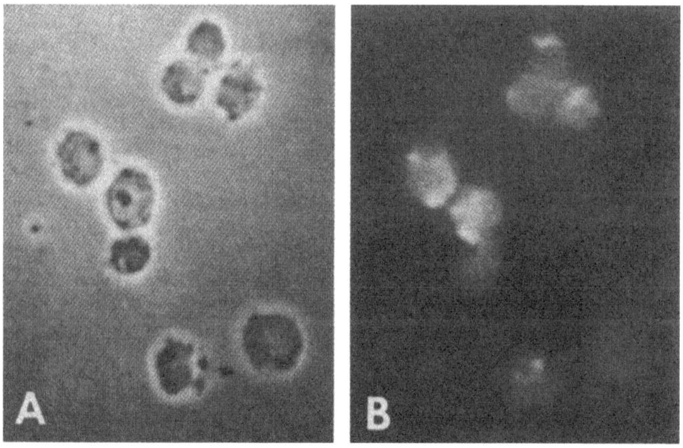

Figure 6: Oyster hemocytes stained with rat anti-oyster serum lectin plus FITC-goat anti-rat Ig: (A) phase contrast; (B) fluorescence excitation for FITC.

When analyzed by flow cytofluorometry (Figure 7) the oyster hemocyte preparations were homogeneous in their light scattering properties. When analyzed relative to their fluorescence, although certain heterogeneity in the fluorescence intensity was observed, no discrete differences reflected as distinctive peaks in the histograms could be determined in the fluorescence of the cells tested, i.e., no clear-cut subpopulations could be identified. Figure 7 shows the flow cytofluorometry diagrams for comparisons of light scattering (A) and fluorescence distribution (B) histograms for hemocytes incubated with Rt anti-OSL (a) vs hemocytes incubated with HBC (b) and Rt anti-OSL (a) vs NRtS (c). The light scattering diagrams (A) indicate that in all experiments the cells examined exhibited homogeneous light scattering properties, suggesting that their viability remained unmodified by the labeling procedure when compared to the controls. Fluorescence intensity distribution ratios between the pairs of curves considered from their intersecting points indicated by the bar (B) are shown in the same figure. The fluorescence distribution ratios indicate that a low background fluorescence can be ascribed to nonspecific binding of rat normal serum Ig since the rat serum ratios Rt anti-OSL/NRtS and Rt anti-OSL/HBS are both 5.02.

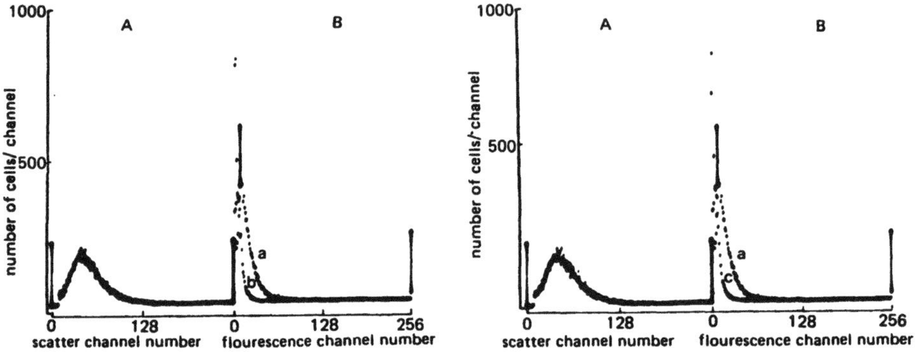

Figure 7: Light scattering (A) and fluorescence intensity distributions (B) obtained by flow cytometry of oyster hemocytes stained with rat antisera plus FITC-goat anti-rat Ig. Histograms show distribution of hemocytes incubated with (a) rat anti-oyster serum lectins (dil 1/20), (b) normal rat serum (dil 1/20) and (c) HBS. The bar placed at the intersection of the histograms indicates the channel number at which the background fluorescence intensity was assigned.

Conclusions

Our results show that a fraction of the serum lectins which we studied exhibits a carbohydrate specificity similar to the hemocyte membrane lectin, although the membrane lectin does not portray a unique carbohydrate specificity (9,10). The membrane lectin is serologically related to the soluble serum lectins as revealed by the immunofluorescence experiments and the blocking of microsomal binding to RBC. In addition, the hemocyte membrane lectin appears to be an integral membrane protein since it is copurified through the preparation of microsomes and hence, it is unlikely that it is incorporated from the serum as cytophilic proteins. This evidence suggest that the oyster hemocyte membrane-associated lectin may be a true integral membrane protein, and therefore may function as a membrane receptor in non-self recognition by molluscan hemocytes (9,10). The serological cross-reactivity of the oyster serum lectin with C-reactive protein is interesting, and once obtained the complete amino acid sequence, their comparison might reveal stretches of significant homology that may account for the observed cross-reactivity. The purification of oyster hemocyte membrane lectins is now in progress in our laboratory in order to compare their physicochemical properties and molecular structure with lectins purified from oyster serum.

Acknowledgements: This research was supported in part by Fogarty Institute Fellowship Award (National Institutes of Health) 1F OSTW 031 155 to G.R. Vasta and Grants PCM 820-8016 to T.C. Cheng and G.R. Vasta (National Science Foundation) and GM 30672 to J.J. Marchalonis and G.R. Vasta (National Institutes of Health). I thank Dr. Dominick DeLuca for expert advice in the FACS measurements, Mrs Teresa Vasta for the preparation of the figures and Ms. Lesley Secker for typing the manuscript.

References

1. Prokop, O., Uhlenbruck, G. and Kohler, W. (1968) Vox Sang 14: 321.
2. Hildermann, W.H., Johnson, I.S. and Jokiel, P.L. (1979) Science 204: 420.
3. Lackie, A.M. (1980) Parasitology 80: 393.
4. Lackie, A.M. (1983) Cell Sci. 63: 181.
5. Lehtinen, P., Pikkarainen, J. and Kulonen, E. (1976) Biochim. Biophys. Acta 439: 393.
6. Monsigny, M., Kieda, C. and Roche, A.C. (1979) Biol. Cellulaire 33: 289.
7. Amirante, G.A. and Mazzalai, F.G. (1978) Dev. Comp. Immunol. 2: 723.
8. Renwrantz, L. and Stahmer, A. (1983) J. Comp. Physiol. 149: 535.
9. Vasta, G.R., Sullivan, J.T., Cheng, T.C., Marchalonis, J.J. and Warr, G.W. (1982) J. Invert. Path. 40: 367.
10. Vasta, G.R., Cheng, T.C. and Marchalonis, J.J. (1984) Cell. Immunol. 88: 475.
11. Acton, R.T., Bennett, J.C., Evans, E.E. and Schrohenloher, R.E. (1969) J. Biol. Chem. 244: 4128.

Author Index

Accumulated Subject Index for Volumes 1-5

Bei Fragen zur Produktsicherheit wenden Sie sich bitte an:
If you have any questions regarding product safety,
please contact:

Walter de Gruyter GmbH
Genthiner Straße 13
10785 Berlin
productsafety@degruyterbrill.com